# International Studies in Entrepreneurship

Volume 42

**Series Editors**
Zoltan J. Acs, George Mason University, Fairfax, VA, USA
David B. Audretsch, Indiana University, Bloomington, IN, USA

More information about this series at http://www.springer.com/series/6149

Charles W. Wessner • Thomas R. Howell

# Regional Renaissance

How New York's Capital Region Became
a Nanotechnology Powerhouse

Charles W. Wessner  
Georgetown University  
Washington, DC, USA

Thomas R. Howell  
Dentons, LLP  
Washington, DC, USA

ISSN 1572-1922        ISSN 2197-5884   (electronic)  
International Studies in Entrepreneurship  
ISBN 978-3-030-21193-6        ISBN 978-3-030-21194-3   (eBook)  
https://doi.org/10.1007/978-3-030-21194-3

© Springer Nature Switzerland AG 2020  
This work is subject to copyright. All rights are reserved by the Publisher, whether the whole or part of the material is concerned, specifically the rights of translation, reprinting, reuse of illustrations, recitation, broadcasting, reproduction on microfilms or in any other physical way, and transmission or information storage and retrieval, electronic adaptation, computer software, or by similar or dissimilar methodology now known or hereafter developed.  
The use of general descriptive names, registered names, trademarks, service marks, etc. in this publication does not imply, even in the absence of a specific statement, that such names are exempt from the relevant protective laws and regulations and therefore free for general use.  
The publisher, the authors, and the editors are safe to assume that the advice and information in this book are believed to be true and accurate at the date of publication. Neither the publisher nor the authors or the editors give a warranty, express or implied, with respect to the material contained herein or for any errors or omissions that may have been made. The publisher remains neutral with regard to jurisdictional claims in published maps and institutional affiliations.

This Springer imprint is published by the registered company Springer Nature Switzerland AG  
The registered company address is: Gewerbestrasse 11, 6330 Cham, Switzerland

# Endorsements

How can an ailing, declining region be transformed into a vibrant economic success? The answer has eluded thought leaders in policy and business for decades. With the publication of this path-breaking new book, Wessner and Howell provide a resolute and compelling answer. Based on their penetrating analysis of the resurgence of the New York state Capital Region, the authors provide a startling clear roadmap of how policy can trigger a breathtaking economic turnaround. Thanks to this book, there are no more reasons or excuses why any particular region cannot dramatically enhance its economic performance.

David B. Audretsch, *Distinguished Professor and Ameritech Chair of Economic Development, Indiana University*

This is a thoroughly researched and well-rendered account of one of the most dramatic regional economic turnarounds in modern US history. New York's large and sustained public investments in universities and research infrastructure in partnership with industry led to the creation of thousands of new, well-paying high-tech manufacturing jobs its Capital Region. Significantly, this effort enjoyed the support of a long line of Republican and Democratic leaders, from governors to the local level, as well as business and academic leaders who worked together to achieve a shared vision—the creation of New York's 'Tech Valley.' This study suggests best practices that can guide manufacturing and technology development efforts by federal and state policymakers across the nation.

Philip Singerman, *the Associate Director for Innovation and Industry Services at the NIST (National Institute of Standards and Technology) in the Department of Commerce*

# Executive Summary

This book describes how the State of New York has revitalized the economy of its Capital Region, creating a center of nanotechnology research and manufacturing that has come to be known as "Tech Valley," a corridor running along the Middle Hudson Valley from East Fishkill through Albany, Schenectady, and Troy to Saratoga Springs. This effort, which spanned several governorships, required substantial investment on the part of the state and included the creation of new institutions, which in turn attracted advanced manufacturing facilities. The initiative has been successful beyond expectations, creating nearly 10,000 manufacturing jobs in the region and tens of thousands of indirect and induced jobs, helping to reverse the area's long economic decline. This positive outcome and the policies that enabled it are especially relevant in an era in which Americans are seeking ways to revive US onshore manufacturing and to create viable long-term, well-paid career pathways for young people.

In the mid-twentieth and late twentieth century, Upstate New York was in a steep economic decline, reflecting a seemingly intractable combination of demographic, geographic, and structural disadvantages. Beginning with Governor Nelson Rockefeller and continuing down through Governor Andrew Cuomo, successive Republican and Democratic governors have worked to reverse this trend, emphasizing the practical application of scientific knowledge to promote innovation and restore job growth. Importantly, the governors enjoyed sustained bipartisan support from the state's principal legislative leaders in this effort. Despite its rough-and-tumble politics, New York leaders have been able to maintain the policy continuity necessary for the successful pursuit of a long-term strategic goal.

## Building on the Universities

The foundation of New York's developmental effort has been its educational system, which every governor from Nelson Rockefeller onward has worked to improve and expand. Governor Rockefeller drove the buildout of the State University of

New York (SUNY), transforming it from a loose aggregation of struggling undergraduate institutions into an educational and research powerhouse. In the early 1980s, Governor Hugh Carey and state legislative leaders supported the establishment of a Center for Industrial Innovation at Rensselaer Polytechnic Institute (RPI) in Troy, New York, with a then-unprecedented outlay of $30 million in state funds. Governor Mario Cuomo established Centers for Advanced Technology (CATs) at New York research universities, strengthening graduate-level research, upgrading research equipment and infrastructure, and fostering research collaborations between universities and local companies.

Governor George Pataki (R) took New York's technology-driven economic development effort to a new level, supported by a bipartisan duo of legislative leaders, Senate Majority Leader Joseph L. Bruno (R) and House Speaker Sheldon Silver (D). At the beginning of his tenure in 1995, Governor Pataki made an all-out, successful effort to dissuade IBM from relocating its headquarters outside the state, with the eventual result that IBM invested in a new, state-of-the-art semiconductor manufacturing facility in East Fishkill. He secured legislation creating the New York State Office of Science, Technology, and Academic Research (NYSTAR) with a then-unprecedented first-year budget of $156 million and created Centers of Excellence at state universities, thematic applied research teams focusing on technological issues of particular interest to industry. Perhaps most importantly, together with his allies in the legislature, he oversaw successive major investments in nanotechnology research and research infrastructure at SUNY Albany, enabling that institution to emerge as the foremost academic center of applied nanotechnology research in the world.

## Betting on Nanotechnology

Nanotechnology is the manipulation of matter with at least one dimension sized at 1–100 nm, a nanometer being one billionth of a meter, the equivalent of eight hydrogen atoms lined up side by side. Nanotechnology is applicable across a wide range of human activity and is leading to revolutionary new technologies in electronics, materials science, medicine, biotechnology, and renewable energy. New York State has been a leader in nanotechnology since its inception, with IBM and New York universities pioneering in the investigation of nanotech themes. At the end of the 1990s, the state began investing in nanotechnology for purposes of economic development in the Capital Region. Because the semiconductor industry was the first major industry to cross the nanotechnology threshold in manufacturing, the state's initial developmental effort focused on that industry.

## IBM's Contribution

IBM, headquartered in Armonk, New York, played a key role in the evolution of Tech Valley. Although the US semiconductor industry is most commonly associated with the West Coast and the southwestern United States, New York State, where IBM is headquartered, has been the site of semiconductor research and some manufacturing since the 1960s, an industrial legacy that enabled future success. IBM has conducted semiconductor-related R&D virtually from the inception of the industry, and it operated semiconductor manufacturing facilities producing devices for its own internal consumption at a site at East Fishkill, New York, which became operational in 1963. In the 1980s, IBM worked through US semiconductor associations and consortia to steer industry-supported research projects to New York universities, collaborated with university-industry research projects, and acted as an advocate for New York state within the US semiconductor industry. Reflecting its regional commitment, IBM was a persistent advocate of upgrading the state's educational system and research infrastructure.

## Center for Economic Growth

In addition to IBM's unique contributions, the state's public policy measures have been paralleled by an ambitious private-sector effort to forge a regional approach to development in the Albany area, which was traditionally characterized by parochialism and squabbles between competing jurisdictions and governmental units. The Center for Economic Growth (CEG), formed under the auspices of the Albany-Colonie Chamber of Commerce in 1987, emerged as a think tank and advocacy organization promoting regional developmental efforts. Beginning in 1997, CEG launched an effort to attract semiconductor manufacturing to the Capital Region and, in an effort to rebrand the region, promoted the expression "Tech Valley," a term which was initially derided by some but gradually taken more seriously as the region's remarkable transformation unfolded. CEG's development efforts received sustained support from National Grid, a power transmission company committed to the long-term economic development of the Capital Region.

## The Emergence of the Capital Region as a Center of Nanotechnology Research

In the 1980s, New York universities, supported by the state, participated in consortia established by the US semiconductor industry. These included the Semiconductor Research Corporation (SRC), founded in 1982, which conducts commercially relevant basic research, and SEMATECH, established in 1988, which focused on cooperative R&D to enhance US semiconductor manufacturing. SUNY Albany Professor Alain Kaloyeros, working with IBM and other companies, was conducting research of particular interest to US semiconductor producers, involving connectivity between semiconductor memory cells, and in 1988, SRC and SEMATECH created a New York "Center of Excellence" in conjunction with RPI and SUNY Albany. In the 1990s, nanotechnology research programs at SUNY Albany began to attract major industry funding, and he emerged as an advocate for building on such research to bring additional semiconductor manufacturing to the Capital Region.

## Nanotechnology Research at SUNY Albany

During his tenure, Governor Pataki became convinced that large state investments in research infrastructure at universities could enable the achievement of his economic development goals. In the 1990s and early 2000s, major state investments in nanotechnology research facilities were undertaken at SUNY Albany to attract investment on a much larger scale by semiconductor companies seeking access to what was rapidly becoming a leading center of applied nanotechnology research. To capture this potential, in 2001, IBM offered $100 million, and the state committed $50 million forming a partnership to create an actual state-of-the-art 300 mm semiconductor wafer fabrication facility at SUNY Albany, the only such resource on a university campus anywhere in the world. The 300 mm research production line enabled participating semiconductor companies to test new tools, materials, and processes in a cutting-edge manufacturing environment, a major competitive advantage.

## Creation of College of Nanoscale Science and Engineering

In 2004, Governor Pataki announced the formation of the College of Nanoscale Science and Engineering (CNSE or the "NanoCollege") at SUNY Albany, the first nanotechnology college in the United States. CNSE recruited faculty with industry and engineering experience and offered students a curriculum emphasizing nanotechnology's practical applications and commercial potential. The CNSE was created as a "research in Switzerland," a neutral site where competitors could work together to address shared technological problems, sharing both the cost of the work and the knowledge it generated.

Executive Summary ........ xi

## *The Industry Influx*

As the NanoCollege grew, many semiconductor producers concluded that it was important to establish a major local presence in the Albany area. In 2002, SEMATECH, then based in Austin, Texas, announced that it would establish a $400 million research center adjacent to the SUNY Albany 300 mm pilot manufacturing facility, and, following an intensive recruiting effort, in 2010, SEMATECH moved all of its operations to Albany. A snowballing influx of high-tech companies followed, including semiconductor makers (Advanced Micro Devices, Toshiba, Infineon, Samsung), toolmakers (Applied Materials, Tokyo Electron, ASML), and supply chain firms (M+W Zander). Investments by these firms usually involved consortia and frequently were supported by state incentives. At present, the CNSE operates four large, specialized facilities housing 300 mm manufacturing systems and tools and featuring as industrial partners some of the leading semiconductor firms in the world.

## Establishing a Foundation for Semiconductor Manufacturing

New York policymakers' long-term objective was not simply to promote research in the Capital Region but to leverage the research activities and infrastructure to attract nanotechnology-based manufacturing to the region. This, it was hoped, would help to offset the employment effects of the decline of traditional manufacturing industries in the area. Despite widespread and continuing skepticism, this approach ultimately worked. A large, sustained, well-informed, and well-executed team effort by state and regional leaders succeeded in persuading one of the world's leading semiconductor manufacturers, Advanced Micro Devices (AMD), to establish a manufacturing presence in the Capital Region's Saratoga County. Today, a corporate successor operates there as GlobalFoundries, one of the most advanced semiconductor foundries in the world.

## *Learning from Setbacks*

In the decade-long effort to attract semiconductor manufacturing, state policymakers were forced to learn from failure. In 1999, an ambitious effort by state development officials to secure "pre-permitting" regulatory approval for a generic semiconductor manufacturing facility at a site chosen by officials at RPI's Technology Park in North Greenbush failed when the local Town Board, reflecting concerns about unfounded environmental hazards, voted against the project. Regrouping, policymakers and business leaders concluded that despite the setback, pre-permitting remained a useful policy tool, enabling the securing of necessary regulatory approvals ahead of time at prospective sites, so that the site could be offered to an investor on a "shovel-ready" basis. Going forward, communities would self-select in a bottom-up approach, with interested local communities demonstrating their interest and support for local semiconductor manufacturing, manifested in prior regulatory zoning approvals by local authorities for generic manufacturing operations. The leaders of several communities advanced proposed sites for semiconductor fabrication plants, with Saratoga County emerging as the leading contender.

## *Saratoga Economic Development Corporation*

Saratoga County's bid for a chip manufacturing plant was spearheaded by the Saratoga Economic Development Corporation (SEDC), a small, private nonprofit corporation staffed with economic development professionals and funded by local businesses for the purpose of developing the local economy. Under the leadership of President Ken Green and Vice President Jack Kelly, in the mid-1990s, SEDC began laying the conceptual, practical, and regulatory foundations for establishment of a semiconductor manufacturing plant in Saratoga County. They identified a site in Luther Forest, in the towns of Malta and Stillwater, which enjoyed extraordinary natural advantages, and, with seed funding from National Grid, the regional electric power utility, noted above, commissioned engineering and planning studies for the site. The CEG backed SEDC's efforts as well as other similar initiatives in the Capital Region, most notably in Marcy, near Utica. The SEDC acquired purchase options for property in Luther Forest and funded travel by local officials and first responders to semiconductor manufacturing sites elsewhere in the United States to allay the environmental concerns that had stopped the North Greenbush project.

## *Regulatory Approvals*

In 2002, the SEDC and CEG jointly submitted plans for establishment of a "Technology Campus" in Luther Forest, to be the site of a generic semiconductor manufacturing plant or plants, to the Town Boards of Malta and Stillwater, requesting creation of a pre-approved "Planned Development District" (PDD) for the site. Technical specifications for 300 mm semiconductor fabrication were used to assess impacts such as noise, soil displacement, environmental impact, and effects on nearby traffic patterns. Over a 2-year period, SEDC made abundant information about semiconductor manufacturing available to the public in the two towns, and the proposed PDD received a thorough public review in multiple hearings and town meetings. An Environmental Impact Statement (EIS) was prepared and made public. Significant changes were made in the plans to respond to public concerns such as traffic, noise, and air pollution. This open review process paid off, and in 2004, both town boards voted in favor of the PDD, enabling SEDC and its allies to present the site to semiconductor manufacturers as "shovel-ready."

## *Outreach to the Industry*

Following the towns' approval of the PDD, SEDC assembled a deep team of industry experts to develop proposals for presentation to semiconductor manufacturers with respect to a manufacturing plant in Luther Forest. The team included environmental and transportation engineers: M+W Zander, an experienced builder of semiconductor wafer fabrication plants, experts in vibration and RF measurements, and power transmission experts provided by National Grid. Abbie Gregg, a semiconductor plant design expert, was part of SEDC's team of engineers and, reflecting her contacts in the industry, was able to provide entree to Hector Ruiz, at that time the CEO of AMD, one of the preeminent semiconductor manufacturers in the world. In 2006, AMD chose the Luther Forest site for construction of its next 300 mm wafer fabrication plant. Ruiz said that the site was superior to alternatives outside the United States and credited state and local officials for "the most well-crafted economic development package he could remember seeing."

## *State Leadership Engagement and Financial Commitment*

State leaders from Governor Pataki on down were closely engaged in the industry outreach effort which ultimately secured an investment commitment from AMD. New York delegations participated in trade shows, conferences, and other major semiconductor industry events and offered tours of CNSE's facilities. Crucially, the state offered AMD a very substantial incentive package valued at $1.2 billion, outbidding Dresden in Germany by $100 million. Ruiz said that the incentive package was "the key" in the company's final choice of New York State.

## *Significance of Research Ties*

An important factor underlying AMD's choice of New York was also the company's existing research ties with the state, which had been bolstered by state investments. At the time of its locational decision, AMD, along with three other semiconductor makers, was already engaged in a major, $600 million 7-year partnership with the CNSE to develop nanolithography technology, a partnership called INVENT. AMD was also collaborating with IBM on semiconductor design R&D at the latter's facility in East Fishkill. AMD reportedly valued the availability of RPI's Computational Center for Nanotechnology, featuring a $100 million supercomputer, which would "help AMD in its race against rival Intel" to design smaller and more powerful devices.

## The Infrastructure Buildout

The transportation, water, and electric power infrastructure necessary to support semiconductor manufacturing in Saratoga County did not exist at the time AMD committed to build a wafer fabrication plant in Luther Forest. A wide-ranging effort to secure regulatory approvals and build new infrastructure was required. Although this process encountered delays and setbacks, each obstacle was surmounted by ad hoc actions by Governor Pataki, Senate Majority Leader Bruno, and Speaker Silver, by individual communities, by development organizations like the SEDC and CEG, and by commitments and actions by AMD and then GlobalFoundries. At the same time, federal officials played a significant role. Senator Chuck Schumer continually weighed in with his support for the project. And local Congresswoman Kirsten Gillibrand and her office played a lead facilitative role at a critical time in the project's development in working with the State Senate Majority leader's office and convening federal, state, and local agencies and others required to obtain regulatory approvals and fill funding gaps. Indeed, the role the congressional office played in advancing the project may be seen as a best practice as congressional offices in general are uniquely positioned to play a facilitative role.

The infrastructure projects necessary to support the chip fabrication plant were generally supported by local authorities, reflecting the fact that the improvements brought ancillary benefits to communities in the vicinity. Construction of waterlines to the factory site involved the regulatory approval and construction of roughly 30 miles of new waterline connecting a water source in Moreau, New York, with the fab site in Malta/Stillwater, a scheme which also enhanced water supply availability to communities along the route. Road improvements to ease traffic flow near the factory enabled the village of Round Lake to secure a long-sought bypass around its center, easing traffic, and provided a vehicle for realization of the vision by Saratoga County to modernize and upgrade its trail system. New power lines and substations built by National Grid improved power availability and reliability for local communities. By the time the fab was built and became operational, the necessary infrastructure to support it was in place.

## The Launch of GlobalFoundries

Between 2010 and 2017, GlobalFoundries invested $15.786 billion at the Malta site. The construction of the GlobalFoundries wafer fabrication plant in Luther Forest, begun in 2011, was one of the largest building projects ever undertaken in the United States, requiring an effort comparable to that of the construction of the Empire State Building. The original plans were revised on a number of occasions to further expand the scope of the enterprise. The task was unusually challenging because of the sophistication and precision associated with semiconductor manufacturing, which requires handling of exotic materials, meticulous alignment of heavy machinery, and the creation of a particle-free clean room environment. The construction was completed on time, notwithstanding numerous challenges. A Project Labor Agreement (PLA) with local construction trades ensured good wages and labor peace.

## Economic and National Security Impact of New York's Nanotechnology Investments

The economic benefits for the region of the state's investments in nanotechnology have exceeded all forecasts.

### *Direct and Supply Chain Jobs*

Initial studies predicted that the AMD/GlobalFoundries wafer fabrication plant would result in direct employment of roughly 1500 people—estimates that were criticized as too optimistic. In fact, GlobalFoundries has created over 3500 "direct jobs" (employees and vendors' employees working full-time onsite) at its Luther Forest site in Saratoga County while preserving nearly 2000 jobs at IBM's former operation in East Fishkill. The state and private investments in the CNSE created another 4000 jobs within the CNSE and its on-site industrial partners in Albany. The figure of roughly 9500 direct jobs has been complemented by large numbers of "indirect jobs" (jobs created by a manufacturer when it buys supplies or services) within the GlobalFoundries supply chain.

### *Construction Jobs*

Construction industry employment also exceeded forecasts. Initial forecasts predicted that construction of the AMD/GlobalFoundries facility would result in employment of about 1500 workers for a temporary project of under 2 years duration. In fact, since the startup of construction in 2009, the GlobalFoundries facility in Malta/Stillwater has been a more or less continuous construction site, with as many as 3500 construction workers employed in the winter of 2014–2015 and 900–1100 workers present at the site through 2015, long after the completion of the original fab. The secondary construction projects attributable to the presence of GlobalFoundries—new housing, hotels, healthcare facilities, and retail—have generated hundreds of additional construction jobs. The construction job boom in the Capital Region coincided with, and eased the impact of, the deepest recession the United States has experienced since the Great Depression, in which the construction industry was particularly hard hit.

## *Induced Jobs*

The high salaries associated with semiconductor industry employment, with average pay of $92,000 a year at GlobalFoundries Malta fab, have had major ramifications for the growth of the regional economy, creating thousands of "induced jobs" (jobs created by employees of the manufacturer spending money locally) in sectors as diverse as restaurants, healthcare, banking, and retail sales such as automobiles and household goods. Similarly, the many visitors, technicians, engineers, and vendors associated with GlobalFoundries have boosted the region's hotels and contributed to substantial new hotel construction. While it is difficult to quantify with precision the number of induced jobs attributable to the state's investments in nanotechnology, local economies near Malta and Albany clearly derive a substantial number of jobs indirectly from the nanotechnology complex. The US Semiconductor Industry Association estimates that for each direct job in the industry, 4.89 induced and indirect jobs are created, a figure that correlates with estimates by the European semiconductor industry and recent academic work on the subject. On this basis, the GlobalFoundries facility alone would account for over 15,000 indirect and induced jobs. This is in addition to the 3500 direct jobs that substantially exceeded investment requirements as well as the many indirect jobs directly linked to the supply chain.

## *The National Security Dimension*

Semiconductors now comprise the core of all major US defense platforms, from satellites and aircraft to naval vessels, communications systems, and support vehicles. For a number of years, US defense planners have been concerned that the globalization of semiconductor research and production posed a fundamental threat to the security of US defense systems. The Pentagon established a program pursuant to which certain critical devices are fabricated by "trusted" companies at secure sites physically located in the United States. One such "trusted" site is the former IBM semiconductor plant in East Fishkill, New York, now owned and operated by GlobalFoundries, which is engaged in a variety of projects supporting US defense needs. The CNSE also performs work for the US defense agencies. The New York nanocluster now comprises an important part of the industrial foundation of US national security.

# Educating and Training a High-Tech Work Force

The Capital Region has been able to attract major inward investments by high-technology companies in significant part because its numerous and strong educational institutions ensured not only a pool of engineering graduates but also—at least potentially—the ability to generate workers with skills suitable for positions in manufacturing plants.

## *Four-Year Institutions*

The drive to create Tech Valley has been spearheaded by the science and engineering departments of local universities, particularly SUNY Albany's College of Nanoscale Science and Engineering (which now forms part of SUNY Polytechnic) and those at RPI. These are among over 20 colleges and universities in the Capital Region offering a broad range of curricula—what one local educator characterized as a "rich stew of higher education institutions that offer virtually anything that economic development specialists or corporate relocation specialists look at when they want to locate their plants." State and private investments in these institutions are substantial and continuing, including the establishment of a new engineering school at SUNY Albany since the transfer of the CNSE to SUNY Poly.

Such investments reflect New York State's, and SUNY's, long-standing belief that education is a key economic development tool and that education should support the workforce demands of employers. SUNY's 64 universities and community colleges represent the largest system of its type in the United States. These institutions are engaged with the private sector on multiple fronts to ensure that their curricula and program align with the needs of employers. SUNY has a senior vice chancellor tasked with oversight of all SUNY community colleges, which play a particularly important role in preparing students for the demands of high-tech manufacturing.

## *Community Colleges*

Though often overlooked, the principal workforce needs of semiconductor manufacturers are not engineers and managers with advanced degrees—although these are important—but technicians and operators who comprise roughly two-thirds of a wafer fabrication plant's labor force. These are what former New York Senate Majority Leader Joseph L. Bruno terms the "blue collar workers of the future," often holding only a high school diploma but possessing an array of specialized skills that enable them to function in a high-tech manufacturing environment. Community colleges in the Capital Region have mobilized to establish curricula relevant to high-tech manufacturing, working closely with companies. Hudson Valley Community College (HVCC) in Troy has been involved in the state's outreach efforts to the semiconductor industry from the beginning and has developed degree programs specifically designed to meet the needs of semiconductor manufacturers. One HVCC initiative is an extension facility, TECH-SMART, adjacent to the GlobalFoundries facility in Malta/Stillwater, which features simulated semiconductor manufacturing environments. Semiconductor manufacturing and advanced manufacturing curricula have also been introduced at Schenectady County Community College (SCCC), Fulton-Montgomery Community College (FMCC), and SUNY Adirondack.

## *K–12 Education*

The public K–12 education system in New York outperforms those of most other states, and the Capital Region's schools collectively outperform New York state averages. In the 2015–2016 school year, 61 of the region's school districts spent 70% more per pupil than the national average, and 10 of the region's districts spent more than double the national average. Despite generally strong performances by the schools, however, weaknesses have become observable in the K–12 system's STEM education programs and preparation of students for life after graduation. A number of model programs have emerged to address these problems and are described in this study. These include the establishment of Tech Valley High School in Troy by a coalition of regional school districts and the development of "pathways to higher education, integrating various educational levels and post-degree, workplace environments," spearheaded by the Ballston Spa Central School District.

## *Skills Gap*

The growth of the region's technology intensive industries has exceeded forecasts, and local tech firms warn of a "skills gap," a shortfall in available workers with the requisite knowledge and skill sets for high-tech manufacturing. Across the region, educational institutions are scrambling to respond with new investments, programs, and initiatives. A particularly difficult challenge has proven to be scaling up model education and training programs which have produced good outcomes but which do not produce nearly enough graduates to meet the needs of local manufacturers. Other concerns include the difficulty in forecasting industry's manpower needs, the time required to secure state approval for new curricula in the community colleges, and the difficulty associated with constantly adapting course offerings to maintain relevance to the rapidly evolving, frequently mercurial industry. Collectively, these factors, plus a tangle of other practical problems and institutional anomalies, combine to produce the skills gap, which may well represent Tech Valley's single most important challenge.

## Emerging Challenges

While the creation of Tech Valley is indisputably a success story, it will not necessarily remain one. In addition to the emergence of the skills shortage, a number of significant developments have occurred during the past decade that raise the question of whether the Capital Region can sustain its momentum.

### *Changing Political Leadership*

The emergence of Tech Valley benefitted from a confluence of top state leaders—Governor Pataki, Senate Majority Leader Bruno, and Speaker Silver—for a decade after 1995. These three men, while not always in agreement, saw to it that the effort received the resources and senior political level attention it needed to succeed. All three have by now passed from the scene, and the Capital Region finds itself in competition with other New York regions which—ironically—are applying some of the same methods and models used in the creation of Tech Valley.

Governor Andrew Cuomo is seeking to adapt and apply best practices derived from the Tech Valley experience across Upstate New York. He has succeeded in wrestling a substantial degree of control over economic development funding from the legislature. As an alternative to pork-barrel projects ("member items"), he has divided the state into Regional Economic Development Councils (REDCs) which develop strategic regional plans driven by the private sector—similar to those pioneered within the Capital Region by CEG—and which are allocated state economic funds based on the merits of those plans as assessed by state policymakers. Governor Cuomo has attempted to replicate the CNSE model in other Upstate cities, beginning with Buffalo. The model is arguably working where it has been closely followed (Buffalo/Niagara) and not working in other areas (such as an attempted film hub near Syracuse) where the model was not followed.

### *The Procurement Scandal*

Since mid-2015, the enterprise known as Tech Valley has been overshadowed by a scandal arising out of procurement irregularities involving the Former President of SUNY Poly, Professor Alain Kaloyeros, and others in connection with the Buffalo projects. Kaloyeros was indicted on federal bid-rigging charges in September 2016, and a number of major CNSE-related projects under way at the time of the indictments stalled or fell apart. Kaloyeros was found guilty in July 2018. Subsequently, Empire State Development (ESD), under the direction of Howard Zemsky, moved to install appropriate oversight, implement institutional changes, and restructure stalled projects. Former GlobalFoundries CEO Doug Grose was appointed to run

SUNY Poly's research and economic development activities through a newly established entity, NY CREATES. While this effort has not yet been finalized at the time of this writing, recent developments are promising. In 2018, New York and Applied Materials announced plans for an $880 million semiconductor equipment and materials center at SUNY Poly's NanoFabX, and in 2019, New York disclosed a new collaboration with IBM to develop component technologies for artificial intelligence, a $300 million project that will be housed at SUNY Poly. Regardless of what may be failures of judgment on the part of some individuals, the fact remains that the CNSE—and, to a greater degree, SUNY Poly—have proven to be important institutions in the development and ongoing support for the advanced manufacturing and semiconductor development ecosystem in the state and nation.

## *Challenges Facing the Luther Forest Technology Campus*

SEDC's plan for the Luther Forest Technology Campus envisioned that the campus' administering organization, Luther Forest Technology Campus Economic Development Corporation (LFTCEDC), could fund itself over time through sale of parcels of land within the campus to companies planning industrial sites. That has not happened, reflecting zoning restrictions and the disappearance of state incentives for the campus, leaving LFTCEDC without the financial resources to improve and maintain the infrastructure for a semiconductor manufacturing cluster. The problem has been complicated by divisions within the Saratoga County economic development community. While recent events suggest movement toward resolution of these issues, questions remain of whether the infrastructure necessary to support an expanding cluster of semiconductor supply chain firms in Luther Forest will be put in place in a relevant timeframe.

## *Global Competition*

In the semiconductor industry, global competition and accelerating technological change will always constitute a major and potentially destabilizing challenge for regions in which the industry is present. Many countries view the industry as strategic, and a number of them, most notably China, are committing unparalleled resources to enhance the competitive position of their indigenous producers. The location of future semiconductor research and manufacturing facilities will be determined, to a substantial degree, by the level of commitment in terms of incentives, infrastructure, and workforce development put in place by competing national and regional governments.

## *Fostering Startups*

An important aspect of the long economic malaise of Upstate New York, in general, and the Capital Region, in particular, has been the relative scarcity of innovative startup companies and the fact that successful startups often leave the region. The advent of nanotechnology has seen the emergence of a number of promising startups originating in local universities, such as BessTech (lithium-ion batteries), Glauconix (nanostructures with medical applications), and ThermoAura (thermoelectric nanocrystals). A support network for tech-oriented startups is emerging, including sources of angel and venture funding and curricula at local universities emphasizing entrepreneurialism. The experience of innovative centers such as North Carolina's Research Triangle and Austin, Texas, has shown that the existence of a cluster of large, established tech-oriented firms—now an established fact in the Capital Region—eventually leads to and supports the proliferation of innovative startups. Nonetheless, the need to foster more tech-oriented startups in the region remains a significant challenge.

## The Tech Valley Model

A compelling feature of New York's creation of Tech Valley is its demonstration that it is possible to reverse long-term economic decline in an old industrial region through the right combination of public policies and private-sector engagement. Although it can be argued that New York's experience is sui generis, reflecting factors such as the state's large financial resources, its location, and the pre-existing industrial foundation of tech-oriented firms such as GE, IBM, and Kodak, these advantages were offset by daunting disadvantages which led many experts to write Upstate's economic situation off as hopeless—such as high taxes, high labor costs, fragmented local governments, and a long-standing out-migration of young people. With some difficulty, these challenges were surmounted, and many of the practices and techniques instrumental in the creation of Tech Valley are arguably "best practices" which can be successfully adapted and applied in other regions that confront similar challenges.

## *Best Practices*

Best practices are techniques that are widely accepted as superior to alternative methods because they are likely to deliver better outcomes. Such methods, when adapted to particular local circumstances and conditions, offer the realistic prospect of success. New York's experience suggests a number of best practices:

- **Policy continuity**. The sustained character of the investments and policies by New York's leaders with respect to innovation-based economic development—spanning nearly half a century—is one of the most remarkable features of this effort. The scale and continuity of the investments made the success of Tech Valley possible.
- **Leadership in depth**. While the commitment of New York's senior leadership was essential, the creation of Tech Valley was driven by local officials, academic leaders, and business executives who demonstrated initiative, tenacity, and teamwork in pursuit of the regional vision. Significantly, such local leaders were backed by the state with financial and other resources at numerous key junctures.
- **Preservation of the industrial legacy**. Tech Valley was built, in substantial part, upon expertise, know-how, financial resources, and continued presence in the region of older tech-oriented firms, most notably IBM and GE, which, in the absence of continued outreach by the state, would have exited the region.
- **Public investments in research infrastructure**. New York made large and sustained investments in nanotechnology research facilities and equipment at universities in the Capital Region, most notably at SUNY Albany and RPI, enabling the Albany area to emerge as the leading center of applied nanotechnology research in the world. The existence of research facilities and expertise that were literally available nowhere else proved a powerful draw for major semiconductor firms to establish a local presence.
- **Formation of university-industry consortia**. The formation of numerous university-industry nanotechnology research consortia in the Capital Region was an effective institutional mechanism for maximizing the value and minimizing the cost of commercially relevant research undertaken in university facilities. Locating the state's investment in the university enabled the state to support facilities without favoring a particular firm. The consortium model also enabled participating semiconductor companies to minimize the risks and share the costs associated with the adoption of new equipment, materials, and processes while capturing the full technological benefit of the research.
- **Use of intermediary organizations**. A key advantage was that SUNY Albany's research collaborations with industry were undertaken through not-for-profit corporate intermediary organizations not bound by academic rules and protocols, a structure which facilitated engagement with private firms. The newly established NY CREATES, which will restructure and operate SUNY Poly's research and economic development activities, is expected to play this role in the future.
- **Regional approach to economic development**. Creation of "Tech Valley" was enabled by a collaborative effort driven by business organizations to forge a single regional strategy for economic development, sidestepping local rivalries and political gridlock that had previously impeded economic growth.
- **Pre-permitting industrial sites**. The regional development organizations seeking to attract semiconductor manufacturers to a local site undertook the considerable effort required to secure regulatory approvals for a generic semiconductor

manufacturing facility. This enabled outreach to semiconductor makers based on a "shovel-ready" location.
- **Partnering with existing industry research consortia**. From the beginning of the Tech Valley effort, New York's academic and business leaders reached out to established, external semiconductor research consortia, building partnerships with the Research Triangle-based Semiconductor Research Corporation, which funded relevant basic R&D, and Austin-based SEMATECH, which conducted manufacturing R&D. Over time, this effort familiarized the semiconductor industry with the research resources available in New York and helped facilitate SEMATECH's eventual relocation from Austin to Albany.
- **Building a professional development team**. The Saratoga Economic Development Corporation, which led the region's outreach to semiconductor manufacturers, assembled an elite team of professionals with deep semiconductor industry expertise and contacts within the industry. This enabled the region to put together a winning proposal to present to manufacturers.
- **Competitive incentives package**. The state government showed the vision and the will to put forward an internationally competitive incentives package (initially valued at $1.2 billion) able to draw a semiconductor producer to the region, a package which proved superior to those of other countries and regions.
- **Abiding commitment to education and training**. The state has demonstrated long-term commitment to STEM education at all levels, from K–12 through graduate programs and to relevant skills training, centered on community colleges. New York's emphasis on human resources development and the high quality of its educational institutions and programs has dramatically enhanced its competitiveness relative to other regions and other countries.

## Conclusion

New York has overcome numerous obstacles to create a large and growing cluster of nanotechnology-based research, manufacturing, and supply chain enterprises and educational institutions in "Tech Valley." Its success also underscores the possibilities for other US cities and regions to attract and retain high-tech manufacturing. At the same time, the region's continued success is not guaranteed. The semiconductor industry's unique needs and fierce global competition raise serious challenges, underscoring the need for ongoing political and financial commitment, institutional teamwork, and sustained effort from the region's leadership.

# Acknowledgments

We would like to express our appreciation to the following individuals in the Capital Region who made themselves available for interviews in connection with this study:

| Individual | Affiliation |
|---|---|
| Todd Alhart | GE Global Research |
| Peter Altenberg | National Grid |
| Sean Atkinson | GlobalFoundries |
| Heather Briccetti | President, The Business Council of New York State, Inc. |
| Dennis Brobston | Saratoga Economic Development Corporation |
| David Catalfamo | Park Strategies |
| Ed Cody | GlobalFoundries |
| Greg Connors | GlobalFoundries; Former Supervisor, Town of Stillwater |
| Carolyn Curtis | Academic Vice President, Hudson Valley Community College (HVCC) |
| Richard Dewey | New York Independent System Operator |
| James Dias, PhD | Vice President for Research, University at Albany, State University of New York |
| Jonathan Dordick, PhD | Vice President for Research, Rensselaer Polytechnic Institute |
| Joseph Dragone | Former Superintendent, Ballston Spa Central School District |
| Ronald Epstein | CFO, New York State Department of Transportation |
| Michael Fancher | SUNY Polytechnic Institute; Director, New York State Center for Advanced Technology in Nanomaterials and Nanoelectronics |
| Kimberly Finnigan, Esq. | GlobalFoundries |
| Doug Grose | NY CREATES (Former CEO of GlobalFoundries) |
| Pradeep Haldar, PhD | Vice President of Entrepreneurship, Innovation, and Clean Energy Programs, SUNY Polytechnic Institute |
| Robert Hanks | GlobalFoundries |
| Brian Hannafin | John W. Danforth Company, Inc. |
| James Harris | President, Janitronics |

| Individual | Affiliation |
|---|---|
| Catherine Hill, Esq. | Clean Energy Leadership Institute, Skidmore College |
| Linda Hill | Formerly of National Grid |
| Penny Hill | Associate Dean, TEC-SMART, Hudson Valley Community College |
| Karen Hitchcock, PhD | Former President, University at Albany, State University of New York |
| Cynthia Hollowood | General Manager, Holiday Inn Saratoga |
| Michael Izdebski | GE Power & Water |
| Kristina Johnson | Chancellor, State University of New York |
| Jack Kelley | Former Vice President, Saratoga Economic Development Corporation |
| Paul C. Kelly | Interim Vice President for Finance and Research, SUNY Polytechnic Institute |
| Andrew Kennedy | President, Center for Economic Growth (CEG); Former Deputy Director of State Operations for Governor Andrew Cuomo |
| Edward Kinowski | Supervisor, Town of Stillwater |
| Michael Liehr | Executive Vice President of Innovation and Technology, SUNY Polytechnic Institute |
| Mark Little | Former Head, GE Global Research |
| Jeff Lovell | Park Strategies; Former Senior Policy Advisor to Governor George Pataki |
| Andrew Matonak, EdD | President, Hudson Valley Community College |
| Angelo Mazzone | President, Mazzone Hospitality |
| Robert Megna | Senior Vice Chancellor and Chief Operating Officer, State University of New York |
| Robert Miller, Esq. | CEO, Windsor Development Group |
| Stephanie Montag | PeroxyChem |
| Katie Newcombe | National Grid |
| Ken Petranoski | Supervisor, Town of Stillwater |
| Daniel Pickett | Chairman and CEO, nfrastructure |
| Irv Plotnik | KLA-Tencor |
| Emily Reilly | GlobalFoundries |
| Michael Relyea | Former Director, Luther Forest Technology Campus |
| David Rooney | Former Vice President, Center for Economic Growth |
| Ray Rudolph, PE | Chairman, CHA Companies |
| Mike Russo | GlobalFoundries |
| Laura Schultz, PhD | Assistant Professor of Nanoeconomics, College of Nanoscale Science and Engineering |
| Robert Snyder | FPI Mechanical |
| Jeffrey Stark | New York State Building and Construction Trades |
| Darren Suarez | Director of Government Affairs, Business Council of New York State |
| F. Michael Tucker | Former President, Center for Economic Growth |
| Jason Van Buren | GlobalFoundries |
| Martin Vanags | Saratoga County Prosperity Partnership |
| Michele Vennard | President, Albany County Convention & Visitors Bureau |

# Acknowledgments

Upstate New York is dotted with small- and medium-sized cities, most of which continue to support local newspapers. Over the years, the story of the emergence of Tech Valley has been thoroughly covered by strong local reporting, often featuring interviews with principal actors as key events have unfolded. We acknowledge and express our appreciation for the invaluable archive of "first drafts of history" which these regional newspapers have produced. In particular, we thank the *Times Union* (Albany), *The Daily Gazette* (Schenectady), the *Albany Business Review* (Albany), *The Post-Star* (Glens Falls), *The Record* (Troy), *The Saratogian* (Saratoga Springs), *The Buffalo News* (Buffalo), the *Observer-Dispatch* (Utica), the *Daily Messenger* (Canandaigua), and *The Post-Standard* (Syracuse). In addition, we would like to acknowledge the in-depth reporting and commentary of the *Gotham Gazette*, which covers New York State policymaking and politics.

This study of the origins, growth, achievements, and challenges of the Tech Valley nanocluster was made possible through the financial support and cooperation of a broad array of institutions. The Center for Economic Growth was instrumental in initiating the study and providing direct financial contributions and facilitating the contributions and cooperation of others. The contributors to the study encompass a number of regional development authorities including the Capital Region Chamber, Saratoga County Prosperity Partnership, Saratoga Economic Development Corporation, Rensselaer County Regional Chamber of Commerce, Washington County Local Development Corporation, Greene County Industrial Development Agency, and Columbia Economic Development Corporation. The financial support and insights into the region's development were also provided by major manufacturing companies such as GlobalFoundries and IBM and significant funding from National Grid, the regional electricity utility. This support was essential in providing the means to undertake the study and the local knowledge and diverse perspectives to better understand the many contributions to the growth of the Tech Valley nanocomplex.

Reflecting the national significance of microelectronics and the policies deployed to support its growth here in the United States, the study also attracted support from a variety of federal departments and agencies. These include the Office of the Secretary of Defense, the Economic Development Administration and National Institute of Standards and Technology of the Department of Commerce, and the National Cancer Institute of the National Institutes of Health. Institutionally, the work was supported at Georgetown by the Security and Software Engineering Research Center ($S^2$ERC) funded by the National Science Foundation. This broad-based support provided the resources that were essential to complete the extensive field research that underpins the study. It also reflects both the regional and the national interest in understanding New York's accomplishment and its potential application elsewhere in the nation.

The authors would like to recognize a special debt to two individuals: Mike Russo of GlobalFoundries was instrumental in providing early encouragement for the research, urging the authors to take a comprehensive view of the region's accomplishments and its current challenges. Without his leadership and support, the study would not have taken place. Michael Tucker of the Center for Economic Growth and now of Tucker Strategies played a crucial role throughout the study, identifying interviewees, facilitating meetings, and sharing his many insights and deep knowledge of the region.

# Disclaimer

The statements, findings, conclusions, and recommendations in this report are those of the authors and do not necessarily reflect the views of any of the sponsors.

# Contents

1  Introduction ................................................. 1
2  Upstate New York: Reversing Economic Decline
   Through Innovation ......................................... 21
3  Nanotechnology Research in Albany, 1980–2016 ............... 49
4  Establishing a Foundation for Nanotechnology Manufacturing .... 93
5  The Infrastructure Buildout: A Detailed Look ................ 133
6  The Launch of GlobalFoundries .............................. 163
7  Economic Impact of New York's Nanotechnology Investments ..... 181
8  Educating and Training a High-Tech Workforce ............... 217
9  The Changing Landscape of Tech Valley ...................... 277
10 Conclusion ................................................ 337

Appendices .................................................... 363

Bibliography .................................................. 395

Index ......................................................... 411

# About the Authors

**Charles W. Wessner** teaches Global Innovation Policy at Georgetown University (United States) and is a powerful advocate of effective innovation policies. Previously, he served for two decades as a National Academies Scholar where he founded and directed the National Academy of Sciences Technology, Innovation, and Entrepreneurship Program (United States). He is recognized nationally and internationally for his expertise on innovation policy, including public-private partnerships, entrepreneurship, early-stage financing for new firms, twenty-first-century universities and manufacturing, and the special needs and benefits of high-technology industry. As an outgrowth of his work with the US government, he advises technology agencies, universities, and government ministries in Europe and Asia. In addition, he cooperates closely with international organizations and lectures at major universities in the United States and abroad. The overarching goal of his work is to develop a better understanding of how we can bring new technologies forward to address global challenges in health, climate, energy, water, infrastructure, and security. Reflecting his commitment to international cooperation, he was recently named an Officer of the Order of Merit by the President of France. He can be reached at cw826@georgetown.edu.

**Thomas R. Howell** is known for his work in the international trade field and for his publications and research in high-tech sectors such as microelectronics and advanced manufacturing. As part of his legal work, he developed a specialization in field research and analytic studies for clients. He examines factual, policy, and legal issues utilizing research in primary sources, foreign language sources, and extensive interviews to capture the tacit knowledge often not otherwise available. His subjects include industrial, research, innovation, and science policies outside the United States, foreign markets and business practices, cartels, and innovation-based economic development. The National Academy of Sciences (United States) has commissioned and published a number of his works. He has also contributed as a writer/researcher under contract to the National Academies producing chapters and book-length reports which must pass a rigorous peer review. He has published numerous articles in academic journals and law reviews and has performed research under contract to the Semiconductor Industry Association, and the National Academy of Sciences. In recognition of his expertise, he has been invited to serve on policy committees undertaking studies for both the National Academy of Sciences and the Defense Science Board. He can be reached at thomas.howell@dentons.com.

# Chapter 1
# Introduction

In the past half century, fundamentally new technologies—microelectronics, digital computing and communications, biotechnology—have revolutionized human economic endeavor and everyday life, spawning entirely new industries, and, in the United States, bringing unprecedented prosperity to regions in which those industries have become concentrated. But the same era has coincided with the erosion or disappearance of vast swaths of the US manufacturing base, the displacement of millions of workers, and the economic decline of formerly prosperous regions. The epicenter of this phenomenon, the old industrial regions of the Northeast and Upper Midwest, long ago became known colloquially as the "Rust Belt."[1] In these hard-hit areas, the loss of manufacturing jobs has meant fewer opportunities that offer "good wages for workers who lack advanced education," steeply declining population, and an array of social maladies including rising crime, broken families, substance abuse, and declining educational attainment.[2] The disappearance of well-paying manufacturing jobs underlies much of the increase in income inequality that emerged during the latter half of the twentieth century.[3] In the 2016 presidential election, the

---

[1] The term "Rust Belt" has its origin in the 1984 presidential campaign of Walter Mondale. It is an imprecise expression and is sometimes used to embrace the coal-producing regions of Appalachia and former iron ore mining areas, the industrial regions in the upper South, as well as the old industrial Northeast and Midwest. "Midnight in the Rust Belt," *Beltmag.com* (September 21, 2013).

[2] According to a 2013 study by the Federal Reserve Bank of Cleveland, four Rust Belt cities—Cleveland, Buffalo, Pittsburgh, and Detroit—collectively lost 45% of their population between 1970 and 2006 and experienced major declines in level of household income and educational attainment. David Hartley, "Economic Decline in Rust Belt Cities," *Economic Commentary* (May 20, 2013).

[3] Martin Neil Baily and Barry P. Bosworth, "US Manufacturing: Understanding its Past and Potential Future," *Journal of Economic Perspectives* 28(1) (2014). In 2013, New York State had the most disproportionate top-to-bottom ratio of income disparity of any state in the United States. The top 1% of earners averaged income of $2,006,652 compared with the average of $44,163 for the bottom 99%, or a top-to-bottom ratio of 45.4. The smallest ratio was observable in Alaska, 13.2. "Income Inequality in the US by State, Metropolitan Area, and Country," *Economic Policy Institute* (June 16, 2016).

economic and social pain felt in these regions and the sense that they have been left behind moved to the center of the national political discussion.

Upstate New York (New York north and west of Westchester County) has been one of the hardest-hit areas of the Rust Belt, experiencing a longstanding, demoralizing hemorrhage of manufacturing companies, jobs, and people to other states.[4] The *New York Times* ran a feature in 1997, during a time of surging national economic growth, observing that Upstate New York was stagnating, with "one of the weakest economies of any significant region of the country."[5] Two decades later, former New York Lieutenant Governor Robert Duffy observed that if Upstate New York broke away from Downstate, "the economic indicators of Upstate would be among the lowest in the country."[6] More recently, President Donald Trump suggested that residents should consider leaving a blighted "upper New York" for jobs in other regions.[7] The region has not escaped the social problems closely associated with industrial decline.[8]

Rust Belt states and regions have struggled not only with a seemingly intractable tangle of economic and social ills, but with the widespread assumption that over the long run their situation is essentially hopeless, a product of sweeping global economic changes that individual communities and regions are powerless to combat. A 2016 economic study of the erosion of US manufacturing observed that in old industrial cities, the inevitable consequences of sustained population loss include higher rates of poverty and crime coupled with diminished fiscal capacity for local authorities to address these and other problems, leading to "vicious cycles that perpetuate further decline."[9] That perspective frequently characterizes assessments of Upstate New York's long economic malaise.[10]

---

[4] See *generally* Kansar Hamdami, Richard Deitz, Ramon Garcia, and Margaret Cowell, "Population Out-Migration from Upstate New York," in Federal Reserve Bank of New York (Buffalo Branch), *The Regional Economy of Upstate New York* (Winter 2005). Binghamton, New York, has lost roughly three out of four manufacturing jobs in the past three decades. Rochester's manufacturing employment has plummeted by 53% since 1990. Reflecting the fact that manufacturing jobs pay $15,000–20,000 more per year than the average wage in the private sector, the wholesale loss of such jobs means that "upstate residents, as a whole have less disposable income to spend on cars, homes and vacations." "Made in NY? Forget It, as State Loses to Others," Rochester, *Democrat & Chronicle* (March 10, 2017).

[5] "As US Economy Races Along, Upstate New York is Sputtering," *The New York Times* (May 11, 1997).

[6] "The Upstate Economy is One of the Worst in the Country," *Politifact.com* (September 16, 2016).

[7] "Upstate New Yorkers to Trump: 'Hello, it's not the 1940s and 1950s,'" Schenectady, *The Daily Gazette* (July 27, 2017).

[8] "Watchdog Report: Upstate Sinks in a Sea of Legal Opioids," *Pressconnects* (December 16, 2016).

[9] Michael Manville and Daniel Kuhlmann, "The Social and Fiscal Consequences of Urban Decline: Evidence from Large American Cities, 1980–2010," *Urban Affairs Review* (November 11, 2016).

[10] "Upstate New York's persistently slow economic growth is often viewed as the result of local disadvantages, such as the region's heavy reliance on a declining manufacturing sector, elevated business costs, or lack of high-tech business services." Kansar Hamdami, Richard Deitz, Ramon Garcia, and Margaret Cowell, "Population Out-Migration from Upstate New York," in Federal Reserve Bank of New York (Buffalo Branch), *The Regional Economy of Upstate New York* (Winter

1 Introduction

**Table 1.1** Percent change in manufacturing employment by decade for metropolitan areas in the State of New York

|  | 1970–1980 | 1980–1990 | 1990–2000 | 2001–2010 | 2010–2014 |
|---|---|---|---|---|---|
| New York City | −22.63 | −26.00 | −27.48 | −37.15 | −0.99 |
| Albany/Troy/Schenectady | −14.71 | −22.82 | −15.04 | −24.62 | 20.89 |
| Buffalo/Niagara | −19.02 | −25.19 | −13.13 | −35.67 | 5.96 |
| Rochester | −4.20 | −16.48 | −18.86 | −37.86 | −3.50 |
| Syracuse | −5.00 | −13.37 | −11.23 | −32.95 | −7.34 |
| Utica/Rome | −23.30 | −23.80 | −16.56 | −31.82 | −0.44 |
| Binghamton | −1.31 | −13.78 | −26.99 | −32.38 | −17.56 |
| Median | −14.71 | −22.62 | −16.56 | −32.95 | −0.44 |

Source: Compiled by John Bacheller in "The Decline of Manufacturing in New York and the Rust Belt," *Policy by the Numbers* (October 4, 2016), using data from U.S. Department of Commerce, Bureau of Economic Accounts, Regional Economic Accounts, and http://www.headhunterseconomics.com/

This book offers a rejoinder. It describes how local business, academic, and government leaders in New York's Capital Region—embracing the cities of Albany, Troy, and Schenectady and a number of smaller cities and towns—stubbornly refused to accept the inevitability of economic decline and the disappearance of manufacturing and, backed by a succession of capable and committed state leaders, mounted a decades-long effort to renew and restore the economy of their region and their communities. The effort to create a technology-intensive research and manufacturing "Tech Valley" along the middle Hudson River has involved multifaceted, sustained initiatives mounted under a succession of Republican and Democratic governors and centered on the region's research universities. The development effort has been based primarily on nanotechnology, the new science of manipulating materials at the molecular or atomic scale to build microscopic devices, materials, and physical structures with practical applications.

The employment impact of the Tech Valley effort is observable in Table 1.1, which depicts changes in manufacturing employment, by decade, in New York State's metropolitan areas. As can be seen, these figures are unrelievedly bleak for Upstate New York for the forty years between 1970 and 2010, with almost all metro areas experiencing double-digit percentage declines decade after decade. However, in the 5-year period from 2010 to 2014, during the worst recession since the Great Depression, the Albany/Troy/Schenectady metro area saw a surge in manufacturing

---

2005). In 2001, an academic expert on technology-driven economic development, Stuart W. Leslie, published what amounted to a postmortem verdict on an effort in the 1980s by the then-president of Rensselaer Polytechnic Institute, George Low, to replicate Silicon Valley in New York's economically struggling Capital Region. Noting that Low's effort fell short of its objectives, he concluded that his initiatives "could not overcome the regional disadvantage that kept them from competing effectively with emerging high technology centers in other parts of the country. [They] illustrate the limits of local actions in the face of large corporate restructuring and regional economic decline." Stuart W. Leslie, "Regional Disadvantage: Replicating Silicon Valley in the Capital Region," *Technology and Culture* (2001), p. 238.

employment of nearly 21%, the only increase of such magnitude recorded by any of the Upstate metro areas in nearly half a century. The Buffalo/Niagara Falls area—where the Tech Valley model is in early stages of replication—saw a smaller but striking increase of about 6%, contrasting sharply with a 36% decline in the preceding decade. In August 2016 the president and CEO of the Federal Reserve Board of New York acknowledged that the Capital Region, in contrast to most of the rest of the state, was enjoying "sturdy growth, with significant manufacturing job gains in industries, such as nanotechnology, that are benefitting from partnerships with the region's higher education institutions."[11]

The basic premise upon which the Tech Valley effort rests is that scientific knowledge generated in universities and research centers can be applied locally in a manner that stimulates economic growth, creates manufacturing jobs, and substantially increases prosperity. That proposition has arguably been validated—albeit unevenly and episodically—by commercially oriented research universities like MIT and Stanford and land-grant colleges and universities in a number of states.[12] However, notwithstanding such examples, innovation-based economic development has proven difficult to implement systematically, reflecting factors such as local culture, academic resistance to commercially oriented research, political discontinuities, inadequacy of resources committed, and the occasionally shattering impact of international competitive challenges, ranging from dumping to the abrupt offshore relocation of manufacturers and research centers.[13]

## 1.1 Innovation Clusters

Since the days of Governor Nelson Rockefeller, New York political, academic, and business leaders struggling to revive the Upstate economy have been aware of experiential reference points such as Silicon Valley, Boston's Route 128, and North Carolina's Research Triangle, where great research universities drove economic growth and regional prosperity—areas that have come to be known as "innovation clusters." They were also aware of earlier local examples of knowledge-based economic growth, such as the establishment of the General Electric electrochemical

---

[11] "Feds: Upstate New York Job Growth 'Flat,'" *Syracuse.com* (August 18, 2016).

[12] See *generally* David Kaiser (ed.), *Becoming MIT: Moments of Decision* (Cambridge, MA, and London: The MIT Press, 2010); C. Stewart Gillmore, *Fred Terman at Stanford: Building a Discipline, a University, and Silicon Valley* (Stanford CA: Stanford University Press, 2004); N. Rosenberg and R. R. Nelson, "American Universities and Technical Advance in Industry," *Research Policy* (1994) 23:326.

[13] For example, a dynamic innovation cluster of photovoltaic energy enterprises which grew up in Toledo, Ohio, driven by technological support from the University of Toledo, was devastated by Chinese dumping of photovoltaic cells in and after 2012. National Research Council, Charles W. Wessner (ed), *Best Practices In State and Regional Innovation Initiatives: Competing in the 21st Century* (Washington, DC: The National Academies Press, 2013), pp. 135–140.

laboratories at the turn of the twentieth century in Schenectady by an MIT professor, Willis Whitney, which eventually spawned a huge manufacturing complex[14] and the innovation-based industrial development around RPI in Troy, New York, sometimes described as a nineteenth century Silicon Valley.[15] Fredrick Terman, the Stanford University provost known as the father of Silicon Valley, consulted with New York leaders in the 1960s as part of his broader effort to propagate his innovation-based development model in other regions.[16]

In addition to such real-world examples, state policymakers have been able to draw more recently on a growing body of learning developed in academia which systematically examines the dynamics and geography of knowledge-based regional economic development.[17] These studies emphasize the importance of what has become known variously as public–private partnerships (PPPs), "innovation clusters," and the "Triple Helix," mutually reinforcing collaborations of geographically concentrated research universities, government agencies, and private companies systematically promoting innovation and its practical application within a region. New York Governor Andrew Cuomo summarized this model in 2016 in the following way:

> *Business development plans are best when focused on synergistic economic clusters. Clusters are a number of businesses that are involved in the same general enterprise, usually including research and development, manufacturing or related supply chain companies. Clusters often occur in concert with an institution of higher learning. The new*

---

[14] George Wise, *Willis R. Whitney: General Electric and the Origins of US Industrial Research* (New York: Columbia University Press, 1985).

[15] Thomas P. Carroll, "Designing Modern America in the Silicon Valley of the Nineteenth Century," *RPI Magazine* (Spring 1999).

[16] Stuart W. Leslie and Robert H. Karagon, "Selling Silicon Valley: Fredrick Terman's Model for Regional Advantage," *Business History Review* (Winter 1996).

[17] The phenomenon of the industry cluster, in which enterprises in a given sector group themselves together in a particular location, thus enhancing their collective competitiveness, was first examined by the great nineteenth century British economist Alfred Marshall, who examined the Sheffield industrial district in Great Britain. (See Fiorenza Belussi and Katia Caldon, "At the Origin of the Industrial District: Alfred Marshall and the Cambridge School," *Cambridge Journal of Economics* (2009)). Marshall identified the elements of a successful cluster, which have come to be known as "Marshall's trinity"—supply chain linkages, a pool of skilled labor, and "knowledge spillovers" reflecting the availability in the cluster of market intelligence, new designs and applications, and improvements in manufacturing technique. Paul Krugman, a winner of the Nobel Prize in economics, stressed the abiding importance of Marshall's trinity in *Geography and Trade* (Cambridge, MA: The MIT Press, 1991). Marshall's ideas were carried forward, refined, and popularized by Michael Porter, who argued in his influential 1990 book, *The Competitive Advantage of Nations*, that in advanced economies, regional clusters of related firms and industries, rather than individual firms or sectors, were the principal source of economic competitiveness as well as rising regional employment and per-capita income levels. (See Michael Porter, "Clusters and the New Economics of Competition," *Harvard Business Review* (December 1998)). The role played by research universities and supportive government organizations in the development of regional technology-intensive industry clusters has been articulated in academia and refined by Henry Etzkowitz, Loet Leydesdorff, and others, into a model known as the "Triple Helix." Henry Etzkowitz, *The Triple Helix: University-Industry-Government Innovation in Action* (New York and London: Routledge, 2008).

> *economic clusters often build from that region's existing assets. For example, Rochester, which has a long history of engineering and academics coming from Kodak and the Xerox era, is now developing a cluster economy in the photonics area. Often the key to the future is updating the past.*[18]

The formula described by Governor Cuomo is straightforward, but its successful implementation is not. Terman, for example, consulted widely with leaders of other regions on the dynamics of Silicon Valley, but despite occasional successes (most notably Taiwan's Hsinchu Science-Based Industrial Park), the model proved very difficult to adapt in other regions.[19] Since then numerous "Triple Helix" initiatives have failed more or less completely.[20] Governor Cuomo warns that there can be "no copycat economic blueprints," that "what works for one region may not necessarily work for another region," and that a region's cluster strategy should be "built on that region's assets, institutions and advantages."[21] There is ample academic and empirical support for this cautionary perspective.[22] At the same time, with appropriate qualifiers and caveats, the saga of New York's Tech Valley arguably *does* represent a model applicable elsewhere in Upstate New York as well as in other US states and regions and around the world. It would not be the first occasion in which the Empire State served as a model for the rest of the United States with respect to institutional—and engineering—innovations.[23]

---

[18] Andrew M. Cuomo, "Rebuilding the Upstate Economy," *Huffington Post* (September 1, 2016). In the European Union, economic development strategies which seek to identify and build upon a region's strengths, including the industrial legacy described by Governor Cuomo, have come to be known as "smart specialization." See generally Slavo Radosevic, et al. (eds), *Advances in the Theory and Practice of Smart Specialization* (London, San Diego, Cambridge, MA, and Oxford: Elsevier, 2017).

[19] Stuart W. Leslie and Robert H. Karagon, "Selling Silicon Valley: Fredrick Terman's Model for Regional Advantage," *Business History Review* (Winter 1996). See also Vivek Wadwha, "Silicon Valley Can't Be Copied," *MIT Technology Review* (July 3, 2013). Hsinchu Science Park was founded by Kuo-Ting Li, a Taiwanese policymaker known retrospectively as the "architect of Taiwan's economic miracle." He consulted with Terman as to how Taiwan could create its own version of Silicon Valley and applied the Terman model with spectacular results. "Fred Terman, the Father of Silicon Valley," *Net Valley* (October 21, 2010).

[20] Andrew M. Cuomo, "Rebuilding the Upstate Economy," *Huffington Post* (September 1, 2016).

[21] Franz Todtling and Michaela Tripple, "One Size Fits All? Towards a Differentiated Regional Innovation Policy Approach," *Research Policy* (2005).

[22] "What We Can Learn from £100 m and 10 Years Wasted on the Technique Programme," *WalesOnline* (June 1, 2013). *The Economist* observed in 2013 that "the world is littered with high-tech enclaves that fail to flourish. Malaysia's biotech valley has been nicknamed 'Valley of the Bioghosts.'" "Crazy Diamonds," *The Economist* (July 20, 2013). For case studies of failure of the Triple Helix model, see Denis Gray, Eric Sundstrom, Louis G. Tomasky, and Lindsey McGowen, "When Triple Helix Unravels: A Multi-Case Analysis of Failures in Industry-University Cooperative Research Centers," *SAGE Journals* (October 1, 2011).

[23] The construction of the New York Thruway, completed in 1956, served as a model and established standards for the construction of the federal interstate highways, a transformational project which began soon afterward. New York's Thruway Authority "foreshadowed the creation of the Highway Trust Fund, the legislation that ensured that all federally—collected, motorist-generated revenue would be earmarked for highway construction." Michael R. Fein, *Paving the Way: New York Roadbuilding and the American State, 1880–1956* (Lawrence, KA: University of Kansas

## 1.2 A Manufacturing Revival?

Tech Valley is not an isolated phenomenon, but a particularly dramatic example of a trend becoming observable in other Rust Belt regions (and comparable old industrial regions of Europe) that are improbably becoming "hotspots of innovation."[24] In addition, although it is premature to draw definitive conclusions, there are indicators—noted by surveys conducted by institutions such as MIT and the Boston Consulting Group—that paralleling the emergence of new centers of innovation, a significant "re-shoring" of manufacturing and manufacturing jobs is under way in North America, driven by factors such as US advances in factory automation and sharply lower US natural gas prices.[25]

## 1.3 Advantages of Geographic Co-Location

One factor guiding high-tech companies' locational decision-making is the advantages associated with having at least some production functions in close geographical proximity to process R&D.[26] It follows that if sophisticated centers of relevant R&D are present in the United States, they constitute an incentive for the co-location of related manufacturing operations. New York policymakers gambled that large investments in local research infrastructure would eventually draw manufacturing activity and jobs to the Capital Region. This bet has paid off.

Physical proximity of applied research and manufacturing offers significant advantages. "Tacit knowledge," or "know-how," is the real-world ability to make things, fix problems, and adjust operations on the basis of actual experience rather than from knowledge derived from textbooks or theory. While much scientific and technical knowledge can be written down and stored or transmitted digitally around the world, know-how is gained primarily through hands-on operational experience,

---

Press, 2008), p. 182.

[24] This phenomenon is explored in a recent book by Antoine van Agtmael and Fred Bakker, *The Smartest Places on Earth: Why Rustbelts are Emerging as Hotspots of Global Innovation* (New York: Public Affairs, 2016).

[25] "Reshoring: A Boost in American Manufacturing," *Machine Design* (August 30, 2017); "Manufacturers Bring Back Jobs to Central Mass," *WBJournal* (September 4, 2017); "Manufacturing Bringing the Most Jobs Back to America," *USA Today* (April 23, 2016); Lindsay Oldenski, *Reshoring By US Firms: What Do the Data Say?* (Peterson Institute for International Economics, September 2015); David Simchi-Levi, "US Re-Shoring: A Turning Point," *MIT Forum for Supply Chain Innovation 2012 Annual Re-Shoring Report* (Cambridge, MA: The MIT Press, 2012); TD Economics, "Onshoring, and the Rebirth of American Manufacturing," (October 15, 2012); "US Manufacturers 'Relocating' from China," *Financial times* (September 23, 2013); "Overseas Jobs are Coming Home—S.C. Business," Columbia, SC, *The State* (September 8, 2013).

[26] Dieter Hagmann of Stanton Chase, quoted in "Reshoring Is an Issue for Europe Too," *Finanz & Wirtschaft* (October 23, 2013).

close observation of actual results, and trial-and-error experimentation. It is transmitted through on-site demonstration and mentoring, much as masters have passed skills to apprentices over generations.[27] Physical presence in the geographic locales where know-how is generated powerfully enhances its transmission and absorption. Thus in the development and refinement of highly complex manufacturing techniques, there is usually no substitute for the creation of an actual factory environment in which tools, materials, and processes can be tried out, defects identified, and fixes developed and improved by people on the spot.[28] "The key experimental tool of the…engineer is…the pilot plant, and inferences drawn from experimental data provided by such plants."[29]

Most of the large expenditures made by the State of New York in the creation of Tech Valley have involved acquisition of facilities and equipment necessary to create such genuine manufacturing environments for research purposes, in some cases creating capabilities that do not exist anywhere else on earth. An influx of high-tech companies and skilled engineers, scientists, technicians, and managers has been the result.

## 1.4 Betting on Nanotechnology

Nanotechnology is the manipulation of matter with at least one dimension sized at 1 to 100 nanometers (nm), a nanometer being one-billionth of a meter or 0.000000001 m. One nanometer is equivalent in size to eight hydrogen atoms lined

---

[27] Michael Polanyi, a scientist who closely studied this phenomenon, observed in 1958 that much knowledge cannot be transmitted by prescription, only by example, and that "this restricts the range of diffusion to that of personal contacts… Craftsmanship tends to survive in closely circumscribed local traditions." Michael Polanyi, *Personal Knowledge: Toward a Post-Critical Philosophy* (Chicago: University of Chicago Press, 1958), p. 52. Eugene S. Ferguson, an engineer by training, observed in 1992 that "an engineer's intelligent first response to a problem that a worker brings in from the field is 'Let's go see.' It is not enough to sit at one's desk and listen to an explanation of a difficulty. Nor should the engineer refer immediately to drawings or specifications to see what the authorities say. The engineer and the worker must go together to the site of the difficulty if they expect to see the problem in the same light. There and only there can the complexities of the real world, the stuff that drawings and formulas ignore, be appreciated." Eugene S. Ferguson, *Engineering and the Mind's Eye* (Cambridge MA: The MIT Press, 1992), p. 56.

[28] Nathan Rosenberg and Edward Steinmueller noted the example of experiments conducted at Stanford University between 1916 and 1926 which subjected aircraft propellers to wind-tunnel testing. The tests were conducted because "there was no way in which the body of scientific knowledge would permit a more direct determination of the optimal design of a propeller given the fact that the propeller operated in combination with both the engine and the airframe… and it must be compatible with the power-output characteristics of the former and the flight requirements of the latter." The tests represented "the development of a specialized methodology that could not be deduced from scientific principles, although it was obviously not inconsistent with those principles." Nathan Rosenberg and Edward Steinmueller, "Engineering Knowledge," *Industrial and Corporate Change* (October 2013).

[29] Nathan Rosenberg and Edward Steinmueller, "Engineering Knowledge," *Industrial and Corporate Change* (October 2013), p. 21.

up side by side, and a human hair is 80,000–100,000 nm wide. The term embraces a broad range of activities that occur at or below this size threshold, engaging fields such as semiconductor manufacturing, molecular biology and engineering, microfabrication, and organic chemistry. By the end of the 1990s, the dramatic potential of nanotechnology was becoming widely apparent, and in 2000 President Clinton launched the National Nanotechnology Initiative (NNI) to coordinate federal efforts in the field, doubling federal outlays on nanoscale science and engineering.[30] Since the origins of nanotechnology as a concept, New York State has been a leader in relevant research, with IBM, Rensselaer Polytechnic Institute, and the State University of New York at Albany pioneering the investigation of nanotech themes.[31] The high-tech industrial renaissance that is now plainly evident in the middle Hudson Valley has been built on this scientific and engineering legacy.

### 1.4.1 The Semiconductor Industry: Pioneer in Nanotechnology

The semiconductor industry is the first major industry in which much of the technology involved in research and manufacturing has crossed the nanoscale threshold, with feature sizes in semiconductor devices currently ranging as small as 7 nm.[32] As a result, although nanotechnology research ranges across many fields and disciplines, the first decades of the effort to create a nanotechnology-based Tech Valley in the Capital Region naturally centered around outreach to the semiconductor industry. In this way, the state came to embrace an industry of enormous strategic and national security importance with strong potential for creating high-value, high-compensation jobs but which also faced daunting and, at times, existential risks.

#### 1.4.1.1 A Vital Industry

Semiconductors are the basic building blocks and enabling technology of all advanced information, communications, automation and electronics systems which comprise the core of the modern global economy, and for this reason, in East Asia

---

[30] National Research Council, *A Matter of Size: Triennial Review of the National Nanotechnology Initiative* (Washington, DC: The National Academies Press, 2006).

[31] "If You Build It They Will Come," *The Chronicle of Higher Education* (February 7, 2013); "IBM Team Makes Atomic-Scale Circuitry Breakthrough," *Watertown Daily Times* (February 3, 2000); "RPI Creates Center for Nanotechnology Studies," Albany, *The Times Union* (March 30, 2001).

[32] GlobalFoundries' wafer fabrication facility in Malta/Stillwater, New York, is currently manufacturing semiconductor wafers with technology nodes as small as 7 nanometers (GlobalFoundries, May 2017). See generally Jan G. Korvink and Andreas Greiner, *Semiconductor for Micro- and Nanotechnology: An Introduction for Engineers* (Wileys, 2002).

they are sometimes called "the rice of industries."[33] (See Box 1.1.) Their progressively increasing performance capability has enabled technological advances and reductions in cost across a broad range of other industries, including automobiles, robotics, telecommunications, energy generation and transmission, healthcare, and lighting. Semiconductor technology makes possible such basic features of contemporary life as the Internet, the Global Positioning System, and the iPhone.

> **Box 1.1: Semiconductors and the Dawn of Integrated Circuits**
> "Semiconductor" is a generic term for devices that are capable of controlling the flow of electrical signals via conductive impurities that are introduced into a pure element, usually silicon, or a compound such as gallium arsenide. Integrated circuits (ICs) are semiconductor devices in which multiple electronic functions are fabricated and connected on a single chip (as opposed to discrete semiconductor devices which exist and are connected separately). The first integrated circuit was created in 1958 by Jack Kilby of Texas Instruments, who demonstrated that by incorporating numerous connected electronic functions on a single block of silicon, the practical problems associated with wiring together numerous separate electronic components could be avoided. Months later Robert Noyce, who later co-founded Intel, found a way to interconnect all components on a chip that would facilitate mass production. Called "solid state" technology, it revolutionized the modern world and heralded the advent of the information age. Today integrated circuits incorporate on a single device a vast array of logic, memory, communications, and sensor functions. Advances in semiconductor technology continue to enable new forms of human endeavor and to destabilize established economic and social structures.

Semiconductors play a pervasive role in national security, forming the core of all smart weapons systems, defense communications networks, aerial and satellite surveillance systems, and every major military platform from aircraft to warships to armored vehicles. In twenty-first century warfare, offensive and defensive capability is increasingly a function of semiconductor-based electronic measures and countermeasures. Cybersecurity of strategic systems can be enhanced—or compromised—based on the quality and security of semiconductor devices.

---

[33] G. Dan Hutcheson, "Economics of Semiconductor Manufacturing," in Yoshio Nishi and Robert Doering (eds.), *Handbook of Semiconductor Manufacturing Technology (2d Ed.)* (Boca Raton, London, and New York: CRC Press, 2017), p. 1137.

[34] Recent academic work supports the proposition that semiconductor manufacturing results in much higher local employment multipliers than other manufacturing and services activities. Enrico Moretti of the University of California at Berkeley concludes, based on an analysis of 11 million American workers in 320 metropolitan areas, that high-tech manufacturing, of which semiconductor fabrication is arguably the most advanced expression, supports dense clusters of supply chain and services firms which themselves pay above-average salaries and require specialized services.

Semiconductor research and manufacturing activities are also widely recognized as sources of major economic benefits in regions in which they are located.[34] In the United States, the semiconductor manufacturing workforce earns more than twice the average for all US manufacturing, and local expenditures by these workers contributes to a higher-than-average number of indirect and induced local jobs[35] The economic and national security importance of semiconductors is so fundamental that established and aspiring global and regional powers are deploying extensive resources to create and attract semiconductor research and manufacturing activities to locations within their own borders, a process that has led to competition between countries for inward semiconductor investment. New York's effort to attract semiconductor investment has thus pitted it against national governments in East Asia and Europe.

### 1.4.1.2 A Volatile Industry

Competition in the semiconductor industry is resource-intensive, unpredictable, and risky, characterized by spectacular returns on the best innovations and periodic, punishing shakeouts. In the first decade after the team led by Robert Noyce perfected the first commercially viable integrated circuits, new companies introduced commercial electronic components based on integrated circuit technology.[36] Companies raced each other to create newer, faster, and more sophisticated chips and to keep up with or exceed Moore's Law, an industry rule of thumb holding that the number of transistors on a chip doubles roughly every 2 years.[37] Competition in product design, price, service, and quality was ferocious and firms emerged, disappeared, or fragmented as the technology advanced.[38]

---

High-tech manufacturers and supply chain firms require additional local services, including sophisticated information technology, graphic design, business consultancy and legal and security services, and restaurants, hotels, healthcare, and other services. Moretti concludes that for each new high-tech job in a metropolitan area, five additional local areas are created outside of high tech over the long run. Enrico Moretti, *The New Geography of Jobs* (Boston and New York: Mariner Books, 2013). This is consistent with the Semiconductor Industry Association's estimate—that every job in the semiconductor manufacturing sector fosters nearly five indirect jobs nearby. Other estimates are even higher. Semiconductor Industry Association, *US Semiconductor Industry Employment* (January 2015).

[35] In 2015, semiconductor manufacturing workers earned an average of $138,100 per year, compared with $64,305 for US manufacturing workers generally. Average wage date from Bureau of Labor Statistics, cited in Michaela D. Platzer and John F. Sargent, *US Semiconductor Manufacturing Trends, Global Competition, Federal Policy* (Washington, DC: Congressional Research Service, June 27, 2016), p. 8.

[36] "Tales of Silicon Valley Past: Legendary Founders Talk About Early Days at Fairchild," *San Jose Mercury News* (May 13, 1995); "Growth of Silicon Empire: Bay Area's Intellectual Ground Helped Sprout High Technology Industry," *The San Francisco Chronicle* (December 27, 1999).

[37] See generally "Moore's Law: Past, Present and Future," *IEEE Spectrum* (June 1997).

[38] National Research Council, *Competitive Status of the U.S. Electronics Industry: A Study of the Influences of Technology in Determining International Competitive Advantage* (Washington, DC: National Academy Press, 1984), p. 43.

The industry's rapid growth in the 1960s was fueled by federal government procurement of integrated circuits for the Minuteman and Apollo programs, and by the curriculum of Stanford University's Department of Electrical Engineering, which kept close pace with the rapid evolution of integrated circuit design and fabrication technology.[39] In contrast to many US manufacturing sectors, which were at the time largely content to serve the domestic market, in the 1960s and 1970s US makers of integrated circuits established manufacturing facilities in Europe and assembly and test operations in the Far East.[40] In the US semiconductor industry's first decades, most firms were vertically integrated, keeping in-house such varied activities as research, design, fabrication, assembly testing, and packaging. (See Box 1.2.)

**Box 1.2: The Complexity of Semiconductor Fabrication**
Semiconductor fabrication is quite possibly the most complex manufacturing activity in human history, involving over 300 process sequences, exotic materials, and multiple, interacting pieces of complex precision equipment. From the industry's inception, it has tended to conduct research into new and refined production methods ("process R&D") directly on the factory floor rather than in separate research laboratories. When Noyce and Gordon Moore established Intel, they recalled the problems their former firm, Fairchild, had experienced transferring technology from its central research lab to its manufacturing lines. They created Intel with no central lab. Process R&D and manufacturing were co-located, enabling process designers to make empirical choices and try them out on the spot to see if they worked. This approach, which proved key to Intel's success, was eventually carried over into the Sematech consortium, dramatically accelerating the developmental pace of the US semiconductor industry.[41] A 2009 survey of US semiconductor producers concluded that process R&D requires proximity to manufacturing operations.[42]

---

[39] Nathan Rosenberg, "America's Entrepreneurial Universities," in David M. Hurt (ed.), *The Emergence of Entrepreneurship Policy; Governance and Growth in the US Knowledge Economy* (Cambridge: Cambridge University Press, 2003); Stuart W. Leslie, "The Biggest Angel of All: The Military and the Making of Silicon Valley," in Martin Kenney (Ed.), *Understanding Silicon Valley: The Anatomy of an Entrepreneurial Region* (Stanford: Stanford University Press, 2000).

[40] In Europe, national governments initially sought to promote "national champions"—large, vertically integrated electronics firms producing chips primarily for internal consumption—but over time this approach was largely abandoned in favor of strategies building on Europe's superb research institutions and its tradition of research collaborations. European firms emerged as leaders in niche areas of design, toolmaking, and specialized materials.

[41] Elias C. Carayannis and James Gover, "The Sematech-Sandia National Laboratories Partnership: A Case Study," *Technovation* (2002).

[42] Semiconductor Industry Association, *Maintaining America's Competitive Edge: Government Policies Affecting Semiconductor Industry R&D and Manufacturing Activity* (March 2009).

## 1.4 Betting on Nanotechnology

The industry is notoriously unstable. In the 1970s, the first East Asian country to create a major indigenous semiconductor industry—Japan—began to establish a competitive position in world semiconductor markets, entering the market for memory devices in the early 1980s with shattering impact on US producers. Japan's come-from-behind developmental effort benefited from comprehensive government support measures and resulted in a series of high-profile trade disputes with the United States. Japanese semiconductor makers surpassed the US industry in some key product areas including semiconductor equipment and materials and memory devices as well as in manufacturing competitiveness, and for a time it looked as though Japan would dominate the world market. At that point, the US government—recognizing the strategic importance of the industry—deployed unprecedented policy measures on behalf of domestic chip makers, most notably trade retaliation and the formation of the Sematech manufacturing R&D consortium. While the US–Japan relationship in microelectronics eventually stabilized, the industry itself did not. During and after the 1980s, the global industry was periodically destabilized by sharp cyclical downturns and the advent of aggressive, disruptive new players (Korea, Taiwan, China), new business models (foundries, vertical disaggregation), and new technologies (digital imaging, the iPhone).[43]

The global semiconductor industry is undergoing another profound upheaval, driven by looming, fundamental physical challenges to further miniaturization. The industry's traditional assumption that chip complexity would increase according to Moore's Law simply cannot be sustained as circuit sizes approach the molecular and atomic or subatomic level. Designing and manufacturing semiconductors at the technological cutting edge is becoming so complex and so costly that it is forcing a dramatic restructuring on the industry.[44] At present a few industry giants, such as Intel and Samsung ("integrated device makers" or IDMs) continue to operate vertically integrated semiconductor production chains performing design, wafer fabrication, assembly, test, and packaging functions themselves; however, this business model is proving unsustainable, with a more disaggregated approach emerging in which separate firms specialize in different stages of semiconductor production such as design, fabrication, and packaging.

The disaggregation of the semiconductor industry has been driven by the advent of the foundry business model. An increasing number of semiconductor producers have abandoned manufacturing altogether, becoming "fabless" firms which design

---

[43] In an extraordinary demonstration of technological catch-up and eventual leadership in the mid-1990s, South Korea's Samsung surpassed the United States and Japan in a key semiconductor memory device category—dynamic random access memories (DRAMs)—and continues to dominate this product segment today. Linsu Kim, "The Dynamics of Samsung's Technological Learning in Semiconductors, *California Management Review* (Spring 1997). The government of Taiwan supported technology transfer and creation of new companies which culminated in the achievement of technological parity in semiconductor manufacturing with US, Japanese, and Korean firms. Interview with Ding-Hua Hu, "Taiwanese IT Pioneers: Ding-Hua Hu," recorded February 10, 2011 (Computer History Museum, 2011).

[44] See Laszlo B. Kich, "End of Moore's Law: Thermal (Noise) Death of Integration in Micro and Nano Electronics," *Physics Letters* (December 2002).

semiconductors to be fabricated by foundries, which specialize in manufacturing chips designed by other firms. By concentrating solely on production functions, foundries have been able to achieve better yields, update processes and systems, and incorporate next-generation tools more rapidly than most other manufacturers. Similarly, a significant number of fabless firms have enjoyed dramatic success by concentrating solely on design, marketing, and sales.[45] In 2016, foundries accounted for well over 25% of total global semiconductor production,[46] and the number of IDMs is shrinking, trends which have directly affected New York. The two IDM semiconductor makers with which the state had its principal relations a decade ago—IBM and AMD—have now gone fabless, and their manufacturing operations have been acquired and consolidated by GlobalFoundries.

#### 1.4.1.3 The National Security Dimension of a Changing Industry

The disaggregation of the semiconductor industry has national security implications. At the end of the 1990s, as a growing proportion of semiconductor fabrication was outsourced to foundries in Asia, the Pentagon began worrying about the security of chips fabricated offshore which were incorporated in US defense systems.[47] The result was a succession of US defense policy initiatives pursuant to which the Department of Defense contracted with designated "trusted" producers physically located in the United States to fabricate devices for use in military systems. IBM participated in this effort and fabricated chips for the military in its facilities in East Fishkill, New York, and Burlington, Vermont, until it divested its manufacturing assets in 2015. GlobalFoundries has taken over those facilities as well as the responsibility for carrying out IBM's defense commitments. In effect, GlobalFoundries' New York operations have become an important part of the industrial foundation upon which US national security rests.[48]

## 1.5 New York's Challenging Path Forward

The success which the Tech Valley effort has enjoyed to date was far from inevitable and over time has experienced major setbacks and near-disasters. An ambitious 1987 bid by the state to be the site of the newly forming Sematech microelectronics

---

[45] Major fabless firms include Qualcomm, Broadcom, Nvidia, Mediatek, and Xilinx.

[46] Based on TrendForce estimate of 2017 global foundry revenue of $57.3 billion and World Semiconductor Trade Statistics (WSTS) estimates of total global semiconductor revenues for 2017 of $409 billion. "Top Ten Foundries 2017," *Electronics Weekly* (December 1, 2017); http://www.wsts.org/76/Recent-News-Release.

[47] Defense Science Board, *Task Force on High Performance Microchip Supply* (Washington, DC: Department of Defense, February 2005).

[48] "U.S. Paves Roads to Trusted Fabs," *EE Times* (July 11, 2017), http://www.trustedfundryprogram.org.

research consortium ended in failure. In 1995, the state narrowly avoided what might well have been the death knell of its aspirations in nanotechnology when IBM was persuaded, at the eleventh hour, to abandon plans to move its headquarters out of New York. In 1999, an effort which was gaining momentum to establish a semiconductor manufacturing center in North Greenbush, New York, was halted in its tracks when the local Town Board voted against the project. Outbreaks of intramural infighting among local authorities have periodically jeopardized progress. In 2016, one of the figures associated with the Capital Region's nanotechnology effort, Professor Alain Kaloyeros of the SUNY Polytechnic Institute (SUNY Poly), was indicted on state and federal bid-rigging charges and, coincidentally or not, several major nanotechnology projects have been halted, at least temporarily, including a proposed semiconductor manufacturing facility at Marcy, New York.

The effort to create Tech Valley and more recently the attempt to expand the model across Upstate New York have been widely disparaged and criticized. *The Wall Street Journal* derided the Albany initiative in 1999 as the "umpteenth" attempt by a community to "package itself as the next ... Silicon Valley."[49] As the state's investments in nanotechnology research grew, critics charged that the money was being wasted and that hoped-for new manufacturing jobs would never materialize. At this writing, while the success of the Albany-area developmental effort is increasingly regarded as an established fact, Governor Cuomo's initiative to replicate that success in Buffalo, Utica, Rochester, and other Upstate urban centers is coming under critical fire, notwithstanding promising indicators in Buffalo, where that effort was initiated.[50]

Put in proper perspective, New York's nanotechnology initiative is a recent expression of the state's traditional approach to economic development, which has been characterized by bold, risky projects of outsized scale which, when successful, have proven transformational. Not infrequently, reflecting the Hudson Valley's role as the cradle of American engineering, the projects have involved bravura and precedent-setting feats of engineering.[51] Some of them provided technical and insti-

---

[49] "Even Bad Publicity is Good," Albany, *The Times Union* (November 28, 1999).

[50] "Buffalo Boondoggle," *New York Post* (March 6, 2016). "Spending Billions While Killing Jobs," *New York Post* (March 28, 2017). Former state Assemblyman Richard Brodsky, whose committee in the legislature oversaw state economic development projects, said with respect to Governor Cuomo's programs in 2016 that "what I'd say is, change course. If [Cuomo] insists on maintaining these kinds of policies, more and more the evidence will show they're not working. And he'll be forced to defend the indefensible." EJ McMahon, with the fiscal oversight group The Empire Center, commented that "I would hope the public would ask, 'Hey, wait a minute. Why are you spending all this money?'" See "Cuomo Manages the Fallout from Corruption Scandal," WBFO (September 29, 2016).

[51] Daniel H. Calhoun, *The American Civil Engineer: Origins and Conflict* (Cambridge, MA: Technology Press, 1960.) The first technical and engineering college in the United States was the Military Academy at West Point. Rensselaer Polytechnic Institute (RPI), founded in 1824, is the oldest degree-conferring engineering school in the English-speaking world. Union College established an engineering program in 1845. The engineers who built America's railroads, canals, and bridges in the nineteenth century were overwhelmingly graduates of these three institutions. Frederick Rudolph, *Curriculum: A History of the American Undergraduate Course of Study Since 1636* (San Francisco: Josey Bass, 1977), p. 63.

tutional models for subsequent federal initiatives implemented on a nationwide scale.⁵² All of these projects hit snags along the way, and, given the costs and risks involved, were controversial, requiring strong leadership to move them forward. Most famously, Governor DeWitt Clinton's effort to connect the Great Lakes with the Atlantic seaboard via a 363-mile canal, involving unprecedented engineering challenges, was mocked by critics and the newspapers as "Clinton's Ditch" and "Clinton's Folly." But at a stroke, completion of the Erie Canal in 1825 ensured New York City's ascension to status of the leading commercial center in the world.⁵³ More recent examples of transformational New York projects are the construction of the first Tappan Zee Bridge and the New York Thruway.⁵⁴

---

⁵²The creation of the New York State Thruway, completed in 1955, "provided an important model for the nation's Interstate Highway System. Its experience in executing public works projects on a grand scale was instrumental in opening the way and setting the standards for the federal interstate highways." The Thruway Authority provided an important institutional model for the federal Highway Trust Fund. Michael R. Fein, *Paving the Way: New York Road Building and the American State, 1880–1956* (Lawrence: University Press of Kansas, 2008), p. 231.

⁵³Evan Cornog, *The Birth of Empire: Dewitt Clinton and the American Experience* (Oxford: Oxford University Press, 1998).

⁵⁴The Tappan Zee Bridge, opened in 1955, bridged the Hudson at one of its widest points, between Tarrytown and Nyack, "a clear expression of engineering confidence," and was based on a novel design utilizing eight floating caissons. During its construction, it was lampooned as a "basic error," "costly," and of "freak design." *The New York Times* advocated an alternative site at a narrower crossing point, a suggestion that was rejected by the Dean of the Rensselaer Polytechnic Institute, a design engineer, who said that the important thing was "not the width of the body of water you want to cross but the number of people who want to cross there." In fact, when the bridge opened it "prompted a general reordering of economic life" all along the lower Hudson. The west shore city of Kingston, previously viewed as being on the "wrong shore" of the river (e.g., the side opposite New York City), abruptly found itself to be on the "right" shore going forward. The city experienced an explosive economic boom. Among other things, IBM established a plant in Kingston, eventually bringing over 7000 jobs. Michael R. Fein, *Paving the Way: New York Road building and the American State, 1880–1956* (Lawrence: University of Kansas Press, 2008), pp. 215–216; "Kingston: The IBM Years Gives a Peek into Tech Giant's History," *Hudson Valley Magazine* (June 4, 2014).

**Image 1.1** SUNY Poly Colleges of Nanoscale Science and Engineering, Albany, New York

# Bibliography[55]

As US Economy Races Along, Upstate New York is Sputtering. (1997, May 11). *The New York Times*.
Bacheller, J. (2016, October 4). The Decline of Manufacturing in New York and the Rust Belt. *Policy by the Numbers*.
Baily, M. N., and Bosworth, B. P. (2014, Winter). US Manufacturing: Understanding its Past and Potential Future. *Journal of Economic Perspectives*. 28(1).
Belussi, F., and Caldon, K. (2009). At the Origin of the Industrial District: Alfred Marshall and the Cambridge School. *Cambridge Journal of Economics*.
Buffalo Boondoggle. (2016, March 6). *New York Post*.
Calhoun, D. H. (1960). *The American Civil Engineer: Origins and Conflict*. Cambridge, MA: Technology Press.
Carayannis, E. C., and Gover, J. (2002). The Sematech-Sandia National Laboratories Partnership: A Case Study. *Technovation*.
Carroll, P. T. (1999, Spring). Designing Modern America in the Silicon Valley of the Nineteenth Century. *RPI Magazine*.
Cornog, E. (1998). *The Birth of Empire: Dewitt Clinton and the American Experience* Oxford: Oxford University Press.
Crazy Diamonds. (2013, July 20). *The Economist*.

---

[55] As noted in the front matter of this book, the study also drew on interviews carried out by the authors and numerous articles from *The Times Union* (Albany), *The Daily Gazette* (Schenectady), the *Albany Business Review* (Albany), the *Post-Star* (Glens Falls), *The Record* (Troy), *The Saratogian* (Saratoga Springs), *The Buffalo News* (Buffalo), The *Observer-Dispatch* (Utica), The *Daily Messenger* (Canandaigua), and the *Post-Standard* (Syracuse). These are not individually included in the bibliography.

Cuomo, A. M. (2016, September 1). Rebuilding the Upstate Economy. *Huffington Post*.
Cuomo Manages the Fallout from Corruption Scandal. (2016, September 29). WBFO.
Defense Science Board. (2005, February). *Task Force on High Performance Microchip Supply*. Washington, DC: Department of Defense.
Etzkowitz, H. (2008). *The Triple Helix: University—Industry—Government Innovation in Action* (New York and London: Routledge).
Feds: Upstate New York Job Growth 'Flat.' (2016, August 18). *Syracuse.com*.
Fein, M. R. (2008). *Paving the Way: New York Roadbuilding and the American State, 1880–1956*. Lawrence, KA: University of Kansas Press.
Ferguson, E. S. (1992). *Engineering and the Mind's Eye*. Cambridge, MA: The MIT Press.
Fred Terman, the Father of Silicon Valley. (2010, October 21). *Net Valley*.
Gray, D, Sundstrom, E., Tomasky, L. G., and McGowen, L. (2011, October 1). When Triple Helix Unravels: A Multi-Case Analysis of Failures in Industry-University Cooperative Research Centers. *SAGE Journals*.
Growth of Silicon Empire: Bay Area's Intellectual Ground Helped Sprout High Technology Industry. (1999, December 27). *The San Francisco Chronicle*.
Hamdami, K., Deitz, R., Garcia, R., and Cowell, M. (2005, Winter). Population Out-Migration from Upstate New York. In Federal Reserve Bank of New York (Buffalo Branch), *The Regional Economy of Upstate New York*.
Hartley, D. (2013, May 20). Economic Decline in Rust Belt Cities. *Economic Commentary*.
Hutcheson, G. H. (2017). Economics of Semiconductor Manufacturing. In Y. Nishi and R. Doering (Eds.), *Handbook of Semiconductor Manufacturing Technology* (2nd), p. 1137. Boca Raton, London, and New York: CRC Press.
IBM Team Makes Atomic-Scale Circuitry Breakthrough. (2000, February 3). *Watertown Daily Times*.
Income Inequality in the US by State, Metropolitan Area, and Country. (2016, June 16). *Economic Policy Institute*.
Kaiser, D. (Ed.) (2010). *Becoming MIT: Moments of Decision*. Cambridge, MA, and London: The MIT Press.
Kich, L. B. (2002, December). End of Moore's Law: Thermal (Noise) Death of Integration in Micro and Nano Electronics. *Physics Letters*.
Kim, L. (1997, Spring). The Dynamics of Samsung's Technological Learning in Semiconductors. *California Management Review*.
Kingston: The IBM Years Gives a Peek into Tech Giant's History. (2014, June 4). *Hudson Valley Magazine*.
Korvink, J. G., and Greiner, A. (2002). *Semiconductor for Micro- and Nanotechnology: An Introduction for Engineers*. Wileys.
Krugman, P. (1991). *Geography and Trade*. Cambridge, MA: The MIT Press.
Leslie, S. W. (2000). The Biggest Angel of All: The Military and the Making of Silicon Valley. In M. Kenney (Ed.), *Understanding Silicon Valley: The Anatomy of an Entrepreneurial Region*. Stanford: Stanford University Press.
Leslie, S. W. (2001). Regional Disadvantage: Replicating Silicon Valley in the Capital Region. *Technology and Culture*.
Leslie, S. W., and Karagon, R. H. (1996). Selling Silicon Valley: Fredrick Terman's Model for Regional Advantage. *Business History Review*.
Made in NY? Forget It, as State Loses to Others. (2017, March 10). Rochester, *Democrat & Chronicle*.
Manufacturers Bring Back Jobs to Central Mass. (2017, September 4). *WBJournal*.
Manufacturing Bringing the Most Jobs Back to America. (2016, April 23). *USA Today*.
Manville, M., and Kuhlmann, D. (2016, November 11). The Social and Fiscal Consequences of Urban Decline: Evidence from Large American Cities, 1980–2010. *Urban Affairs Review*.
Midnight in the Rust Belt. (2013, September 21). *Beltmag.com*.
Moore's Law: Past, Present and Future. (1997, June). *IEEE Spectrum*.
Moretti, E. (2013). *The New Geography of Jobs*. Boston and New York: Mariner Books.

National Research Council. (1984). *Competitive Status of the U.S. Electronics Industry: A Study of the Influences of Technology in Determining International Competitive Advantage*. Washington, DC: National Academy Press.

National Research Council. (2006). *A Matter of Size: Triennial Review of the National Nanotechnology Initiative*, Washington, DC: The National Academies Press.

National Research Council. (2013). *Best Practices In State and Regional Innovation Initiatives: Competing in the twenty-first Century*. C. W. Wessner (Ed.) (Washington, DC: The National Academies Press.

Oldenski, L. (2015, September). *Reshoring By US Firms: What Do the Data Say?* Peterson Institute for International Economics.

Overseas Jobs are Coming Home—S.C. Business. (2013, September 8). Columbia, SC, *The State*.

Platzer, M. D., and Sargent, J. F. (2016, June 27). *US Semiconductor Manufacturing Trends, Global Competition, Federal Policy*. Washington, DC: Congressional Research Service.

Polanyi, M. (1958). *Personal Knowledge: Toward a Post-Critical Philosophy*. Chicago: University of Chicago Press.

Porter, M. (1998, December). Clusters and the New Economics of Competition. *Harvard Business Review*.

Radosevic, S., et al. (Eds.) (2017). *Advances in the Theory and Practice of Smart Specialization*. London, San Diego, Cambridge, MA, and Oxford: Elsevier.

Reshoring: A Boost in American Manufacturing. (2017, August 30). *Machine Design*.

Reshoring Is an Issue for Europe Too. (2013, October 23). *Finanz & Wirtschaft*.

Rosenberg, N. (2003). America's Entrepreneurial Universities. In D. M. Hurt (Ed.), *The Emergence of Entrepreneurship Policy; Governance and Growth in the US Knowledge Economy*. Cambridge: Cambridge University Press.

Rosenberg, N., and Nelson, R. R. (1994). American Universities and Technical Advance in Industry. *Research Policy* 23:326.

Rosenberg, N., and Steinmueller, E. (2013, October). Engineering Knowledge. *Industrial and Corporate Change*.

Rudolph, F. (1977). *Curriculum: A History of the American Undergraduate Course of Study Since 1636*. San Francisco: Josey Bass.

Semiconductor Industry Association. (2009, March). *Maintaining America's Competitive Edge: Government Policies Affecting Semiconductor Industry R&D and Manufacturing Activity*.

Semiconductor Industry Association. (2015, January). *US Semiconductor Industry Employment*.

Simchi-Levi, D. (2012). US Re-Shoring: A Turning Point. *MIT Forum for Supply Chain Innovation 2012 Annual Re-Shoring Report*. Cambridge, MA: The MIT Press.

Spending Billions While Killing Jobs. (2017, March 28). *New York Post*.

Taiwanese IT Pioneers: Ding-Hua Hu. (2011, February 10). Recorded interview. Computer History Museum.

Tales of Silicon Valley Past: Legendary Founders Talk About Early Days at Fairchild. (1995, May 13). *San Jose Mercury News*.

TD Economics. (2012, October 15). Oenshoring, and the Rebirth of American Manufacturing.

Todtling, F., and Tripple, M. (2005). One Size Fits All? Towards a Differentiated Regional Innovation Policy Approach. *Research Policy*.

Top Ten Foundries 2017. (2017, December 1). *Electronics Weekly*.

The Upstate Economy is One of the Worst in the Country. (2016, September 16). *Politifact.com*.

US Manufacturers 'Relocating' from China. (2013, September 23). *Financial Times*.

U.S. Paves Roads to Trusted Fabs. (2017, July 11). *EE Times*.

van Agtmael, A., and Bakker, F. (2016). *The Smartest Places on Earth: Why Rustbelts are Emerging as Hotspots of Global Innovation*. New York: Public Affairs.

Wadwha, V. (2013, July 3). Silicon Valley Can't Be Copied. *MIT Technology Review*.

Watchdog Report: Upstate Sinks in a Sea of Legal Opioids. (2016, December 16). *Pressconnects*.

What We Can Learn from £100 m and 10 Years Wasted on the Technique Programme. (2013, June 1). *WalesOnline*.

# Chapter 2
# Upstate New York: Reversing Economic Decline Through Innovation

**Abstract** For a half century, New York's leaders have worked to reverse the state's decline relative to other states and regions in manufacturing, particularly in upstate areas experiencing an erosion of companies and jobs. This effort, based on the promotion of innovation driven by the state's universities and colleges, has been sustained by a succession of governors and legislative leaders of both political parties. In the Capital Region, New York's developmental effort, drawing on best practices from Silicon Valley and other dynamic regions, was sufficient to enable the state to make a strong albeit unsuccessful bid to attract Sematech (1988) and to persuade IBM to reverse a decision to move its headquarters out of the state (1995). These policies provided the foundation for further growth in the decade ahead.

Throughout its history, New York State has unabashedly embraced the view of Alexander Hamilton (a New Yorker) that government should actively and directly promote economic growth through measures such as infrastructure improvements, subsidies to industry, and workforce training.[1] New York's leaders have implemented a succession of monumental initiatives to spur economic growth—the Erie Canal, the infrastructure projects of Robert Moses to transform New York City, the construction of the turnpike system, and the creation of Olympics infrastructure at Lake Placid. It is altogether unsurprising that the state has responded to the challenge of eroding manufacturing in the upstate region with the massive and sustained public investments in high-tech manufacturing described in this study. Although Hamilton's views have remained controversial at the federal level down to the present day, there has been relatively little controversy within the state as to whether the

---

[1] Hamilton's *Report on Manufactures*, which recommended subsidies, workforce training and trade protection, was not implemented by Congress. Thomas Jefferson stated that Hamilton's proposed industrial policy "flowed from principles adverse to liberty, and was calculated to undermine and demolish the republic." Jefferson to Washington, September 1792, *cited in* Douglas A. Irwin, "The Aftermath of Hamilton's Report on Manufactures," *The Journal of Economic History* (September 2004), p. 813. For a survey of recent US policy debates on industrial policy, *see* Wendy H. Schact, *Industrial Competitiveness and Technological Advancement: Debate Over Government Policy* (Washington, DC: Congressional Research Service, December 3, 2013).

© Springer Nature Switzerland AG 2020
C. W. Wessner, T. R. Howell, *Regional Renaissance*, International Studies in Entrepreneurship 42, https://doi.org/10.1007/978-3-030-21194-3_2

upstate investments in high technology are appropriate. As Governor Mario Cuomo said of his economic policies,

> We have gone ahead with an economic development strategy that has nothing to do with ideologies and everything to do with common sense.[2]

The story of New York's "Tech Valley" strongly reinforces the proposition that public investments in university education and research infrastructure not only lead to economic growth but also to improved standards of living and quality of life. Tech Valley also illustrates how coherent state industrial policies can buffer and even counteract the powerful international competitive forces that have been unleashed by globalization. During the Great Recession, as US manufacturers were moving offshore, outsourcing production and supply chain functions to foreign manufacturers, or in some cases simply shutting down, New York's policies produced large new investments in high-tech manufacturing, thousands of new manufacturing jobs, thousands more construction jobs, and the beginnings of local supply chains supporting the new factories. As this study makes clear, these developments did not just happen in a market (or environment) free of government intervention.

## 2.1 The Challenge

A 2010 study of New York State politics observed that "[Postwar] New York has the dubious distinction of leading the nation in industrial decline."[3] In the years after World War II, all of the old industrial areas of the Northeast experienced deteriorating economic conditions, but as former Governor Hugh Carey expressed it in the mid-1970s, whereas the rest of the Northeast had a "common cold," New York had "a case of pneumonia."[4] After 1946, New York State failed to match the average rate of economic growth for the nation as a whole by virtually any economic measure, weakness which "persisted through the 1950s and 1960s and then intensified sharply in the 1970s."[5] New York experienced an out-migration of manufacturing

---

[2] Morton Schoolman and Alvin Magid (eds.) *Reindustrializing New York State: Strategies, Implications, Challenges* (Albany: SUNY Press, 1986), pp. 27–29. Economist Erik Reinert observed in 2007 that "…since its founding fathers, the United States has always been torn between two traditions, the activist policies of Alexander Hamilton (1755–1804) and Thomas Jefferson's (1743–1826) maxim that 'the government that governs least, governs best'. With time and usual American pragmatism, this rivalry has been resolved by putting the Jeffersonians in charge of the rhetoric and the Hamiltonians in charge of policy." Erik Reinert, *How Rich Countries Got Rich and Why Poor Countries Stay Poor* (London: Constable, 2007), p. 23.

[3] Edward V. Schneier, John Brian Murtaugh, and Antoinette Pole, *New York Politics: A Tale of Two States* (Armonk and London: M.E. Sharpe, 2010), p. 12.

[4] Timothy B. Clark, "The Frostbelt Fights for a New Future," *Empire State Report II* (October–November 1976), p. 332.

[5] Peter D. McClelland and Alan L. Magdovitz, *Crisis in the Making: The Political Economy of New York State Since 1945* (Cambridge: Cambridge University Press, 1981), p. 61.

## 2.1 The Challenge

companies, skilled workers, and young, educated adults, reflecting factors such as the availability of lower cost land and labor in other states, the advent of air conditioning in the Sun Belt, and what one survey euphemistically termed local firms "desire to escape a negative situation in their operating environment."[6]

Economic decline was particularly severe in Upstate New York, which experienced chronically stagnant or negative growth rates, depressed wages, and a hemorrhage of young adults leaving the region. By the mid-1990s, the region's principal manufacturing firms, including IBM, Xerox, Bausch & Lomb, and Eastman Kodak, were shedding thousands of jobs, and between 1995 and 1997, "departures exceeded arrivals in upstate New York by nearly 169,000 people."[7] Brookings scholar Rolf Pendall observed in 2003 that if Upstate New York were a separate state, it would be the third slowest growing state in the Union and would rank 48th out of 50 with respect to a number of economic indicators.[8] A 2012 SUNY study observed that "the upstate economy has been deteriorating for more than 50 years."[9]

It is beyond the scope of this study to examine in detail the factors underlying the economic decline of Upstate New York, except to note that similar experiences have been observable to a greater or lesser degree throughout the Northeast and Upper Midwest, the so-called Rust Belt, for nearly half a century. This study examines the emergence of "Tech Valley" in the Capital Region of Upstate New York, a manifestation of efforts by a generation of state leaders to reverse long-run economic decline through systematic promotion of innovation and technology-driven economic development. Along the way, some observers, citing specific setbacks, have characterized this effort as doomed, with state leaders said to be "unable to overcome the regional disadvantage that kept them from competing effectively with emerging high-technology centers in other parts of the country."[10] Such judgments have been confounded by the events described in this study. Improbable as it might seem, a recent report prepared for the U.S. Department of Commerce states that in nanotechnology, "Perhaps, New York State, and specifically the Albany area, may be the best role model in the U.S. for job creation from laborer to scientist."[11]

---

[6] Peter D. McClelland and Alan L. Magdovitz, *Crisis in the Making: The Political Economy of New York State Since 1945* (Cambridge: Cambridge University Press, 1981), p. 59.

[7] National Research Council, Charles W. Wessner (ed), *Competing in the 21st Century: Best Practices In State and Regional Innovation Initiatives: Competing in the 21st Century* (Washington, DC: The National Academies Press, 2013), p. 144.

[8] Rolf Pendall, *Upstate New York's Population Plateau: The Third-Slowest Growing State* (Washington, DC: The Brookings Institution, 2003).

[9] Robert F. Pecorella, "Regional Political Conflict in New York State," in Robert F. Pecorella and Jeffrey M. Stonecash, (eds) *Governing New York State* (Albany: SUNY Press, 2012), p. 14.

[10] Stuart W. Leslie, "Regional Disadvantage: Replicating Silicon Valley in New York's Capital Region," *Technology and Culture* (2001), p. 237.

[11] Robert D. McNeil, et al., "Barriers to Nanotechnology Commercialization," Report prepared for U.S. Department of Commerce Technology Administration (Springfield: The University of Illinois, September 2007).

While it is premature to conclude that New York's technology-driven economic development policies will succeed in fully revitalizing the upstate economy, certain signal successes are indisputable. These include retention of, and local reinvestment by, mainstay companies like IBM and GE; creation of the world's leading center of nanotechnology research and education in Albany; the establishment in Saratoga of the world's most advanced semiconductor manufacturing site; the in-migration of scores of supply chain firms; and the creation of thousands of high-skill, high-wage jobs. As will be shown, the positive economic, infrastructural, and quality-of-life impacts of these events are not always evident or necessarily evenly distributed. But they are real.

## 2.2 Upstate Advantages

Upstate New York and the state as a whole have long been criticized for intrinsic competitive disadvantages including "bad business climate," "high taxes," "high labor costs," and "expensive energy." Looking back on the emergence of Tech Valley from the perspective of time, however, it becomes evident that Upstate New York also has enjoyed crucial advantages which have played a role in enabling recent success. These include an established base of world-leading high-technology companies, numerous strong educational institutions, transportation advantages, and leadership at the state level that has been able to maintain policy stability with respect to economic development notwithstanding periodic changes in administration.

### 2.2.1 The Strength of the Industrial Legacy

Upstate New York was the site of a number of large technology-intensive firms that had been established in the late nineteenth or first half of the twentieth century, including IBM, GE, Corning, and Eastman Kodak. These firms were characterized by one study as:

> a collection of now priceless economic resources, the seeds for which had been sown at different times since the turn of the century and for much different purposes and which had matured at various rates of development, eventually provided a fertile groundwork for high-tech firms to flourish.[12]

These legacy companies made contributions of incalculable value to New York's evolution into a high-tech state. They developed, manufactured, and commercialized generations of proprietary, advanced-technology products; trained workers and

---

[12] Martin Schoolman, "Solving the Dilemma of Statesmanship: Reindustrialization Through an Evolving Democratic Plan," in Morton Schoolman and Alvin Magid (eds.) *Reindustrializing New York State: Strategies, Implications, Challenges* (Albany: SUNY Press, 1986), p. 18.

attracted educated talent from out of state; and provided valuable support for school systems and universities. The companies demanded that the state make improvements in education and infrastructure that benefitted the broader public. In addition, they advised generations of New York policymakers on the nature and realities of international competition in high-technology sectors, enabling political leaders to develop and implement enlightened—and increasingly effective—public policies.

### 2.2.2 Transportation

New York City occupies one of the world's most superb natural harbors, an advantage which was dramatically enhanced by the completion of the Erie Canal in the 1820s, making New York "the first state to burst through the Appalachian Mountain barrier that divided coastal cities from the farms and mineral resources of the Midwest."[13] The water route along the Hudson-Mohawk corridor was augmented by railroads (beginning in the 1840s) and eventually by the New York State Thruway. The value of these transportation links in economic development terms is sufficiently pronounced that, excluding Long Island, over 80% of New York's citizens still "reside in counties that lie along this inverted L-shaped route."[14]

The upstate region's transportation resources and its location with nearby access to industrial suppliers and customers in New York City, Boston, Montreal, and Buffalo is routinely cited by regional economic development professionals seeking to attract new businesses to the area.[15] The largest concentration of population in North America—132 million people and 56% of all skilled workers—resides within one day's shipping time (850 miles) from New York's Capital Region.[16]

### 2.2.3 Policy Continuity

Since colonial days, New York has had a tradition of strong governors, based on the constitutional powers they enjoy, the fact that governors of New York are almost automatically regarded as potential presidents, and the force of personality of many individual governors.[17] New York governors spearheaded economic development projects

---

[13] Edward V. Schneier, John Brian Murtaugh, and Antoinette Pole, *New York Politics: A Tale of Two States* (Armonk and London: M.E. Sharpe, 2010), p. 41.

[14] Edward V. Schneier, John Brian Murtaugh, and Antoinette Pole, *New York Politics: A Tale of Two States* (Armonk and London: M.E. Sharpe, 2010), p. 8.

[15] "Campus Poses Transport Challenge," Albany, *The Times Union* (May 10, 2004).

[16] E. Michael Tucker, "The Rise of Tech Valley," *Economic Development Journal* (Fall 2008), pp. 35–36.

[17] Edward V. Schneier, John Brian Murtaugh, and Antoinette Pole, *New York Politics: A Tale of Two States* (Armonk and London: M.E. Sharpe, 2010), p. 24.

and institutional innovations of dramatic scale and impact. Governor DeWitt Clinton was responsible for the construction of the Erie Canal, which transformed New York City into the commercial center of the world; Governor Alfred E. Smith created the precursor of the modern US welfare state; and Governor Nelson Rockefeller enabled and empowered the State University of New York to develop into one of the greatest public university systems in the world. While it is too early to tell whether the creation of Tech Valley will rank with these earlier achievements, it differs in one respect—Tech Valley is attributable not to one but to a bipartisan succession of governors, spanning over half a century, who have consistently pursued policies aimed at fostering technology-driven economic growth (see Table 2.1). As a result, with respect to Tech Valley, to date New York has avoided destabilizing policy reversals commonly associated with the democratic process and changes of administration.

A 1986 study published by SUNY of economic revitalization efforts in New York observed that the state's high-tech economic development efforts had been characterized by "administrative continuity" amidst "political change." With the transition from Hugh Carey to Mario Cuomo's governorship in 1982–1983, the study

**Table 2.1** A Series of Governors and Initiatives

| Governor | Tenure | Initiatives |
|---|---|---|
| Nelson Rockefeller (R) | 1959–1973[a] | Built SUNY into a major research institution |
| Hugh Carey (D) | 1975–1982 | Supported creation of Rensselaer Polytechnic Institute RPI Technology Park |
| Mario Cuomo (D) | 1983–1994 | Created Centers for Advanced Technology (CATs)<br>Led first New York bid for Sematech<br>Supported Graduate Research Initiative |
| George Pataki (R) | 1995–2006 | Led effort to retain IBM in New York<br>Created NYSTAR and Empire State Development<br>Created Centers of Excellence<br>Made unprecedented investments in research infrastructure<br>Led successful bids for Sematech and Advanced Micro Devices (GlobalFoundries) |
| Eliot Spitzer (D) | 2007–2008 | Oversaw incentive package for relocation of Sematech headquarters to New York |
| David Paterson (D) | 2008–2010 | Supported $1.5 billion expansion by IBM<br>Brokered accord between GlobalFoundries and construction unions |
| Andrew Cuomo (D) | 2011–present | Applied the "Albany model" of innovation-based economic development across Upstate New York<br>Established Regional Economic Development Councils to foster strategic planning on a regional basis |

[a]Malcolm Wilson served as governor for a year after Rockefeller's resignation in 1973. He did not undo or reverse Rockefeller's initiatives with respect to SUNY

observed that because Carey's economic strategies had not been fully developed or implemented, "it would have been a relatively simple matter for the new governor to ignore politely the accomplishments of the Carey administration and begin afresh." Instead, "to his great credit … Cuomo retained Carey's economic strategies intact," reflecting a shared value of the two governors—"avoidance of ideology and a depoliticization of the policy process," and the need for coherent economic development policy.[18] The CEO of an Albany engineering firm, noting the value of stable and consistent economic policies, said that with respect to Tech Valley, state government "is to be applauded, period. Pataki picked up where Mario left off. He made it as nonpartisan as I've seen an issue."[19]

**Box 2.1: Three Men in a Room**
"Most major issues in New York are negotiated at some point in face-to-face discussions among the governor, the majority leader of the state senate, and the speaker of the assembly," discussions which have frequently required bipartisan consensus. Conventional wisdom holds that these "three men in a room" (the offices have never been held by women) ultimately set policy directions for the state as a whole.[20] Thus in 1981, Democratic Governor Hugh Carey, Democratic Assembly Speaker Stanley Fink, and Republican Senate Majority Leader Warren Anderson collaborated in an effort to create Centers for Advanced Technology (CATs) and to provide $30 million in state-backed funding for a major new research center, the Center for Industrial Innovation (CII) at Rensselaer Polytechnic Institute.[21] In the formative years of Tech Valley, when CNSE was created and the state succeeded in attracting semiconductor manufacturing investment, the "three men in a room" were George Pataki, Joseph Bruno, and Sheldon Silver:

| Position | Individual | Tenure in Office |
| --- | --- | --- |
| Governor | George Pataki (R) | 1995–2006 |
| Speaker of the Assembly | Sheldon Silver (D) | 1994–2015 |
| Senate Majority Leader | Joseph Bruno (R) | 1994–2008 |

---

[18] Morton Schoolman and Alvin Magid (eds.) *Reindustrializing New York State: Strategies, Implications, Challenges* (Albany: SUNY Press, 1986), p. 29.

[19] Interview with Ray Rudolph, Chairman, CHA Companies (September 15, 2015).

[20] Edward V. Schneier, John Brian Murtaugh, and Antoinette Pole, *New York Politics: A Tale of Two States* (Armonk and London: M.E. Sharpe, 2010), pp. 27–28.

[21] Michael Black and Richard Worthington, "The Center for Industrial Innovation at RPI," in Morton Schoolman and Alvin Magid (eds.) *Reindustrializing New York State: Strategies, Implications and Challenges* (Albany: SUNY Press, 1986), pp. 261–265.

> Governor Pataki, Leader Bruno, and Speaker Silver disagreed with each other at many points during their tenure, and former staffers for the three men tend to attribute Tech Valley's success to whichever man they served, downplaying the contributions of the other two.[22] However, for Tech Valley's most formative decade, this trio of leaders consistently displayed the unity of purpose necessary to enable technology-driven economic development to succeed in Upstate New York.

## 2.3 First State Efforts in Innovation-Based Economic Development

A recent academic study notes that according to many historical accounts, Upstate New York's descent into "rust belt" status occurred in the 1970s. In fact, Governor Rockefeller's files reveal that as long ago as the 1950s he was "keenly aware of a declining industrial base that was draining upstate counties of revenue, employment opportunity, and population in the early Cold War era." Moreover, many residents of those distressed counties, as well as Rockefeller himself, saw a potential remedy in a massive expansion of the state university system, which at the time was struggling and shackled with restrictions but which represented a potential "economic lifeline."[23] If the story of Tech Valley can be said to have a starting point, it was Rockefeller's subsequent efforts to build SUNY into an educational and research powerhouse, which many regard as his greatest achievement.

What one study of the New York economy termed "drumbeats from Dixie" had their origins at the turn of the twentieth century, as southern states began efforts to recruit northern-based manufacturers with tax incentives, industrial sites, and employer-friendly labor laws, a process which accelerated after World War II.[24] Although the observation by *Business Week* that "Southern growth achieved a critical mass, turning orderly [company] migration into a flood" perhaps exaggerates the gravitational pull from the Northeast to the Sun Belt, from the end of World War II onward, New York faced unprecedented competition from other states and countries for investment and jobs.

---

[22] "Legislature Stakes Out New Priorities as Convention Continues," Troy, *The Record* (January 4, 2008).

[23] Elizabeth Tandy Shermer, "Nelson Rockefeller and the State University of New York's Rapid Rise and Decline," (Rockefeller Archive Center Research Reports Online, 2015), p. 3.

[24] Peter D. McClelland and Alan L. Magdovitz, *Crisis in the Making: The Political Economy of New York State Since 1945* (Cambridge: Cambridge University Press, 1981), p. 104; Nicholas Lowe, "Southern Industrialization Revisited: Industrial Recruitment as a Strategic Tool for Local Economic Development," in Daniel P. Gitterman (Ed.), *The Way Forward: Building a Globally Competitive South* (Chapel Hill: Global Research Institute, 2011); National Research Council, Charles W. Wessner (ed), *Best Practices In State and Regional Innovation Initiatives: Competing in the 21st Century* (Washington, DC: The National Academies Press, 2013).

## 2.3 First State Efforts in Innovation-Based Economic Development

In 1960, Governor Rockefeller established an Advisory Council for the Advancement of Industrial Research and Development comprised of executives from the state's high-tech companies and academics from New York's research universities. Among other activities the Council consulted with Stanford University's provost and former dean of engineering, Frederick Terman, widely known as the "father of Silicon Valley," who was a strong advocate of university-driven, innovation-based regional economic development.[25] (See Box 2.2.) Rockefeller, impressed, established the New York State Science and Technology Foundation (NYSSTF) in 1963 with a mandate to spur economic growth through state-sponsored R&D projects. NYSSTF was run by a board appointed by the governor. From the mid-1960s through the early 1980s, NYSSTF made small research grants to local universities for research equipment, to support graduate students engaged in research, and to attract top researchers to New York.[26]

The state's initial efforts to promote university-driven economic development necessarily centered on its private universities, which had resisted the creation of a public university system so successfully that New York was the last state in the union to create a public university.[27] Moreover, the 1948 launch of the SUNY system "began with an abject surrender to the interests of higher education," with SUNY proscribed from offering graduate education in the arts and sciences and prohibited from engaging in research.[28] These restrictions were dismantled in the wake of the 1957 Sputnik shock and Rockefeller's efforts to transform SUNY into a true university system as well as an instrument for economic development. Between the last year in office of Rockefeller's predecessor and Rockefeller's last year—a 15-year interval—state outlays for higher education increased by a multiple of 24, from $43 million to $1.052 billion, and SUNY had become the largest system

---

[25] Stuart W. Leslie and Robert H. Kargon, "Selling Silicon Valley: Frederick Terman's Model for Regional Advantage," *Business History Review* (Winter 1996).

[26] Stuart W. Leslie, "Regional Disadvantage: Replicating Silicon Valley in New York's Capital Region," *Technology and Culture* (2001).

[27] At the end of World War II, all of New York's universities were supervised by the University of the State of New York, whose powerful Board of Regents had propelled New York into a national leadership position in primary and secondary education. With respect to higher learning, the Regents championed the state's private universities and fought creation of competing state institutions. Syracuse University's chancellor warned at war's end that while temporary state institutions might be opened to serve returning GIs, "we ought to guard against the danger of temporary agencies becoming permanent institutions. We do not want an embryo of a state university . . . which would be difficult to liquidate." John B. Clark, W. Bruce Leslie, and Kenneth P. O'Brien (eds.), *SUNY at Sixty: The Promise of the State University of New York* (Albany: SUNY Press, 2010), p. xvii.

[28] Roger L. Geiger, "Better Late Than Never: Intentions, Timing and Results in Creating SUNY Research Universities," in John B. Clark, W. Bruce Leslie, and Kenneth P. O'Brien (eds.), *SUNY at Sixty: The Promise of the State University of New York* (Albany: SUNY Press, 2010), p. 172. SUNY was created under the leadership of Governor Thomas E. Dewey to provide for the education of returning GIs and in response to exposés of discrimination and anti-Semitism in New York's private universities. Tod Ottman, "Forging SUNY in New York's Political Cauldron" in John B. Clark, W. Bruce Leslie, and Kenneth P. O'Brien (eds.), *SUNY at Sixty: The Promise of the State University of New York* (Albany: SUNY Press, 2010), pp. 15–29.

of higher education in the country.[29] However, "the Regents became more aggressive in stifling the development of SUNY, including an organized effort to suppress fledgling doctoral programs." For decades after its creation, the research capabilities of the SUNY system were affected by "a belated start and continuing vacillation in Albany."[30]

> **Box 2.2: The Example of Silicon Valley**
> In the 1960s, New York State leaders, including Governor Rockefeller and Rensselaer Polytechnic Institute's president George Low, were strongly influenced by the ideas and experience of Frederick Terman, Stanford University's provost and its former dean of engineering, widely known as the "father of Silicon Valley."[31] At the time, Terman was serving as an advisor and consultant to states seeking to replicate Silicon Valley and was engaged by Governor Rockefeller in 1968 to study the state of engineering education in New York as it related to economic development.[32]
>
> In most accounts of how Silicon Valley came to be, Stanford University occupies a central position.[33] Founded in 1891, Stanford's founders looked to MIT as a model for encouraging technology-intensive company formation from academic knowledge, a counterpoise to the region's exploitation by eastern economic interests.[34] In the 1930s, Frederick Terman, then a professor of engineering, encouraged his students to consider the commercial potential of electronics technology and to visit nearby technology-oriented firms. He supported two of

---

[29] P. D. McClelland and A. L. Magdovitz, *Crisis in the Making: The Political Economy of New York State Since 1945*, Cambridge: Cambridge University Press (1981), p. 182.

[30] Roger L. Geiger, "Better Late Than Never: Intentions, Timing and Results in Creating SUNY Research Universities," in John B. Clark, W. Bruce Leslie, and Kenneth P. O'Brien (eds.), *SUNY at Sixty: The Promise of the State University of New York* (Albany: SUNY Press, 2010), p. 176.

[31] "The Father of Silicon Valley," *TechHistoryWorks* (September 21, 2016). http://www.techhistoryworks.com/silicon-valley-history/2016/9/21/the-father-of-silicon-valley.

[32] Stuart W. Leslie and Robert H. Kargon, "Selling Silicon Valley: Frederick Terman's Model for Regional Advantage," *Business History Review* (Winter 1996).

[33] National Research Council, C. Wessner (ed.), *Best Practices in State and Regional Innovation Initiatives* (Washington DC: The National Academies Press, 2013), pp. 219–27; Timothy J. Sturgeon, "How Silicon Valley Came to Be," in Martin Kenney (Ed.), *Understanding Silicon Valley: The Anatomy of an Entrepreneurial Region* (Stanford: Stanford University Press, 2000); "Upstarts and Rabble Rousers … Stanford Fetes 4 Decades of Computer Science," *San Francisco Chronicle* (March 20, 2006).

[34] Henry Etzkowitz, "Silicon Valley: The Sustainability of an Innovation Region," Triple Helix Research Group, 2012.

## 2.3 First State Efforts in Innovation-Based Economic Development

his students, William Hewlett and David Packard, in the founding of a company to commercialize an audio-oscillator that Hewlett had developed through academic work at Stanford.[35] Later reflecting on the success of Hewlett-Packard and other companies founded on the basis of technologies developed at Stanford, he observed that—

> Industry is finding that, for activities involving a high level of scientific and technological creativity, a location in a center of brains, is more important than a location near markets, raw materials, transportation or factory labor.[36]

In the 1950s, Terman was responsible for three initiatives that accelerated the transformation of the region around Stanford into the foremost concentration of high-technology industry in the world, including:

- **The Honors Cooperative Program** facilitated enrollment by engineers from nearby electronics companies in Stanford graduate courses, enabling them to remain abreast of current technology.
- **The Stanford Research Institute** (SRI) conducted defense-related and other practical research relevant to regional businesses.
- **The Stanford Industrial Park**, the first university-owned industrial park in the world, facilitated research cooperation between industry and the university and was the site of Fairchild Semiconductor, whose alumni founded Intel, AMD, LSI Logic, and National Semiconductor.[37]

### 2.3.1 RPI's George Low

State planners' early-stage efforts to use research universities as economic drivers necessarily focused on private institutions, given SUNY's late developmental start.[38] The most significant early efforts centered on the Rensselaer Polytechnic Institute (RPI) in Troy, New York, founded in 1824, which was the United States' first school of civil engineering. RPI graduates and faculty played a major role in the industrialization of the country in the nineteenth century, designing and constructing the transcontinental railroad, subways, and bridges (including the Brooklyn Bridge),

---

[35] AnnaLee Saxenian, *Regional Advantage: Culture and Competition in Silicon Valley and Route 128* (Cambridge, Mass: Harvard University Press, 1994), p. 20.

[36] Stuart W. Leslie and Robert H. Karagon, "Selling Silicon Valley: Fredrick Terman's Model for Regional Advantage," *Business History Review* (Winter 1996), p. 437.

[37] AnnaLee Saxenian, *Regional Advantage: Culture and Competition in Silicon Valley and Route 128* (Cambridge, Mass: Harvard University Press, 1994), pp. 22–24; C. Stewart Gillmore, *Fred Terman at Stanford: Building a Discipline, a University, and Silicon Valley* (Stanford: Stanford University Press, 2004).

[38] John W. Kalas, "Reindustrialization in New York: The Role of the State University," in Morton Schoolman and Alvin Magid (eds.) *Reindustrializing New York State: Strategies, Implications, Challenges* (Albany: SUNY Press, 1986).

and RPI-trained engineers contributed substantially to innovation in machinery and steelmaking. After World War II, RPI concentrated on providing the best undergraduate training in engineering and largely as a result was eclipsed by younger institutions such as MIT, Stanford, Carnegie-Mellon, and CalTech, which established large graduate departments and secured extensive external funding for research. In 1968, Stanford's Fred Terman characterized RPI as a respected regional institution with a good undergraduate program but lacking graduate programs of national stature.[39] In 2000, Warren H. Bruggeman, a member of RPI's Board of Trustees observed that RPI—

> has been very much diminished in its importance [relative to other engineering schools]. I suspect that over the years RPI has concentrated more on putting out fine engineers at the undergraduate level and has not kept pace with schools of notable recognition.[40]

In 1975, RPI's Board of Trustees recruited a new president, George Low, who had conceived, developed, and managed the Apollo program, "The man who put a man on the moon."[41] Low, an admirer of Terman's role in the development of Silicon Valley, sought to transform RPI into a driver of regional economic growth based on three new research centers he called "steeples of excellence":

- **The Build Program** was designed to upgrade RPI's "outdated and outmoded" laboratory equipment, particularly in the area of information technology. In 1977, RPI established the **Center for Interactive Computer Graphics** featuring advanced computers and CAD-CAM technology.
- **The Center for Manufacturing Productivity and Technology Transfer**, established in 1979, was a research collaboration with manufacturers such as Boeing, GE, and GM to combine emerging technologies like robotics and CAD-CAM into automated integrated systems that would comprise the "factories of the future."

---

[39] Stuart W. Leslie, "Regional Disadvantage: Replicating Silicon Valley in New York's Capital Region," *Technology and Culture* (2001); "RPI will Devote More Attention to Research—Locally, University at Albany Has Stolen Much of the Spotlight," Schenectady, *The Daily Gazette* (April 2, 2000). Terman's 1968 survey of engineering schools in New York found that enrollment in Ph.D. programs in engineering were "surprisingly weak for a state as industrialized as New York," and that in general, New York institutions were "nearly all lacking in top leadership." Terman advised that "faculty, not buildings or improved equipment, should be the state's highest priority." C. Stewart Gilmore, *Fred Terman at Stanford: Building a University, a Discipline, and Silicon Valley* (Stanford: Stanford University Press, 2014), p. 459.

[40] "RPI Will Devote More Attention to Research—Locally, University at Albany Has Stolen Much of the Spotlight," Schenectady, *The Daily Gazette* (April 2, 2000). Reflecting RPI's preeminence in undergraduate engineering education, GlobalFoundries' New York-based semiconductor manufacturing operation recruits more of its workforce from RPI than anywhere else. Keynote Address by RPI President Shirley Ann Jackson, National Research Council Symposium, "New York's Nanotechnology Model: Building the Innovation Economy," Albany, New York, April 4, 2013.

[41] Low, serving as NASA's Chief of Manned Space Flight, wrote a memo to President Kennedy suggesting that it was technologically possible to put a man on the moon. Based on that memo, Kennedy made the dramatic statement in 1961 that man would land on the moon and return safely to earth by the end of the 1960s. "The Legacy of George Low," *Albany Business Review* (November 26, 2003).

## 2.3 First State Efforts in Innovation-Based Economic Development

- RPI founded the ***Center for Integrated Electronics*** in 1981 to develop very large scale integration (VLSI) semiconductor technology with the support of local manufacturers such as IBM and GE.[42]

Following the launch of the "three steeples of excellence," Low embarked on an effort to establish a Center for Industrial Innovation (CII), which would seek to integrate the emerging fields of integrated electronics, advanced manufacturing, and computer graphics. At the recommendation of GE CEO Jack Welch, Low sought major financial support for this project from the State of New York. In late 1981, joined by Welch, IBM CEO John Opel, Kodak's Walter Fallon, and other state business leaders, successfully pitched the CII idea to Governor Hugh Carey. The companies assisted Low in lobbying the state legislature for the necessary funding, an effort which yielded $30 million in state funding for the new center.[43]

Low also sought to replicate Terman's creation of Stanford Industrial Park at RPI, providing the infrastructure for a high-tech cluster closely linked to the university's research programs.[44] He began in 1980 by establishing a business incubator under RPI supervision to house innovative startups. (See Box 2.3.) The RPI incubator launched 16 startups in its first decade, employing 700 people, although one of its most successful companies, computer graphics firm Raster Technologies, moved to Boston's Route 128. A 2010 retrospective on the RPI business incubator observed that—

---

[42] Stuart W. Leslie, "Regional Disadvantage: Replicating Silicon Valley in New York's Capital Region," *Technology and Culture* (2001).

[43] Stuart W. Leslie, "Regional Disadvantage: Replicating Silicon Valley in New York's Capital Region," *Technology and Culture* (2001). The legislation establishing CII provided that the state's Urban Development Corporation would let $30 million in bonds and lend the funds to RPI to finance construction of the facility, with the state holding a lien on the property. Upon completion of the construction, the state would lease the facility from RPI and lease it back for $600 thousand per year until the principle was paid off. The state paid the interest, and RPI assumed responsibility for the estimated $35 million cost of equipping the facility. Michael Black and Richard Worthington, "The Center for Industrial Innovation at RPI: Critical Reflections on New York's Economic Recovery," in Morton Schoolman and Alvin Magid (eds.) *Reindustrializing New York State: Strategies, Implications and Challenges* (Albany: SUNY Press, 1986), p. 261.

[44] Terman oversaw the creation of Stanford Industrial Park on land owned by the university after World War II. His goal was to create a university-linked center of high technology. Varian Associates became the first industrial tenant in 1951, followed by Hewlett Packard, GE, Kodak, and others, including the Shockley Semiconductor Laboratory, a corporate seedbed whose offspring and descendants included Fairchild Semiconductor, Intel, and Advanced Micro Devices. Terman used his former student, David Packard, co-founder of Hewlett-Packard, to explain to companies the benefits of physical proximity to a cooperative research university. By 1977 the park hosted 75 companies and 19,000 employees. C. Stewart Gilmore, *Fred Terman at Stanford: Building a University, a Discipline, and Silicon Valley* (Stanford: Stanford University Press, 2014), p. 328; Carolyn Tajnai, "From the Valley of Heart's Delight to Silicon Valley: A Study of Stanford University's Role in the Transformation," Stanford: Stanford University Department of Computer Science, 1996).

It is not hyperbole to say there would be no Tech Valley without the business incubator program at Rensselaer Polytechnic Institute .... Many of the companies that incubated at Rensselaer—companies such as AMRI, originally known as Albany Molecular Research, MapInfo, Vicarious Visions, entransmedia, ReQuest Inc.—are icons of the Capital Region's private sector.[45]

---

**Box 2.3: An Early Incubator Program at the Rensselaer Polytechnic Institute**

One of the United States' first business incubators was established at Rensselaer Polytechnic Institute (RPI) in 1980 as part of George Low's effort to transform the institute into a global technology industry leader. Based directly on Stanford's example, the RPI administration wanted to "develop dynamic consulting opportunities by generating new firms." The incubator was directed by the RPI administration but was "not under the academic purview of the university."[46]

The incubator was characterized by a National Academies study group as a "living laboratory for applied research." RPI faculty consulted with industry tenants and in some cases started companies themselves. Tenants benefitted from "easy access to faculty consultants in technical, financial and management areas" as well as access to laboratory facilities, libraries, databases, and office support services. The university received rent for use of space, fees for use of facilities, and a 2% equity interest in all companies except those started by RPI faculty. As a private institution, RPI could legally accept stock from start-ups in lieu of rent and could invest in companies. RPI faculty were encouraged and given incentives to engage in entrepreneurial projects.

After 5 years of operation, the RPI incubator had spawned several dozen companies and had an occupancy of 16 companies. The National Academies study group concluded that—.

> The [RPI] Incubator Program has been successful in meeting the goals for which it was established. It provides job training opportunities for students, consulting for faculty, and a source of potential future donors to the university and the community.

Source: National Research Council, *New Alliances and Partnerships in American Science and Engineering* (Washington, DC: National Academy Press, 1986), pp. 73–74.

---

[45] "Incubating Tech Valley," *Albany Business Review* (February 8, 2010).

[46] Stuart W. Leslie, "Regional Disadvantage: Replicating Silicon Valley in New York's Capital Region," *Technology and Culture* (2001).

## 2.3 First State Efforts in Innovation-Based Economic Development

In 1981, Low launched the Rensselaer Technology Park (RTP) commenting that "Hewlett-Packard and Stanford—that's the model."[47] RTP achieved moderate success in its first decade, hosting successful companies like MapInfo, a software company founded by four RPI students which in 12 years grew into a publicly traded firm with 400 employees—"No Hewlett-Packard, perhaps, but nonetheless a model for would-be entrepreneurs in the Capital Region."[48] One of the first commercial internet service providers, PSINet, had its origin in the RPI park.[49]

Low died in 1984, but before his death he made observations about obstacles confronting the State of New York in developing and retaining high-tech manufacturing. He was "amazed by the division and roadblocks he saw here [New York]: among local municipalities, among entities that make up the region—businesses, school and government—and even on his own campus, where various disciplines were kept separate and apart."[50] In 1983, New York had lost in a competition with Texas to attract the newly formed computer consortium, the Microelectronics and Computer Technology Corporation (MCC). New York had made a late start in the bidding; other states had already set up task forces to win MCC "before anyone in New York's state government had apparently even heard of the new consortium." Low pointed out that Texas had committed to $45 million in incentives plus $23 million to endow relevant chairs at the University of Texas, accompanied by active engagement by the governor and other state officials and professionally prepared pitch presentations. Low observed that "we were about four years behind the competition."[51]

Low's basic answer to the challenges facing the state was the concept of public–private partnerships convening government, academia, and business, something he experienced firsthand at NASA, and he was one of the first in the region to promote the concept although at the time the state had begun to go build such partnerships with private universities and private companies. Mike Wacholder, recruited by Low to work at the RPI park, commented in retrospect:

> *How do you put a man on the moon? You couldn't do it without building partnerships between industry, educational institutions and government. The educational institutions had to solve the fundamental problems and do the research. The industry had to build the solution. And the government had to define and direct the mission.*[52]

---

[47] Stuart W. Leslie, "Regional Disadvantage: Replicating Silicon Valley in New York's Capital Region," *Technology and Culture* (2001).

[48] Stuart W. Leslie, "Regional Disadvantage: Replicating Silicon Valley in New York's Capital Region," *Technology and Culture* (2001).

[49] "The Legacy of George Low," *Albany Business Review* (November 26, 2003).

[50] "The Legacy of George Low," *Albany Business Review* (November 26, 2003).

[51] George Low, Memorandum for the Record, "Texas' Quest for MCC," August 11, 1983, RPI Archives, cited in Stuart W. Leslie, "Regional Disadvantage: Replicating Silicon Valley in New York's Capital Region," *Technology and Culture* (2001).

[52] "The Legacy of George Low," *Albany Business Review* (November 26, 2003).

## 2.4 Efforts Under Governor Mario Cuomo

Governor Carey's Lieutenant Governor, Mario Cuomo, was an advocate for technology-driven industrialization in New York State.[53] In 1982, as he pursued the Democratic nomination for governor, he argued that "we have to work very hard in the development of high technology and manufacturing," which required maintaining a superior educational base, including vocational training. Noting how Germany and Japan subsidized the acquisition of high-tech capital equipment, Mario Cuomo touted the Carey administration's decision to provide $30 million in financial assistance to RPI's Center for Industrial Innovation, declaring "High technology is a great opportunity for us."[54]

### 2.4.1 Centers for Advanced Technology

In the mid-1980s, under the direction of Governor Cuomo, NYSSTF began to concentrate on deepening university collaborations with industry, speeding up commercialization and developing two sectors—information technology and biotechnology. Governor Cuomo's administration established Centers for Advanced Technology (CATs) funded through NYSSTF at universities on the basis of prior academic excellence and an institution's willingness to work with, and secure funding from, private industry.[55] Each of the first seven CATs was located at a different university with a specific technology theme as shown in Table 2.2.

A 1986 assessment of the CATs indicated that they strengthened graduate programs, created new faculty positions, upgraded universities' scientific equipment, and made university science and technology resources available to small businesses that lacked comparable R&D resources.[56]

---

[53] "Cuomo Pushes Technology to Kickstart State's Engine," *Watertown Daily Times* (January 27, 1994); "Hi-Tech Funding Extended—Su Cater, Six Others to Benefit," *Syracuse Herald-Journal* (March 16, 1987).

[54] "Cuomo Tells IUE: High Tech Manufacturing Needed," Schenectady, *The Daily Gazette* (May 22, 1982).

[55] John B. Clark, W. Bruce Leslie, and Kenneth P. O'Brien (eds.), *SUNY at Sixty: The Promise of the State University of New York* (Albany: SUNY Press, 2010), p. 142.

[56] Morton Schoolman, "Solving the Dilemma of Statesmanship: Reindustrialization Through a Democratic Plan," Morton Schoolman and Alvin Magid (eds.), *Reindustrializing New York State: Strategies, Implications and Challenges* (Albany: SUNY Albany Press; 1986).

**Table 2.2** The Seven Centers for Advanced Technology (CATs)

| CAT University | Technology focus |
| --- | --- |
| Columbia University | Computers and information systems |
| Cornell University | Agricultural biotechnology |
| SUNY Stony Brook | Medical biotechnology |
| SUNY Buffalo | Medical instruments |
| Syracuse University | Computer applications and software |
| University of Rochester | Advanced optical technology |
| Polytechnic Institute of New York | Telecommunications |

## 2.4.2 New York State and the Semiconductor Industry, 1982–1988

At the beginning of the 1980s, New York-based IBM was the world's largest producer and largest purchaser of semiconductors. However, at that time most of the semiconductor industry's manufacturing and research activities were located in other states and foreign countries. At the beginning of the decade, the US semiconductor industry, facing a growing competitive challenge from Japan as well as rapidly rising R&D and capital costs, began to organize unique institutional arrangements in response. New York State was able to capitalize on the industry's initiatives by securing industry and government support for semiconductor-relevant research in the state's universities. IBM, a major player in the Semiconductor Industry Association (SIA) and in the consortia which SIA spawned, played an important role in steering semiconductor R&D projects to New York institutions.[57]

In 1982, SIA formed the Semiconductor Research Corporation (SRC) to provide industry funding for microelectronics R&D through research contracts with universities. IBM's Erich Bloch was the primary mover behind, and founding Chairman of the SRC. At the time SRC was formed, "there were only a few isolated pockets of integrated circuit (IC) research in universities, mostly related to design," and fewer than 100 graduate students nationwide were engaged in IC-related research. Within a few years of SRC's formation, well over 1000 graduate students were conducting IC-related research.[58] SRC made a dramatic contribution to the US industry's human capital, putting "thousands of highly qualified students into the industry."[59]

---

[57] "IBM Pushes for Chip Consortium," Ft. Lauderdale, *Sun Sentinel* (January 7, 1987).

[58] Robert M. Burger, *Cooperative Research: The New Paradigm* (Durham: Semiconductor Research Corporation, 2001).

[59] See summary of remarks by John E. Kelly in National Research Council, Charles W. Wessner (rapporteur), *The 5/14/2019 in the United States* (Washington, DC: The National Academies Press, 2011), p. 73. Between its inception in 1983 and 2011, SRC invested $1.3 billion, supported 7500 graduate students through 3000 research contracts, 1700 faculty, and 241 universities. SRC support resulted in more than 43,000 technical documents, 326 patents, 579 software tools, and work on 2315 research tasks and products. See summary of remarks by Larry Sumney in National Research Council, Charles W. Wessner (rapporteur), *The Future of Photovoltaic Manufacturing in the United States* (Washington, DC: The National Academies Press, 2011), pp. 184–185.

SRC's initial research contracts to universities in 1982—

*did not reflect the industry's needs or goals. Instead they reflected interests and capabilities of the university research community in 1982....*[60]

In order to focus university-based R&D on themes relevant to the industry's needs, SRC originated the first industry-wide technology "roadmap" in 1984–1985, setting 10-year research targets to organize university research to hit the milestones set in the roadmap.[61] SRC found that numerous universities around the United States were eager to collaborate but "only a few were well equipped to do so." Necessarily SRC narrowed its focus to roughly three dozen institutions "with clear capabilities and the required facilities."[62] As a testament to the existing microelectronics infrastructure of several private New York institutions, during the decade after SRC's formation, it made very substantial research investments in New York State. Among the early projects were—

- **Cornell University**, which was awarded the very first SRC research contract, received $30.2 million in support, originally focusing on advanced devices, multilevel interconnect, CAD, and lithography.[63] "A significant factor in the original selection of Cornell was the existence of an NSF-supported nanofabrication center that had positioned itself to contribute to device investigations."[64]
- **Rensselaer Polytechnic Institute** received $20.3 million in support for R&D in themes such as electron beam lithography, copper interconnect technology, and metallization.

The existence of these early SRC research projects created an awareness of New York's research resources and provided a foundation for the state to bid for additional research projects and inward investment by the semiconductor industry. (See Table 2.3.)

---

[60] Robert M. Burger, *Cooperative Research: The New Paradigm* (Durham: Semiconductor Research Corporation, 2001), p. 59.

[61] Robert R. Schaller, *Technological Innovation in the Semiconductor Industry: A Case Study of the International Technology Roadmap for Semiconductors,* Ph.D. Dissertation (Arlington, VA: George Mason University, 2004), pp. 442–443.

[62] Robert M. Burger, *Cooperative Research: The New Paradigm* (Durham: Semiconductor Research Corporation, 2001), pp. 67–68.

[63] Robert M. Burger, *Cooperative Research: The New Paradigm* (Durham: Semiconductor Research Corporation, 2001).

[64] Robert M. Burger, *Cooperative Research: The New Paradigm* (Durham: Semiconductor Research Corporation, 2001), p. 70.

## 2.4.3 The Bid for Sematech

In the mid-1980s, US semiconductor makers, recognizing they had fallen behind Japan in manufacturing began forming a manufacturing research consortium, Sematech, and in 1987 sought federal funding for the project.[65] In May 1987, the semiconductor industry's trade association, SIA, invited RPI to submit a concept paper with respect to the potential location of the newly formed consortium at RPI in New York, with a facility to be completed by mid-1988. RPI and New York State officials felt that the Capital Region had strong selling points for a Sematech bid, which included the "RPI Technology Park, RPI's faculty and multiple disciplines, skilled labor force, contamination-free rooms and a number of companies involved in semiconductor research."[66]

In July 1987, Governor Cuomo proposed legislation that would provide a $40 million interest-free 30-year state loan to finance the construction of the Sematech research facility at the Rensselaer Technology Park.[67] During the summer of 1987, Sematech received roughly 100 initial proposals from 36 states for the site of the research facility, including RPI's technology park in North Greenbush.[68] In contrast to the failed 1983 effort to attract the MCC computer consortium, the state made a unified and powerful push to land Sematech:

> Gov. Mario M. Cuomo, state legislative leaders and State Economic Development Director Vincent Tese are working as a closely knit team, in concert with RPI and a number of leading university officials in the Northeast to entice Sematech to locate at the RPI site.... The State Legislature has approved an interest-free, 30-year, $40 Million loan as part of the bait. This could be the deciding factor because as far as is known it is the largest subsidy offered by any of the other 13 states bidding to be the chosen site.[69]

The front-runner state in the competition, Texas, was sufficiently concerned about the challenge from New York that the chairman of the Texas Department of Commerce, in a statement that did not mention any other competing states, questioned why New York, if it were such a competitive location, needed to offer Sematech $40 million in incentives in its bid.[70]

In November 1987, Sematech selected 12 finalist states in the site selection process. Governor Cuomo upped the state's incentive bid from $40–80 million, and

---

[65] See generally the summary of comments by Gordon Moore of Intel Corporation in National Research Council, Charles W. Wessner (ed.) *Securing the Future: Regional and National Programs to Support the Semiconductor Industry* (Washington, DC: The National Academies Press, 2003).
[66] "Industry Group Plans Cooperative—RPI Eyed as Semiconductor Manufacturing Site," Albany, *The Times Union* (May 22, 1987).
[67] "Cuomo Urges Bill to Draw High Tech," Albany, *The Times Union* (July 1, 1987).
[68] "State Fights Silicon Valley for Computer Chip Site," Albany, *Knickerbocker News* (July 29, 1987).
[69] "A Vigorous Effort," Syracuse, *The Post Standard* (September 1, 1987).
[70] Vetter stated "I have to ask why New York has to offer $40 million (in incentives): is it because Troy, N.Y. . . . is such a lousy place to locate? Have you ever tried to get to Troy, N.Y.? It's not an easy job." "Officials outraged by Texas Shot at Troy as Lousy Site for Project," Albany, *The Times Union* (September 3, 1987).

**Table 2.3** SRC Research Contracts with New York Universities, 1982–1993

| Start date | Theme | Principal investigator | Institution |
|---|---|---|---|
| 1982 | Microscience and technology | Frey, MacDonald | Cornell University |
| 1983 | CAD for VLSI layout | Kinnen | University of Rochester |
| 1983 | Growth kinetics of thin insulator and interface defects | Raj | Cornell University |
| 1983 | Advanced beam systems | Steekl, Muraska | Rensselaer Polytechnic Institute |
| 1985 | Self-testing VLSI circuits | Albicki | University of Rochester |
| 1986 | Electronic packaging | Li | Cornell University |
| 1989 | Ion projection lithography | Wolf | Cornell University |
| 1991 | W-Cu MOCVD | Kaloyeros | SUNY Albany |
| 1991 | Parallel E-beam lithography | MacDonald | Cornell University |
| 1993 | Stress voicing and electromigration | Li | Cornell University |

Source: Robert M. Burger, *Cooperative Research: The New Paradigm* (Durham: Semiconductor Research Corporation, 2001)

RPI offered additional square footage and access to the George M. Low Center for Industrial Innovation.[71] Despite these last-minute efforts, Sematech chose Austin, Texas, as the site for its research facility. Reportedly, the deciding factor in the selection of Austin was the city's available 300,000 square foot Data General Corp. plant, a dense existing semiconductor industry concentration, and the state's central location.[72] Although disappointed, New York leaders recognized that the state was emerging as a real contender for semiconductor investments. A spokesman for the State Department of Economic Development, Harold Holzer, commented that:

> What's good is that New York state is finally in the big leagues for projects like this. It wasn't long ago that New York wouldn't have ever been considered. Just like the lottery, you can't win if you're not in it.[73]

While the competition with other states for Sematech was the subject of headlines, a less-noticed program initiated by Governor Cuomo pursued a different kind of interstate competition in innovation. Governor Cuomo's Graduate Research Initiative, launched in 1987, allocated state funds to the SUNY system to enable SUNY to pay higher salaries to attract leading university faculty to positions at SUNY.[74] The initiative also funded the acquisition of equipment for use in university-based scientific research. One of the faculty recruited under this program, involving, among other things, a personal interview with Governor Cuomo himself, was a

---

[71] "State to Up Ante in Sematech Pot to $80 M," Albany, *Knickerbocker News* (December 3, 1987).

[72] "Sematech Decision Tipped by Existing Building, State Aid," Albany, *The Times Union* (January 7, 1988).

[73] "Sematech Decision Tipped by Existing Building, State Aid," Albany, *The Times Union* (January 7, 1988).

[74] "Funding Roller-Coaster Derails Research," *The Buffalo News* (June 10, 1992).

young physicist, Alain Kaloyeros, a Greek-Lebanese émigré, who was hired by the University at Albany (SUNY Albany[75]) in 1988, in the immediate aftermath of New York's Sematech setback,[76] who would play a significant role in nanotechnology research at the university.[77] Later, from the perspective of 2011, Governor Andrew Cuomo, reflecting on New York's emerging preeminence in nanotechnology, commented that—

> Success has many fathers. I think a founding father of this success happened to be my father. All of the buzzwords you hear today, they're in the documents from 1988. They knew it was going to take time. They didn't want a rocket to go straight up, straight down.[78]

### 2.4.4 Emergence of a Regional Approach to Economic Development

The region around Albany was historically comprised of "self-contained urban centers (e.g., Schenectady, Troy, Albany, Cohoes), fragmented local government structures and a culture of machine-style politics." The region's parochialism and balkanized governmental units were seen as obstacles to economic growth, and in the early 1980s the business community began systematic efforts to foster entrepreneurialism, a broad regional approach to development, and, as one observer noted, "more collaborative, flexible governance arrangements."

> At the forefront of this effort was the Center for Economic Growth (CEG), a regionally-based consortium of Chambers of Commerce and economic stakeholders, and their invention of the Tech Valley moniker.[79]

---

[75] The University at Albany is also sometimes referred to as "UAlbany" or "UA."

[76] Kayloyeros earned his Ph.D. in physics from the University of Illinois at Champaign–Urbana. His research area, methods of applying metal coatings to manufactured objects, was "such a hot area of research that he was able to pick and choose among the many universities that recruited him." "Love of Teaching Brought Scientist to Capital District," Albany, *The Times Union* (December 11, 1990).

[77] Susan V. Herbst, provost and acting president of SUNY Albany, in "A Raise for the Record Books," *Inside Higher Ed* (June 5, 2007).

[78] "Nano's Seeds Planted Long Ago," Albany, *The Times Union* (October 3, 2011).

[79] Michael Buser, "The Production of Space in Metropolitan Regions: A Lefebvrian Analysis of Governance and Spatial Change," *Planning Theory* (March 21, 2012), pp. 288–289; "A Special Executive is Needed," Albany *The Times Union* (May 10, 1987). A 1988 editorial in the Albany *Times Union*, noting CEG's formation, stated that "For too long, local politicians and executives have kept their distances, each fearing that closer ties might be against their best interests. That has led to a parochial approach in which neither side has been a winner. Executives viewed politicians as more interested in votes than in helping the economy grow. Politicians feared executives might spell trouble if their plans for expansion offended special interest groups. The price of parochialism has been high—a drop of 15 points in the national standings." "More Airport Study? Yes," Albany, *The Times Union* (May 12, 1988).

CEG was created by the Albany-Colonie Chamber of Commerce in 1987.[80] It quickly emerged as a strong advocate for regional collaboration to combat economic decline. "CEG drove the bus, and was the think tank" that created the intellectual basis for state leaders to support high-tech development policies.[81] In 1993, CEG President Kevin O'Connor wrote that—

> The Capital Region has less than one-half of 1 percent of the nation's population, but more than 1 percent of all the local governments. There's something wrong here.... [We need] reorganization, restructuring, in short, regionalization. I'm not talking about one big regional government and I'm not talking about giving up local control. I am talking about creating larger entities in some cases.. about consolidating services among local governments and about agreements among municipalities for sharing equipment and services.[82]

CEG played a major role in securing state funding for the rebuilding of Albany International Airport (1996) and the Albany-Rensselaer train station (2002), part of a broader effort to rebrand the region as relevant in high technology.[83] In 1997, CEG launched a major effort to attract semiconductor manufacturing to the region (see Chap. 4).

Regional business leaders worked to "re-brand" the Capital Region. The term "Tech Valley," embracing New York's Capital Region and the Hudson Valley north of New York City, is attributed to Wallace Altes, former president of the Albany-Colonie Regional Chamber of Commerce, who coined the term in 1998 in conjunction with a broader effort to market the region to high-technology industries.[84] The term caught on quickly and within a year it was being used by numerous regional businesses and economic development organizations.[85] In 2000, Capital Region motorists were given an option to obtain "Tech Valley" license plates, with a special logo.[86] "For many of those involved Tech Valley was primarily concerned with the perceived culture of defeatism and expressing an alternative idea of what the region could become.... Tech Valley not only served as a marketing ploy to court global capital, it was a statement to the citizens of the Capital District that they were now part of a vibrant culture of entrepreneurialism, growth and technology."[87]

---

[80] "Million Dollar Drive Getting Under Way Economic Development Fund-Raiser," Albany *The Times Union* (March 26, 1987).

[81] Interview with former Executive Director of NYSTAR Michael Relyea (January 8, 2016).

[82] "The Argument for Regionalization," Albany, *The Times Union* (April 11, 1993).

[83] Michael Buser, "The Production of Space in Metropolitan Regions: A Lefebvrian Analysis of Governance and Spatial Change," *Planning Theory* (March 21, 2012), p. 289; "Business Group Pushes $390 M Airport Expansions," Albany, *The Times Union* (August 4, 1989).

[84] "Not Yet Tech Valley, But Getting Closer," Albany, *The Times Union* (March 7, 1999); "Obama Nods to Tech Valley," Albany, *The Times Union* (September 20, 2009).

[85] "Tech Valley Image Has Winning Edge," Albany, *The Times Union* (October 7, 1999).

[86] "New on Tech Valley Roads," Albany, *The Times Union* (January 23, 2000).

[87] Michael Buser, "The Production of Space in Metropolitan Regions: A Lefebvrian Analysis of Governance and Spatial Change," *Planning Theory* (March 21, 2012), p. 289.

2.4 Efforts Under Governor Mario Cuomo 43

"Tech Valley" was also "greeted with laughter and the sneers of naysayers."[88] In November 1999, the *Wall Street Journal* featured the Capital Region in an article on cities' efforts to attract companies. The article observed that, "Albany is just the umpteenth community trying to package itself as the next (or new and up-and-coming) Silicon Valley. Like many of the others it has a long way to go."[89] Nevertheless, at the time of the *Journal* article, Tech Valley already had over 700 high-technology companies employing 45,000 people, and regional cooperation was growing. The Albany *Times Union* observed that—

> *[Tech Valley] companies are more open to sharing talent and in presenting a united front in selling the region to the rest of the country. Organizations like the Software Alliance, the Center for Economic Growth, and the Albany-Colonie Regional Chamber of Commerce are pooling resources to attract top-notch talent into the Capital Region work force.*[90]

### 2.4.5 Retaining IBM

In the early 1990s, IBM was the mainstay of the Hudson Valley economy and its largest employer.[91] In 1992, IBM reported an operating loss of nearly $5 billion, and its chairman indicated that the company would take "aggressive actions" to turn its performance around.[92] Among other measures, the company indicated it would abandon its longstanding no-layoffs practice and cut up to 3500 jobs at its Upstate New York plants in Poughkeepsie, Kingston, and East Fishkill.[93] In February, the workforce reduction figure for the upstate facilities was increased to 6000. The IBM semiconductor plant in East Fishkill would stop making semiconductors, eliminate 4000 jobs, and focus on making ceramic packages for semiconductors.[94] Ultimately, IBM laid off over 8000 workers in the Hudson Valley between December 1992 and December 1993.[95] These events sent a shock wave through Upstate New York, threatening local economies and small businesses that supplied goods and services to the three IBM plants.[96]

[88] "Region Thrives Amid High-Tech Revolution," Albany, *The Times Union* (March 19, 2013).
[89] "Even Bad Publicity is Good," Albany, *The Times Union* (November 28, 1999).
[90] "Tech Valley Living Up to its New Name," Albany, *The Times Union* (January 30, 2000).
[91] IBM's employment in the Hudson Valley peaked in 1984 at 31,300 jobs. "The Town IBM Left Behind," *Business Week* (September 10, 1995).
[92] "IBM Posts Biggest Loss in U.S. Corporate History," Albany, *The Times Union* (January 20, 1993).
[93] "IBM to Cut 3500 Jobs From New York Plants," Albany, *The Times Union* (January 7, 1993).
[94] "IBM Braces for Additional Cuts, Plans to Implement First Layoffs," *Watertown Daily Times* (February 25, 1993).
[95] "State Offers $40 M in Loans in Effort to Help IBM Grow," Albany, *The Times Union* (January 30, 1994).
[96] "IBM Country Under Siege," Albany, *The Times Union* (March 26, 1993); "In IBM Country, A Kick in the Teeth," *Watertown Daily Times* (April 4, 1993); "Small Manufacturers Feeling Squeeze—Shrinking of Industrial Giants Like IBM, Kodak Hurting State's Smaller Firms," *Watertown Daily Times* (May 17, 1993).

In response to IBM's retrenchment, in May 1993 Governor Cuomo announced the formation of 11 working groups convening representatives of the state, Hudson Valley local governments, and the private sector "to respond to the unprecedented regional economic crisis." These groups undertook a number of initiatives aimed at buffering the local economy from the IBM shock, including dislocated worker adjustment, establishment of a small business loan fund by a consortium of regional banks, job training, and preparation of plans for business incubators.[97] In January 1994, the Cuomo administration announced a $40 million loan program to support a joint venture between Cirrus Logic Corp. and IBM[98]; from IBM's perspective such measures, while welcome, did not offset disadvantages associated with New York State as a location for manufacturing operations.[99]

## 2.5  Efforts Under Governor George Pataki

Mario Cuomo was succeeded as governor in 1995 by George Pataki. By that time IBM's restructuring had enabled the company to return to profitability.[100] However, IBM's new CEO, Louis Gerstner, who had implemented a thoroughgoing overhaul of the company's operations, was nearing a decision to move the company's headquarters out of New York State altogether. Gerstner later recalled that the company "was on its way to another state. We even had the building picked out."[101] Governor Pataki, who was in communication with IBM as soon as he took office, recalls that—

> [W]hen I got elected, I went and visited IBM officials. I went and visited the corporate executives across the state, even before I took office, because our economic climate was so bad. I remember the IBM leadership telling me they had not invested in New York in a decade. And not only that, wherever they had to make a decision as to where to disinvest, New York was at the top of the list. And they had taken tens of thousands of jobs out of New York.[102]

Governor Pataki made a flurry of commitments to IBM to forestall the move of its headquarters. Gerstner commented that "I've never seen a government move as fast as this one has in the past two weeks." IBM reversed its plans and stated that it would build a new headquarters in Armonk, "because we see a shift in policy toward

---

[97] Peter Fairweather and George A. Schnell, "Reinventing a Regional Economy: The Mid-Hudson Valley and the Downsizing of IBM," *Middle States Geographer* (Vol. 28, 1995).

[98] "State Offers $40 million in Effort to Help IBM Grow," Albany, *The Times Union* (January 30, 1994). A $15 million loan from the state Urban Development Corporation and a $25 million loan from the state Job Development Authority would be made available to purchase equipment, make renovations, train employees, and cover energy costs. Ibid.

[99] "Pataki Hails IBM's New Chip Plant—$700 million Deal to Create 400 Jobs," *The Buffalo News* (November 18, 1997).

[100] "IBM Reports Solid Quarterly Results," *The Buffalo News* (October 21, 1994).

[101] "Pataki Hails IBM's New Chip Plant—$700 million Deal to Create 400 Jobs," *The Buffalo News* (November 18, 1997).

[102] "Pataki Headlines Fox's 'NanoNow'," *Albany Business Review* (September 18, 2006).

## 2.5 Efforts Under Governor George Pataki

pro-economic growth on the part of this government."[103] During his tenure in office (1995–2006), Governor Pataki would oversee extraordinary and unprecedented state investments in infrastructure for applied, commercially relevant research, and IBM would be the state's foremost industrial research collaborator.

Upon assuming the governorship in 1995, George Pataki implemented a series of initiatives to promote innovation-based economic growth in the state. Soon after taking office, he consolidated a number of state economic development organizations to form Empire State Development (ESD), an entity which enjoyed authority to provide tax credits, loans, and grants to companies. His attempt to merge the NYSSTF with ESD was rebuffed by the legislature, which enacted legislation in 1999 creating the New York Office of Science, Technology and Academic Research (NYSTAR), an institution primarily answerable to the legislature rather than to the governor, to whom NYSSTF reported.[104] NYSTAR assumed the responsibilities of NYSSTF, including the section of CAT sites, and NYSSTF itself was dissolved. In contrast to ESD, which had a primary mission of retaining existing businesses in the state, NYSTAR's focus was on the use of university-based research to foster start-ups and to attract high-tech companies from outside the state.[105]

The 1999 legislation creating NYSTAR, the "Jobs 2000 Act" or J2K, was sponsored by Senate Majority Leader Joseph Bruno, a longstanding proponent of technology-based economic development, who was so closely associated with the legislation that some dubbed it "Joe-2 K."[106] The same legislation gave NYSTAR a first-year budget of $156.5 million, or about six times the former budget of NYSSTF and substantially more than ESD's annual budget of roughly $120 million. The law was "hailed by university officials and business leaders as a sea change in state policy that promises to invigorate academic research, stimulate business development and revive the upstate economy." Kevin O'Connor, president of the Albany-based Center for Economic Growth, observed that—

> *The magnitude of this is something I don't think anybody has focused on. This is the most radical shift in economic development policy that I've seen in this state in the last 10 years.*[107]

In his 2001 State of the State message, Governor George Pataki announced plans to establish "Centers of Excellence" across the state to link university research with high-tech companies, with a proposed investment of $283 million[108] in contrast to

---

[103] Governor Pataki committed to acquire vacant IBM facilities in the mid-Hudson region for $13 million and use them to consolidate New York's data processing operations, at that time scattered in 49 locations. "Big Blue to Stay in State, Build New Headquarters," *Watertown Daily Times* (February 17, 1995).

[104] "NYSTAR is Fostering High Tech Business," *The Buffalo News* (February 16, 2000).

[105] "NYSTAR is Fostering High Tech Business," *The Buffalo News* (February 16, 2000).

[106] "State Investing $522 million to Create High-Tech Jobs," *Watertown Daily Times* (November 11, 1999).

[107] "Jobs 2000 Gets Backing from Pataki, Bruno—State Sets Aside Half Billion Dollars for High-Tech R&D," Schenectady *The Daily Gazette* (November 11, 1999).

[108] "The State of the State," *The Buffalo News* (January 4, 2001).

the CAT centers, which "did not hold university researchers accountable for an immediate impact on economic development." One of the first of three of these would focus on nanotechnology and would be located at SUNY Albany; the others were in Buffalo (bioinformatics) and Rochester (optics).[109] Pataki's Centers of Excellence policy effectively vested responsibility for economic development in local university officials—

> [U]niversity officials at the three public universities [where Centers of Excellence were established] were expected to collaborate with industry by building partnerships and research centers. It was expected that these research centers, and their partners, would result in job creation for their regions.[110]

An official in Governor Pataki's office recalled afterward that those who were most surprised by the choice of SUNY Albany for the Center of Excellence for nanotechnology were "officials at New York's more prestigious research institutions who were already conducting nanoscale research." A key decisional factor was the perception by the governor's office that SUNY Albany's leadership was more willing to serve industry's requirements. One Pataki administration official said that "universities will always take the money" but the governor was looking for universities willing to "constantly, consistently and clearly focus on industry relations."[111]

## Bibliography[112]

Big Blue to Stay in State, Build New Headquarters. (1995, February 17). *Watertown Daily Times*.

Black, M., and Worthington, R. (1986). The Center for Industrial Innovation at RPI. In M. Schoolman and A. Magid (Eds.) *Reindustrializing New York State: Strategies, Implications and Challenges* (pp. 261–265). Albany: SUNY Press.

Burger, R. M. (2001). *Cooperative Research: The New Paradigm*. Durham: Semiconductor Research Corporation.

Buser, M. (2012, March 21). The Production of Space in Metropolitan Regions: A Lefebvrian Analysis of Governance and Spatial Change. *Planning Theory*. 288–289.

Clark, J. B., Leslie, W. B., and O'Brien, K. P. (Eds.) (2010). *SUNY at Sixty: The Promise of the State University of New York*. Albany: SUNY Press.

Clark, T. B. (1976, October-November). The Frostbelt Fights for a New Future. *Empire State Report II*. p. 332.

---

[109] "NanoTech Idea Created a Decade Ago," Schenectady, *The Daily Gazette* (September 15, 2011).

[110] Robert W. Wagner, *Academic Entrepreneurialism and New York State's Centers of Excellence Policy*, Ph.D. dissertation, (SUNY Albany, 2007), p. 6.

[111] Robert W. Wagner, *Academic Entrepreneurialism and New York State's Centers of Excellence Policy*, Ph.D. dissertation, (SUNY Albany, 2007), p. 84, citing interviews conducted by the author.

[112] As noted in the front matter of this book, the study also drew on interviews carried out by the authors and numerous articles from *The Times Union* (Albany), *The Daily Gazette* (Schenectady), the *Albany Business Review* (Albany), *The Post-Star* (Glens Falls), *The Record* (Troy), *The Saratogian* (Saratoga Springs), *The Buffalo News* (Buffalo), The *Observer-Dispatch* (Utica), The *Daily Messenger* (Canandaigua), and the *Post-Standard* (Syracuse). These are not individually included in the bibliography.

Cuomo Pushes Technology to Kickstart State's Engine. (1994, January 27). *Watertown Daily Times*.

Etzkowitz, H. (2012). *Silicon Valley: The Sustainability of an Innovation Region*. Triple Helix Research Group.

Fairweather, P., and Schnell, G. A. (1995). Reinventing a Regional Economy: The Mid-Hudson Valley and the Downsizing of IBM. *Middle States Geographer*. 28.

The Father of Silicon Valley. (2016, September 21). *TechHistoryWorks*.

Geiger, R. L. (2010). Better Late Than Never: Intentions, Timing and Results in Creating SUNY Research Universities. In J. B. Clark, W. B. Leslie, and K. P. O'Brien, (Eds.), *SUNY at Sixty: The Promise of the State University of New York* (p. 172). Albany: SUNY Press.

Gillmore, C. S. (2004). *Fred Terman at Stanford: Building a Discipline, a University, and Silicon Valley*. Stanford CA: Stanford University Press.

Hi-Tech Funding Extended—Su Cater, Six Others to Benefit. (1987, March 16). *Syracuse Herald-Journal*.

IBM Braces for Additional Cuts, Plans to Implement First Layoffs. (1993, February 25). *Watertown Daily Times*.

IBM Pushes for Chip Consortium. (1987, January 7). Ft. Lauderdale, *Sun Sentinel*.

In IBM Country, A Kick in the Teeth. (1993, April 4). *Watertown Daily Times*.

Irwin, D. A. (2004, September). The Aftermath of Hamilton's Report on Manufactures. *The Journal of Economic History*.

Kalas, J. W. (1986). Reindustrialization in New York: The Role of the State University. In M. Schoolman and A. Magid (Eds.), *Reindustrializing New York State: Strategies, Implications, Challenges*. Albany: SUNY Press.

Leslie, S. W., and Karagon, R. H. (1996). Selling Silicon Valley: Fredrick Terman's Model for Regional Advantage. *Business History Review*.

Lowe, N. (2011). Southern Industrialization Revisited: Industrial Recruitment as a Strategic Tool for Local Economic Development. In Daniel P. Gitterman (Ed.), *The Way Forward: Building a Globally Competitive South*. Chapel Hill: Global Research Institute.

McClelland, P. D., and Magdovitz, A. L. (1981). *Crisis in the Making: The Political Economy of New York State Since 1945*. Cambridge: Cambridge University Press.

McNeil, R. D. et al. (2007, September). Barriers to Nanotechnology Commercialization. Report prepared for U.S. Department of Commerce Technology Administration. Springfield: The University of Illinois.

National Research Council. (1986). *New Alliances and Partnerships in American Science and Engineering*. Washington, DC: National Academy Press.

National Research Council. (2003). *Securing the Future: Regional and National Programs to Support the Semiconductor Industry*. C. W. Wessner (Washington, DC: The National Academies Press.

National Research Council. (2011). *The Future of Photovoltaic Manufacturing in the United States*. C. W. Wessner (rapporteur). Washington, DC: The National Academies Press.

National Research Council. (2013). *Best Practices In State and Regional Innovation Initiatives: Competing in the 21st Century*. C. W. Wessner (Washington, DC: The National Academies Press.

Ottman, T. (2010). Forging SUNY in New York's Political Cauldron. In J. B. Clark, W. B. Leslie, and K. P. O'Brien, (Eds.), *SUNY at Sixty: The Promise of the State University of New York* (pp. 15–29). Albany: SUNY Press.

Pecorella, R. F. (2012). Regional Political Conflict in New York State. In R. F. Pecorella and J. M. Stonecash (Eds.) *Governing New York State* (p. 14). Albany: SUNY Press.

Pendall, R. (2003). *Upstate New York's Population Plateau: The Third-Slowest Growing State*. Washington, DC: The Brookings Institution.

A Raise for the Record Books. (2007, June 5). *Inside Higher Ed*.

Reiert, E. (2007). *How Rich Countries Got Rich and Why Poor Countries Stay Poor*. London: Constable.

Saxenian, A. (1994). *Regional Advantage: Culture and Competition in Silicon Valley and Route 128.* Cambridge, MA: Harvard University Press.

Schact, W. H. (2013, December 3). *Industrial Competitiveness and Technological Advancement: Debate Over Government Policy.* Washington, DC: Congressional Research Service.

Schaller, R. R. (2004). *Technological Innovation in the Semiconductor Industry: A Case Study of the International Technology Roadmap for Semiconductors,* Ph.D. Dissertation. Arlington, VA: George Mason University.

Schneier, E. V., Murtaugh, J. B., and Pole, A. (2010). *New York Politics: A Tale of Two States.* Armonk and London: M.E. Sharpe.

Schoolman, M. (1986). Solving the Dilemma of Statesmanship: Reindustrialization Through an Evolving Democratic Plan. In M. Schoolman and A. Magid, *Reindustrializing New York State: Strategies, Implications, Challenges.* Albany: SUNY Press.

Schoolman, M., and Magid, A. (1986). *Reindustrializing New York State: Strategies, Implications, Challenges.* Albany: SUNY Press.

Shermer, E. T. (2015). Nelson Rockefeller and the State University of New York's Rapid Rise and Decline. Rockefeller Archive Center Research Reports Online.

Small Manufacturers Feeling Squeeze—Shrinking of Industrial Giants Like IBM, Kodak Hurting State's Smaller Firms. (1993, May 17). *Watertown Daily Times.*

State Fights Silicon Valley for Computer Chip Site. (1987, July 29). Albany, *Knickerbocker News.*

State Investing $522 Million to Create High-Tech Jobs. (1999, November 11). *Watertown Daily Times.*

State to Up Ante in Sematech Pot to $80M. (1987, December 3). Albany, *Knickerbocker News.*

Sturgeon, T. (2000.) How Silicon Valley Came to Be. In M. Kenney (Ed.), *Understanding Silicon Valley: The Anatomy of an Entrepreneurial Region.* Stanford, CA: Stanford University Press.

Tajnai, C. (1996). From the Valley of Heart's Delight to Silicon Valley: A Study of Stanford University's Role in the Transformation. Stanford: Stanford University Department of Computer Science.

The Town IBM Left Behind. (1995, September 10). *Business Week.*

Tucker, F. M. (2008, Fall). The Rise of Tech Valley. *Economic Development Journal.*

Upstarts and Rabble Rousers … Stanford Fetes 4 Decades of Computer Science. (2006, March 20). *San Francisco Chronicle.*

Wagner, R. W. (2007). *Academic Entrepreneurialism and New York State's Centers of Excellence Policy,* Ph.D. dissertation. Albany: SUNY.

# Chapter 3
# Nanotechnology Research in Albany, 1980–2016

**Abstract** Large state investments in universities in the Capital Region, most notably in the University at Albany (SUNY Albany) and Rensselaer Polytechnic Institute, transformed the region into one of the most formidable centers of nanotechnology in the world. Most notably, the state underwrote expansion of the nanotechnology research infrastructure at SUNY Albany, culminating in the creation in 2004 of the College of Nanoscale Science and Engineering (CNSE). In 2002, Sematech began a process of relocation from Austin to Albany which was completed a decade later, followed by an influx of other semiconductor companies seeking joint research projects at the NanoCollege.

The Albany NanoCollege (currently part of SUNY Polytechnic Institute [SUNY Poly]) is arguably the foremost center for nanotechnology studies in the world. It had its origins in the 1980s when the US semiconductor industry began funding university-based R&D in microelectronics. The subsequent growth and spectacular success of the NanoCollege flowed from a steady stream of industry, state, and federal investments, with each new initiative building on prior achievements. Success was substantially attributable to the leadership, the research culture, and the institutional innovations which characterized the SUNY Albany site.

## 3.1 RPI-SUNY Albany Research Collaboration

In the early 1980s, Rensselaer Polytechnic Institute (RPI) President George Low partnered with SUNY Albany President Vincent O'Leary to secure state support for local economic development initiatives. The partnership increased the clout of the two universities in the legislature and led to research collaborations between the two institutions in applied physics.[1] In 1987, RPI and SUNY Albany formed the Joint Laboratories for Advanced Materials (JLAM) to convene scientists from the two

---

[1] Creso M. Sá, "Redefining University Roles in Regional Economies: A Case Study of University-Industry Relations and Academic Organization in Nanotechnology," *Higher Education* (2011) 61:193–208.

institutions to collaborate on thin film materials and process technologies. One of JLAM's founders was a former GE scientist, physics professor James W. Corbett, who had served at RPI as an adjunct faculty member while still at GE.[2] Another major figure was SUNY Albany's Alain Kaloyeros, who soon demonstrated an aptitude for raising research funding.[3]

> **Box 3.1: Recognizing Nanotechnology's Potential**
>
> "Nanotechnology" is the "science, engineering and technology related to the understanding and control of matter at the length of scale of approximately 1 to 100 nanometers," which commonly involves dealing with materials at the molecular or atomic level.[4] (By means of comparison, the diameter of a human hair is about 80,000–100,000 nm.) Throughout the 1990s, a growing number of public and private studies called attention to the revolutionary potential of nanotechnology in diverse fields including electronics, materials, medicine, and energy.[5] In January 2000, President Clinton announced the National Nanotechnology Initiative (NNI) to coordinate the R&D investments of federal departments and agencies in nanotechnology. The President's speech was accompanied by a doubling of the federal budget for nanoscale science and engineering from $270 million in FY 2000 to $495 million in FY 2001. More importantly, the advent of NNI "triggered a wave of primarily positive media coverage of nanotechnology and eventually led to increased investment in nanoscience and nanotechnology by universities, states, venture-backed startups, global 1000 companies, and foreign governments."[6]

---

[2] Creso M. Sá, "Redefining University Roles in Regional Economies: A Case Study of University-Industry Relations and Academic Organization in Nanotechnology," *Higher Education (2011)*, 61:197. The RPI and SUNY Albany research efforts were seen as complementary. RPI had research programs under way in sputtering and ion beam deposition physical analysis and semiconductor process applications of thin film metals and insulators. SUNY Albany had ongoing research programs in chemical vapor deposition of metals and superconductors, ultra-high vacuum (UHV) surface analysis, and nuclear reaction analysis. Michael Fury and Alain E. Kaloyeros, "Metallization for Microelectronics Program at the University of Albany: Leveraging a Long-Term Mentor Relationship," *IEEE Explore* (1990), pp. 59–60; "Love of Teaching Brought Scientist to Capital District," Albany, *The Times Union* (December 11, 1990).

[3] In 1989, Kaloyeros received a $50,000 grant from the State of New York for superconductivity research seeking a technique for achieving better electrical current characteristics in complex shapes and forms. In 1991, he received a National Science Foundation grant of $62,500 per year for 5 years to conduct research in advanced electronic materials, matched by $37,000 per year by corporate sponsors. "Superconductivity Study Sparks Grants," Albany, *The Times Union* (January 9, 1989), "SUNY Physicist Gets 5-Year Grant," Albany, *The Times Union* (May 16, 1989). By 1993 his work was being supported by about $4 million per year in grants and donated equipment. "SUNY Gets High Tech Designation," Albany, *The Times Union* (May 15, 1993).

[4] President's Council of Advisors on Science and Technology (PCAST), *The National Nanotechnology Initiative at Five Years: Assessment and Recommendations of the National Nanotechnology Advisory Panel* (Washington, DC: Executive Office of the President, 2003).

[5] Philip Shapira and Jue Wang, "Case Study: R&D Policy in the United States: The Promotion of Nanotechnology R&D," (Atlanta: Georgia Institute of Technology, November 2007.)

[6] Neal Lane and Thomas Kalil, "The National Nanotechnology Initiative: Present at the Creation," *Issues in Science and Technology* (Summer 2005).

> From the beginning New York was a leader in nanotechnology. IBM pioneered nanoscale research, and in 1989 its scientists used a scanning tunneling microscope to arrange 35 individual xenon atoms into the letters "IBM." SUNY Albany was working on nanoscale themes from the early 1990s onward. In 1997, officials from SUNY Albany and Rensselaer Polytechnic Institute (RPI) asked New York State lawmakers for $15 million to support nanoscale R&D, and "to the credit of the state, they recognized the need."[7] In 2000, IBM scientists demonstrated that molecular atomic-level electronic circuits can work, enabling "an era of ubiquitous computing in which everyday objects will have profound intelligence and the ability to anticipate and adjust to human needs," a concept now embodied in the so-called internet of things.[8] A year after Clinton's speech announcing NNI, RPI announced a plan to create a center for nanotechnology studies.[9] By 2002, Albany had "emerged as a world capital in nanotechnology."[10]

In 1988, SRC and Sematech chose RPI as the New York State Sematech Center of Excellence, with SUNY Albany included as an affiliated institution responsible for developing chemical vapor deposition (CVD) technology for copper metallization, a topic of considerable interest in the semiconductor industry. IBM was particularly interested in SUNY Albany's thin film CVD copper research and provided separate funding for a number of specific subtopics, as well as mentoring for the research, complementing mentors from other companies.[11] A senior IBM engineer, Michael Fury, served as a mentor, and it became evident that his research interests and those of SUNY Albany's Kaloyeros were "strongly aligned." The two recalled later in a joint article that, "it is this alignment which contributed to the subsequent expansion of our interactions." The geographic proximity of SUNY Albany to IBM research and manufacturing facilities in East Fishkill, New York, and Burlington, Vermont, enabled frequent personal interactions. IBM research mentors assumed

---

[7] "If You Build It, They Will Come," *The Chronicle of Higher Education* (February 7, 2003).

[8] "IBM Team Makes Atomic-Scale Circuitry Breakthrough," *Watertown Daily Times* (February 3, 2000).

[9] "RPI Creates Center for Nanotechnology Studies," Albany, *The Times Union* (March 30, 2001).

[10] "NY's High-Tech Hope," *New York Post* (September 18, 2002).

[11] At the time, semiconductor devices were made from a silicon substrate coated with silicon oxide, and aluminum was used to connect a device's memory cells and as a connector between devices. Kaloyeros explained that "you have two regions in the device and you want them to talk to each other. This is where the metals come in." Aluminum, however, had high electrical resistance, which slowed the flow of electrons and increased the heat created by the current passing through, and was subject to electromagnetic degradation, which limited the thinness with which aluminum connectors could be made. The "metal of choice" to replace aluminum was copper, which had lower electrical resistance and other advantages. However, copper interacts with silicon, requiring a barrier to isolate copper from the silicon substrate. While this posed considerable technological challenges, as Kaloyeros put it, "We don't have a choice any more. If we are going to have higher speed, we are going to have to learn how to use copper." "Scientists Explore the Future of Computer Technology," Albany, *The Times Union* (December 11, 1990).

roles which included adjunct professor, industry advisor to an SRC fellow, and submission of research proposals with industrial mentor as co-principal investigator and Associate Director.[12]

In addition to CVD copper research, SUNY Albany worked with industrial partners on CVD deposition techniques involving aluminum, tungsten, titanium, and titanium nitride. Although SUNY Albany supplied some of the equipment used in the research effort, this basic infrastructure was substantially augmented by donated tools from industry partners which in some cases were modified by graduate students to address technological challenges particular to the research. Industrial partners also donated proprietary chemicals, provided modeling services, and furnished test wafers for trial deposition runs. (See Table 3.1.)

SUNY Albany's CVD metallization research focused on a variety of chemical approaches to thin film deposition before all of the properties of the processes and the resulting films were known, with the expectation that the diversity of approaches increased the prospect "that a truly commercializable process will result." Collaboration significantly diminished the costs and risks that the industrial partners would have faced conducting the same trial-and-error research separately.[13] By 1996, "Kaloyeros and students at the Center [had] developed an advanced process for depositing high-purity metals on silicon wafers."[14] As a result, Varian Associates, a California-based maker of semiconductor manufacturing equipment, entered into a $4.5 million research collaboration with the Center, with Varian's chairman observing that SUNY Albany's researchers had "made the most progress in the technology of chemical vapor deposition, which can place materials on a silicon wafer more precisely than the [then] current method of physical vapor deposition."[15]

## 3.2 CAT Designation for SUNY Albany

By the early 1990s, Kaloyeros' semiconductor research was drawing about $4 million a year in grants and donated equipment, most of it from private industry. In 1993, Governor Mario Cuomo designated SUNY Albany as a State Center for Advanced Technology (CAT) in thin films and coatings, creating the Center for Advanced Thin Film Technology and making the institution eligible for state funding of $1 million per year for 10 years.[16] The Thin Film Center was housed in

---

[12] Michael Fury and Alain E. Kaloyeros, "Metallization for Microelectronics Program at the University of Albany: Leveraging a Long-Term Mentor Relationship," *IEEE Explore* (1990), pp. 59–60.

[13] Michael Fury and Alain E. Kaloyeros, "Metallization for Microelectronics Program at the University of Albany: Leveraging a Long-Term Mentor Relationship," *IEEE Explore* (1990), p. 63.

[14] "UAlbany Hopes to Bring in Semiconductor Center," Albany, *The Times Union* (August 31, 1996).

[15] "UAlbany, Varian Unveil Manufacturing Line," Albany, *The Times Union* (September 13, 1996).

[16] Thin films are extraordinarily thin layers of materials with ubiquitous applications in the semiconductor industry, and these film deposition techniques are a major subject of research.

3.2 CAT Designation for SUNY Albany

**Table 3.1** Contributions of Industrial Partners—SUNY Albany Metallization Program

| Metal | Tools | Chemistry | Modeling | Devices |
|---|---|---|---|---|
| Copper | WJ 7000<br>MKS | SUNYA<br>Schumacher | RPI<br>Essential Research | IBM<br>Motorola<br>Intel<br>National Semiconductor |
| Al-Cu | Drytek<br>Drytek Quad<br>MKS | Air Products | Essential Research | IBM |
| Ti | SUNYA | NASA Lewis | | IBM<br>Motorola<br>Intel |
| TiN | SUNYA | NASA Lewis<br>Gelest<br>Schumacher | | IBM<br>Motorola<br>Intel |
| W | SUNYA | Gelest<br>Schumacher | | Motorola<br>National Semiconductor |

Source: Michael A. Fury and Alain E. Kaloyeros, "Metallization for Microelectronics Programs at the University of Albany: Leveraging a Long Term Mentor Relationship," *IEEE Xplore* (1993), p. 61

SUNY Albany's physics building. By 1995, in addition to the $1 million per year from the state, the center was bringing in $7 million per year in industry grants and equipment donations.[17]

In 1995, SUNY Albany disclosed that it would build a 75,000 square foot Center for Environmental Sciences and Technology Management (CESTM, also known as NanoFab200) at the university's uptown campus, with the state Urban Development Corporation providing $10 million in funding and the university pledging another $2 million.[18] According to Kaloyeros, the purpose of the new center would be "to develop technology and commercialize it" and to "help existing companies with manufacturing and competitiveness."[19] One specific function of the new structure would be to house the Center for Advanced Technology in Thin Films.[20] In addition, it would house the Atmospheric Sciences Research Center, a National Weather Service forecast office, and a business incubator large enough to support 10–20

---

[17] "Technology Carries Ball as Collegiate Moneymaker," Albany, *The Times Union* (February 25, 1996).

[18] According to one account, Governor Cuomo initially opposed funding CESTM, preferring proposals advanced by Cornell and Columbia universities. However, SUNY Albany President H. Patrick Swygert determined to secure a CAT center at his institution, moved two highly regarded SUNY Albany assets to CESTM as a lure—the National Weather Service and the Atmospheric Sciences Research Center. Swygert also sought help from Victor Riley, at the time CEO of Key Bank and a leading Capital Region community leader. Riley agreed to chair a committee of local business leaders to pitch SUNY Albany to the governor and Assembly Speaker Sheldon Silver. This effort proved successful. "Others Deserve Credit for Nano College's Success, Too," Albany, *The Times Union* (October 8, 2011).

[19] "UAlbany Reveals Blueprints for High Tech Center," Albany, *The Times Union* (February 15, 1995).

[20] "Work on High Tech Site Means Business," Albany, *The Times Union* (November 11, 1995).

spinoff companies.[21]

By the mid-1990s, SUNY Albany's CAT for thin film technology was achieving distinction with respect to microelectronics research with industrial applications. By the end of 1996, after 3 years of operations, the Center had worked with 50 companies, attracted $32 million in private funding, and fostered over 20 new high-tech products.[22] In 1996, Varian Associates, a maker of semiconductor manufacturing tools, selected SUNY Albany as the site for a $4.5 million semiconductor manufacturing line to test newly developed equipment. Varian chose SUNY Albany because the Center had "made the most progress in the technology of chemical vapor deposition, which can place materials on a silicon wafer with more precision than the current level of physical vapor deposition."[23] Other industry research partners included IBM and Motorola. Rich Saburro, an executive at the Center, said in 1996 that it had gained an "international reputation for expertise in microelectronics," particularly in the areas of interconnects, optoelectronics, and specialized hard coatings. Kaloyeros commented that—

> We're going from not just doing the science in a typical research center, but carrying it all the way through to pre-manufacturing. We feel that our job, as a center sponsored by New York State, is to help bring jobs to New York State and the area.[24]

In 1998, SUNY Albany was chosen as a partner in another newly created CAT, to be established at SUNY Stony Brook, to specialize in emerging electronics, photonics, and materials technologies for diagnostic tools and sensor systems. The SUNY Albany/Stony Brook proposal was reviewed by the National Research Council, the research arm of the National Academies of Sciences, and Engineering, which singled out the proposal because of the combined technical capacity of the two institutions and the potential economic significance of the topics they proposed to study.[25] SUNY Stony Brook emphasized the importance of SUNY Albany's partnership in the successful bid, noting that "the Albany CAT develops and manufactures sensors that are used in computer chips—the sensors are not manufactured at Stony Brook."[26]

The Center was supported in its efforts to grow nanotechnology research by Karen Hitchcock, who became President of SUNY Albany in 1995. Hitchcock, who held degrees in biology and anatomy, oversaw the "creation of cutting-edge facilities and a slew of business partnerships in nanotechnology as well as in biotechnol-

---

[21] "Work on High-Tech Site Means Business," Albany, *The Times Union* (November 11, 1995).

[22] "Pataki Budget Threatens Three Technology Programs," Albany, *The Times Union* (January 17, 1997).

[23] "UAlbany, Varian Unveils Manufacturing Line," Albany, *The Times Union* (September 13, 1996).

[24] "High-Tech Center at RPI, SUNY-A Work Together to Bring in Big R&D Dollars," *Albany Business Review* (October 7, 1996). A key feature of the new line was a "cluster tool" designed to integrate as many as eight stand-alone tools, reducing the number of clean rooms required and significantly reducing costs. Ibid.

[25] "SUNY Albany, Stony Brook Named Partners for Latest Center for Advanced Technology," Schenectady, *The Daily Gazette* (July 30, 1998).

[26] "Joins in Partnership with Stony Brook for New Endeavors," Schenectady, *The Daily Gazette* (August 3, 1998).

**Table 3.2** Not-for-profit Corporate Intermediaries Established by SUNY Albany

| Organization | Function |
|---|---|
| Fuller Road Management Corporation | Property management, serve as conduit for state funding |
| Albany Nanotech Inc. | Operation of research facilities |
| Nanotech Resources Inc. | Support for SUNY Albany Center of Excellence |

ogy." Hitchcock became aware of the nanotechnology research in 1993, when serving as SUNY Albany's Vice President of Academic Affairs. She later recalled that Kaloyeros shared her view that "universities need to develop strong, broadly based research that includes the study of fundamental problems as well as work that develops practical applications."[27] In 1998, Hitchcock made the seemingly improbable prediction that "New York's high-technology corridor will compete with Silicon Valley in California, North Carolina's Research Triangle Park and Route 128 outside Boston.[28]

The creation of the CAT, the Focus Center, and CESTM enabled SUNY Albany to "shop its faculty, but especially its capital assets, to attract public and private [funding] support." Moreover, SUNY Albany—

> Caught the attention of New York's top policymakers, including the governor (Pataki) and speaker of the Assembly. State officials were especially eager to capitalize on the attention of corporations, like those in the semiconductor industry, were paying to UA's research initiatives. As one state official described it, UA opened the eyes of the governor that investment in New York's science and technology research capacity at universities could help him achieve his economic development goals, especially upstate.[29]

## 3.3 Intermediary Organizations

SUNY Albany established a number of not-for-profit corporate intermediaries to serve as interfaces between the university and the private sector. Reportedly, these entities were created because "IBM wanted to deal with a 501-C3" to ensure that "they were not just dealing with an academic campus."[30] Three incorporated entities were established which were governed by corporate boards and largely staffed with individuals with backgrounds in industry or government. (See Table 3.2.)

These entities presented a "more corporate presence relating to potential industry partners than the persona of an academic institution could provide." In contrast to an academic organization, industry partners were "concerned about three metrics, cost, performance and time." The corporate intermediaries demonstrated to industry that

---

[27] "If You Build It, They Will Come," *The Chronicle of Higher Education* (February 7, 2003).
[28] "SUNY Albany, Stony Brook Named Partners for Latest Center for Advanced Technology," Schenectady, *The Daily Gazette* (July 30, 1998).
[29] Robert W. Wagner, *Academic Entrepreneurialism and New York State's Centers of Excellence Policy,* Ph.D. dissertation, (SUNY Albany, 2007), p. 81.
[30] Robert W. Wagner, *Academic Entrepreneurialism and New York State's Centers of Excellence Policy,* Ph.D. dissertation, (SUNY Albany, 2007), pp. 101–102.

SUNY Albany understood and appreciated these concerns.[31] SUNY Albany's strategy involved "building an academic institution that has the appearance of a corporate structure thereby appeasing the needs of business officials who might be uncomfortable negotiating with an academic institution."[32]

The corporate entities were founded in collaboration with the SUNY Research Foundation, the principal institutional entity through which SUNY promotes public–private partnerships with CNSE.[33] The SUNY Research Foundation "does not have university-type rules" governing subjects such as human resources policies and faculty tenure.[34] Like other university research foundations, the SUNY Foundation provides university administrators "funds for programs and supplies that are not within the bounds of their regular budget authority."[35] Kaloyeros characterized the foundation in 2011 as a "flexible, proactive and innovative partner."[36] He was able to avoid the constrictions of university rules with respect to faculty by building SUNY Albany's nanotech complex under Foundation auspices. The faculty serving the complex were employees of the Foundation's not-for-profit corporate entities. Without this autonomy, some observers believe the Albany nanotech experience "never could have happened."[37]

Timothy Killeen, president of the SUNY Research Foundation, commented in a 2013 symposium convened by the National Research Council that the corporate intermediaries were designed to achieve goals for CNSE that were beyond the direct reach of SUNY or its research foundation:

> These corporations are able to increase flexibility of CNSE in ways that academia is not equipped to do. In particular, they provide a dedicated corporate structure that can ensure alignment with SUNY's not-for-profit mission of research and education. One of them is the Fuller Road Management Corporation, which manages a land lease with SUNY, designs and constructs facilities, provides financing for construction, and issues debt for facility

---

[31] Robert W. Wagner, *Academic Entrepreneurialism and New York State's Centers of Excellence Policy*, Ph.D. dissertation, (SUNY Albany, 2007), p. 103.

[32] Robert W. Wagner, *Academic Entrepreneurialism and New York State's Centers of Excellence Policy*, Ph.D. dissertation, (SUNY Albany, 2007), p. 155.

[33] The Fuller Road Management Corporation (FRMC), for example, was co-founded on an equal basis by the Research Foundation of the State of New York and the University at Albany Foundation. FRMC is private 501(c) not-for-profit real estate holding corporation formed in 1993 to "plan, design, develop, construct, own, operate, and lease facilities supporting the technical programs, strategic partnerships and business consortia" of SUNY Albany's nanotechnology activities. The state continued to own the land under CESTM and the Center of Excellence but leased it to FRMC, which assumed responsibility for the land. New York State Office of the Comptroller General, *Fuller Road Management Corporation*, Report 2012-S-26, (January 2013), p. 5.

[34] Interview with Professor Catherine Hill, Saratoga Springs, New York (September 15, 2015).

[35] Edward V. Schneier, John Brian Murtaugh, and Antoinette Pole, *New York Politics: A Tale of Two States* (Armonk and London: M.E. Sharpe, 2010), p. 46.

[36] SUNY Buffalo State, "Research Foundation of SUNY Celebrates 60th Anniversary," Press Release, February 15, 2011.

[37] Interview with Catherine Hill, Saratoga Springs, New York (September 15, 2015).

*construction, with the research foundation as the credit tenant—an important backstop. It also provides access to research programs and facilities and owns and operates the facility, leasing office space to industry.*[38]

## 3.4 Center for Advanced Interconnect Science and Technology (CAIST)

By 1996, SIA had identified interconnect technology—the pathways between transistors in semiconductor devices—as "one of the key road map technologies for future developments in the industry... interconnects [were] at the top of the list as the most critical for the future of the industry."[39] The industry was "rapidly approaching the current limit of interconnect technology," a bottleneck which was "slowing the overall performance of computer chips."[40] The fact that SUNY Albany had developed expertise in this area benefitted the region in competing with other regions. Rick Saburo, a former employee of the Center for Economic Growth (CEG) who had moved to a business development position at the SUNY Albany Thin Film Center, observed in 1996 that specializing in interconnects was "giving the Capital region an edge now in vying for the other areas in research finding." He added—

> *It's one of those technologies that have been identified by SIA as critical to the growth and future of the semiconductor industry. The people who got our center started had the foresight to develop specialization in that particular area. And it's worked well for us; we've grown very quickly—definitely one of the fastest-growing CATs.*[41]

In 1997, the US semiconductor industry launched its Focus Center Research Program. The Microelectronics Advanced Research Corporation (MARCO) was formed as a wholly owned subsidiary of SRC. MARCO was staffed and operated separately from SRC but used the latter's resources and infrastructure. MARCO established R&D Focus Centers in collaboration with US universities. Each Focus Center is comprised of a team of US universities given the task of exploring research themes over an 8–9 year time horizon that addressed anticipated "gaps and barriers" in the semiconductor research roadmap. The Focus Center program was initially funded by Semiconductor Industry Association (SIA) members (50%), by suppliers (25%), and by the US government (25%) with a 6-year budget of $300 million.[42]

---

[38] National Research Council, Charles W. Wessner (rapporteur), *New York's Nanotechnology Model: Building the Innovation Economy* (Washington, DC: The National Academies Press, 2013), p. 75.

[39] "High-tech Centers at RPI, SUNY-A Work Together to Bring in Big R&D Dollars," *Albany Business Review* (October 7, 1996).

[40] "Semiconductor Industry Eyes Schools," Albany, *The Times Union* (October 3, 1996).

[41] "High-tech Center at RPI, SUNY-A Work Together to Bring in Big R&D Dollars," *Albany Business Review* (October 7, 1996).

[42] National Research Council, Charles W. Wessner (ed.) *Securing the Future: Regional and National Programs to Support the Semiconductor Industry* (Washington, DC: The National Academies Press, 2003), pp. 203–204.

While the Focus Center program was still in its planning stages, Albany would make a bid to become the Focus Center for interconnects.[43]

In late 1997, MARCO released an RFP for proposals for a major interconnects Focus Center, and the State of New York committed $5 million a year for 5 years to the center if SUNY Albany and RPI were successful in a bid to land the center.[44] SUNY Albany and RPI prepared a bid in collaboration with Georgia Tech, Stanford, and MIT, with the two New York institutions bringing strength to the bid given their expertise in materials processing and manufacturing.[45] In August 1998, Governor Pataki announced that the semiconductor industry and DARPA had chosen the consortium for the establishment of a MARCO Focus Center specializing in interconnects. It would be called the Center for Advanced Interconnect Science and Technology (CAIST). Although Georgia Tech was assigned the lead role in negotiating the details of the consortium with the industry and government, the primary research site was designated "Focus Center—New York" at SUNY Albany.[46] Governor Pataki reiterated the state's commitment of $5 million/year for 5 years and told a news conference that "the word is out, New York can compete with Silicon Valley and Austin, Texas, for high-tech jobs of the future and win."[47] Focus Center—New York was formally launched in December 1998 at an initial 3-year budget level of $45 million.[48]

CESTM's 70,000 square foot complex was completed in June 1997.[49] The completion of the futuristic-looking building enabled SUNY Albany faculty members involved with the Focus Center, the CAT, and the Institute for Materials to relocate from a 30-year-old physics building to the new CESTM structure. Virtually from its inception, the new center was fully occupied and the university was planning a $10 million, 50,000 square foot expansion to house a pilot semiconductor manufacturing line.[50] The building itself was said to "suggest all the technological promise of the coming millennium"—

> From Fuller Road, the curved green glass and aluminum tower rises into view like the prow of a ship, the almost inevitable comparison goes, ushering in a gargantuan, gleaming white building that is the new Center for Environmental Sciences and Technology Management.[51]

---

[43] "Semiconductor Industry Eyes Schools," Albany, *The Times Union* (October 3, 1996).

[44] "UAlbany, RPI Ready Research Center Bid," Albany, *The Times Union* (November 6, 1997); "State Pays to Go High-Tech," Albany *The Times Union* (December 2, 1997).

[45] "Semiconductor Research in Place," Schenectady, *The Daily Gazette* (November 23, 1997).

[46] "Albany Draws Microchip Center," Albany, *The Times Union* (August 12, 1998).

[47] "UAlbany, RPI, Join Microchip Network," Schenectady, *The Daily Gazette* (August 12, 1998).

[48] Focus Center—New York was to receive $5.85 million from MARCO, $15 million from the state and $25 million in cash and equipment from industry. "Computer Technology Initiative a Reality," Albany, *The Times Union* (December 10, 1998).

[49] CNSE, http://www.sunycnse.com/AboutUs/History.aspx

[50] "New UAlbany Science Center Thriving," Albany, *The Times Union* (May 20, 1997).

[51] "Millennium Modern Airily Impressive Science Center Offers a High-Tech Alternative to Areas," Albany, *The Times Union* (March 23, 1997).

## 3.5 Unprecedented Investments in Research Equipment

The Establishment of a 300 mm Pilot Manufacturing Facility. SUNY Albany's microelectronics research programs were distinguished from those of most other US universities by the high caliber of their research equipment. An executive from the U.S. Semiconductor Industry Association who visited in 1996 commented that "this university is somewhat unique in that it has some very high-tech, state-of-the-art equipment."[52]

In the mid-1990s, the state of the art in semiconductors was based in 200 mm (8-in.) wafer technology, but the industry was looking ahead to the introduction of 300 mm (12-in.) wafer technology, which would enable manufacture of 2.5 times as many chips at 1.7 times the cost of the 200 mm process. In 1997, SUNY Albany was reportedly seeking state funds to establish a 300 mm pilot manufacturing line on its campus as part of the CESTM, citing the economic growth that Austin, Texas, had experienced after establishing a then-state-of-the-art 200 mm pilot line in the late 1980s. The cost of the pilot line was estimated at $150 million. Senate Majority Leader Bruno indicated that he supported the provision of state funds for the pilot line and that he thought Governor Pataki and Assembly Speaker Sheldon Silver did also.[53]

### 3.5.1 A Powerful Rationale for Pilot Manufacturing Facilities

From an industry perspective, the existence of a pilot semiconductor manufacturing line featuring state-of-the art equipment is of incalculable value and exerts a powerful gravitational pull on device and equipment companies that aspire to global leadership. Cutting-edge semiconductor manufacturing equipment can be very costly—in 2006, one ASML-made EUV lithography tool acquired by SUNY Albany weighed 44,000 pounds and carried a price tag of $65 million.[54] However, the first generation of newly developed equipment, the so-called alpha tools, are often characterized by defects and quirks which only become evident when the tools are employed in actual manufacturing operations. Toolmakers eventually responded by developing "beta tools," versions of the same equipment that addressed and overcome the problems observed during initial production runs. But a device manufacturer that has already made huge investments in "alpha" tools is stuck with capital investments in a plant that, in the best case, is less competitive than beta variants and in the worst case, is unusable.

---

[52] "Semiconductor Industry Eyes Schools," Albany, *The Times Union* (October 3, 1996).
[53] "UAlbany Banks on $10 M to Fund a Dream," Albany, *The Times Union* (October 10, 1997).
[54] "$65 Million Chip Etching Tool Arrives," Schenectady, *The Daily Gazette* (August 29, 2006).

A research pilot manufacturing line enables industrial collaborators to spread the cost of initial production runs among themselves, and to identify problems in "alpha tools" before investing in their own manufacturing facilities. A team of engineers from IBM, Micron Technologies, Infineon, AMD, and ASML presented a paper in 2006 describing how SUNY Albany's 300 mm research production line mitigated the costs and risks associated with introduction of several generations of ASML lithography tools. At the 157 nm node, the technology was "not exactly a success story for the semiconductor industry," but costs were shared across the program, SUNY Albany covered much of the cost, and company financial exposure was therefore "low." In the more successful 193 nm research program, involving ASML "alpha" lithography tools, the pilot line enabled consortium members to avoid becoming saddled with expensive-but-unusable early-generation equipment—

> [E]ven when a technology development program is successful, as it was with 193nm lithography, there is still a high cost of early adoption and alpha tool evaluation. Most of the early 193 alpha tools were used for only a short time before they were replaced by higher NA, more capable scanners. Most of them ended up listed for sale on equipment re-marketing web sites, for pennies on the dollar, and there weren't many takers. Most of them are still there. Yet the need for active customer involvement early in the development phase of this equipment cannot be overlooked. The early learning for the user in developing processes, and the feedback required by the supplier to enable improvements in the equipment technology is critical. But, the cost of developing new technology is only going up, along with the risk. The best approach is through a consortium of users, equipment suppliers, universities and government. In an environment where costs and risks are shared, where the possibility exists to provide funding for equipment development and no one company bears the sole financial burden.[55]

From the university's perspective, large-scale investment in research equipment offered a strategy for competing for research funding with leading-edge research universities like MIT and Stanford. A study of New York's Centers of Excellence policies cited interviews with SUNY Albany officials who explained the rationale underlying the huge state investments in research equipment—

> They knew that they could not compete against the more traditional, and powerful, research institutions for limited federal research grants. However, if they could attract industry funds by leveraging state support for capital projects, they could expand their research capacity. One official close to Albany's development described it as a "new game"—rather than competing academically for research support, UA [SUNY Albany] would attract and partner with industry by giving companies what they need (research infrastructure capacity), while partnering with the state by emphasizing economic development. Thus UA could have a win-win strategy for everyone involved.[56]

---

[55] Michael Tittnich, et al., "A Year in the Life of an Immersion Lithography Alpha Tool at Albany Nano Tech," in Proceedings of SPIE, Vol. 6151, *Emerging Lithographic Technologies* (2006), pp. 1–3.

[56] Robert W. Wagner, *Academic Entrepreneurialism and New York State's Centers of Excellence Policy*, Ph.D. dissertation, (SUNY Albany, 2007), p. 84.

Kaloyeros, describing the 300 mm project, stressed that the university would offer manufacturers access to the pilot line to develop their own technology and "they won't incur the costs of the infrastructure."[57]

From the standpoint of the state's policymakers, large investments in university research facilities offered an attractive alternative to traditional incentives offered to encourage companies to locate in the state. The risk of traditional incentives, as New York had learned first-hand, was that the recipient might leave the state after a few years, having benefitted from state subsidies. But when investments were undertaken in university research infrastructure, the assets were owned by the state and would remain in the state regardless of the comings and goings of individual industrial participants. The state could continue to control the assets and use them to attract additional companies.[58]

### 3.5.2 IBM Renews Its Commitment to New York

The 300 mm project got a boost from IBM's 1997 disclosure that it would invest $700 million to build a commercial 300 mm wafer fabrication facility in East Fishkill, Dutchess County. The estimated cost of IBM's 300 mm fab escalated to $2.5 billion by late 2000, representing what was then the largest single private investment in the history of the state.[59] The new pilot line at the university would enable worker training to "feed into IBM's manufacturing needs."[60] Kaloyeros credited the research infrastructure at SUNY Albany with IBM's decision to build its 300 mm fab in New York. "These guys were leaving and going to California, and we're keeping them in New York state."[61] A state official "close to IBM decision makers" observed that in deciding to build its 300 mm fab in New York:

> *The company was especially interested in New York state's strong support of UA's 300mm research lab in the newly-announced facility adjacent to CESTM. The company was currently using its own 200mm facility, but was ... particularly 'allured' by the opportunity to use the next generation 300mm facility with public financial support. This would allow IBM to be involved in R&D without having to front the cost to build the facility themselves.*[62]

---

[57] "Researcher Outlines 21st Century Incubator," Schenectady, *The Daily Gazette* (November 19, 1998).

[58] Interview with Catherine Hill, Counsel to Albany Chamber of Commerce, 1995–1996, Saratoga Springs, New York (September 15, 2015).

[59] "IBM Plant Likely a Magnet," Albany, *The Times Union* (October 11, 2000).

[60] "UAlbany Center Hopes for High Tech Boost," Albany, *The Times Union* (November 18, 1997). IBM's Vice President John Kelly concurred that the SUNY Albany pilot line would benefit his company "because of the school's proximity to the new [IBM] plant." See "UAlbany High Tech Plans Get a Boost," Albany, *The Times Union* (December 2, 1997).

[61] "Plant's Benefit to Area Touted—IBM'S Facility May Spark Local Opportunities," Schenectady, *The Daily Gazette* (October 11, 2000).

[62] Robert W. Wagner, *Academic Entrepreneurialism and New York State's Centers of Excellence Policy,* Ph.D. dissertation, (SUNY Albany, 2007), p. 83.

At the end of 1997, the state committed $10 million to the SUNY Albany 300 mm project, and SUNY Albany indicated that it had secured commitments totaling $25 million from three of the world's seven largest manufacturing equipment suppliers.[63] In 2000, New York State's pledge of $10 million toward the cost of the 300 mm pilot manufacturing facility was increased to $28 million, while project leaders reported that negotiations with semiconductor companies were likely to raise $200 million in the form of donated equipment.[64] The 300 mm research pilot line would closely support IBM's 300 mm manufacturing facility:

> Not only will the specifications of the UAlbany [SUNY Albany] facility and the IBM plant mesh seamlessly, but the university and the company already have partnered on research and projects, starting out small and increasing to an average of $5 million a year in funding and infrastructure support.[65]

In 1998, SUNY Albany was reportedly considering raising and investing $45 million in private and public funds to construct 15,000 square feet of clean rooms to be built in conjunction with the 300 mm pilot manufacturing facility. The idea was to lease clean room space in the incubator facility associated with the pilot line to high-tech manufacturers, particularly semiconductor firms, to support their on-site research efforts. Kaloyeros observed that in microelectronics, "access to clean rooms is the No. 1 hurdle for start-up companies."[66] Under the proposal, the clean room space would be established at a cost estimated at $20 million, $4 million would be spent on the so-called dry space around clean room areas, $20 million would be allocated to support infrastructure, and $1 million would go toward a building linking the clean room space to the pilot manufacturing plant.[67] Assembly Speaker Sheldon Silver strongly backed the idea, pointing out that in Austin, creation of a previous-generation of clean room technology had led to the creation of 10,000 jobs directly and 40,000 jobs in related industries.[68]

### 3.5.2.1 Establishment of the State Center of Excellence in Nanoelectronics and Nanotechnology (CENN)

In 2001, Governor Pataki designated SUNY Albany as one of the first three New York Centers of Excellence, specializing in nanotechnology. The Center of Excellence in Nanoelectronics and Nanotechnology (CENN) would be housed at SUNY Albany's Center for Environmental Sciences and Technology Management and would comprise part of the project establishing a 300 mm pilot facility at CESTM. The Center of Excellence program would also involve establishment of an incubator for small companies and collaboration with local community colleges to

---

[63] See "UAlbany High Tech Plans Get a Boost," Albany, *The Times Union* (December 2, 1997).

[64] "UAlbany Chip Site to Get $18 M More—Pataki and Bruno to Announce Funding," Schenectady, *The Daily Gazette* (February 15, 2000).

[65] "Plant Opens Door to Future," Albany, *The Times Union* (October 15, 2000).

[66] "'Clean Rooms' are Eyed," *Albany Business Review* (December 28, 1998).

[67] "'Clean Rooms' are Eyed," *Albany Business Review* (December 28, 1998).

[68] "Silver Pledges Support for Chip Plant Clean Room," Albany, *The Times Union* (February 24, 1999).

3.5 Unprecedented Investments in Research Equipment    63

provide relevant training utilizing semiconductor manufacturing equipment that the colleges could not afford themselves.[69]

#### 3.5.2.2 IBM's Pledge of $100 Million Toward 300 mm Research Fab

Concurrently with the Governor's announcement of the Center of Excellence for nanotechnology, IBM pledged $100 million in investment in the 300 mm research fab, with the State of New York contributing another $50 million. IBM would enjoy access to the 300 mm facility for its research and development projects. (See Box 3.2.) The company provided 20 internships for SUNY Albany students to study at IBM facilities and pledged to support the university's research programs with grants and equipment donations. One observer commented that "the governor's creation of the Center of Excellence will guarantee a tenfold expansion in the research, development, prototyping and work force training programs between IBM and Albany."[70]

> **Box 3.2: Rationale for IBM Investment in CESTM**
> When New York created a Center of Excellence at SUNY Albany in 2001, IBM pledged to invest $100 million at the CESTM's planned 300 mm wafer fabrication facility at CESTM, with the state contributing another $50 million. Under the terms of the deal, SUNY Albany would lease equipment from IBM pursuant to a 3-year agreement, and IBM would have access to the 300 mm wafer facility for its R&D projects. From IBM's perspective, this investment made sense because of its plan to establish a $2 billion, high-volume 300 mm manufacturing line in East Fishkill. Access to the research line would enable IBM to test, prove, and refine the equipment, materials, and processes ultimately intended for use at the East Fishkill site in a shared-cost manufacturing environment. This would enable IBM to accelerate ramp-up of the East Fishkill site and to sidestep costly investments in unproven new equipment and processes that fail when operational.[71] Thus, although IBM's $100 million commitment represented a very large investment, by reducing the risks associated with the much-larger East Fishkill investment, it promised to save money for the company.

### 3.5.3 Completion of the 300 mm Research Fab

In 2003, the first 300 mm research manufacturing facility was completed at NanoFab 300 South, with 17,000 square feet of cleanroom space, and another 14,000 square feet of cleanroom was added in February 2004.[72] Operated by Albany Nanotech,

---

[69] "Research Initiative Will Benefit Area Universities," Schenectady, *The Daily Gazette* (January 4, 2001); "Playing a Big Role in a Tiny World," Albany, *The Times Union* (January 5, 2001).

[70] "$100 M Boosts Tech Center," Albany, *The Times Union* (April 24, 2001).

[71] "$100 M Boosts Tech Center," Albany, *The Times Union* (April 24, 2001).

[72] Michael Tittnich, et al., "A Year in the Life of an Immersion Lithography Alpha Tool at Albany Nano Tech," in Proceedings of SPIE, Vol. 6151, *Emerging Lithographic Technologies* (2006).

Inc., one of the intermediary organizations established to serve as an interface between the university and the private sector, the 300 mm facility featured some of the most advanced semiconductor manufacturing equipment available. In 2004, Albany NanoTech began using the world's first 193 nm pre-production immersion lithography system, valued at $26 million, donated by the Dutch producer ASML Holding NV. TEL donated an associated tool, a $5–6 million coater/developer system. One industry expert characterized the ASML system as the "bleeding edge," a machine that was not even on the market yet. Acquisition of the new system, which was being used to produce 300 mm wafers, gave Albany NanoTech, one of the "fullest complements of [lithography] machines in the world." Shonna Keogan, a spokesperson for Albany NanoTech, commented that "basically we have every single lithography tool being used for research into chip development for commercial markets." This included an extreme ultraviolet tool using a technology that she characterized as "way, way off in the distance."[73]

In December 2003, the Albany NanoTech 300 mm research manufacturing line fabricated its first 300 mm silicon wafer, representing the first-ever such achievement by a university. LaMar Hill, director of business development at Albany NanoTech, commented that the achievement by a team of university and industry scientists, was unprecedented: "other universities don't even dream of doing something like this, let alone do it. We're doing historic stuff here."[74] In late 2004, IBM disclosed that it had used the 193 nm immersion lithography tool at SUNY Albany to create IBM's first 64-bit power microprocessor.[75] The company's researchers used the tool to develop the immersion portions of the microprocessor, then transferred the device to IBM's fab in East Fishkill to complete the work.

## 3.6 Sematech Revisited

The Sematech consortium was originally formed by the US semiconductor industry to enhance the industry's competitiveness in manufacturing through joint research on leading-edge process technology.[76] Although New York State had bid unsuccessfully to be the site of the newly formed Sematech in 1987, the state continued to show interest in developing a local presence by the consortium. In September 2001, when semiconductor industry leaders, including Sematech President Robert Helms, were present in Albany for a symposium on nanotechnology, they were persuaded to make a detour, first to the SUNY Albany nanotechnology research facilities and then to the symposium, where Governor Pataki was present. Governor Pataki and Robert

---

[73] "Albany NanoTech Fills Toolbox," Albany, *The Times Union* (August 26, 2004).

[74] "New Chips Mark UAlbany Milestone, Schenectady, *The Daily Gazette* (January 15, 2004).

[75] IBM's John Kelly commented on the acquisition of the 193-nanometer immersion system: "That's the first in the world and the most advanced lithography tools and it's sitting in Albany NanoTech." "Brain Power Will Win Nanotech Wars," *Albany Business Review* (September 16, 2004).

[76] See generally National Research Council, Charles W. Wessner (ed.) *Securing the Future: Regional and National Programs to Support the Semiconductor Industry* (Washington, DC: The National Academies Press, 2003).

Helms reportedly talked at some length about a potential partnership. Ten months of "hush hush" discussions followed between Sematech leadership and "a small group of players" on the New York side, including Governor Pataki himself and his higher education aide. It became evident in these talks that Sematech's real need was a research center focusing on extreme ultraviolet light (UVL) lithography.[77]

### 3.6.1 Establishment of a Sematech Research Center in New York

In July 2002, Sematech announced that it would establish its next research center, "International Sematech North," a $400 million project focusing on extreme UVL lithography, at the SUNY Albany 300 mm wafer cleanroom complex. The project would focus on three themes, mask blanks, resist, and extreme UVL extensions.[78] Sematech agreed to contribute $40 million in cash to the project over 5 years and another $60 million in in-kind contributions. The State of New York pledged $160 million and the Governor's office another $50 million toward completing construction work. IBM contributed equipment for research.[79] By 2006, International Sematech North was operating the EUVL Mask Blank Development Center, the leading R&D center of its kind in the world, and the Sematech Resist Test Center, which focused on EUV and Hyper NA 193 nm photoresist evaluation and development.[80]

Sematech's selection of Albany as the site for this research center surprised many industry observers, one of whom remarked that "Albany certainly wouldn't have been on any list of places to go. I'm just not familiar with what's going on there." Governor Pataki observed that when the announcement was made, "There are a lot of places where jaws have been dropped." One factor favoring Sematech's choice was the fact that the consortium's most influential member, IBM, was based in New York and favored the prospect of having the research facility nearby. The region's lower costs and concentration of nearby universities and colleges were factors. But—

> *chip makers said the most important factor was the caliber of work already being done at SUNY Albany, and the enthusiastic backing it had from the governor and the legislature.*[81]

---

[77] "If You Build It, They Will Come," *The Chronicle of Higher Education* (February 7, 2003); "Fall Meeting Planted Seed for Deal," Albany *The Times Union* (July 18, 2002).

[78] "Sematech, SUNY Seal EUV Lithography Program," *Solid State Technology* (January 29, 2003).

[79] "Sematech Touts the Benefits of its New York Alliance," *Austin American-Statesman* (July 19, 2002).

[80] Michael Tittnich, et al., "A Year in the Life of an Immersion Lithography Alpha Tool at Albany Nano Tech," in Proceedings of SPIE, Vol. 6151, *Emerging Lithographic Technologies* (2006); Vibhu Jindal, "Getting up to Speed with Roadmap Requirements for Extreme UV Lithography," *SPIENewsroom* (January 2013).

[81] "Albany No Longer A Secret in High-Tech Chip World," *New York Times* (July 19, 2002).

Semiconductor industry observers also commented on the impact of the comparative incentives offered in Texas and New York. An Austin economist, Angelos Angelou, hired by New York in 2002, said that Governor Pataki had "provided a unique set of circumstances, including economic incentives, which Texas did not offer, to encourage Sematech to come to New York. Patrick Shaughnessy, a spokesman for the Texas state economic development agency, said that high-tech firms come to Texas "without any additional prompting from the state," and that while individual communities might offer incentives, "the state stands clear from that." Angelou countered that such thinking was "getting Texas in trouble," and that "there's not been a semiconductor company in recent history that's gone somewhere on their own." Saralee Tiede, Vice President of the Greater Austin Chamber of Commerce, commented that:

> *We are concerned that Texas is losing its competitive advantage in high tech . . . [T]he fact is, money talks, and companies go places where there are incentives.*[82]

### 3.6.2 Relocation of Sematech Headquarters to Albany

Reports circulated in Texas as long ago as 2003 that Sematech was considering relocating its headquarters from Austin to New York although Kaloyeros denied that either he or Governor Pataki had ever approached Sematech with such a proposal.[83] However, in January 2006 Sematech announced the layoff of 80 people (15% of the total staff) at its main office in Austin, while at the same time Sematech members were adding 50 to 60 employees to the Albany NanoTech Complex.[84] In May 2007, International Sematech announced it would move its headquarters from Austin to Albany. The letter of intent between the state and the consortium set forth the terms of the deal:

- International Sematech would be headquartered at CNSE and would bring or create 450 jobs in the Albany area.
- Sematech would invest $150 million in cash and $150 million in "cash equivalents" over a 7-year period, plus $5 million to universities in the state.
- Sematech would add two New York board members, one from SUNY Albany and one from the Fuller Road Management Corporation.
- The state will contribute $300 million to the "strategic alliance."[85]

---

[82] "Momentum for New Hope," Albany, *The Times Union* (November 22, 2002).
[83] "Getting All of Sematech Never Part of Area Plan," Schenectady, *The Daily Gazette* (June 5, 2003).
[84] "Sematech Expands Presence at Albany With $50 M R&D Center," *Albany Business Review* (January 26, 2006).

In 2010, Sematech disclosed that it would move the "bulk of its remaining operations" from Austin to Albany. Sematech CEO Dan Armbrust commented that "New York state has put tremendous backing behind this initiative.... [w]e've crossed the tipping point where there are enough entities that are investing here that we have to be here too."[86]

## 3.7 Arrival of Tokyo Electron

In November 2002, Tokyo Electron, Ltd. (TEL), a Japanese maker of semiconductor manufacturing equipment, announced that it would establish a $300 million R&D center at SUNY Albany, and that it expected to employ 300 researchers at the Center of Excellence in Nanoelectronics.[87] TEL pledged to contribute $200 million to the university over 7 years, with the State of New York committing an additional $100 million.[88] The TEL Technology Center of America (TTCA), TEL's first research center outside of Japan, was established in 2003, focusing on semiconductor equipment and process development.[89] TEL, which at the time was the world's second largest maker of semiconductor manufacturing equipment, indicated that it had chosen Albany for its research center because it was the only university-owned R&D center focusing on 300 mm research, and that it gave companies access to "equipment that would otherwise be prohibitively expensive."[90] Tetsuro "Terry" Higashi, TEL's president and CEO, commented in December 2002 that he was pleased with "the level of technology at UAlbany [SUNY Albany], which was why TEL was coming to the region."[91] He indicated that:

> By participating in this center, we will significantly enhance our internal development efforts, ultimately allowing us to shorten the time required to bring critical technology from the research lab to the production floor. I was extremely impressed with the potential of this facility.[92]

---

[85] "Details of Agreement Emerge in Contract Documents," Albany, *The Times Union* (May 11, 2007); International Sematech Move Expected to Transform Albany Economy," *Austin Business Journal* (May 10, 2007); "Legislation for Int'l Sematech Funding Approved by Assembly," *Albany Business Review* (May 14, 2007).

[86] "Sematech Moving Operations, 100 Jobs to Albany," *Albany Business Review* (October 12, 2010).

[87] "UAlbany Lands R&D Center," Albany, *The Times Union* (November 21, 2002). TEL make machines for a number of steps in the semiconductor manufacturing process, including coating wafers with light-sensitive chemicals, circuit etching, and wafer cleaning. "Momentum for New Hope," Albany, *The Times Union* (November 2, 2002).

[88] "If You Build It, They Will Come," *The Chronicle of Higher Education* (February 7, 2003).

[89] Michael Tittnich, et al., "A Year in the Life of an Immersion Lithography Alpha Tool at Albany Nano Tech," in Proceedings of SPIE, Vol. 6151, *Emerging Lithographic Technologies* (2006).

[90] "Tech Valley's Love," Albany, *The Times Union* (November 22, 2002).

[91] "Momentum for Hope," Albany, *The Times Union* (November 22, 2002).

[92] "Tokyo Electron Plugging $300 M R&D Center Into Albany, NY," *Site Selection* (December 2002).

## 3.8 Albany Nanotech

In 2001, SUNY Albany established "Albany NanoTech" as an umbrella designation under which the University's nanotechnology programs operated, including the CAT and the Center of Excellence. The School of Nanoscience and Nanoengineering was established—the forerunner of the College of Nanoscale Science and Engineering (CNSE). The school, jointly funded by state grants and contributions from endowments and private sources, was the first institution of its kind in the United States, dedicated entirely to the study of atomic-level sciences. Open to graduate and doctoral candidates, the school awarded degrees in a number of nanotechnology disciplines, including thin film materials, in which SUNY Albany had already achieved distinction.[93]

In 2001, SUNY Albany also created an entity designated as the Albany NanoTech Complex, located at Fuller Road/Washington Avenue in Albany, to house facilities used for research, development, and commercialization of nanotechnologies. The complex would incorporate CESTM's existing facility and others yet to be built and would consist of research facilities used for educational purposes, shared facilities which can be leased by private industry, and proprietary facilities leased by individual companies.

In 2003, the Albany NanoTech Complex on Fuller Road/Washington Avenue was opened to house industrial tenants collaborating with the university in research projects. The complex incorporated CESTM's original facility, which had opened in 1997 and housed an older 200 mm research center. A second building, the 120,000 square foot NanoFab300 South building—named in reference to its ability to make semiconductors on 300 mm wafers—housed International Sematech, TEL, IBM, Infineon, and GE. A third structure, the 228,000 square foot NanoFab300 North, was completed in 2005.[94] By the beginning of 2004, Albany NanoTech housed 350 researchers with another 100 expected by June, and the complex was nearing the limits of its existing capacity.[95] (See Table 3.3.)

## 3.9 The Flow of State Funds to Nanotechnology Research, 2000–2009

Between 2000 and 2009, New York State budgeted nearly $900 million in funding for nanotechnology research at SUNY Albany, allocated to a bewildering array of programs (See Table 3.4). State funding to these programs was administered by the Research Foundation of the State University of New York via the Fuller Road Management Corporation, one of the intermediary organizations established to

---

[93] "School of Nanosciences Planned," Albany, *The Times Union* (April 14, 2001).
[94] "Brand Name Tenants Fill Vacancies," Albany, *The Times Union* (August 9, 2003).
[95] "New Chips Mark UAlbany Milestone," Schenectady, *The Daily Gazette* (January 15, 2004).

**Table 3.3** The Albany NanoTech Complex in 2013

| Building | Year completed | Cost (millions of dollars) | Overall size/clean room size (thousands of square feet) |
|---|---|---|---|
| NanoFab 200 | 1997 | 16.5 | 70/4 |
| NanoFab South 300 | 2004 | 50 | 150/32 |
| NanoFab North 300 | 2005 | 170 | 228/35 |
| NanoFab East/ Central | 2009 | 150 | 350/15 |
| NanoFab Xtension | 2013 | 365 | 320/50 |

Source: Jason Chernock and Jan Youtie, "State University of New York at Albany Nanotech Complex," in Georgia Tech Enterprise Innovation Institute, *Best Practices in Foreign Direct Investment and Exporting Based on Regional Industry Clusters* (Atlanta: Georgia Tech Research Corporation, February 2013, Prepared for the Economic Development Administration, U.S. Department of Commerce)

serve as an interface between the university and the private sector. A 2010 audit of these outlays by the New York Office of the State Comptroller found that "the State funding had been spent as intended to construct and equip various nanotechnology research facilities at the University and to support research conducted at those facilities."[96]

Although the state's investments in nanotechnology have been very substantial, those investments have catalyzed much-larger private sector investments in the state. In 2013, the Empire State Development (ESD) President and CEO Ken Adams indicated that New York had invested roughly $1.3 billion in the semiconductor sector over the years but that "if you think about that $1.3 billion investment, this has had a leveraging effect in attracting or supporting over $20 billion from world leaders in the industry."[97]

## 3.10 IBM, The Anchor Tenant

Professor Laura Schultz of the College of Nanoscale Science and Engineering observes that IBM played the role of "anchor tenant" in fostering CNSE's success. An anchor tenant in innovation terms is—

---

[96] Office of the New York State Comptroller, *Fuller Road Management Corporation & The Research Foundation of the State of New York: Use of State Funding for Research into Emerging Technologies at the State University of New York at Albany: Nanotechnology* (2010-S-4) http://www.osc.state.ny.us/audits/allaudits/093010/10s4.pdf, page 13.

[97] National Research Council, Charles W. Wessner (rapporteur), *New York's Nanotechnology Model: Building the Innovation Economy* (Washington, DC: The National Academies Press, 2013), p. 77.

Table 3.4 New York State Funding of Nanotechnology Research, 2000–2009

| Project | State funding Source | Recipient entity | Fiscal year | Budgeted amount (millions of dollars) |
|---|---|---|---|---|
| CESTM Building | State | SUNY Foundation | 2000–2001 | 10.0 |
| CATN2 | NYSTAR | SUNY Foundation | 2001–2002 | 9.7 |
| State Center of Excellence (Construction) | ESD | FRMC | 2003–2004 | 50.0 |
| Sematech Facilities | ESD | SUNY Foundation | 2003–2004 | 160.0 |
| Tokyo Electron | ESD | SUNY Foundation | 2004–2005 | 100.0 |
| CNSE | State | University | 2005–2006 | 8.3 |
| IMPLSE | ESD/SUCF | FRMC | 2005–2006 | 75.0 |
| Power Substation/Albany NanoTech | SUCF | FRMC | 2005–2006 | 5.0 |
| CENN operations | ESD | FRMC | 2006–2007 | 3.7 |
| INDEX—capital | ESD | FRMC | 2006–2007 | 75.0 |
| INDEX—operations | ESD | FRMC | 2006–2007 | 3.0 |
| INVENT | SUCF | FRMC | 2006–2007 | 75.00 |
| Sematech—machinery and equipment | ESD | FRMC | 2008–2009 | 300.00 |
| Total | | | | 876.1 |

Notes: *ESD* Empire State Development; *FRMC* Fuller Road Management Corporation; *NYSTAR* New York State Foundation for Science, Technology and Innovation; *State* Direct appropriation from state budget; and *SUCF* State University Construction Fund
Source: Office of the New York State Controller, *Fuller Road Management Corporation & The Research Foundation of the State University of New York: Use of State Funding for Research into Emerging Technologies at the State University of New York at Albany: Nanotechnology* (2010-S-4)

> A firm traditionally heavily engaged in R&D with research interests in a technology being developed in the geographic area. Anchor tenants can aid the development and commercialization of university research through the direct sponsorship of faculty work, the hiring of graduates, and collaboration with professors. . . . [T]he existence of an anchor tenant significantly [contributes] to patenting in a region both for the anchor and the non anchor firm.[98]

---

[98] Laura I. Schultz, "Nanotechnology's Triple Helix: A Case Study of The University of Albany's College of Nanoscale Science and Engineering," *Journal of Technology Transfer* (2011), p. 560.

IBM was the largest tenant at Albany NanoTech and made massive investments in research and research infrastructure at the site on Fuller Road. Its presence catalyzed investments in Albany NanoTech by other major industry players like ASML, the world's largest semiconductor equipment maker, which invested $325 million at Albany NanoTech—co-funded by IBM—to create its first R&D center outside of Europe.[99] IBM's presence was an "intangible incentive" for Sematech's decision to invest in an R&D center at Albany NanoTech.[100] IBM's presence on Fuller Road was also "a strong draw for the [nanotechnology] school."[101] In addition, IBM was instrumental in the creation of a computational center for nanotechnology at RPI (see Box 3.3).

In 2008, with IBM beginning to shed local jobs in the slumping semiconductor business, Governor David Paterson concluded a new agreement with the company to bolster its operations in the state, retain jobs, and create 1000 new jobs. IBM would undertake three projects in the state:

- IBM would expand its research operations at Albany NanoTech, a $325 million effort that would create 325 new jobs.
- The company would undertake a $1 billion upgrade at its existing semiconductor plant at East Fishkill and would not lay off any of the 1400 employees at that site for the remainder of the 2008.
- IBM would establish a new $125 million, 675-employee R&D center for advanced semiconductor packaging at a to-be-determined site in upstate New York under the ownership and management of Albany NanoTech.

The state committed $140 million in incentives to the deal. The existing and planned research facilities at Albany NanoTech played a key role in IBM's renewed commitment to the state:

> Holding IBM's expansion announcement inside Albany NanoTech underscored one of the deal's central elements. A $3.5 billion, 450,000 sq. ft. (40,500 sq. m.) complex, Albany NanoTech will be the key research partner in the two IBM projects creating the 1,000 new jobs.[102]

## 3.11 Arrival of the M + W Group

The M + W Group is one of the world's leading design, engineering and construction firms for high-technology manufacturing facilities, operating in over 30 countries. Formerly, known as M + W Zander, an Austrian firm with US offices in Plano, Texas, the company designed IBM's 300 mm fab at East Fishkill and helped construct SUNY Albany's nanotechnology research facilities. In 2004, M + W Zander established a $6 million job-training program at the Watervliet Arsenal with

---

[99] "IBM, Partners Creating 1000+ Jobs With $2.7 Billion in New York Projects," *Site Selection* (January 2005).
[100] "Upstate New York Gets Nod For Sematech's $403 M R&D Center," *Site Selection* (July 2002).
[101] "IBM's Big New York Compute: $1.5 B Investment, 1000 Jobs," *Site Selection* (August 2008).
[102] "IBM's Big New York Computer: $1.5 B Investment, 1000 Jobs," *Site Selection* (August 2008).

financial support from the state, the US Army, and a nonprofit group, the Arsenal Business & Technology Partnership. The training center planned to work with 120 union members per year teaching semiconductor-related construction skills. M + W had 70 workers in New York at the time of the announcement and planned to add 50 more by the end of 2004.[103] M + W's decision to open the training center was "largely because of the work being done at UAlbany [SUNY Albany]."[104]

In 2010, following its selection as the general contractor for the new GlobalFoundries fabs, M + W announced that it would move its North American headquarters from Texas to Watervliet, bringing an additional 190 jobs to the region.[105] A company spokesperson said that M + W had shifted its US headquarters several times since first setting up in Silicon Valley. Its subsequent moves reflected its assessment of where the future of nanotechnology was located—

> *Albany ... is known worldwide as a leader in the (development) of nanotech and nanoscience, as evidenced by the companies that have centered here. It is the center of the universe now in nanotechnology.*[106]

M + W was "moving its center of gravity [to Albany] to be closer to its major semiconductor industry clients that are doing research at the NanoCollege."[107]

---

**Box 3.3: Supercomputing at RPI**
In 1999, Dr. Shirley Ann Jackson became President of the Rensselaer Polytechnic Institute. A physicist by background, Jackson launched a strategic effort to remake RPI, the Rensselaer Plan, creating state-of-the-art research platforms and hiring 325 new tenure-track faculty. She focused on "signature thrusts" in a number of disciplines, including computational science and engineering, nanotechnology and advanced materials, and biotechnology.[108] In 2001, RPI created a center for nanotechnology studies with $2 million in support from IBM, Kodak, Philip Morris, Albany International Corp., and several federal agencies.[109]

---

[103] "New Deal in Works at Arsenal," Albany, *The Times Union* (February 10, 2014).
[104] "UAlbany Project Raises Hopes," Albany, *The Times Union* (January 6, 2005).
[105] "Firm Moving Headquarters to Arsenal," Schenectady, *The Daily Gazette* (February 10, 2010).
[106] "GlobalFoundries Construction Moving HQ from Texas to Watervliet," Saratoga Springs, *The Saratogian* (February 9, 2010).
[107] "M + W Move a Sign of High Tech to Come," Albany, *The Times Union* (February 10, 2010).
[108] Rensselaer Polytechnic Institute, http://rpi.edu/president/profile.html
[109] "RPI Creates Center for Nanotechnology Studies," Albany, *The Times Union* (March 30, 2001).

In May 2006, Jackson announced the formation of a partnership between RPI and IBM establishing a new Computational Center for Nanotechnology (CCNI) based on the RPI campus. The site featured the establishment of $100 million IBM "Blue Gene" supercomputer system, the most powerful system on any university campus in the world.[110] IBM contributed $33 million (largely comprised of hardware, software, and support) to the project, the state contributed $33 million, and RPI and partner firms contributed $34 million.[111]

Jackson said that the new supercomputer directly played into the semiconductor manufacturing research under way at SUNY Albany and that between the two institutions, scientists in the Capital Region would have access to a full spectrum of chip making that was "impossible elsewhere." John Kelly of IBM observed that with the pattern of processing speed doubling every other year, "we are simply outrunning our capabilities" and that CCNI's system would "allow companies large and small to simulate complex miniaturization scenarios and effectively reduce the research time and cost of developing nanoscale electric systems."[112] Dr. Jackson observed in 2013 that—

> *CCNI has allowed companies of all sizes to improve their products and processes by tapping the expertise of Rensselaer scientists and engineers and the power of high-performance computing for simulation, modeling, and the manipulation of big data.*[113]

## 3.12 Establishment of the College of Nanoscale Sciences and Engineering

In his 2004 State of the State address, Governor Pataki outlined the formation of the College of Nanoscale Sciences and Engineering at SUNY Albany, the first nanotechnology college in the United States. CNSE would be housed at the three-building cluster at Fuller Road and the Washington Avenue extension in Albany. CNSE would have a separate budget and governance structure from SUNY Albany itself, a fact which reportedly had "the potential to irk UAlbany [SUNY Albany] faculty members who maintain that the school's nanotechnology efforts are siphoning resources from other programs there." Governor Pataki indicted that CNSE

---

[110] "RPI to Get Supercomputer—System Expected to Create 300–500 Jobs," Schenectady, *The Daily Gazette* (May 11, 2006).

[111] "A 'Magical' Moment for Tech Valley—Many Await Chance to Use RPI Supercomputers Which Will Link Region to a Powerful Network," Albany, *The Times Union* (May 12, 2006).

[112] "RPI to Get Supercomputer—System Expected to create 300–500 Jobs," Schenectady, *The Daily Gazette* (May 11, 2006).

[113] National Research Council, Charles W. Wessner (rapporteur), *New York's Nanotechnology Model: Building the Innovation Economy* (Washington, DC: The National Academies Press, 2013), p. 69.

would train the high-technology workforce that companies like IBM, Sematech, and Tokyo Electron (TEL) needed.[114] Kaloyeros became the vice president and chief administrative officer of CNSE.[115] CSNE Associate Vice President Jim Castrane observed in the fall of 2004 that nanotechnology at SUNY Albany had evolved from an institute to a department to a school, and ultimately a full-fledged college. "We have transformed the research staff into faculty."[116]

At the inception of CNSE, its faculty was drawn from other universities and from companies. (See Box 3.4.) In addition, CNSE brought in some scientists (including several from IBM and SEMATECH) who worked on-site but did not have teaching assignments.[117] Laura I. Schultz, Assistant Professor of Nanoeconomics at CNSE, recalls that when CNSE was formed, the roles of faculty and staff were "not constrained by traditional academic expectations," but were defined to "maximize technology transfer and economic development." As of 2010, over one-third of CNSE's tenure-track faculty had industrial experience, a background which enabled them to "better understand the needs of corporate partners," to identify potential collaborators, to develop university-industry alliance, and to attract "research funding from corporate partners."[118]

> **Box 3.4: Profiles from CNSE'S First Faculty**
> The College of Nanoscale Sciences and Engineering (CNSE) at SUNY Albany began its first graduate-student classes in the fall of 2004 with an enrollment of 73. The absence of an established academic program in nanotechnology created the "unique opportunity to build a pioneering academic and research program from the ground up," enabling flexibility, creativity, and "out-of-the-box" thinking.[119] At the outset, CNSE had 25 faculty, who cited their motivations to come to the NanoCollege as the promise of something new and to gain access to state-of-the-art tools. They included "a mix of people with new doctoral degrees, several with patents under their belt, and people with experience in industry."

---

[114] "UAlbany to Have Nanotech College," Albany, *The Times Union* (January 8, 2004).

[115] "Nano College Propels UAlbany Program," Albany *Business Review* (April 22, 2004); "Nanoscience College Destined for Albany," *Albany Business Review* (April 20, 2004). The SUNY Albany School of NanoSciences and NanoEngineering was absorbed by CNSE.

[116] "It's More Than a Tech Transfer," *Albany Business Review* (September 2, 2004).

[117] Interview with Karen Hitchcock, Troy New York (April 4, 2013), in National Research Council, Charles W. Wessner (ed), *Best Practices In State and Regional Innovation Initiatives: Competing in the 21st Century* (Washington, DC: The National Academies Press, 2013), p. 153.

[118] Laura I. Schultz, "Nanotechnology's Triple Helix: A Case Study of the University of Albany's College of Nanoscale Engineering," *Journal of Technology Transfer* (2011), p. 553.

[119] Laura I. Schultz, "Nanotechnology's Triple Helix: A Case Study of the University of Albany's College of Nanoscale Science and Engineering," *Journal of Technology Transfer* (2011), p. 560.

## 3.12 Establishment of the College of Nanoscale Sciences and Engineering

- **James Ryan**, a Rensselaer Polytechnic Institute (RPI) graduate and a former professor of nanoscience, worked at IBM for 25 years before joining CNSE, Ryan is the author of over 100 publications, holder of 47 US patents, and between 2003 and 2005 served as the site executive for IBM at Albany Nanotech. He went on to become the founding dean of the Joint School of Nanoscience and Nanoengineering of North Carolina A&T State University and the University of North Carolina at Greensboro.[120]
- **Brad Thiel** left a faculty position at Cambridge University to come to CNSE, citing the latter's newness as a draw. Of Cambridge he said "The older the place and the more prestigious the program, the more entrenched it tends to be.... My work on the academic side of things here is focused on basic science, but the structure of the place is such that we're able to take individually critical problems and address the basic science issues that under pin them."
- **Eric Lifshin**, an expert on scanning electron microscopy and author of several books, was a 38-year veteran of the GE Global Research Center who rose to the position of manager of 75 researchers. He said that "I wanted to teach and do some research. I wanted to offer something to [CNSE] and they could offer something to me—to do some research."
- **Gregory Denbeaux** left a research position at Lawrence Berkeley National Laboratory to join the CNSE faculty. He said that "All of the equipment I could possibly want to do nanoscience experiments is all right here in Albany. This is bigger than that—the scale of the number of programs, the quality of the tools. This is not a stagnant place."
- **Kathleen Dunn**, an expert in microscopy techniques, said of joining the CNSE faculty, "This was a different path. I could be part of that—to be part of the beginning, to help define how a college developed. That was very appealing to me."[121]

CNSE's curriculum was based around four "constellations of scholarly excellence in research and development, education, technology development, and economic outreach":

- **Nanoscience.** The study, experimental investigation, and theoretical interpretation of nanoscale phenomena.
- **Nanoengineering.** The application of nanoscience to practical applications, including the atomic scale design, manufacture, and operation of efficient and functional structures, machines, processes, and systems.

---

[120] Joint School of Nanoscience and Nanoengineering, *http://jsnn.ncat.uncg.edu/*
[121] "It's More Than a Tech Transfer," *Albany Business Review* (September 2, 2004).

- **Nanoeconomics.** The study of economic and business principles underlying the development and use of nanoscale technology, products, and systems.
- **Nanobioscience.** (Added in 2006.) The application of nanoscale concepts to biological and medical procedures, practices, structures, systems, and organisms.[122]

CNSE's Pradeep Haldar, Head of Nanoeconomics at the College, commented in 2013 that—

> The way we approach it is truly interdisciplinary. Typical academic institutions tend to be organized by silos: engineering guys don't talk to the science guys; within engineering the chemical engineering guy will not talk to the electrical engineering guy. They have nothing in common. We have mixed it all up to make sure the students get the cross-collaboration they need to understand the field.[123]

CNSE developed two separate curricula for graduate students leading to two separate sets of M.S. and Ph.D. degrees in nanoscale science and nanoscale engineering. CNSE also implemented dual cross-disciplinary programs:

- M.S. Nanotechnology management, partnering with SUNY Albany's School of Business
- M.S. and Ph.D. in nanomedicine, in collaboration with selected SUNY medical schools

CNSE also offered separate undergraduate degrees in nanoscale science and nanoscale engineering.[124]

The research infrastructure at CNSE was unsurpassed in the nanotechnology field. In 2010, CNSE had over four times the research facilities than at the second largest center in Austin and those facilities were staffed by twice as many researchers. CNSE's cleanrooms were built to accommodate "300mm silicon wafers, the industry standard, instead of 200mm, the academic research standard." CNSE had one of only two extant EUV lithography tools in the world. As CNSE leadership observed—

> CNSE is the most advanced research enterprise in the academic world. The presence of cutting-edge, one-of-a-kind equipment and state-of-the-art cleanrooms allows scientists to conduct advanced research that is simply impossible without those capabilities. The large scale attracts corporate partners and a highly skilled workforce (Table 3.5). [125]

---

[122] CNSE, *A Proposal for Undergraduate Academic Programs Leading to the B.S. in Nanoscale Science and B.S. in Nanoscale Engineering* (Submitted to SUNY Albany Senate, May 5, 2008), p. 7.

[123] Pradeep Haldar, "Pioneering Innovation to Drive an Educational and Economic Renaissance in New York State," in National Research Council, Charles W. Wessner (rapporteur), *New York's Nanotechnology Model: Building the Innovation Economy* (Washington, DC: The National Academies Press, 2013), p. 81.

[124] SUNY Working Group Report, *The SUNY College of Nanoscale Science and Engineering* (June 13, 2013), p. 5 ("SUNY Working Group Report on CNSE").

[125] Laura I. Schultz, "Nanotechnology's Triple Helix: A Case Study of the University of Albany's College of Nanoscale Science and Engineering," *Journal of Technology Transfer* (2011), p. 561.

3.12 Establishment of the College of Nanoscale Sciences and Engineering 77

**Table 3.5** Growth of the NanoCollege, 2007–2016

|  | Facilities square footage | | Capital expenditures (millions of dollars) | |
| --- | --- | --- | --- | --- |
| Year | Total | Class 1 clean room | Facilities expansion | Facilities acquisition |
| 2007–2008 | 445,000 | 67,000 | 41 | 175 |
| 2008–2009 | 795,000 | 82,000 | 52 | 60 |
| 2009–2010 | 795,000 | 82,000 | 33 | 75 |
| 2010–2011 | 800,000 | 80,000 | 11 | 100 |
| 2011–2012 | 800,000 | 85,000 | 207 | 100 |
| 2012–2013 | 800,000 | 85,000 | 10 | 100 |
| 2013–2014 | 1,300,000 | 135,000 | 55 | 100 |
| 2014–2015 | 1,300,000 | 135,000 | 195 | 100 |
| 2015–2016 | 1,654,000 | 135,000 | 42 | 100 |
| Total |  |  | 646 | 910 |

Source: SUNY Poly

In May 2006, 2 years after it began operations, *Small Times* magazine ranked CNSE first overall in the United States among colleges and universities offering curricula in nanotechnology and microtechnology. It was also ranked number one in the field with respect to educational facilities and industrial outreach.[126] A SUNY working group summarized CNSE's education performance in 2013, observing that "over 90% of the CNSE Ph.D. and M.S. graduates remain in New York and are employed by the 'Who's Who' in the nanotechnology industry in very high paying jobs across the state." At the undergraduate level, entering students had average SAT scores above 1350, "on a par with top science and engineering schools such as MIT and Stanford," and that the undergraduate population had grown rapidly and was expected to reach 500 by 2017. "By design, all of the CNSE undergraduate students are New Yorkers."[127].

## 3.12.1 Arrival of Advanced Micro Devices (AMD)

In June 2006, Governor Pataki disclosed that one of the world's foremost semiconductor makers, Advanced Micro Devices (AMD), had signed an agreement with New York State to build a 300 mm semiconductor fabrication plant in Saratoga County. This agreement, and its evolution into the present-day GlobalFoundries operation in New York, is described in Chap. 6. While AMD's decision was based

---

[126] "Small Times Magazine Ranks UAlbany College Tops in Nanotech," *Albany Business Review* (May 12, 2006).
[127] SUNY Working Group Report, *The SUNY College of Nanoscale Science and Engineering* (June 13, 2013), p. 5 ("SUNY Working Group Report on CNSE"), p. 5.

**Table 3.6** NanoCollege Research Spending by Source (Millions of Dollars)

| Year | Industry | New York State | Federal | Foundation/nonprofit | Total |
|---|---|---|---|---|---|
| 2010–2011 | 80.8 | 107.9 | 9.9 | 10.0 | 208.6 |
| 2011–2012 | 82.9 | 113.7 | 17.3 | 4.0 | 217.9 |
| 2012–2013 | 201.6 | 96.3 | 24.1 | 2.3 | 324.3 |
| 2013–2014 | 170.1 | 170.4 | 21.2 | 2.6 | 364.3 |
| 2014–2015 | 187.2 | 27.5 | 14.6 | 26.4 | 255.7 |
| 2015–2016 | 183.7 | 76.6 | 17.4 | 13.1 | 290.8 |
| 2016–2017 | 141.0 | 27.6 | 38.2 | 5.0 | 211.8 |
| Total | 1047.3 | 620.0 | 142.7 | 63.4 | 1873.4 |

Note: Total may not add due to rounding
Source: National Science Foundation, Higher Education R&D Survey

on a number of factors, one of the most important was the presence of CNSE in Albany, where AMD researchers were already deeply engaged in cutting-edge research. AMD's decision, which ultimately led to the creation of thousands of manufacturing and construction jobs in the Capital Region, was seen as an affirmation of the state's investments in CNSE. A 2013 white paper by the Georgia Tech Enterprise Innovation Institute observed that—

> The attraction of GlobalFoundries validated the state's and the region's plans to build a nanotechnology cluster not based solely on research and development, but on creating high-paying, quality jobs for area residents.[128]

### 3.12.2 CNSE–Industry Collaborations

CNSE has developed an extensive array of industry research collaborations engaging the entire semiconductor supply chain, including device makers, materials suppliers, construction firms, and equipment vendors. Collaborations include a number of on-site company research centers based on "customized" agreements with each company or consortium. CNSE's leadership stressed the importance of flexibility in these arrangements:

> Agreements can be tailored to reflect the particular strengths and advantages of each collaboration, as well as to reflect the goals of each participating entity. The intended outcome is one in which each participating entity contributes to the eco-system, while at the same time utilizing the partnership to attain its desired goal in the most cost-effective manner possible. Agreements are designed to offer a degree of flexibility to address a variety of possibilities, from changing market conditions to new technology innovations, while continuing to respect the desired outcomes of each entity.[129]

---

[128] Jason Chernock and Jan Youtie, "State University of New York at Albany Nanotech Complex," in Georgia Tech Enterprise Innovation Institute, *Best Practices in Foreign Direct Investment and Exporting Based on Regional Industry Clusters* (Atlanta: Georgia Tech Research Corporation, February 2013, Prepared for the Economic Development Administration, U.S. Department of Commerce), p. 65.

[129] Laura I. Schultz, "Nanotechnology's Triple Helix: A Case Study of the University of Albany's College of Nanoscale Science and Engineering," *Journal of Technology Transfer* (2011), p. 553.

## 3.12 Establishment of the College of Nanoscale Sciences and Engineering

CNSE's corporate partners enjoy "a virtual one-stop shop" for overcoming barriers to commercialization of technology, including technology incubation, prototyping, and test-bed integration support (see Box 3.5):

> *This support by Albany NanoTech has generally resulted in accelerated deployment of nanotechnology-based products. Proof of concept technology incubation is provided through 450,000 sq. ft. of on-site office, laboratory and clean-room incubation facilities... . Product qualification support is provided via access to a unique state-of-the-art industry standard semiconductor fabrication facility, serving as a technology test-bed leading to the development, demonstration, integration and qualification of advanced fabrication technologies for the semiconductor industry.*[130]

Table 3.6 depicts research outlays by the NanoCollege by funding source. As can be seen, the largest source of funding is private industry, which accounts for about 56% of the total. In addition, a portion of each annual federal outlay consists of federally funded industry-directed projects.

---

**Box 3.5: A Research "Switzerland"**

The semiconductor research facilities at the College of Nanoscale Science and Engineering (CNSE) comprise a neutral research "Switzerland" where competing companies collaborate on shared technological problems, with the university serving as the neutral intermediary, "without worrying [companies'] about their technology falling into their competitors' hands."[131] Proprietary technologies developed on the CNSE's premises belong to CNSE. Industrial partners, however, typically lease adjacent facilities, and technologies developed on their leased property become their intellectual property. The proprietary process technology developed on the CNSE site can be licensed or acquired as a service by parties in their own manufacturing operations. In addition, on-site employees of industrial partners gain hands-on experience and know-how. Toolmakers who participate have the opportunity to refine their machines based on operating experience gained on the CNSE pilot lines and could work alongside device makers whom they hoped would be customers.[132]

---

### 3.12.2.1 Center for Semiconductor Research (CSR)

The Center for Semiconductor Research, founded in 2005 at Albany Nanotech, is a research center valued at over $1 billion, conducting R&D on semiconductor technology beginning at the 32 nm node. It is the world's only university-based R&D center which integrates device design, fabrication, modeling, testing, and pilot

---

[130] Harpal Dhillon, Salahuddin Qazi, and Sohail Anwar, "Mitigation of Barriers to Commercialization of Nanotechnology: An Overview of Two Successful University-Based Initiatives." *Proceedings of the ASEE 2008 Annual Conference* (Pittsburgh, Pennsylvania, 2008).

[131] "High Tech Companies Team Up on Chip Research," *Wall Street Journal* (August 27, 2012).

[132] Interview with Catherine Hill, former counsel to Albany County Chamber of Commerce (September 16, 2015).

manufacturing. Original industrial partners include IBM, SONY, Toshiba, AMD, Applied Materials, and Tokyo Electron.[133] At this writing, IBM, Samsung, and Tokyo Electron continue as industrial participants.

### 3.12.2.2 International Venture for Nanolithography (INVENT)

In 2003, a partnership was formalized between SUNY Albany and a consortium comprised of IBM, Advanced Micro Devices (AMD), Micron Technology, and Germany's Infineon Technologies to develop nanolithography technology. The consortium was publicly announced in 2005, when it was disclosed that the State of New York would contribute $180 million to the consortium over 7 years and the industry participants another $420 million.[134] The rationale for the consortium was the reality that advanced semiconductor research was becoming too expensive for individual firms—even very large ones—to conduct on their own.[135] The research would be conducted at CNSE and was expected to employ a total of 500 researchers, over 300 of them at the SUNY Albany campus.[136]

### 3.12.2.3 Institute for Nanoelectronic Discovery and Exploration (INDEX)

In 2006, the Semiconductor Industry Association and the Semiconductor Research Corporation launched two nanotechnology institutes, one of them in Silicon Valley and the other at SUNY Albany's Center for Excellence, to be managed by CNSE. The new institute, called INDEX, partnered university researchers from Yale, Harvard, RPI, MIT, Georgia Tech, and Purdue with researchers from IBM, AMD, Intel, Texas Instruments, Micron Technology, and Freescale Semiconductor. The State of New York contributed $80 million toward the $435 million cost of the new institute, which would feature a 250,000 square foot facility.[137] Governor Pataki said that 10 years previously, INDEX would have been located on the West Coast but that SUNY Albany was "becoming the academic center for nanoelectronics in America."[138]

---

[133] College of Nanoscale Science and Engineering, "Landing Edge Research and Development Research Centers," http://www.sunycnse.com/LandingEdgeResearchandDevelopmentResearchCenters

[134] "More Chips in Tech Jackpot," Albany, *The Times Union* (July 19, 2005).

[135] "More Chips in Tech Jackpot," Albany, *The Times Union* (July 19, 2005).

[136] "Huge Prize Awaits Tiny Science—UAlbany to Play Role in Nanotech," Schenectady, *The Daily Gazette* (January 4, 2016). "Better Microchips Sought by Alliance," *Kansas City Star* (July 19, 2005).

[137] "New York Gets Nanotech Institute," *Albany Business Review* (January 3, 2006).

[138] "Albany Leads in Tiny Realm," Albany, *The Times Union* (January 3, 2006).

### 3.12.2.4 The IBM/ASML Project

ASML, the maker of the 193 nm immersion lithography tool used successfully by IBM in 2004, is a Dutch maker of semiconductor manufacturing equipment and the largest maker of photolithography equipment in the world. In 2005, Governor Pataki disclosed in his State of the State address that ASML would collaborate with IBM and other industrial partners in a $2.7 billion research project centered on Albany Nanotech. The project had several elements:

- IBM and ASML would each establish new R&D centers at SUNY Albany, with New York State contributing $225 million for research equipment.
- ASML, working with IBM, would invest $325 million at its own research center at SUNY Albany, the International Multiphase Partnership for Lithography Science and Engineering (IMPLSE).
- IBM, together with Applied Materials and TEL, would invest $450 million and would use new clean rooms at the Albany Nanotech Complex to conduct long-term research into 32- and 22-nanometer semiconductor features.
- IBM and six industry partners (SUNY, AMD, Infineon, Samsung, Chartered Semiconductor, and Toshiba) would invest $1.9 billion in a new 380,000-square foot semiconductor lab in East Fishkill.[139]

### 3.12.2.5 Partnership with Applied Materials

Applied Materials is a leading provider of semiconductor manufacturing equipment and services and the foremost provider of materials engineering solutions for the semiconductor, photo voltaic, and flat panel display industries. In 2005, Albany NanoTech and IBM announced a $300 million R&D partnership with Applied Materials to study leading-edge nanomaterials and technologies. The collaboration was expected to create 100 jobs at Albany NanoTech, where the project would be housed in NanoFab300 East.[140]

In 2018, New York and Applied Materials announced that Applied would enter into a partnership with the state to create an $880 million semiconductor equipment and materials research center at SUNY Poly's NanoFabX, expected to open in 2019. In addition, the SUNY system will collaborate with Applied to conduct semiconductor R&D at various SUNY campuses across the state. ESD and Applied will also contribute to a venture fund to foster startups and to attract established companies to the semiconductor supply chain.

---

[139] "$2.7 Billion Boost for Tech Valley," Albany, *The Times Union* (January 5, 2005); "UAlbany Project Raises Hopes," Albany, *The Times Union* (January 6, 2005).
[140] "$300 M Partnership for Nano Tech," Albany, *The Times Union* (September 27, 2005); "More than 100 High Tech Jobs to be Created," Troy, *The Record* (September 27, 2005).

### 3.12.2.6 M + W Expansion

In 2015, CNSE and M + W announced that M + W would relocate and expand its US headquarters at CNSE's NanoTech Complex in Albany. A new 30,000 square foot facility would house 160 new and existing M + W employees, together with Gehrlicher Solar America Corporation, an M + W division. It was also announced that SUNY Polytechnic Institute (SUNY Poly),[141] M + W US, and Gehrlicher would collaborate on a 5-year, $105 million solar power plant construction project, creating up to 400 jobs within the state.[142]

### 3.12.2.7 AIM Photonics

In 2015, Governor Cuomo and Vice President Joe Biden jointly announced that New York had been chosen as the site for the American Institute for Manufacturing Integrated Photonics (AIM Photonics). Total public and private investment in the institute was forecast to exceed $600 million, including a commitment of $250 million by New York State to "equip, install and make operational" a leading-edge photonics prototyping operation. The project was also supported by a $110 million federal grant.[143] Administrative and operational functions would be located at SUNY Polytechnic's Albany Nanotech Campus and in Rochester. Partners in the AIM Photonics consortium included SUNY Poly, the University of Rochester, the Rochester Institute of Technology, NASA, NSF, the Departments of Energy and Defense, the University of California at Santa Barbara, and the University of Arizona.[144] Industrial partners include GE, IBM, Cisco Systems, Mentor, Cadence, Infinera, Boeing, Corning, and Analog Photonics. The operational plan calls for photonics chips to be fabricated in CNSE's Center for Semiconductor Research and shipped to Rochester where they will be tested, assembled, and packaged at AIM Photonics' new testing, assembly, and packaging (TAP) facility at the Eastman Business Park.[145]

---

[141] In September 2014, CNSE was transferred to the State University of New York at Utica-Rome (SUNY IT), merging with it to form a new entity, SUNY Polytechnic Institute (SUNY Poly).

[142] SUNY POLY, "SUNY Poly CNSE Announces Milestone as M + W Group Opens US Headquarters at Albany Nanotech Complex and Research Alliance Begins $105 M Solar Power Initiative," Press Release, October 21, 2015.

[143] "Governor Cuomo and Vice President Biden Announce New York State to Lead Prestigious National Integrated Photonics Manufacturing Institute," New York State Press Release, July 27, 2015.

[144] "Teaming Up to Get Workers Ready for the Tech of the Future," Columbia, South Carolina *The State* (September 12, 2015).

[145] "Photonics: The Next Big Thing," Albany *The Times Union* (May 12, 2018).

3.12 Establishment of the College of Nanoscale Sciences and Engineering

### 3.12.2.8 Small- and Medium-Sized Supply Chain Firms

While large makers of semiconductor manufacturing equipment and materials like TEL, ASML, and Applied Materials attracted the most attention when they made commitments for R&D at SUNY Albany, small- and medium-sized equipment and material makers were also establishing a presence in or near the Albany NanoTech Complex. These firms had a strong interest in seeing their equipment incorporated into CNSE's 300 mm manufacturing line. Jim Castracane, Associate Vice President of CNSE, commented in 2006 that—

> *The key here is that this facility acts as a clearing house not only for research and development; but also for tool development and tool qualification and demonstration to the hundreds of companies and organizations that come through here. The publicity and visibility that is presented to toolmakers to have their equipment part of the cleanroom—that's the value of toolmakers bringing their tools here.*[146]

One toolmaker also stressed that participation in the manufacturing line helped his company improve the quality of its equipment. "To demonstrate your tools in an actual manufacturing environment, you can develop complete solutions rather than one-off solutions." In addition, the manufacturing line was beneficial to small start-ups which could "utilize the equipment that normally would be very difficult for them to acquire."[147]

In 2006, Vistec Lithography, a small UK-based maker of semiconductor lithography equipment, announced that it would move its global headquarters from Cambridge in the United Kingdom to Watervliet, New York. Vistec would receive $18 million from the Assembly's capital appropriation fund and a $12 million appropriation earmarked in the state budget. Vistec would move 20–25 people from England to New York and was committed to creating 80 jobs at its own operation and 50 jobs among suppliers.[148]

### 3.12.2.9 The EUV Lithography Project

In February 2016, SUNY Poly and GlobalFoundries announced creation of a new Advanced Patterning and Productivity Center to be housed at CNSE in Albany. The $500 million, 5-year project was designed to accelerate the introduction of extreme ultraviolet (EUV) lithography into the semiconductor manufacturing process. EUV lithography is a next-generation manufacturing process that utilizes short wavelengths of light (14 nm or less) to create microscopic patterns on semiconductor wafers (7 nm node and beyond). The project was expected to benefit from ASML

---

[146] "Albany NanoTech is Magnet for Companies That Make the Tools That Make the Chips," *Albany Business Review* (December 26, 2005).

[147] "Albany NanoTech is Magnet for Companies That Make the Tools That Make the Chips," *Albany Business Review* (December 26, 2005).

[148] "Region Draws British Company," Albany, *The Times Union* (October 19, 2006); "U.K. Transplant Seeks 60 Employees for R&D, High-Tech Manufacturing," *Albany Business Review* (January 22, 2007).

NXE:3300B EUV scanner already in place at CNSE and installation of a more advanced model. The project, which was to convene materials and equipment supplies as participants, was intended to prepare for implementation of EUV lithography at GlobalFoundries' site in Malta/Stillwater.[149] The subsequent developments with respect to this project are described in Chap. 8.

#### 3.12.2.10 Artificial Intelligence Center

In 2019, New York disclosed a new collaboration with IBM to develop component technologies for use in artificial intelligence and cloud computing. The $300 million project will be housed at SUNY Poly and is projected to create 326 new semiconductor-related jobs over 5 years. IBM also committed to invest $2 billion over 5 years on R&D at the Center for Semiconductor Research at SUNY Poly and at its main research labs in Yorktown, New York.[150]

### *3.12.3 Defense-Related Research*

Although the highest visibility R&D collaborations between SUNY Albany's emerging nanotechnology center and industry involved the civilian sector, the university also undertook projects with national defense implications. In 1998, Lockheed Martin Federal Systems chose SUNY Albany to develop G18 semiconductors for satellite, telecommunications, and possibly defense applications. Lockheed sent its own researchers to work on-site at SUNY Albany's Center for Advanced Thin Film Technology, providing $1 million in cash and $2 million worth of equipment and personnel. Lockheed was interested in developing devices with radiation-hardened systems ("rad-hard") for use in extreme environments, including outer space. Of particular interest to Lockheed was SUNY Albany's interest in copper interconnect technology, which had the potential to operate at twice the speed of the current industry standards utilizing aluminum.[151]

In 2004, Albany NanoTech received two federal grants for defense-related R&D:

- $1.5 million was allocated for the development of semiconductors for use by the US military
- $1.3 million was directed to Albany NanoTech for the development of highly efficient electronic devices for the U.S. Navy's All-Electronic Ship program.[152]

---

[149] GlobalFoundries, "SUNY Poly and GlobalFoundries Announce New $500 M R&D Program in Albany to Accelerate Next Generation Chip Technology," Press Release (February 9, 2016).
[150] "IBM Promises 326 New Jobs at SUNY Poly," Albany *The Times Union* (February 20, 2019).
[151] "UAlbany Picked for Chip Research," Albany, *The Times Union* (November 1, 1998).
[152] "Defense Bill Has Funds for Region," Schenectady, *The Daily Gazette* (July 24, 2004).

The research for the Navy involved, among other things, development of cryogenic power control systems for Navy ships, submarines, and aircraft. In 2004, an official involved with the Navy's research programs, Captain David Schubert, toured Albany NanoTech's facilities and commented afterward that "the investment the state has made in this facility is incredible. I've never seen anything of this scale at another university."[153]

In 2008, the U.S. Army Research Laboratory (ARL) and CNSE announced creation of a research partnership to accelerate the development and commercialization of nanotechnology-enabled sensors and electronic devices for applications in Army combat and support systems. ARL and CNSE created the Center for National Nanotechnology Innovation & Commercialization (NNICC), headquartered at CNSE, to conduct joint research on themes such as "sensor-on-a-chip" systems for remote sensing, nanomaterial coatings offering lighter but stronger protection against chemical, thermal, and environmental challenges, and multi-functional, low-power consumption sensor networks, and power electronic devices.[154]

## 3.13 The Creation of SUNY Polytechnic

In the spring of 2013, a SUNY plan to spin off CNSE from SUNY Albany became public knowledge.[155] This development was the most recent in a decade-long pursuit of independence by the NanoCollege. CNSE had been created by the SUNY Board of Trustees in 2004 as an "autonomous administrative, programmatic and budgetary structure" tasked with "strategic education" and R&D, and serving as an "economic outreach engine" for the state.[156] In 2008, the Trustees granted full administrative, academic, and fiscal authority over CNSE to CEO Alain Kaloyeros, who thereafter reported directly to SUNY Chancellor Nancy Zimpher.[157] The 2013 plan arose out of a working group established by Zimpher to review the relationship between SUNY, SUNY Albany, and CNSE.[158] When the plan became public, the Albany *Times Union* observed that "Albany Nano is already something of an island, with its CEO, Alain Kaloyeros, reporting directly to State University of New York Chancellor Nancy Zimpher rather than UAlbany [SUNY Albany] President Robert Jones."[159]

---

[153] "Naval Research Leader Likes UAlbany's Nanotechnology," Schenectady, *The Daily Gazette* (August 25, 2004).

[154] "U.S. Army and UAlbany NanoCollege Sign Agreement to Establish Unique Research Partnership," *Nanowerk* (May 20, 2008).

[155] "Plans to Spin off College in Play," Albany, *The Times Union* (March 14, 2013).

[156] SUNY Board of Trustees Resolution No. 2004–41, adopted April 20, 2004.

[157] SUNY Board of Trustees Resolution No. 2008–165, adopted November 18, 2008.

[158] CNSE Working Group, *The SUNY College of Nanoscale Science and Engineering: A Vibrant Engine for Innovation, Education, Entrepreneurship and Economic Vitality for the State of New York* (June 13, 2013). The working group included representatives of CNSE, the governor's office, SUNY Albany, the SUNY Research Foundation, and the SUNY Board of Trustees.

[159] "Nano U, Big Questions," Albany, *The Times Union* (March 15, 2013).

The SUNY Working Group report concluded that CNSE's independence from SUNY Albany would "allow CNSE's further growth and impact by enhancing its role and ability to move quickly and nimbly to take advantage of the numerous opportunities." The report cited the governor's mandate to CNSE to "link local academic, business and economic resources for each region of upstate with complementary CNSE assets and capabilities to establish a '21$^{st}$ Century High Technology Erie Canal' that hosts vertically integrated supply chain partnerships that stabilize and expand Upstate's business foundation and industrial base." Independence from SUNY Albany would reduce "administrative complexity for corporate stakeholders who seek simplified and authoritative systems."[160]

In response to the Working Group's report, the Trustees endorsed the separation of CNSE from SUNY Albany and the establishment of "a new degree-granting structure" that included CNSE. Zimpher trusted Implementation Teams with developing action plans to support creation of a new entity. As a result of these deliberations, which considered establishment of CNSE both as a stand-alone entity or merged with the Marcy-based SUNY Institute of Technology (SUNYIT), the Trustees ultimately endorsed a merger of CNSE with SUNYIT, creating a "new science, engineering, and technology research and education institution with co-principal locations in Albany and Utica-Rome." The two institutions would "build on each other in a mutually-beneficial fashion, while analyzing intellectual cross-fertilization and free exchanges of new ideas."[161]

In July 2013, the SUNY Board of Trustees voted 13-3 to separate CNSE from SUNY Albany, with the separation taking place in early 2014. One of the dissident trustees warned that "CNSE is on a path of moving beyond the control of the Board."[162] The creation of the new stand-alone entity (SUNY Polytechnic Institute [SUNY Poly]) coincided with a mandate from Governor Cuomo to Kaloyeros and CNSE to "export" the NanoCollege model to other regions in Upstate New York, an effort that is described in Chap. 9.[163]

The "export" effort had already begun with respect to Utica, with the CNSE-SUNYIT merger simply the latest manifestation of growing ties between the nanotechnology communities in Utica and Albany. Utica-based microelectronics companies had been working with Kaloyeros' in SUNY Albany research facilities since the mid-1990s, characterizing the institution as "unique in the R&D services

---

[160] SUNY Working Group Report, *The SUNY College of Nanoscale Science and Engineering* (June 13, 2013), p. 5 ("SUNY Working Group Report on CNSE"), pp. 1, 6, and 12.

[161] UAlbany/CNSE/IT Implementation Teams, *Report to the SUNY Board of Trustees on the Potential Merger of the SUNY Institute of Technology and the SUNY College of Nanoscale Science and Engineering* (2013).

[162] "CNSE to Spin Off on its Own," Albany, *The Times Union* (July 17, 2013).

[163] "A Nano Model Ripe for Export," Albany, *The Times Union* (October 24, 2013).

[164] Comment by Tom Clynne, President of Infrared Components Corp, of Utica (later named Critical Imaging), a maker of infrared components and cameras, in "Playing a Big Role in a Tiny World," Albany, *The Times Union* (January 5, 2001); "Imaging Firm to Work With NanoTech," Schenectady, *The Daily Gazette* (February 24, 2005). In 2004, with the collaboration of CNSE's

it provides companies."[164] In 2008, on the first occasion a state nanotechnology official promoted initiatives in the Utica area, Kaloyeros supported the establishment of a semiconductor packaging facility near SUNYIT in Marcy, stating that "we are focused on building the relationship to make sure that the Utica-Rome area hopefully gets a significant portion, if not the lion's share, of the contractor and supplier jobs that will support the facility."[165] In July 2009, Assembly Speaker Sheldon Silver and Governor Paterson announced the formation of a cross-regional partnership between SUNYIT and CNSE, the Computer Chip Hybrid Integration Partnership (CHIP).[166] The subsequent progress of CNSE's Utica initiatives is described in Chap. 8.

## 3.14 Global 450 Consortium (G450C)

In 2011, Governor Andrew Cuomo announced that the state had entered into agreements with IBM, GlobalFoundries, Samsung, Intel, and TSMC to develop the next-generation semiconductor technology required for the transition to 450 mm wafers. This first-of-its-kind developmental project, budgeted at $4.8 billion, was headquartered and housed at CNSE. The state planned to invest $400 million over a 5-year period in this project, including $100 million for energy efficiency and low-cost energy allowances. All state funds were to be directed to CNSE and all tools and equipment will be owned by CNSE. Private investment included a commitment to purchase $400 million worth of tools from New York State companies. Global 450 was comprised of two projects:

- IBM and its partners focused on developing the next generation of semiconductor devices.
- IBM, Intel, TSMC, GlobalFoundries, and Samsung conducted research to facilitate the transition from 300 mm to 450 mm wafer sizes.[167]

---

Professor Bai Xu, an expert in micro-electro-mechanical structure, Utica-based Critical Imaging was able to fabricate prototype 200 mm wafers for Silicon-based infrared sensor devices "Imaging Firm to Work With NanoTech," Schenectady, *The Daily Gazatte* (February 24, 2005).

[165] "Utica Region in Running for Nanotech Support Jobs. Utica, *Observer-Dispatch* (July 22, 2008); "Nanogeek, the Man Behind SUNYIT's Nanotech Project, "Utica *Observer-Dispatch* (November 28, 2009); "SUNYIT Tech Partnership to Bring 475 Jobs," Utica *Observer-Dispatch* (July 15, 2009).

[166] "Major Technology Partnership Announced," *US Fed News* (July 21, 2009).

[167] Office of the Governor, "Governor Cuomo Announces $4.4 Billion Investment by International Technology Group Led by Intel and IBM to Develop Next Generation Computer Chip Technology in New York," Press Release, (September 27, 2011) *http://www.governor.ny.gov/press/092711chip technologyinvestment*

The 5-year Global 450 Consortium expired at the end of 2016 without being extended with a new agreement.[168] The Consortium was credited with "a number of technological breakthroughs," but no device maker had a present plan to build a 450 mm fab, with all producers seeking instead to maximize 300 mm technology for the foreseeable future.[169] The industry has reportedly developed "cold feet" with respect to 450 mm, reflecting a number of technological and cost factors (see Box 3.6). "That, along with the fallout from the September arrest of SUNY Poly founder Alain Kaloyeros on state and federal bid-rigging charges, appears to have made the G450C a low priority for its one-time members."[170]

---

**Box 3.6: Wafer Size—From 300 mm to 450 mm**

For a half century, the semiconductor industry has followed Moore's Law, a rule of thumb holding that, reflecting technological advance, the number of transistors on integrated circuits will double roughly every 2 years, improving performance while reducing costs. An important aspect of this relentless technological process has been increasing the size of wafers on which integrated circuits are made, a transition which has occurred roughly every 10 years. The industry is currently utilizing 300 mm (12 in.) wafers at the cutting edge, with the next step being a move to 450 mm (18 in.) wafers. Larger wafer sizes enable production of larger volumes of chips and a greater rate of throughput, reducing costs. "If the cost to process a wafer stays the same, but the wafer contains more devices, then the cost per device goes down."[171] Retooling in order to make the transition to larger wafer sizes has required massive investments by toolmakers and semiconductor device manufacturers. However, each recent generational transition in wafer size has enabled about a 30% reduction in the cost per area of silicon and thus the cost per device.[172]

---

[168] "State Funding Isn't Why Chip Group Died," Albany, *The Times Union* (January 14, 2017).

[169] "Chip Makers Winding Down SUNY Research Program," *The Buffalo News* (January 12, 2017).

[170] "Chip Giants Leave SUNY Alliance," Albany, *The Times Union* (January 11, 2017).

[171] "EUV is key to 450 mm Wafers," *Semiconductor Engineering* (July 31, 2014).

[172] "Why 450 mm Wafer?" *Semiconductor Engineering* (August 8, 2012).

Wafer Size Across Generations

| Wafer size (mm) | (in.) | Year |
|---|---|---|
| 150 | 6 | 1980 |
| 200 | 8 | 1991 |
| 300 | 12 | 2001 |
| 450 | 18 | ??? |

When the Global 450 Consortium was launched in 2011, "Intel, TSMC and Samsung were aggressively beating the 450mm drum," and semiconductor manufacturers "wanted, if not demanded, 450mm pilot line fabs in place by 2016, with high-volume manufacturing 450mm plants by 2018."[173] However, by 2014, semiconductor makers had "altered course, leaving the next-generation wafer size in limbo." No producer ruled out a move to 450mm, but even Intel, the most bullish, reportedly had come to believe that 450mm was "on hold until the end of the decade," (e.g., 2020).[174]

Device makers confronted the reality that the equipment for a 450 mm fab would cost "a whopping $10 billion or more," but might not deliver the same cost reductions that occurred in prior generational transitions.[175] Instead of the 30%cost reduction experienced in the two preceding transitions, forecasts for the 450 mm shift are for a maximum of 20% cost reduction, and perhaps as little as 10%.[176] To add to the uncertainty, in 2016 "end markets [were] in such

---

[173] "What Happened to 450 mm?" *Semiconductor Engineering* (July 17, 2014). In addition, the European Commission was exploring the feasibility of building a 450 mm fab in Europe. "In the Space of Five Years, It Looks Like 450 mm Manufacturing Has Become Surplus to Current Requirements," *New Electronics* (June 28, 2016).

[174] "450 Mm/Copper/Low-K Convergence Report 2017," *Business Wire* (May 10, 2017).

[175] "The Bumpy Road to 450 mm," *Semiconductor Engineering* (May 16, 2013).

[176] A frequently overlooked phenomenon is that in each generational shift in wafer size, the percentage of wafer costs attributable to lithography increases, and per-chip lithography costs do not drop with changes in wafer size (the cost savings are attributable to other processes, such as deposition and etch). Lithography accounted for 20–25 percent of the cost of making chips on 150 mm wafers. At 200 mm, lithography was 25 percent of total cost, and at 300 mm, 50 percent. "Every time wafer size increases, the importance of lithography to the overall cost of making a chip grows. ... Each wafer size increase affects only the non-litho costs, but those costs are becoming a smaller fraction of the total because of wafer size increases." At 450 mm, in a worst-case scenario, lithography could account for 75 percent of the cost of making a device. "Why 450 mm Wafers?" *Semiconductor Engineering* (August 8, 2012).

> flux that [it was] unclear when there will be enough volume to drive massive investments beyond what is already in the works."[177]
>
> Semiconductor toolmakers were reluctant to fund all of the R&D necessary to develop 450 mm equipment and wanted device makers to share some of that cost—and "for that reason and others, Samsung has completely backed away from 450mm."[178] Equipment makers ruefully recalled the IC device makers' push to make the transition from 200 mm to 300 mm fabs in the mid-1990s, which saw the toolmakers develop the new equipment demanded by the device makers by the late 1990s. At that point, however the chipmakers deferred their plans for 300 mm fabs as the semiconductor market turned down. "Equipment vendors ended up holding the bag and lost a fortune."[179]

## Bibliography[180]

450mm and Other Emergency Measures. (2016, September 22). *Semiconductor Engineering*.
450Mm/Copper/Low-K Convergence Report 2017. (2017, May 10). *Business Wire*.
Albany No Longer A Secret in High-Tech Chip World. (2002, July 19). *The New York Times*.
Better Microchips Sought by Alliance. (2005, July 19). *Kansas City Star*.
The Bumpy Road to 450mm. (2013, May 16). *Semiconductor Engineering*.
Chernock, J., and Youtie, J. (2013, February). State University of New York at Albany Nanotech Complex. In Georgia Tech Enterprise Innovation Institute. *Best Practices in Foreign Direct Investment and Exporting Based on Regional Industry Clusters*. Atlanta: Georgia Tech Research Corporation. Prepared for the Economic Development Administration, U.S. Department of Commerce.
CNSE Working Group. (2013, June 13). *The SUNY College of Nanoscale Science and Engineering: A Vibrant Engine for Innovation, Education, Entrepreneurship and Economic Vitality for the State of New York*.
College of Nanoscale Science and Engineering. Landing Edge Research and Development Research Centers. http://www.sunycnse.com/LandingEdgeResearchandDevelopmentResearchCenters.
College of Nanoscale Science and Engineering. (2008, May 5). *A Proposal for Undergraduate Academic Programs Leading to the B.S. in Nanoscale Science and B.S. in Nanoscale Engineering*. Submitted to SUNY Albany Senate.
Dhillon, H., Qazi, S., and Anwar, S. (2008). Mitigation of Barriers to Commercialization of Nanotechnology: An Overview of Two Successful University-Based Initiatives. *Proceedings of the ASEE 2008 Annual Conference*. Pittsburgh, Pennsylvania.
EUV is key to 450mm Wafers. (2014, July 31). *Semiconductor Engineering*.

---

[177] "450 mm and Other Emergency Measures," *Semiconductor Engineering* (September 22, 2016).

[178] "Is 450 mm Dead in the Water?" *Semiconductor Engineering* (May 15, 2014).

[179] "The Bumpy Road to 450 mm," *Semiconductor Engineering* (May 16, 2013).

[180] As noted in the front matter of this book, the study also drew on interviews carried out by the authors and numerous articles from *The Times Union* (Albany), *The Daily Gazette* (Schenectady), the *Albany Business Review* (Albany), *The Post-Star* (Glens Falls), *The Record* (Troy), *The Saratogian* (Saratoga Springs), *The Buffalo News* (Buffalo), The *Observer-Dispatch* (Utica), The *Daily Messenger* (Canandaigua), and the *Post-Standard* (Syracuse). These are not individually included in the bibliography.

# Bibliography

Fury, M. A., and Kaloyeros, A. E. (1993). Metallization for Microelectronics Program at the University of Albany: Leveraging a Long Term Mentor Relationship. *IEEE Xplore*. 59–60.

GlobalFoundries. (2016, February 9). SUNY Poly and GlobalFoundries Announce New $500 M R&D Program in Albany to Accelerate Next Generation Chip Technology. Press Release.

Governor Cuomo and Vice President Biden Announce New York State to Lead Prestigious National Integrated Photonics Manufacturing Institute. (2015, July 27). New York State Press Release.

Haldar, P. (2013). Pioneering Innovation to Drive an Educational and Economic Renaissance in New York State. In National Research Council, *New York's Nanotechnology Model: Building the Innovation Economy*. C. W. Wessner (rapporteur), p. 81. Washington, DC: The National Academies Press.

High Tech Companies Team Up on Chip Research. (2012, August 27). *Wall Street Journal*.

IBM's Big New York Compute: $1.5 B Investment, 1,000 Jobs. (2008, August). *Site Selection*.

IBM, Partners Creating 1,000+ Jobs With $2.7 Billion in New York Projects. (2005, January). *Site Selection*.

If You Build It, They Will Come. (2003, February 7). *The Chronicle of Higher Education*.

In the Space of Five Years, It Looks Like 450mm Manufacturing Has Become Surplus to Current Requirements. (2016, June 28). *New Electronics*.

Is 450mm Dead in the Water? (2014, May 15). *Semiconductor Engineering*.

Joint School of Nanoscience and Nanoengineering, http://jsnn.ncat.uncg.edu/.

Jindal, V. (2013, January). Getting up to Speed with Roadmap Requirements for Extreme UV Lithography. *SPIENewsroom*.

Lane, N., and Kalil, T. (2005, Summer). The National Nanotechnology Initiative: Present at the Creation. *Issues in Science and Technology*.

Major Technology Partnership Announced. (2009, July 21). *US Fed News*.

National Research Council. (2003). *Securing the Future: Regional and National Programs to Support the Semiconductor Industry*. C. W. Wessner (Ed.) (Washington, DC: The National Academies Press.

National Research Council. (2013a). *Best Practices In State and Regional Innovation Initiatives: Competing in the 21st Century*. C. W. Wessner (Ed.) (Washington, DC: The National Academies Press.

National Research Council. (2013b). *New York's Nanotechnology Model: Building the Innovation Economy*. C. W. Wessner (rapporteur). Washington, DC: The National Academies Press.

New York State Office of the Comptroller General. (2013, January). *Fuller Road Management Corporation*. Report 2012-S-26.

NY's High-Tech Hope. (2002, September 18). *New York Post*.

Office of the Governor. (2011, September 27). Governor Cuomo Announces $4.4 Billion Investment by International Technology Group Led by Intel and IBM to Develop Next Generation Computer Chip Technology in New York. Press Release. http://www.governor.ny.gov/press/092711chiptechnologyinvestment.

Office of the New York State Comptroller. (2010). *Fuller Road Management Corporation & The Research Foundation of the State of New York: Use of State Funding for Research into Emerging Technologies at the State University of New York at Albany: Nanotechnology*. (2010-S-4) http://www.osc.state.ny.us/audits/allaudits/093010/10s4.pdf.

President's Council of Advisors on Science and Technology. (2003). *The National Nanotechnology Initiative at Five Years: Assessment and Recommendations of the National Nanotechnology Advisory Panel*. Washington, DC: Executive Office of the President.

Sá, C. M. (2011). Redefining University Roles in Regional Economies: A Case Study of University-Industry Relations and Academic Organization in Nanotechnology. *Higher Education* 61:193–208.

Schneier, E. V., Murtaugh, J. B., and Pole, A. (2010). *New York Politics: A Tale of Two States*. Armonk and London: M.E. Sharpe.

Schulz, L. I. (2011). Nanotechnology's Triple Helix: A Case Study of the University of Albany's College of Nanoscale Science and Engineering. *Journal of Technology Transfer*.

Sematech, SUNY Seal EUV Lithography Program. (2003, January 29). *Solid State Technology*.
Sematech Touts the Benefits of its New York Alliance. (2002, July 19). *Austin American-Statesman*.
Shapira, P., and Wang, J. (2007, November). Case Study: R&D Policy in the United States: The Promotion of Nanotechnology R&D. Atlanta: Georgia Institute of Technology.
SUNY Board of Trustees Resolution No. 2004-41. (2004, April 20).
SUNY Board of Trustees Resolution No. 2008-165. (2008, November 18).
SUNY Buffalo State. (2011, February 15). Research Foundation of SUNY Celebrates 60th Anniversary. Press Release.
SUNY POLY. (2015, October 21). SUNY Poly CNSE Announces Milestone as M+W Group Opens US Headquarters at Albany Nanotech Complex and Research Alliance Begins $105M Solar Power Initiative. Press Release.
SUNY Working Group Report. (2013, June 13). *The SUNY College of Nanoscale Science and Engineering*.
Teaming Up to Get Workers Ready for the Tech of the Future. (2015, September 12). Columbia, South Carolina *The State*.
Tittnich, M., et al. (2006). A Year in the Life of an Immersion Lithography Alpha Tool at Albany Nano Tech. In Proceedings of SPIE, Vol. 6151, *Emerging Lithographic Technologies*.
Tokyo Electron Plugging $300M R&D Center Into Albany, NY. (2002, December). *Site Selection*.
UAlbany/CNSE/IT Implementation Teams. (2013). *Report to the SUNY Board of Trustees on the Potential Merger of the SUNY Institute of Technology and the SUNY College of Nanoscale Science and Engineering*.
Upstate New York Gets Nod For Sematech's $403M R&D Center. (2002, July). *Site Selection*.
U.S. Army and UAlbany NanoCollege Sign Agreement to Establish Unique Research Partnership. (2008, May 20). *Nanowerk*.
Wagner, R. W. (2007). *Academic Entrepreneurialism and New York State's Centers of Excellence Policy*, Ph.D. dissertation. SUNY: Albany.
What Happened to 450mm? (2014, July 17). *Semiconductor Engineering*.
Why 450mm Wafer? (2012, August 8). *Semiconductor Engineering*.

# Chapter 4
# Establishing a Foundation for Nanotechnology Manufacturing

**Abstract** New York State and Capital Region policymakers' long-range objective for their public investments in nanotechnology was to attract private investment in nanotechnology manufacturing, which would offset the employment effects of the decline of traditional manufacturing in the region. A large, sustained, well-informed, and well-executed team effort by state and regional leaders succeeded in persuading one of the world's leading semiconductor manufacturers to establish a manufacturing presence in Saratoga County.

New York's first and biggest technology-oriented economic development investments were in university-based research infrastructure, but state planners expected that such investment would eventually lead to the establishment of a major high-tech manufacturing presence and the creation of thousands of jobs within the state.[1] To be sure, the state's investments in semiconductor research infrastructure at SUNY Albany and elsewhere represented in substantial part, a successful effort to ensure that established players like IBM and GE kept their existing manufacturing operations in the state. But the state also sought to attract new semiconductor manufacturing operations from outside the region, as Texas and Oregon had done. To realize this vision, beginning in 1998 state and local economic development professionals mounted an extraordinary effort to attract inward investment by semiconductor manufacturers and their associated suppliers. In addition to investments in research infrastructure at SUNY Albany, this effort involved:

- Development of deep expertise on semiconductor manufacturing, including the retention of expert consultants

---

[1] "The ultimate goal of UAlbany's [SUNY Albany's] plan [to build a 300mm pilot line] is to woo a full-scale chip manufacturing facility to the region." See "UAlbany Center Hopes for High Tech Boost," Albany, *The Times Union* (November 18, 1997). In supporting state funding for this initiative, Senate Majority Leader Bruno "said the prize of a manufacturing plant would be lucrative to the area." See "State Pays to Go High Tech," Albany, *The Times Union* (December 2, 1997). In 2002, New York Senator Hillary Clinton said that "All too often in upstate New York we do the research, we get the patent and then the jobs go somewhere else. That's something we cannot permit to continue. We need to translate those applications into jobs." See "Sen. Clinton Wants Research to Result in Jobs for the Region," Schenectady, *The Daily Gazette* (November 21, 2002).

- Extensive, informal, and sophisticated participation in semiconductor industry trade shows, conferences, and other similar events
- Visits by large and often eminent state delegations to individual manufacturers
- Preparation of an extremely sophisticated and detailed proposal for a chip fab to be located at a "shovel-ready" site in New York
- "Pre-permitting" of selected fab sites to enable assurances to be given to manufacturers that regulatory clearances would be forthcoming
- Successful launch of infrastructure projects along a timeframe that would ensure would-be manufacturers that if they built a fab in New York, the necessary power, water, and transportation systems would be in place when needed
- Preparation of incentives packages which were fully competitive with other global semiconductor manufacturing regions (Dresden, Texas, Singapore, Israel)
- Close and effective coordination and sustained collaboration between the semiconductor manufacturer and local government to resolve conflicts and remove regulatory hurdles
- Negotiation between the manufacturer and local construction trade unions of an agreed framework for the terms of employment at the construction site
- Skillful execution of successive construction projects by the engineering and construction firms involved and their workers

New York's high-tech outreach efforts after the late 1990s confronted the state's traditional weaknesses and addressed them, representing a quantum leap forward from prior initiatives in the 1980s which had failed to draw investment by MCC, Sematech, and Samsung. Pre-permitting of fab sites offset the longstanding concern that New York's regulatory environment precluded wafer fabs. New York's infrastructure initiatives demonstrated that the existence of semiconductor infrastructure in areas like Dresden, Singapore, and Austin did not necessarily knock New York out of the running. The sophistication and expertise with respect to the semiconductor business displayed by New York economic development professionals and academic and government leaders eclipsed competing regions in the bidding for Advanced Micro Devices' 300 mm fab, which has evolved into the GlobalFoundries operation in Malta/Stillwater.

Seen in retrospect, New York's success in attracting a major chip fab investment may seem to have been inevitable. In fact, any one of a myriad of legal, regulatory, infrastructural, or political problems could have brought the process to a halt. A 2002 study of the risk factors associated with building semiconductor fabs observed that—

> *Balancing enormous financial risk with cyclical market demands is like a no-limit poker game.... Delayed permits, incomplete tool hookups and similar problems can threaten the schedule and budget of the entire project.*[2]

The following three chapters cover in considerable detail the ways in which these obstacles were surmounted, not only because they form a critical part of the Tech Valley story. The remarkable successes achieved by the participants in this effort required sustained effort, coupled with the willingness and ability to hammer out practical compromises between competing interests and perspectives. A key question for the region is whether this positive environment will continue going forward.

---

[2] Katherine Derbyshire, "Building a Fab—It's All About Tradeoffs," *Semiconductor Magazine* (June 2002).

**Box 4.1: An Abundance of Local Government, with Consequences for Investments**

"New York has one of the most complex networks of local governments of any state."[3] Government subunits include villages, towns, cities, and counties as well as a vast assortment of special-purpose units, often with overlapping jurisdictions, such as parking authorities, sewer districts, fire districts, water authorities, school districts, highway and bridge authorities, and streetlight commissions, some of which have their own taxing authority. In 2007, the State Office of the Comptroller put the total number of local governments at 53,177, plus 6658 special districts.[4] Former Albany *Times Union* columnist Dan Balz has observed that if every single state employee were fired, leaving only local officials, New York would still have more government employees per capita than the neighboring state of Massachusetts.[5]

The dispersed character of governmental authority in New York means that the establishment of large manufacturing plants, such as semiconductor fabs, may require multiple local approvals, not only for the fab itself but for the infrastructure needed to support the fab. This contrasts with foreign companies and some competing regions in the United States such as Austin, Texas, the site of numerous semiconductor fabs. A 2003 opinion piece in the Albany *Times Union* observed that "Austin had the luxury of annexing local communities as it grew, simplifying the planning process [whereas] the Capital Region is a patchwork collection of counties, towns, villages and cities – each with its own agenda."[6] The recalcitrance of a single governmental unit can bring development initiatives to a complete halt, as was demonstrated in 1999, when a 3–2 negative vote by Town Board of North Greenbush (population 12,000) derailed a major effort by the state aimed at attracting a semiconductor manufacturing plant to a site in the town.

This study documents how the challenge posed by fragmented governmental authority was overcome in a collaborative regional effort involving local governments "all pulling together in the same boat," as one participant characterized it. Through teamwork, they achieved one of the greatest economic development successes in the history of New York. This sustained collective effort drew a semiconductor manufacturer to the region and ensured the build-out of infrastructure necessary to enable the new facility to begin operations in a timely fashion.

---

[3] Edward V. Schneier, John Brian Murtaugh, and Antionette Pole, *New York Politics: A Tale of Two States* (Armonk and London: M.E. Sharpe, 2010), p. 24.

[4] Edward V. Schneier, John Brian Murtaugh, and Antionette Pole, *New York Politics: A Tale of Two States* (Armonk and London: M.E. Sharpe, 2010), p. 24.

[5] Edward V. Schneier, John Brian Murtaugh, and Antionette Pole, *New York Politics: A Tale of Two States* (Armonk and London: M.E. Sharpe, 2010), p. 44.

[6] "Work Together, or it Won't Work," Albany, *The Times Union* (February 28, 2003). Semiconductor fabs in Taiwan and China are commonly located in special administrative zones where a single authority is responsible for regulatory approvals.

**Table 4.1** New York Economic Development Organizations Collaborating to Attract AMD/GlobalFoundries

| Organization | Scope |
| --- | --- |
| Empire State Development (ESD) | Statewide |
| Center for Economic Growth (CEG) | 11-County Capital Region |
| Saratoga Economic Development Corporation (SEDC) | Saratoga County |
| Luther Forest Technology Campus Economic Development Corp. (LFTCEDC) | Luther Forest Technology Campus (LFTC) |

## 4.1 The Role of State Development Organizations

New York has over 600 economic development organizations, including public, private, and parapublic entities operating at the state, regional, local, and individual site levels. The state's ultimately successful effort to secure additional semiconductor manufacturing—AMD, later GlobalFoundries—reflected the collaborative efforts of four such organizations—Empire State Development (ESD); the Center for Economic Growth (CEG); the Saratoga Economic Development Corporation (SEDC); and the Luther Forest Technology Campus Economic Development Corporation (LFTCEDC). See Table 4.1.

A private utility with an internal team of economic development professionals, National Grid, worked closely with the development organizations to develop outreach initiatives with respect to semiconductor manufacturers. National Grid supplied technical expertise and funding for marketing the state to the semiconductor industry via intermediary organizations, helping to develop a "branding strategy that established the Capital Region as the epicenter of New York's Tech Valley." National Grid sponsored studies of potential manufacturing sites, conducted site tours, engaged "the best industry consultants" and hosted prospective companies considering locating in the region."[7] The knowledge base which these studies created enabled the region to select one of the best sites for a wafer fab in the world, to engage in extensive site preparatory work, and to create highly sophisticated bids to semiconductor manufacturers. Hector Ruiz, the CEO of Advanced Micro Devices at the time the company decided to build its next fab in Malta/Stillwater, New York, praised New York for "the most well-crafted economic development package he could recall seeing."[8]

---

[7] "Cross Subsidy in NY," *IEEE Power and Energy Magazine* (January/February 2007).

[8] "Tech Valley Vision Pays Off Big—Chip Maker AMD Hopes Rivals Will Also Build Plants in Region," Schenectady, *The Daily Gazette* (June 24, 2006).

## 4.1 The Role of State Development Organizations

> **Box 4.2: Center for Economic Growth**
>
> The Center for Economic Growth (CEG) is a private non-profit economic development organization operating in the 11-county Capital Region of New York. It is funded by its industry members as well as Empire State Development (ESD), National Grid, and the federal Manufacturing Extension Partnership (MEP), part of the National Institute of Standards and Technology (NIST).[9] Founded in 1987 as a spinoff of the Albany-Colonie Regional Chamber of Commerce, CEG was intended to market the entire Capital Region as an alternative to the fragmented and often competing economic development efforts of individual counties and municipalities.[10] This idea was not new, having been tried unsuccessfully in preceding decades, reflecting the patchwork pattern of local governance by cities, towns, and villages.[11]
>
> CEG's first major project was the modernization of the Albany Airport.[12] In 1997, in "a major change in CEG's attitude," the organization began to concentrate on the development of high-technology manufacturing in the Capital Region, including, in particular, "the next generation of billion-dollar computer chip fabrication plants."[13] In 1999, CEG launched a global outreach program that became branded as "NY Loves Nanotech."[14] Beginning in that year, CEG "led the region-wide effort to lure a chip production facility." CEG "worked closely" with the Saratoga Economic Development Corporation (SEDC) to promote SEDC's site in Luther Forest, but it also continued to promote alternative sites in the Capital Region, recognizing that Luther Forest

---

[9] National Research Council, Charles W. Wessner (ed.), *Best Practices In State and Regional Innovation Initiatives: Competing in the 21st Century* (Washington, DC: The National Academies Press, 2013), p. 152.

[10] "Looking Forward—Center for Economic Growth Moving Ahead on New Initiatives," Schenectady, *The Daily Gazette* (April 23, 2000).

[11] Albany Mayor Erastus Corning advocated a regional approach in the 1930s when he was serving as a state senator, without success. Governor Rockefeller created the Hudson Valley Commission in the 1960s without success. In the 1990s, the Rockefeller Institute tried and failed to establish a regional plan for the Capital Region in the 1990s. "Real Regionalism Needs to be Restored," Albany, *The Times Union* (January 10, 2012).

[12] "Gala to Honor Center for Economic Growth," Schenectady, *The Daily Gazette* (October 10, 1998).

[13] CEG President Kevin O'Connor in "Business Development Group Maps Future," Albany, *The Times Union* (December 4, 1997).

[14] F. Michael Tucker, "The Rise of Tech Valley," *Economic Development Journal* (Fall 2008), p. 34.

"isn't a done deal."[15] In 2000, CEG established the Capital Region Semiconductor Task Force, compromised of five committees on industrial outreach, education/workforce, community outreach, site identification, and regional intergovernmental partnerships.[16] CEG was "instrumental in marketing" the College of Nanoscale Science and Engineering, the Luther Forest site, and other Capital Region sites outside of New York State.[17]

**Box 4.3: New York Attracts a Semiconductor Fab: A Timeline**

| Year | Event |
|---|---|
| 1996 | – Samsung chooses Austin, Texas, over New York |
| 1997 | – Center for Economic Growth begins effort to lure chip fab. "Chip Fab '98" launched |
| 1999 | – North Greenbush rejects pre-permitting of a chip fab |
| 2002 | – Saratoga Economic Development Corporation sponsors visits by local leaders to chip fabs to improve knowledge of advantages and operations and applies for approval of Planned Development District for semiconductor manufacturing in Malta/Stillwater |
| 2004 | – Towns of Malta and Stillwater issue generic permits for chip fab in Luther Forest. Infrastructure planning begins |
| 2006 | – Advanced Micro Devices (AMD) announces plan for chip fab in Luther Forest |
| 2009 | – AMD spins off chip manufacturing businesses |
|  | – GlobalFoundries formed. AMD's commitment is adopted by GlobalFoundries and construction of Fab 8 begins |
| 2012 | – Construction on expansion of Fab 8 begins. Plans for R&D Center at site announced. Test production runs of small-volume wafers conducted |
| 2013 | – GlobalFoundries large-scale operations begin |

---

[15] "CEG to Continue Chip Fab Effort," Schenectady, *The Daily Gazette* (June 5, 2002).

[16] "Task Force to Report on Efforts to Lure Chip Fab," Albany, *The Times Union* (September 13, 2000).

[17] Interview with Brian McMahon, executive director, New York State Economic Development Council (October 28, 2015).

Remarkably, the effort to create a chip fab site and to secure a semiconductor manufacturing tenant was led by the local development organization, SEDC, and its affiliate, LFTCEDC. Empire State Development, National Grid, and the Center for Economic Growth (see Box 4.2) played key supporting roles, intervening frequently at important intervals with financial support for the local groups' efforts. ESD and CEG funds paid for environmental and engineering studies, purchases of land, marketing efforts directed at semiconductor manufacturers, preparation of permit applications, and retention of consultants, and ESD was the principal provider of state incentives for semiconductor companies considering locating in New York. (See Box 4.3 for a timeline of the effort to attract a semiconductor fab to New York.)

## 4.2 Creating a Shovel-Ready Site for Semiconductor Manufacturing

The Saratoga Economic Development Corporation was founded in 1978 to work with local government to retain and create jobs in Saratoga County. In the period covered in this study, SEDC was a private non-profit corporation staffed by economic development professionals and funded primarily by local businesses, with a Board of Directors made up of local business people. SEDC emphasized confidentiality in its planning operations, professionalism on the part of its staff, and the "need to keep political and parochial thinking out of the development process."[18] In the first two decades of its existence, it claimed credit for creating over 12,000 new jobs and drawing new companies to the county including Ace Hardware, Target, Northeast Controls, Bell Metal Corporation, State Farm Insurance, and Frito Lay.

SEDC began to establish the foundation for semiconductor fabrication in Saratoga County at least a decade before the first chip manufacturer, Advanced Micro Devices, committed to establish a manufacturing presence in 2006. Based on a thorough study of the requirements of semiconductor manufacturing, SEDC developed what would be characterized as one of the best chip fab sites in the world. This involved a protracted effort to secure regulatory pre-clearance by local governments and initiatives to address the infrastructural and operational needs of semiconductor manufacturing. SEDC's efforts were reinforced at critical junctures by state and federal assistance, and buttressed by the support of Governor Pataki, and Speaker of the Assembly Sheldon Silver, and the engagement of Senate Majority Leader Joseph L. Bruno. When AMD's plans to locate manufacturing at Luther Forest, in Saratoga County, became public in 2006, Bruno commented—

> *They've looked all over the world, looked all over New York state, down the Hudson, out West and settled on Luther Forest. Why? Because we have invested in seven, eight years up there for this event, that's why.*[19]

---

[18] Joe Dalton, SEDC co-founder, "The SEDC Burns the Midnight Oil to Create Job Growth in the County," Albany, *The Times Union* (September 15, 1998).

[19] "Bruno: 2 Chip Plants in Play—Senate Majority Leader Advances Prospects for Bigger AMD Project in Region; Slams Silver Talk," Albany, *The Times Union* (June 21, 2006).

SEDC began looking with interest at an old rocket-testing site on the edge of Luther Forest in Malta/Stillwater as a potential technology-related industrial park as long ago as the mid-1980s.[20] At that time, Luther Forest, a forest of about 7000 acres, was the site of one of the largest planned communities in the United States as well as a light industrial park, buffered by dense second-growth pines.[21] SEDC's Senior Vice President Jack Kelley later recalled that one of the early big advantages of the Luther Forest site was its total lack of federal or state wetlands. In addition, natural ravines on the site divided it into segments which invited incremental development. "The land talks to you. It tells you what can go there."[22] However, prospects for development were impaired in 1986, when 450 acres of Luther Forest were included in the federal Superfund toxic-waste cleanup program, reflecting contamination from rocket and other weapons testing conducted by the U.S. Army in and after 1945.[23] The site was removed from the Superfund list in 1999 following remediation.[24]

In the mid-1990s, SEDC confronted a deteriorating economic and employment environment in Saratoga County as traditional industries downsized. SEDC President Ken Green and Senior Vice President Jack Kelley recalled that they realized at that point "they had to find something new." Their attention was captured by a decision by Samsung, the largest Korean manufacturer of semiconductors, to locate a manufacturing plant in Austin, Texas. Through intensive research and study, they became experts on the semiconductor industry. They began scouting potential sites for chip fabs in Saratoga County, identifying two, the Moreau Industrial Park and Luther Forest.[25] These sites were put in play in 1997 when New York launched an ambitious effort to lure semiconductor manufacturers to the state.

In 1998, National Grid began providing seed funding for the establishment of a chip fab site in Luther Forest. Marilyn Higgins, National Grid Vice President for Economic Development, called the Luther Forest effort a "transformational project

---

[20] "Economic Agency Seized an Opportunity," Albany, *The Times Union* (May 9, 2004). SEDC was negotiating to buy a 280-acre research site in Luther Forest from the New York State Energy Research and Development Authority in 1984, but the deal collapsed when routing testing revealed groundwater contamination from prior rocket fuel testing on the site. "State Plans to Sell Malta Research Site," Schenectady, *The Daily Gazette* (December 30, 1998).

[21] "Luther Forest Planned Development Stands Out from Suburban Sprawl," Albany, *The Times Union* (July 14, 1986).

[22] "New York's Big Subsidies Bolster Upstate's Winning Bid for AMD's $3.2 Billion 300-MM Fab," *Site Selection* (June 10, 2006).

[23] "Building Plan Imperiled, Toxic Cleanup a Luther Forest Kink," Albany, *The Times Union* (May 24, 1986); "Malta Residents Fear Poison from Rocket Site," Albany, *The Times Union* (October 9, 1991).

[24] "From Missiles to Microchips," *Albany, Business Review* (February 3, 2003). In 1999, the U.S. Environmental Protection Agency completed a five-year review of the Malta Rocket Fuel Area, concluding that soil remediation requirements had been satisfied. Drinking water and groundwater sampling continued thereafter. *Luther Forest Technology Campus GEIS: Statement of Findings* (Draft adopted by Stillwater Town Board, June 14, 2004), p. 23.

[25] "Nano Tech Valley: Saratoga County, Capital Region Economy Evolved as Experts Looked to Tech," Troy, *The Record* (June 23, 2013).

for Upstate New York representing a change from the economy of the past to a technology economy of the future."[26] SEDC's Jack Kelly recalls that National Grid was "the number one supporter of SEDC" and that large infusions from National Grid kept [the Luther Forest Technology Campus project] going."[27]

## 4.3 CHIP FAB '98

In 1997, New York state and local economic development officials launched "Chip Fab '98," an innovative initiative to attract semiconductor manufacturing operations to New York State.[28] The state proposed to identify its ten best sites for semiconductor manufacturing and "pre-qualify" them with local permits for hypothetical plants, facilitating entry by companies which fit the pre-qualified profile to begin construction without lengthy regulatory delays.[29] Localities were invited to identify their most promising sites, to be reviewed by a consultant, Industrial Design Corporation (IDC), a leading semiconductor site selection firm, which would select the top ten sites. The state offered up to $50,000 in matching funds to each site.

Pre-permitting would not involve actual semiconductor companies, but would use public input to establish rules for generic semiconductor plants "in advance so that a presented plan can win quick approval if it fits."[30] A spokesperson for Empire State Development indicated that if the initiative drew "significant interest" from semiconductor manufacturers, it would "be willing to talk about substantial enticements."[31] Proposed sites were expected to meet several criteria:

- Total size at least 200 acres
- Have available water, power, and sewage infrastructure
- Be able to supply three million gallons of water per day
- Be close to highways and an airport
- Be within 1 h's drive of a major university[32]

---

[26] John S. Munsey, "Project Case Study: High Tech Land Development," *Civil & Structural Engineer* (May 2006).

[27] Interview with Jack Kelley, Cohoes, New York (October 28, 2015).

[28] Chip Fab '98 was led by Empire State Development and the Governor's Office of Regulatory Reform. It engaged local and regional economic development agencies. "State Hunting for Chip-Fabricating Sites," Albany, *The Times Union* (December 11, 1997).

[29] RPI's George Low had employed pre-permitting as a tool for attracting tenants to the RPI Technology Park in the 1980s. Interview with Skidmore Professor Catherine Hill, Saratoga Springs, New York (September 16, 2015).

[30] "State Hunting for Chip-Fabricating Sites," Albany, *The Times Union* (December 11, 1997). Robert King, Director of the Governor's Office of Regulatory Reform, said that when permitting was completed, "officials could approach companies like IBM and Intel, show them a list of pre-approved sites … and explain that they could break ground in 30 to 60 days." See "State is seeking Pre-Approved Sites for New Factories," Schenectady, *The Daily Gazette* (February 1, 1998).

[31] "Luring Plants is Costly, Experts Say," Albany, *The Times Union* (December 23, 1997).

[32] "State Site Tour Visits Aurelius," Syracuse, *The Post-Standard* (February 5, 1998).

Pre-permitting was an attempt to address New York's reputation for "unusually rigorous environmental rules" which were "scaring [chip] plants away." Under New York's State Environmental Quality Review Act (SEQRA), it could be expected to take 18–24 months to prepare an environmental impact study and hold public hearings, with an uncertain outcome, and Robert King, Director of the Governor's Office of Regulator Reform noted that "the [semiconductor] industry simply refuses to accept this kind of delay." Michael Gerrard, past chairman of New York's environmental law section and co-author of a book on SEQRA compliance, commented that Chip Fab '98 was—

> *a very unusual procedure. But if the plant that someone actually wants to build fits within the parameters that have been submitted and approved, I don't see any problem with it.*[33]

## 4.3.1 Selection of Potential Sites

Fifty-five localities submitted site applications to the state pursuant to Chip Fab '98. In the Capital region, nine proposals were put forward, including proposed sites at North Greenbush, Rensselaer County (at a property owned by RPI), and the property in Malta/Stillwater east of Luther Forest that was owned by New York State Energy Research and Development Authority (NYSERDA).[34] The Saratoga Economic Development Corp., which proposed the Malta/Stillwater site and a site at Moreau Industrial Park in Saratoga County, indicated that "our hopes are not real high" but that Saratoga would throw its support behind whatever site or sites were picked in the Capital region.[35]

IDC ultimately selected 13 sites for potential chip fab locations rather than 10. SEDC's Kelley recalled later that "our [Malta/Stillwater site] was not one of those originally listed, but we continued to push forward.[36] Ironically, with respect to the 13 chosen sites it soon became apparent that designation of potential sites put forward by local development officials did not ensure local regulatory pre-qualification or public support. In October 1999, 6 months after selection of the 13 sites, the

---

[33] "State is Seeking Pre-Approved Sites for New Factories," Schenectady, *The Daily Gazette* (February 1, 1998).

[34] "8 Sites Proposed for New York Computer Chip Plant," Albany, *The Times Union* (January 15, 1998); "State is Seeking Pre-Approved Sites for New Factories," Schenectady, *The Daily Gazette* (February 1, 1998); "State Site Tour Visits Aurelius," Syracuse, *The Post-Standard* (February 5, 1998).

[35] "Chip Plants Require Work," Albany, *The Times Union* (March 1, 1998).

[36] "Nano Tech Valley: Saratoga County, Capital Region Economy Evolved as Experts Looked to Tech," Troy, *The Record* (June 23, 2013).

## 4.3.2 North Greenbush Rejects a Chip Fab

The proposed Rensselaer Technology Park (RTP) site in North Greenbush made the most rapid and comprehensive progress through the initial stages of the permitting process and won the strong endorsement of the five Chambers of Commerce of the Capital region.[38] However, the North Greenbush project was stopped in its tracks on October 14, 1999, when the Town Board voted to block future review of the site as a chip fab location.[39] The key factor underlying the Board's action was mounting local public concern over the potential environmental impact of a chip fab.[40] A spokesperson for the Rensselaer County Greens warned of explosions, fires, and accidents "that could release a 'deadly toxic cloud' on neighboring homes and elementary schools."[41] In the wake of the Board's vote, Rensselaer Polytechnic took steps to rescind its offer of 200 acres for the site.

The state and local economic development officials who had been involved in the North Greenbush episode drew a number of lessons from this setback. Local worries over potential pollution, noise, and adverse effects on rural environments needed to be addressed earlier and in a more comprehensive manner based on "neutral information sources." Nomination of prospective sites was to be left to individual

---

[37] One site, Orchard Park in Erie County, was selected by IDC after local officials had already decided not to proceed with bidding the site. Another, the Airport Industrial Park in Niagara Falls, became "inactive" amidst disputes over who controlled the land and was responsible for pre-permitting. One site in the Capital Region, in the town of Bethlehem, was reportedly "not very active" because the relevant land and access roads were not controlled by a single owner. Another potential site, Merritt Park in East Fishkill, "had already been developed with a 700,000 square-foot warehouse for the Gap retail chain." See "North Greenbush Isn't the Only Site Having Trouble," Schenectady, *The Daily Gazette* (October 24, 1999).

[38] "North Greenbush Chip Fab Permit Review on Track—Site Expected to be Shovel Ready by June," Schenectady, *The Daily Gazette* (March 16, 1999). Rensselaer County budgeted $250,000 to help prepare the site, and RTP's owner, Rensselaer Polytechnic Institute, offered 200 acres of land. *Ibid.*

[39] The board concluded that semiconductor fabrication was "heavy industry," not consistent with RTP's zoning, and that the board would not rezone the property. "RPI Cools Luring Chip Fab Plant—Second Thoughts Come as Schumer Presses Effects," Schenectady, *The Daily Gazette* (November 30, 1999).

[40] Representatives of environmental organizations argued that chip fabs use toxic chemicals in the manufacturing process and consume an average of 3 million gallons of water a day. "Groups Raise Questions About Microchip Plants' Toxins," Schenectady, *The Daily Gazette* (April 21, 1999); "Proposed Chip Making Plant Carries Air Emissions Risk," Albany, *The Times Union* (July 21, 1999).

[41] "North Greenbush Isn't The Only Site Having Trouble," Schenectady, *The Daily Gazette* (October 29, 1999).

communities themselves rather top-down selection by state officials. Before a given site would be presented to potential industrial users, "all of the local officials have to be 'on board.'"[42]

### 4.3.3 The Luther Forest Site

In the wake of the North Greenbush setback, State Senate Majority Leader Joseph L. Bruno, a Republican representing New York State Senate District 43, urged SEDC to take advantage of the opening:[43]

> SEDC's opportunity came on Oct. 14, 1999, when the North Greenbush Town Board killed plans for two chip fabs, clearing the way for the Saratoga County group's Luther Forest proposal to go ahead without competing with another project in the Capital Region.[44]

SEDC and its allies moved discreetly to position the Malta/Stillwater site for chip fab investment in an initiative known by the code name "Project India." Two hundred and eighty acres of Luther Forest were owned by NYSERDA, which began looking for a buyer in 1998.[45] SEDC acquired purchase options on 1350 acres adjoining the NYSERDA site from private owners.[46] In anticipation of the sale, in 1999 the Malta Town Board approved an industrial zone, cleared for light industrial and research and development use, for 440 acres of Luther Forest including the NYSERDA holding.[47] In 2000, the Town Board voted to seek state Empire Zone status for part of NYSEDRA's parcel, which would make businesses locating there eligible for "extensive breaks on the property and sales taxes."[48] In August 2001, Governor Pataki announced that the state would market the NYSERDA land in Luther Forest as a "technology energy park" in collaboration with SEDC and SUNY Albany.[49] SEDC secured the support of Senate Majority Leader Joseph Bruno,

---

[42] Kelly Lovell, President, Center for Economic Growth, in "Task Force Heads West in Order to Lure Chip Fab Plant," Schenectady, *The Daily Gazette* (October 29, 2000).

[43] Interview with former Stillwater Town Supervisor Greg Connors (January 7, 2016). State Senate District 43 includes Saratoga Springs, Stillwater, Ballston Spa, Troy, and Rensselaer.

[44] "Economic Agency Seized an Opportunity," Albany, *The Times Union* (May 9, 2004).

[45] "State Plans to Sell Malta Research Site," Schenectady, *The Daily Gazette* (December 30, 1998).

[46] "Saratoga County Pushes Tech Park," Albany, *The Times Union* (May 30, 2002). The 1,350-acre potential project site was privately owned by the Wright Malta Corporation and the Luther Forest Corporation. At the time the site was primarily a managed second-growth forest joined together by logging roads. After World War II, the portion owned by Wright Malta Corporation was a U.S. Army top secret testing facility for rocket technology. *Luther Forest Technology Campus GEIS: Statement of Findings*, Draft adopted by Stillwater Town Board (June 4, 2004), p. 3.

[47] NYSERDA's parcel had never been previously zoned given its government ownership, which exempted it from local zoning. "Malta Zones Research Site Industrial" (December 28, 1999).

[48] "Luther Forest Site Supported for Empire Zone Program," Schenectady, *The Daily Gazette* (November 1, 2000).

[49] "Pataki Announces Saratoga Technology Energy Park," Schenectady, *The Daily Gazette* (August 21, 2001).

whose district included the Malta/Stillwater site, for a high-tech manufacturing presence at the Luther Forest location.[50]

In February and March of 2002, SEDC funded travel by government officials from the towns of Malta and Stillwater, where the Luther Forest site was co-located, to Chandler, Arizona to inspect Intel's wafer fabrication facility and assess its impact on the community. Malta government officials met with their counterparts from Chandler and Malta/Stillwater firefighters and emergency personnel met with Arizona first responders.[51] "The trips were kept quiet."[52] SEDC justified the secrecy as necessary "to lay the groundwork before facing issues raised by neighbors and officials during subsequent formal review proceedings" and to "avoid tipping off economic developers in other states who might use early notice to prepare a competing proposal."[53]

#### 4.3.3.1 SEDC's Comprehensive Plan

On May 29, 2002, SEDC presented the Malta Town Board with a proposal to establish the Luther Forest Technology Campus on a 1,350-acre site on the border between the towns of Malta and Stillwater. The proposal sought to address one of the site's principal weaknesses, the fact that the most direct highway access to the site ran from I-87 Exit 12 along Dunning Street, a route that ran through the Luther Forest housing development. SEDC proposed two new road links: (1) a new connection to state highway Route 9 bypassing the housing development and (2) a road connecting the proposed campus with Cold Springs Road in Stillwater, which ran past the eastern edge of Luther Forest.[54] SEDC's Ken Green told the Malta and Stillwater Town Boards that while semiconductor companies had expressed "interest" in building at the site, "what we don't have is a ready-to-go site where a company could spend a billion or two."[55]

In June 2002, the SEDC together with the Center for Economic Growth formally submitted plans for the Technology Campus to the Malta Town Board, requesting rezoning of the land as a so-called Planned Development District (PDD).[56] The

---

[50] "Economic Agency Seized an Opportunity," Albany, *The Times Union* (May 9, 2004).

[51] "Malta Leaders' Junket Questioned—Town Officials Went to Arizona to See Chip Manufacturing Plant," Schenectady, *The Daily Gazette* (February 4, 2004).

[52] "Economic Agency Seized Opportunity," Albany, *The Times Union* (May 9, 2004).

[53] "Economic Agency Seized Opportunity," Albany, *The Times Union* (May 9, 2004).

[54] "High-Tech Business Campus Proposed in Malta," Schenectady, *The Daily Gazette* (May 30, 2002).

[55] "Saratoga County Pushes Tech Park," Albany, *The Times Union* (May 30, 2002).

[56] The Town of Malta's zoning regulations provide that PDDs are intended to provide a means for development of entirely new residential commercial or industrial areas "in which certain economies of scale or creative architectural or planning concepts may be used by the developer without departing from the spiritual intent of this [zoning] chapter...." Town of Malta Regulations, Chapter 167, Article VII, § 167–26.

proposal envisioned a plant comprised of four buildings housing up to four silicon water fabricating operations on a 125-acre plot within the 1350-acre campus. Six and a half to ten million gallons of water per day would be pumped from the Hudson River and delivered via pipe to the site. "Four to six" semiconductor companies were said to have expressed interest in the site, and as CEG President Kelly Lovell told the town officials, "We've been working on multiple companies. We just need a site. We need to get approval." Creation of the PDD, he said—

> would give SEDC a shovel-ready site to market to a chip maker, who wouldn't have to jump through all the hoops the agency has already navigated – shaving two years off the time it usually takes to build a fab. The firm would only have to receive town approval for specific building site plans, a process that takes months rather than years.[57]

### 4.3.3.2 Luther Forest Emerges as Front-Runner

The SEDC Luther Forest proposal eclipsed potential rival sites in the region. Following extensive discussion with Albany NanoTech, state officials, and industry representatives, SEDC concluded that no alternative sites had been identified either in New York State or the Northeast United States that were large enough to host an anchor tenant wishing to build four wafer fabrication facilities. "No other sites within this geographic area are large enough, with a minimum of 600 developable acres, that meet the requisite siting requirements for a 'world class' semiconductor manufacturing facility."[58] The geographic position of the site was also fortuitous. Ken Green observed in 2004 that—

> This is an excellent location. It has already a six-lane major interstate [I-87, the Northway] and a four-lane major state highway [Route 9/67]. It's pretty much on the bull's-eye of Northeast America in terms of access to the suppliers needed for these industries in Boston, Buffalo, Montreal and New York City.[59]

In 2005, Gary W. Homonai, an executive at IDC, said, "I think this [Luther Forest] is one of the best greenfield sites I've come across."[60]

### 4.3.3.3 Public Reaction

Initially, the proposed technology campus drew substantial public opposition in Saratoga County. At a 2002 public meeting convened by the Malta Town Board, local citizens, many of them residents of the Luther Forest housing development,

---

[57] "In a Forest, Two Roads Diverge," Albany, *The Times Union* (May 9, 2004).

[58] *Luther Forest Technology Campus GEIS: Statement of Findings*, Draft adopted by Stillwater Town Board (June 4, 2004), p. 46.

[59] "Campus Poses Transport Challenge," Albany, *The Times Union* (May 10, 2004).

[60] "Praise, Cash for Luther Forest," Albany, *The Times Union* (July 7, 2005).

made comments critical of the plan, citing concerns over environmental impact and traffic. A local group opposed to the project, called the Coalition for Responsible Growth, brought in as a speaker Ken Hamidi, a former Intel engineer who campaigned against computer industry practices.[61] Residents of nearby Round Hill worried that the project would increase traffic from Exit 11 on I-87 through the center of town and "would ruin the character of our historic village."[62] Malta Town Board members acknowledged that "traffic and water will be the project headaches."[63] The Schenectady *Daily Gazette* editorialized in July 2002 that—

> The notion of turning 1,350 acres of woodland, in the watersheds of Saratoga and Round lakes into a "green" industrial park is on the face of it a foolish one, in a county that is already losing farmlands and forests to development at an alarming rate and in ways that are permanently degrading its quality of life.[64]

Beginning in 2004, citizens' groups opposed to the project were countered by Tech Valley Capital Region Advocates for Intelligent Growth (TVCRAIG) which participated in public forums arguing that the job creation and increased tax revenue associated with a chip fab outweighed the negatives.[65]

#### 4.3.3.4 Environmental Impact Review

The New York State Environmental Quality Review Act (SEQRA) seeks to ensure consideration of environmental factors in governmental decisions which require planning and approval, extending to actions by political subdivisions, boards, commissions, public benefit corporations, and other public bodies. SEQRA does not itself protect against environmental impacts but exists so that those potentially affected by a decision are made aware of potential environmental economic and social impacts and are thus able to ask informed questions and raise objections. Projects involving significant environmental effects require an Environmental Impact Statement (EIS). SEDC's proposal for a pre-approved Planned Development District for Luther Forest Technology Campus fell under SEQRA and required

---

[61] "New Group Backs Chip Fab Plant—Luther Forest Project Object of Advocacy as Both Sides of Issue," Schenectady, *The Daily Gazette* (April 13, 2004).

[62] "Malta Project Concerns Residents—Many Critical of Nano Technology Campus," Schenectady, *The Daily Gazette* (July 25, 2002); "Technology Campus Met With Opposition," Albany, *The Times Union* (June 28, 2002).

[63] "Computer Chip Plant Eyed for Tech Park," Albany, *The Times Union* (June 4, 2002).

[64] "Where to Put Tech Park," Schenectady, *The Daily Gazette* (July 27, 2002).

[65] "New Group Backs Chip Fab Plant—Luther Forest Project Object of Advocacy on Both Sides of Issue," Schenectady, *The Daily Gazette* (April 13, 2004).

preparation of an EIS. SEDC received roughly $1 million in grant money from the State of New York to finance the preparation of a "Generic Environmental Impact Statement" (GEIS).[66]

The SEDC delivered a draft GEIS to the Malta and Stillwater Town Boards in December 2002, inaugurating what was expected to be a least 6 months of review.[67] Malta's Board was designated as the "lead agency" for SEQRA review purposes and conducted most of the review, in consultation with Stillwater.[68] The GEIS received an extremely thorough public airing. The Town of Malta posted the entire 1,400-page GEIS on its website, as did the SEDC, to enable public review and comment.[69] Both Malta and Stillwater held public hearings in February 2003 and invited written comments through late March 2003.[70] Scores of public meetings were convened. In October 2003, the Malta Town Board declared the environmental review complete. Rejecting complaints by project opponents about the inadequacy of the review, an attorney for the town stated that "We believe this document – and the hard look you've taken – will satisfy any reviewer or court."[71] The SEDC acknowledged the "concerns and fears" of many residents but pointed out that in response to issues raised by opponents, "SEDC has changed 30% of the project," including modifications to traffic preparations and the monitoring of air pollution, and that "the project has been clarified based on residents' responses and questions in the environmental impact reviews."[72]

### 4.3.3.5 Planned Development District Approved

In January 2004, the Malta Town Board began drafting legislation for the proposed Luther Forest PDD which would rezone the site to enable construction of a semiconductor plant. The Town's attorney, Thomas Peterson, presented a rough draft

---

[66] "Past Lessons Shape Tech Proposal," Albany, *The Times Union* (May 31, 2002); "Tech Park Study is Nearing Completion," Albany, *The Times Union* (November 5, 2002). A Generic Environmental Impact Statement (GEIS) is a tool provided under SEQRA to identify and assess potential impacts of development issues with potential effects on the environment and land use in a defined geographical area. Detail is limited to a planning or conceptual level because no specific site plans associated with an actual development exist. The purpose of a GEIS is to identify early in the process potential effects and necessary mitigation measures. *Luther Forest Technology Campus GEIS: Statement of Findings*, Draft adopted by Stillwater Town Board (June 14, 2004), p. 1.

[67] "Malta Northway Exit Seen for Tech Park—Impact Study Eyes 2003 Start for Luther Forest Project," Schenectady, *The Daily Gazette* (December 13, 2002).

[68] Interview with Greg Connors, who served as Stillwater town supervisor when the PDD was drafted (January 7, 2016).

[69] "SEDC Looks to Lure High-Tech Investment," Albany, *The Times Union* (December 20, 2002).

[70] "Malta Tech Park Plan Advances—Board Ready for Comments on Environmental Study," Schenectady, *The Daily Gazette* (January 17, 2003).

[71] "Board's Vote Upsets Tech Park Opponents—Malta Officials Say Environmental Review Sufficient," Schenectady, *The Daily Gazette* (October 17, 2003).

[72] "In a Forest, Two Roads Diverge," Albany, *The Times Union* (May 9, 2004).

addressing issues such as definitions of permitted uses, architectural guidelines, and impact thresholds to conform to the findings in the Environmental Impact Statement.[73] Peterson indicated the rezoning would require a "supermajority" vote of the Town Board (at least 4 of 5).[74] Similarly, the Stillwater Planning Board recommended in March 2004 that the Stillwater Town Board refuse to rezone the Luther Forest site, requiring a 4-out-of-5 supermajority vote by the Town Board to override the recommendation.[75] However, on May 18, 2004, the Malta Town Board voted 5–0 to adopt both the Environmental Impact Statement and the legislation creating the PDD. The Stillwater Town Board voted 5–0 for approval on June 14, 2004.[76] Malta Town Supervisor David R. Meager observed that the town had held 45 meetings since the Luther Forest proposal was unveiled in 2002. "Obviously, not everybody will agree with our decision, but I don't think anybody will be able to say we didn't do our due diligence."[77]

An editorial in the Glens Falls *The Post-Star* commented:

> *Bravo to Malta Town Supervisor David Meager and the Malta Town Board for the meticulous and open manner in which they handled the Luther Forest Technology Park application. Regardless of whether you agree with the Board's decision to approve the legislation or not, you can't fault the Board for its handling of the matter. During the past two years the Board held 45 public meetings and spent hundreds of hours listening to testimony and pouring over detailed documents related to the project. In the end they weren't swayed by emotion or political pressure—just the facts. The public was included every step of the way, and the decision was made with their best interests at heart. That's good government.*[78]

The Malta/Stillwater decision authorized an area of PDD to be used for "nanotechnology manufacturing facilities" up to a maximum of three facilities. Significantly for the future, there was no provision for local incentives to investors because at the time, the PDD qualified for state Empire Zone credits, which were regarded as sufficient.[79] The approvals set forth in provisions of the Towns' Codes, also set forth a number of conditions which would become issues in subsequent years. Among other things, prior to the issuance of a certificate of occupancy for buildings constructed in Luther Forest, (a) a proposed bypass around Round Lake village from I-87 Exit 11 to Route 9 "must be constructed," and (b) the Luther Forest Technology Campus Economic Development Corporation was to construct and maintain paved shared-use trails within the campus to be connected to the exist-

---

[73] "Malta Brining its High Tech Future Into Focus," Albany, *The Times Union* (January 20, 2004).

[74] The Saratoga County Planning Board had recommended in November 2003 that the Malta and Stillwater town boards approve the Luther Forest projects. However, it attached 59 conditions to its approval, some of which the Malta Town Board regarded as "inappropriate." Peterson concluded that for the town to override the planning board's conditions required a supermajority vote. "Supermajority Need for Chip Fab Vote," Schenectady, *The Daily Gazette* (February 11, 2004).

[75] "Luther Forest Proposal Dealt Setback," Schenectady, *The Daily Gazette* (March 17, 2004).

[76] "Malta and Stillwater Town Boards Approve PDD for Luther Forest Technology Campus," *New York Real Estate Journal* (June 3, 2008).

[77] "Rules Open Path for Chip Makers," Schenectady, *The Daily Gazette* (May 19, 2004).

[78] "Boos and Bravos," Glenn Falls, *The Post-Star* (May 24, 2004).

[79] Interview with former Stillwater Town Supervisor Greg Connors (January 7, 2016).

ing trails in Malta and Stillwater, including the Zim Smith Trail, an 8.8-mile path envisioned as the future backbone of countywide trail system.[80]

In August 2004, in the wake of the two towns' approval of the PDD, the Luther Forest Technology Campus Economic Development Corporation, an SEDC subsidiary, paid $4.8 million to acquire 164 acres of land in Luther Forest from Wright Malta Corp.[81] In January 2005, SEDC stated that it expected to receive up to $8 million from New York State to purchase the remaining 1186 acres necessary to complete the site, as well as to cover the cost of engineering work and marketing the site to the semiconductor industry.[82] SEDC closed on the 1186 acres in July 2005.[83]

**Box 4.4: Luther Forest Semiconductor Fabrication Site—Regulatory Milestones**

| Year | Event |
|---|---|
| 2002 | Saratoga Economic Development Corporation (SEDC) submits plans for Planned Development District (PDD) to Malta and Stillwater town boards |
| 2002 | SEDC submits Generic Environmental Impact Statement (GEIS) to town boards |
| 2003 | Town boards declare completion of environmental review |
| 2004 | Town boards enact Planned Development District (PDD) legislation |
| 2008 | Advanced Micro Devices (AMD) concludes Local Development Agreements with towns of Malta and Stillwater |
| 2009 | GlobalFoundries granted soil disturbance permit by Town of Malta |
| 2010–2011 | Town boards approve GlobalFoundries' proposed expansion of original construction plan |
| 2011 | Saratoga Industrial Development Agency approves sales tax exemptions for GlobalFoundries |
| 2012 | Agreement reached between five local taxing entities and GlobalFoundries on assessments for property tax purposes |

---

[80] Town of Stillwater Code, Art. XI § 211–162A(i)(a).

[81] "Developer Pays $4.8M for Luther Forest Land," Albany, *The Times Union* (August 7, 2004).

[82] "State to Help Fund Technology Site," Albany, *The Times Union* (January 28, 2005). "Details [of the state aid package] are being worked out with Gov. George Pataki and state Senate Majority Leader Joseph Bruno." *Ibid.*

[83] "Chip Plant Hopes for Region Remain High," Schenectady, *The Daily Gazette* (July 27, 2005).

### 4.3.3.6 Initial Site Preparation

As state and local economic development officials courted semiconductor manufacturers, county and town officials in Saratoga County moved ahead with initiatives to ensure that the infrastructure necessary to support semiconductor manufacturing would be in place when industrial tenants were secured. SEDC's Ken Green commented in June 2006 that—

> We have step-by-step been working on every aspect of preparing this site. We are very encouraged by the assistance we're receiving from the town of Malta and the town of Stillwater on each detail that gets accomplished. Each piece of the puzzle is going together for the infrastructure to be ready for a project.[84]

Water

The Luther Forest Generic Environmental Impact Statement indicated that the first fab constructed on the campus would require between 1.5 and 2.0 million gallons of water, and that the campus' eventual requirements would be 10 to 15 million gallons. While initially water could be provided by a private water supply company, Saratoga Water Services, Inc., the campus would eventually be supplied with water from the Hudson River through newly constructed water lines pursuant to a proposed regional water plan for Saratoga County.[85] This proposed system would require investments estimated variously at $55–80 million, and the project had been held up for 2 years because low levels of forecast demand appeared to make the project economically unfeasible. The anticipated massive demand from chip fabs at the Luther Forest Technology Campus appeared to change the calculus and make the project viable.[86] In 2004, Congressman John Sweeney secured $10 million in federal grant funds for the new water infrastructure, and Senate Majority Leader Joseph Bruno obtained another $10 million in state funding to support the project.[87] In May 2005, Saratoga County announced that it would use 50% of its budget surplus as an interest-free loan of $15 million to support the water infrastructure buildout.[88]

---

[84] "1B Lure Forms for Chip Fab Site," Albany, *The Times Union* (June 2, 2006).

[85] *Luther Forest Technology Campus GEIS: Statement of Findings*, Draft adopted by Stillwater Town Board (June 14, 2004), p. 8. Under the plan water would be pumped out of the river and treated at a plant in Moreau and piped 27 miles to Malta to support the technology campus.

[86] The county estimated that for the project to be economically feasible, it needed to sell 6 million gallons of water per day. However, the county had only been able to gather commitments of 1.8 million gallons from towns adjacent to the proposed new lines. "Tech Campus May be Key to Water Plan—Saratoga Public Works Chief Still Opposed to River Project," Schenectady, *The Daily Gazette* (January 16, 2004).

[87] "Funding Boosts New Water System," Albany, *The Times Union* (July 15, 2004).

[88] "County Loan to Fund Water System—$15 Million Earmarked to Guarantee Supply for Computer Chip Plant," Albany, *The Times Union* (May 18, 2006).

### Traffic

Perhaps the biggest concern raised by local residents about the proposed Luther Forest development during the pre-clearance proceedings was the anticipated increase in traffic. The site was located east of Interstate 87 (the Northway) between Exits 11 and 12. Traffic destined for the Luther Forest Technology Campus using Exit 11 would pass through Round Lake, threatening its bucolic character.[89] SEDC proposed to address this concern by constructing a bypass road that would route Exit 11 traffic around the village to the Technology Campus. Eventually, after two fabs had been constructed, a new Exit 11A on the Northway between Exits 11 and 12 would be constructed, about a mile north of Round Lake, leading directly to the campus. In addition, traffic elsewhere in the area would be eased by improvements such as roundabouts and turn lanes at over a dozen intersections.[90]

At the time these transportation upgrades were being considered in 2004, the U.S. Department of Transportation (USDOT) was already engaged in several initiatives unrelated to the Luther Forest project that would nonetheless serve to ease anticipated increases in traffic associated with the new chip fabs. The USDOT was already studying plans for a bypass around Round Lake village, a project which the town's mayor was seeking regardless of the presence or absence of a chip fab. The USDOT was also planning improvements around I-87 Exit 12, proposing replacement of five traditional intersections with roundabouts.[91] In December 2005, SEDC state and town officials disclosed that the federal government was prepared to spend $10 million over the next 2 years on road projects around the Luther Forest Technology Campus.[92]

---

[89] "Gov't Allocates $2M for Northway Exit—Sweeney Pushes Effort to Build Tech Park in Malta," Schenectady, *The Daily Gazette* (November 14, 2003). Round Lake with a population of about 1,000 began in the 1860s as a site for Methodist summer camp meetings. The tents that housed the original structures were gradually replaced by cottages, and in the 1880s, lectures based on the Chatauqua Institution were delivered there. In 1975, the Round Lake Historic District was added to the National Register of Historic Places. A village trustee once observed that "the whole village is an antique shop," and another observed characterized it as a "village so rooted in the 19th Century a horse and buggy would fit right in." See "Nothing Easy About Tech Campus," Schenectady, *The Daily Gazette* (February 4, 2006).

[90] "Malta Won't Require Tech Campus Developer to Fund All Traffic Projects," Schenectady, *The Daily Gazette* (May 1, 2004).

[91] Round Lake Mayor Dixie Lee Sacks sought the bypass because "we have as much traffic as we can handle on our small streets. The development going on around us is phenomenal." See "Campus Poses Transport Challenges," Albany, *The Times Union* (May 10, 2004).

[92] Projects included the rebuilding and paving of Cold Springs Road, then a dirt road east of the campus running north-south and the construction of several intersections and/or roundabouts. "Luther Area to Get Road Funding—$10 Million Slated for Paving, New Intersections," Schenectady, *The Daily Gazette* (December 20, 2005).

## Sewers

In April 2004, the Saratoga County Sewer District established a special committee to consider the challenges posed by the proposed Luther Forest chip fab site. The fabs would use as much as 2.5 million gallons of water a day, producing "as much waste water as a small city," which would be discharged into the county sewer system. The Environmental Impact Statement for the project indicated that initially waste water discharge could go into the existing sewer system serving the Luther Forest housing development. However, as the site developed, a new sewer main would be needed connecting the Technology Campus with the country's sewer trunk line.

The county's sewage treatment plant in Halfmoon was seen as adequate "to handle anything that comes along for a while."[93] In February 2006, Saratoga County sewer commissioners committed to provide capacity at the sewer treatment plan for as much as two million gallons per day of industrial wastewater from the Luther Forest Technology Campus, locking in about 40% of the plant's available unused capacity. SEDC's Jack Kelley commented, "That is wonderful. This is an integral, important part of getting the site ready for a prospective client."[94]

The Chip Fab '98 effort led New York to designate 13 sites, including North Greenbush, as potential locations for semiconductor manufacturing.[95] However, 2 years after the launch of Pataki's initiative, there was "still not a single shovel-ready parcel to be offered to a chip manufacturer." Local governments were reportedly "having a hard time getting through the pre-permitting process," reflecting town officials' judgement that the process was "too onerous." Other sites selected by the state proved to be grossly unsuitable, reflecting problems such as intractable land ownership issues and, in the case of Orchard Park, proximity of railroad tracks which would create "too much seismic activity" to allow semiconductor manufacturing. In other areas, the initiatives waned because "no one took charge," local authorities "never heard back from the state" after the initial press release, or "no one applied for the state's matching grant."[96] Over time, vacant sites set aside for Chip Fab '98 were offered to other businesses in the hopes of attracting "anything of consequence."[97]

---

[93] "Panel to Consider Impact of Sewers—Saratoga County Committee Named to Study Effects of Technology Park," Schenectady, *The Daily Gazette* (April 30, 2004).

[94] "Room is Promised at Sewer Plant—Industrial Wastewater Expected from Technology Campus," Schenectady, *The Daily Gazette* (February 23, 2006).

[95] "2 NY Sites on Chip List: Clay and Aurelius Land Spots as State's List of 13 Sites to be Promoted for Microchip Factories," Syracuse, *The Post-Standard* (March 4, 1998).

[96] "North Greenbush Isn't the Only Site Having Trouble," Schenectady, *The Daily Gazette* (October 24, 1999).

[97] In Cayuga County, five years after designation of 220 acres as a Chip Fab '98 site, the only business operating on the land was Oswego Beverage, a distributor of alcoholic and non-alcoholic beverages, and local authorities were pursuing a seed and fertilizer company as a possible occupant. "Cayuga County Chairman Plants Seed for Fertilizer Company," Syracuse, *The Post-Standard* (January 14, 2003).

**Table 4.2** Engineering project team assembled by the Saratoga Economic Development Corporation to develop the Luther Forest site

| Organization | Competencies | Responsibilities |
|---|---|---|
| LA Group | Nationally recognized land-planning firm | Initial site plans |
| C.T. Male | Engineering company | Environmental and geographic analysis; project management infrastructure planning |
| M&W Zander | Builder of 300mm wafer fabs | Master site planning, architecture, engineering, and construction |
| Creighton Manning Engineering | Engineering company | Transportation studies; engineering |
| Abbie Gregg | Clean room consultants | Vibration, electromagnetic, and radiofrequency measurements |
| E/Pro | | Electrical transmission |
| National Grid | Power transmission company | Delivery of electricity and gas |

Source: "Project Case Study: High Tech Land Development," *CE News.com* (May 2006)

## 4.4 Seeking a CHIP FAB Tenant

Having secured local "generic" regulatory approval and acquired ownership of the Luther Forest site, SEDC's efforts to market the site to semiconductor manufacturers intensified. Jack Kelley later recalled that

> Once we acquired the property the heat was really on, because you assume all the responsibility of owning property, such as paying taxes. Intel was very interested. I made 22 visits to meet with them in Chandler, Ariz. They were really our teachers. They were the ones who told us we had one of the two best sites in the world.[98]

As shown in Table 4.2, SEDC assembled a strong team of engineering talent to develop proposals for chip fabs at Luther Forest, a process that was substantially supported financially by National Grid.

Discussions between state officials and prospective investors were secretive. SEDC signed non-disclosure agreements with companies investigating the Luther Forest site and refused to discuss negotiations that were under way.[99] Empire State Development declined to comment on negotiations involving Luther Forest, "saying that the state keeps negotiations with companies confidential."[100] In May 2006, a spokesman for Intel declined to comment on any negotiations that might be under way with respect to the Luther Forest site, stating that—

---

[98] "Nano Tech Valley: Saratoga County, Capital Region Economy Evolved as Experts Looked to Tech," Troy, *The Record* (June 23, 2013).
[99] "$1B Lure for Chip Fab Site," Albany, *The Times Union* (June 2, 2006).
[100] "Tech Valley Dreams Alive After AMD Decision," Albany, *The Times Union* (May 31, 2006).

## 4.4 Seeking a CHIP FAB Tenant

*It is standard operating procedure for Intel to strategically evaluate real estate around the world for potential future expansions. We do this on an ongoing basis. The activities of this strategic process will not be shared publicly until an appropriate time in the process.*[101]

The outreach effort experienced a number of disappointments. In July 2005, Intel announced plans to build a $3 billion plant in Arizona.[102] In 2003, Texas Instruments considered the Luther Forest site but chose Texas for its next fab.[103] In April 2006, "one of the state's main targets," Samsung Electronics, announced that it would build its next fab in Austin.[104] In May 2006, Advanced Micro Devices (AMD) announced that a planned $2.5 billion microprocessor fab would be located in Dresden, Germany.[105]

Behind the scenes, New York officials were working to raise the state's profile with semiconductor manufacturers. SEDC's Ken Green recalled that his team "stirred early interest in Luther Forest by interviewing the consultants who had the ear of AMD and other large chip companies like Intel Corp. and Micron Technology, Inc. Word got back to their clients. Before we ever had any direct meetings with people at AMD, they had heard of us." Fortuitously, one of the consultants working on the Luther Forest project was Abbie Gregg, a semiconductor fab design expert from Phoenix, who was a former colleague of AMD CEO Hector Ruiz. "She used to work with Hector. She could get him to respond to her email. And that got us the opportunity," Green said.[106]

The key selling point for the Luther Forest site was its geographic proximity to the CNSE research infrastructure at SUNY Albany. John Frank, Senior Vice President of M&W Zander, the engineering company specializing in building semiconductor fabs, observed that CNSE was "a critical enabler in the eyes of a chip manufacturer. To be this close to a center of excellence in nanotech research, development, and manufacturing can be a major factor in the success of a new plant."[107]

---

[101] "Tech Valley Dreams Alive After AMD Decision," Albany, *The Times Union* (May 31, 2006).

[102] "Chip Plant Hopes for Region Remain High—Intel to Build $3 Billion Plant at Company's Site in Arizona," Schenectady, *The Daily Gazette* (July 27, 2005).

[103] "New York State's Big Subsidies Bolster Upstate's winning Bid for AMD's $3.2-Billion 300-MM Fab," *Site Selection* (July 10, 2006).

[104] "Chance to Land Chip-Fab Deal," Albany, *The Times Union* (April 16, 2006).

[105] "Tech Valley Dreams Alive After AMD Decision," Albany, *The Times Union* (May 31, 2006).

[106] "Buzz Helped Attract AMD—Saratoga Economic Development Corp. Chief Says Targeting Behind the Scenes People Was Key," Albany, *The Times Union* (September 15, 2007).

[107] John S. Munsey, "Project Case Study: High Tech Land Development," *Civil & Structural Engineer* (May 2006).

## 4.4.1 Advanced Micro Devices

Advanced Micro Devices (AMD) is a storied Silicon Valley-based semiconductor producer formed in 1968 by Jerry Sanders and seven co-workers from Fairchild Semiconductor. AMD makes microprocessors, graphics processors, embedded processors, and chipsets. It is Intel Corporation's principal rival in microprocessors and the two firms have a long history of price warfare and litigation. At the time it was approached by New York representatives, AMD was extremely proficient at semiconductor manufacturing, a fact not well known outside the company.[108] In the second quarter of 2006, the company's fortunes appeared to be ascendant with AMD reporting an eightfold increase in net income over the same period in 2005, and reportedly gaining substantial market share at Intel's expense in microprocessors.[109]

In June 2006, Governor Pataki announced that Advance Micro Devices had entered into a nonbinding agreement with New York State to build a 300mm semiconductor wafer fabrication plant at the Luther Forest Technology Center site in Saratoga County. The deal was the culmination of "months" of negotiations between state officials and AMD management. Construction on the site was to begin at a time of AMD's choosing between July 2007 and July 2009, with the earliest start date for production set for 2012. AMD expected to spend $2 billion at the plant during its first 5 years of operations. The 1.2 million square foot plant would employ 1200 people.[110]

### 4.4.1.1 Factors Underlying AMD's Choice of New York

From AMD's perspective, a variety of unique aspects worked in favor of the New York site. These included a substantial incentives package, strong local talent, major research facilities at CNSE and RPI, previous R&D investments by the company in the region, the natural advantages of the site itself, and the fact that pre-permitting made the site "shovel-ready." Given these multiple advantages, AMD chose New York over potential sites in Germany and East Asia, and the factors underlying its decision have implications for future competitions between regions for high-tech manufacturing investment.

---

[108] Former AMD CEO Hector Ruiz recalls that "AMD was exceptional at making things. We had managed process control at our plant so tightly—more tightly than anyone had imagined was possible—that we were able to patent the process. The science behind it came from the oil industry and employed statistical models. We could control quality and cost better than even Intel. But no one outside the company seemed to know about our success on the factory floor…" Hector Ruiz, *Slingshot: AMD's Fight to Free an Industry from the Ruthless Grip of Intel* (Austin, Texas: Greenleaf Book Group Press, 2013), p. 9.

[109] "Income Surges for AMD—But Threat of Price War With Intel Raises Concern," *San Jose Mercury News* (July 21, 2006).

[110] "$3.2B AMD Plant Big Lift to Region—State Package Key to Luther Forest Success," Albany, *The Times Union* (June 24, 2006).

## 4.4 Seeking a CHIP FAB Tenant

**Table 4.3** New York's successful incentives package for Advanced Micro Devices

| Item | Amount (millions of dollars) |
|---|---|
| State grant for buildings and equipment | 500 |
| State grant for R&D | 150 |
| Empire Zone tax credits/incentives | 250 (est.) |
| Infrastructure (includes some federal funds) | 300 (est.) |
| Total | 1200 |

Note: Commitment by Advanced Micro Devices was to: Create 1205 jobs by 2014; and Maintain 1205 jobs for 7 years (which was surpassed by GlobalFoundries' 3500 jobs)
Source: "New York's Big Subsidies Bolster Upstate's Winning Bid for AMD's $3.2-Billion 300-mm Fab," *Site Selection* (July 10, 2006)

Incentive Package

With respect to incentives, Senate Majority Leader Bruno said that New York offered a package that "outbid Dresden, Germany, by about $100 million."[111] AMD Senior Vice President Doug Grose commented afterward that his company chose New York over Singapore based on New York's "better talent pool and financial incentives."[112] The state pledged a package valued at $1.2 billion, including a $500 million capital grant to AMD to pay for buildings and equipment and a $150 million grant for research and development. AMD would be eligible for tax credits and incentives worth as much as $250 million pursuant to New York's Empire Zone program. Federal, state, and local funds estimated at $300 million would support infrastructural buildout.[113] (See Table 4.3) AMD CEO Hector Ruiz said that the incentive package was "the key" in the company's choice of the New York site.[114] Governor Pataki commented that "philosophically we'd prefer not to have to" offer incentives, but without the package "AMD would have gone elsewhere."[115]

---

[111] Bruno said "either you fold up and walk away, or you deliver a message to the world that New York state is going to step up. And that's exactly what we did." See "Bruno: AMD Bid at $1.3B," Albany, *The Times Union* (April 29, 2008).

[112] "The U.S. $4B Project that Got Away from Singapore," *The Business Times* (March 27, 2009).

[113] "$3.2B AMD Plant Big Lift to Region—State Package Key to Luther Forest Success," Albany, *The Times Union* (June 24, 2006). Publicly funded infrastructure projects included $53 million to expand Saratoga County sewer capacity, $67 million for construction of a water line to serve the Luther Forest site, $22.4 million to build the Round Lake bypass, and $16 million to build transmission lines and a power substation to supply electricity to the fab. "Quest for Chip Factory Hits Paydirt," Albany, *The Times Union* (July 25, 2009).

[114] "Saratoga County Chosen for Multimillion Dollar Microchip Plant," Troy, *The Record* (June 24, 2006).

[115] "Saratoga County Chosen for Multimillion Dollar Microchip Plant," Troy, *The Record* (June 24, 2006).

The Empire Zone program was established by the State of New York in 1986 to offer an array of tax credits and tax refunds to companies that invested and added jobs in blighted areas.[116] To the extent that AMD carried through with its investments and projected job creation, it would be eligible for reductions in state tax payments. The Empire Zone program was controversial because a number of businesses were found to have claimed credits, notwithstanding the fact that they had not created jobs or made investments to support those credits. The program was ended effective June 30, 2010.[117] Companies like AMD/GlobalFoundries that had already qualified for Empire Zone credits continued to receive them, but the program was closed to newcomers.[118] In addition, the entire GlobalFoundries site in Luther Forest remained qualified for Empire Zone benefits, which extend to any future expansions at that site.[119]

The incentive package was important not only because of its size, but because of the unique manner in which it built in time and flexibility to enable AMD to plan the details of its expansion and assess the best time to launch it. AMD was given a two-year period following the signing of the letter of intent to commit to construction and exercise its option for the incentive package. Travis Bullard, a former GlobalFoundries executive who at the time was serving as a member of AMD's site selection team, commented that—

> There is so much investment that goes into bringing one of these facilities on line; New York's willingness to engineer flexibility into its benefit package gave AMD the time it needed to move forward.[120]

Physical Advantages of the Site

In addition to the incentive package, AMD was impressed with the Luther Forest site itself. The nanolithography process which forms circuits on silicon wafers requires absolute quiet, with no vibrations or other disturbances, and the Luther Forest site was extremely "quiet" because its underlying geology, characterized by 60–200 ft of glacial sand, sharply limits the transmissions of vibrations. In addition to the geological advantages, RF and electromagnetic field levels at the site are among the lowest levels found anywhere in North America.[121] In a presentation at CNSE in June 2006, AMD's Hector Ruiz made it clear that the site was "superior to

---

[116] Empire Zone credits were extended for new employees, capital investment and property taxes paid. "Empire Zones Filled With Unknowns—No One Knows How Much They Cost the State or How Many Jobs They Create," Syracuse, *The Post-Standard* (November 23, 2003).

[117] "Officials Glad for Empire Zones' End," Schenectady, *The Daily Gazette* (September 6, 2009).

[118] "Paterson Seeks End of Empire Zones," Albany, *The Times Union* (January 7, 2010).

[119] "Cuomo Balks at Cash for Second Plant," Albany, *The Times Union* (June 30, 2011).

[120] "GLOBAL FOUNDRIES—2010 Gold Shovel Project of the Year," *Area Development* (July 2010).

[121] John S. Munsey, "Project Case Study: High Tech Land Development," *Civil and Structural Engineering* (May 2006).

other locations the company considered in Germany and Asia." He commented to the local New York audience that "you have collected tremendous possible sites for future selections" and praised state and local officials for "the most well-crafted economic development package he could recall seeing."[122] (See Box 4.5 for a description of the semiconductor manufacturing process.)

Previous R&D Investments

Another important factor in AMD's decision was its own growing research presence in New York State.[123] In 2003, it concluded an agreement with IBM to collaborate on semiconductor design R&D at IBM's plant in East Fishkill, New York.[124] In 2004, AMD concluded a deal with CNSE to conduct R&D at the latter's facilities on how to measure performance of transistors for future generation semiconductor devices. AMD transferred workers from its Materials Analysis Laboratory in Dresden to Albany to conduct the research.[125] In 2005, AMD was one of four semiconductor manufacturers to enter into a $600 million, 7-year partnership with CNSE, called the International Venture for Nanolithography (INVENT), to pursue nanolithography research themes and to develop a future workforce for semiconductor production.[126] After AMD's choice of New York as the site for its next fab, semiconductor industry analyst Len Jelinek said that two probable factors underlying the decision were "the University at Albany's College of Nanoscale Science and Engineering as well as the proximity of IBM's chip plant in East Fishkill." He observed that "From an R&D perspective, which is key in this industry, AMD's roots are quite strong in the New York area."[127]

---

[122] "Tech Valley Vision Pays Off Big—Chip Maker AMD Hopes Rivals Will Also Build Plants in Region," Schenectady, *The Daily Gazette* (June 24, 2006).

[123] In a prepared statement the company said in June 2006 that "the location of potential manufacturing operations near joint research and development facilities in New York [will] support faster time to market with more aggressive process technologies." See "Chip Fab seen as First Step on a Path," Albany, *The Times Union* (June 25, 2006).

[124] "IBM Lands Semiconductor Deal Worth Millions," Albany, *The Times Union* (January 9, 2003).

[125] "Straining" or "stretching" silicon accelerates the flow of electrons through transistors, which improves performance and decreases power consumption. Stress levels are determined by shining light on silicon and measuring changes on the wavelengths of the light. AMD chose CNSE as the site for this research because of its expertise and equipment. AMD's director of external research said that "this type of research hinges on having the right facility, and Albany NanoTech has that critical combination of infrastructure and expertise. By joining with Albany NanoTech, we've found a cost-effective way to stay on the cutting edge in this area of nanoscale research." "Shedding Light on a Miniscule Problem," Albany, *The Times Union* (November 10, 2004); "Advanced Micro Devices to Conduct Research at Albany Nano Tech," *Albany Business Review* (November 9, 2004).

[126] "More Chips in the Tech Jackpot," Albany, *The Times Union* (July 19, 2005).

[127] "Spinoff Businesses Likely to Follow," Schenectady, *The Daily Gazette* (June 21, 2006).

Educational Infrastructure

RPI's Computational Center for Nanotechnology, featuring a $100 million supercomputer, also "played a major role" in AMD's decision, according to Senate Majority Leader Bruno, with AMD planning to use the facility for semiconductor R&D.[128] The machine, jointly paid for by the state ($33 million), IBM, and RPI, was the seventh most powerful supercomputer in the world and had an operational capacity of over 90 peak teraflops (one trillion floating point operations per second), or roughly 15,000 calculations per second for every human being on earth.[129] RPI indicated that AMD and IBM were planning to use the supercomputer to design next-generation semiconductors, and the machine was seen as potentially "helping AMD and its race against rival Intel Corporation to make smaller and more powerful chip components." An IBM spokesman commented that with respect to semiconductor "design tests that used to take hours or days to complete will now take a few minutes, greatly accelerating the pace of development. That is a big competitive advantage."[130]

---

**Box 4.5: Semiconductor Manufacturing Process**
Semiconductor manufacturing is the most complex industrial process in the world. It is a highly automated process and requires the use of machines capable of extraordinary precision, a variety of exotic materials, the virtually complete absence of contaminants such as dust and moisture, and protection from external vibration.

**Wafer production**. Semiconductor production begins with the creation of a wafer, usually made of pure silicon, of up to 300mm in diameter, which is polished to create an extremely flat surface.

**Wafer fabrication**. The core manufacturing process involves formation of integrated circuits on the surface of a wafer in a number of process steps in a "clean room" free of dust and other particles:

- ***Deposition*** coats, grows, or transfers material onto the wafer.

---

[128] "RPI's Supercomputer Suits AMD," Albany, *The Times Union* (September 19, 2007); "After AMD, Other Firms Are Interested," Schenectady, *The Daily Gazette* (June 27, 2006).

[129] "RPI's Supercomputer Among the World's Strongest," Schenectady, *The Daily Gazette* (September 8, 2007).

[130] "RPI Supercomputer Suits AMD," Albany, *The Times Union* (September 19, 2007).

4.4 Seeking a CHIP FAB Tenant

- ***Masking or photolithography*** applies photosensitive film to the wafer, a stepper aligns the wafer to a mask and light is shined through the mask and a succession of reducing lenses, exposing the photoresist to the mask pattern.
- ***Etching*** removes the exposed photoresist and the wafer is baked to harden the remaining photoresist.
- ***Doping*** modifies the electrical properties of the silicon through the introduction of impurity atoms in a controllable manner.

The foregoing steps are repeated a number of times until the last desired layer is fabricated. The individual devices are interconnected using various metal depositions.

**Test**. An electrical test system checks to ensure that all chips on the wafer are functioning properly and screens out those which are not.

**Assembly**. The wafer is sliced into individual chips, which are packaged to establish contact leads to the device to which wires are attached by wire-bonding machines. The assembled chip can be encapsulated in plastic or ceramic for protection.

Source: See "How Semiconductors Are Made," *Intersil* <http://rel.intersil.com/docs/lexicon/manufacture.html>.

## *4.4.2 From AMD to GlobalFoundries: AMD's Financial Difficulties*

At the time that the wafer fab deal between AMD and New York was announced, AMD was engaged in a price war with Intel in microprocessors.[131] In the fourth quarter of 2006, AMD reported a net loss of $574 million, reflecting price competition and its $5.4 billion acquisition of graphics semiconductor maker ATI Technologies. An industry analyst commented that—

> *They are getting it on both sides. They have a heavy debt load and interest expense that is going to drain the cash flow. And Intel is inflicting pain. . . . They have some significant headwinds to navigate through.*[132]

---

[131] "AMD, Intel Race to the Bottom," *Forbes* (June 28, 2006).
[132] "AMD's Q4 Retreats Under Competition—ATI By Adds Debt, Intel's New Chip 'Inflicting Pain,'" *San Jose Mercury News* (January 4, 2007).

Between March 2006 and January 2007, one analyst observed that AMD's stock share prices went into "free fall," losing 60% of their value.[133] AMD was reportedly the subject of "a possible takeover or private-equity cash infusion."[134]

Reports of AMD's financial challenges raised questions in New York as to whether and when the company would actually go forward with its chip fab investment. The company indicated in its 2006 annual report that its debt levels might impair its ability to borrow money and pay for $2.5 billion in capital expenditures planned for 2007, most of which was earmarked for Dresden. One investment banker commented that "It's a dilemma – we believe AMD needs to spend the money to build the fabs (chip factories), but they may have to find some additional financing to achieve those goals."[135] The Schenectady *Daily Gazette* commented that—

> The high-flying Advanced Micro Devices Inc. of 2006 has given way to a company in financial peril, saddled with debt and bleeding from a brutal price battle with its larger and suddenly resurgent Silicon Valley archival Intel Corp.[136]

In March 2007, an AMD company spokesman said that "We are still on track with the project. Building a chip fab is not something we go into lightly.[137] The company began work on detailed design of the Luther Forest facility, reportedly involving "80 engineers from all over the world, each sitting at a desk [in AMD's Phoenix headquarters] … [and] all of them … doing Luther Forest work."[138] In a conference call in July 2007, AMD CEO Hector Ruiz was asked point blank by a JPMorgan analyst whether AMD's financial problems might force the company to scrap its plans for Luther Forest. "Not at all," he said. "Our manufacturing strategy has not changed. We're looking forward to benefitting from our plans in New York."[139]

However, the risks associated with investment in a wafer fab were daunting even in the best of circumstances, to say nothing of the financial straits in which AMD found itself in 2007. CEO Hector Ruiz explained the semiconductor business model in a 2013 book, in which he said that a company like AMD would first design a new chip and seek to market it to a major computer make like Dell or HP. Those potential customers would take 6–12 months to build a platform around the chip, designing hardware and incorporating software—but before undertaking that effort, they would require assurances from the chipmaker that it would have the capacity to

---

[133] "AMD, Intel Price War Revisited," <http://www.Peridotcapital.com/2007/01/amd-intel-price-war-revisited.html>.

[134] "Is AMD in Trouble? Money Woes Muddy Plan for Malta Site," Albany, *The Times Union* (March 10, 2007).

[135] "Is AMD in Trouble? Money Woes Muddy Plan for Malta Site," Albany, *The Times Union* (March 10, 2007).

[136] "Saratoga Plant Reported Still a Go Despite AMD Money Woes," Schenectady, *The Daily Gazette* (March 10, 2007).

[137] "Saratoga Plant Reported Still a Go Despite AMD Money Woes," Schenectady, *The Daily Gazette* (March 10, 2007).

[138] "AMD Begins Design Work for Chip Plant," Schenectady, *The Daily Gazette* (March 2, 2007).

[139] "AMD Insists Chip Fab is a Go," Albany, *The Times Union* (July 20, 2007).

manufacture the device. The chip company must answer "yes" or lose the business, and invest in a $2 billion plant that may or may not actually be utilized depending on the decision made by the computer firms. Ruiz quoted an AMD board member who liked to say that building a wafer fab under such circumstances is—

> like Russian roulette [but with a twist]. You pull the trigger and four years later you learn whether you blew your brains out or not.[140]

Box 4.6 describes the increasing cost of building a semiconductor fab.

---

**Box 4.6: The Rising Cost of Building a Semiconductor Fab**

Rapid advances in the development of semiconductor technology have enabled continuous reduction in the cost per function in semiconductor devices, the phenomenon underlying Moore's Law. However, these cost savings are achieved through the continuing deployment of equipment, materials, and processes of progressively increasing sophistication, complexity, and scale. As the cost-per-bit of functionality has fallen, the up-front capital costs required to achieve such gains have ballooned.

In 1991, Sematech estimated that creation of a semiconductor fab capable of starting 20,000 wafers per month cost $400 million to build.[141] A 2003 study by the Semiconductor Industry Association (SIA) which aggregated cost data from its members concluded that the cost of building a 200 mm fab using 130 mm design rules (whether in the United States, Taiwan, or China) was around $1.95 billion.[142] SIA currently estimates that the cost of a new state-of-the-art 300 mm fab ranges from $5 billion to $6.8 billion.[143] Scaling up such facilities can be much more costly: Samsung's new 300 mm fab at Pyeongtaek, capable of producing 300,000 wafers per month, reportedly cost $14.4 billion.[144]

As with prior generations, next-generation 450 mm fabs will achieve major reductions in cost-per-function over 300 mm facilities in the devices which

---

[140] Hector Ruiz, *Slingshot: AMD's Fight to Free an Industry from the Ruthless Grip of Intel* (Austin, TX: Greenleaf Book Press, 2013), p. 8.

[141] Sematech, *Annual Report* (Austin: Sematech,1991).

[142] Semiconductor Industry Association, *China's Emerging Semiconductor Industry: The Impact of China's Preferential Valve Added Tax as Current Investment Trends* (October 2003) Appendix 2.

[143] Semiconductor Industry Association, "Policy Priorities: Tax," *http://www.semiconductors.org/issues/tax/tax*.

[144] "Samsung to Invest Additional $9.2 billion in its $14.44 billion fab," *KitGuru* (April 15, 2015).

> they produce.[145] But these gains will require larger initial investments. A 2012 study prepared for the European Commission by industry analysts projected a cost of "around $10 billion" for a 450 mm fab.[146] In 2012, TSMC announced it would build a 450mm fab and that the estimated cost would be $8–10 billion.[147]

In December 2007, AMD secured an option to purchase land at the Luther Forest site, a move that a company spokesperson characterized as "just getting our ducks in a row."[148] In early 2008, the company retained local attorneys and engineers to assist in forthcoming public review of its plans.[149] In February 2008, AMD began the process of securing the requisite town and environmental approvals to secure a building permit by the end of 2008. It indicated it could begin "pre-construction work" (vegetation removal, grading) on the site as soon as July 2008.[150]

#### 4.4.2.1 A Change in AMD Company Strategy: "Asset-Light"

In April 2007, AMD disclosed a "bold new cost-cutting strategy" termed "asset-light," cutting capital spending by $500 million in 2007. "Asset-Light" was seen by industry analysts as a reference to increased outsourcing of its manufacturing operations to "foundries," which were semiconductor firms that provided manufacturing services to produce the designs of other firms in return for a service fee (see Box 4.7). (Eventually the term was dropped in favor of "asset smart.") One analyst said that "if they were to proceed with this asset-light strategy, then the Albany plant could be at risk."[151] CEO Ruiz and other company executives declined to elaborate on what "asset-light" meant.[152]

---

[145] A 2011 presentation by GlobalFoundries estimated that a 450mm fab with a capacity of 40,000–45,000 wafer struts/month could produce the same volume of die as a 300mm fab with 100k wafer starts per month. Cost savings of 20–25% per die were forecast using 22mm design rules. GlobalFoundries, *Reaping the Benefit of the 450mm Transition* (Semicon West, 2011).

[146] Future Horizons, *Smart 2010/062: Benefits and Measures to Setup 450mm Semiconductor Prototyping and Keep Semiconductor Manufacturing in Europe* (Luxembourg: Office of Official Publications, February 16, 2012).

[147] "TSMC to Spend $10 Billion Building 450mm Wafer Factory," *Reuters* (June 12, 2012).

[148] "AMD Secures Option on Luther Forest Site," Albany, *The Times Union* (December 13, 2007).

[149] "AMD Putting Its Team Together—Local Attorneys, Engineers Would Represent Firm," Schenectady, *The Daily Gazette* (January 30, 2008).

[150] "AMD Plans Gaining Focus—Company Says it Wants to Start Work on Chip Factory by January as Review Process Begins," Albany, *The Times Union* (February 26, 2008).

[151] "AMD Strategy Poses Risk to Chip Fab Plan," Albany, *The Times Union* (April 26, 2007).

[152] "AMD Insists Chip Fab is a Go," Albany, *The Times Union* (July 20, 2007).

**Box 4.7: Semiconductor Foundries**
The "pure play" semiconductor foundry business model, pursuant to which an enterprise provides contract manufacturing services for semiconductors designed by other firms but does not sell devices under its own label, emerged in the mid-1980s as a response to the growing costs and risks associated with semiconductor manufacturing. Carver Mead, a US computer scientist, had argued for years that semiconductor design could be separated from the manufacturing process, but his ideas were generally viewed as commercially unfeasible.[153] The first foundry was created in 1984, when Taiwan's government-supported Industrial Technology Research Institute (ITRI) spun off the Taiwan Semiconductor Manufacturing Corporation (TSMC), led by ITRI head Morris Chang, a veteran of Texas Instruments. The government of Taiwan took a 40% equity stake in TSMC and provided a variety of other forms of support, most notably donated facilities and tax exemptions.[154] The subsequent commercial success of TSMC led to the formation of other foundries in Asia, Taiwan's United Microelectronics Corporation (UMC), Singapore's Chartered Semiconductor, and China's Semiconductor Manufacturing International Corporation (SMIC).

The advent of semiconductor foundries in Asia revolutionized the semiconductor business by addressing the massive and growing capital costs associated with semiconductor manufacturing which increased with each new generation of devices. By the late 1990s an increasing proportion of the US semiconductor industry was becoming "fabless," outsourcing all of the manu-

---

[153] Mead cited the analogy of the printing business and argued that semiconductor design could be separated from the semiconductor manufacturing process, much as the author of a book operates in a separate sphere from the printing company manufacturing books. In the early 1980s, the President of Taiwan's government Industrial Technology Research Institute (ITRI) had a daughter who was one of Mead's students, who suggested that Mead be invited to Taiwan. Mead made the visit and his ideas had a profound impact on ITRI officials. The former head of ITRI, Chintay Shin, recalled later that "I was thrilled the first time I heard about Mead's concept." Interview with Chintay Shin, "Taiwanese IT Pioneers: Chintay Shin," recorded February 21, 2011 (Computer History Museum, 2011), pp. 14–15.

[154] Robert Tsao, the founder of Taiwan's United Microelectronics Corporation (UMC), recalled that ITRI set up an internal semiconductor manufacturing plant and transferred it to TSMC on very concessional terms: "So after [ITRI] spending five years and 100 million U.S. dollars on construction, the second demonstration plant was built, and about 500 to 600 trainees were all sublet to TSMC. Therefore, TSMC had [the] foundry at first and they only needed to spend two million U.S. dollars on subletting the foundry. Apart from this ITRI even gave TSMC seven million U.S. dollars to subsidize the cost of subletting the foundry.... In other words the first 3 and half years of running TSMC was for free and it went very well under the protection of Ministry of Economic Affairs." Interview with Robert Tsao, "Taiwanese IT Pioneers: Robert H.C. Tsao," recorded February 17, 2011 (Computer History Museum), p. 9.

> facturing of their designs to Asian foundries or "fab lite," outsourcing a significant proportion of their manufacturing while retaining some production.
>
> In 2015, nearly 38% of global integrated circuit (IC) sales to systems manufacturers involved products fabricated by foundries, up from 26% in 2010 and 21% in 2005. IC foundry sales to fabless semiconductor firms, systems companies, and integrated device manufacturers (IDMs) accounted for $50 billion in revenues in 2015.[155]

In July 2008, AMD CEO Hector Ruiz stepped down after the company reported a $1.19 billion loss for the second quarter. Ruiz remained in charge of what was called "the company's secretive 'asset smart' strategy" which reportedly included construction of "Fab 4X," as the Luther Forest fab was tentatively designated.[156]

#### 4.4.2.2 Spinning Off AMD's Manufacturing Operations: The Abu Dhabi Deal

In October 2008, AMD announced that in partnership with a new investor, Advanced Technology Investment Company (ATIC), it would commit to build the wafer fab in Luther Forest. ATIC was 100% owned by Mubadala Investment Company, an entity whose sole shareholder was the government of Abu Dhabi. AMD and ATIC would jointly create an entity to be known initially as "The Foundry Company," to be 65% owned by ATIC and 35% owned by AMD. Mubadala, which at the time owned an 8.1% stake in AMD, would increase that stake to 19.3%. AMD would contribute its two existing fabs in Dresden to the Foundry Company, as well as ancillary property rights and other assets. ATIC would invest $1.4 billion in the Foundry Company and would purchase additional shares in the Foundry Company from AMD for $0.7 billion. The Foundry Company would assume around $1.2 billion of AMD's debt.[157]

Through this transaction AMD spun off its manufacturing operations, which had "become a cash drain on a struggling company." Going forward AMD would focus on semiconductor design, "a much less costly—and less risky—business."[158] The Foundry Company would benefit from the combination of ATIC's deep pockets and AMD's superior process technology as well as 3000 AMD employees and a number of senior executives.[159] Hector Ruiz, the former CEO of AMD, would become Chairman of the Foundry Company.[160]

---

[155] "Foundry Sales Defy IC Decline," *EETimes* (October 22, 2015).

[156] "AMD Boss Out Amid $1.19B Loss," Albany, *The Times Union* (July 18, 200); "AMD Factory Still on Track," Albany, *The Times Union* (July 19, 2008).

[157] "Abu Dhabi Money Fuels AMD Project," Troy, *The Record* (October 7, 2008).

[158] "AMD Deal Marks a 'New Dawn,'" Albany, *The Times Union* (October 8, 2008).

[159] "Fabless Future: Struggling AMD Spin-off Factories," *Associated Press* (October 7, 2008).

[160] "AMD Reveals More Details," Troy, *The Record* (October 12, 2008).

4.4 Seeking a CHIP FAB Tenant

The investment plan for the new company envisioned ATIC committing an initial $2.1 billion to be followed by $6 billion more, most of which would be used to upgrade former AMD fabs in Dresden and to build a new $4.2 billion fab at the Luther Forest site in New York.[161] Groundbreaking for the New York fab was anticipated in mid-2009, involving 18 months to construct the building and 18 months to install the tools. Production of wafers was expected to begin in 2012.[162] The deal alleviated the mounting concern in New York over the implications for the fab project of AMD's difficulties. Former Senate Major Leader Bruno commented that—

> Just a few weeks ago, I was getting calls: 'It's never going to happen; it's never going to happen.' Now we're here. Tough decisions, if they're made properly, they lead to benefits like they do here.[163]

Transfer of State Incentives

AMD's original agreement with New York State with respect to the incentives package contained a "no assignment" provision to the effect that AMD could not transfer the incentives to another entity without the approval of Empire State Development. AMD stated in federal filings that if the incentive package were diminished substantially by the state, "the deal won't be acceptable."[164] In December 2008, New York approved the transfer of the $600 million capital grant to the Foundry Company. However, ESD stipulated that if the company did not create 1205 jobs by January 1, 2014, and maintain that employment level for 7 years, ESD could reclaim its grant money.[165]

CFIUS Approval

The spinoff of AMD's manufacturing assets to a company 50% owned by a foreign entity (ATIC) and Mubadala's acquisition of a larger stake in AMD each required approval by the federal Committee on Foreign Investment in the United States (CFIUS), which considers foreign acquisitions from the standpoint of national security. CFIUS concluded in January 2009 that neither transaction raised national security concerns.[166] In February 2009, the last legal obstacle to closure of the deal was surmounted when a quorum of AMD shareholders voted to approve it.[167]

---

[161] "AMD's New Investors Make Big Bet on Chip Plants," Ocala, *Star Banner* (October 8, 2008).
[162] "AMD Reveals More Details," Troy, *The Record* (October 12, 2008).
[163] "Fab Feeling: 'Guess What Guys; It Happened'" Albany, *The Times Union* (October 9, 2008).
[164] "AMD Deal Needs to Vote," Albany, *The Times Union* (October 12, 2008). The deal also required transfer of some German subsidies from AMD to the Foundry Company. "AMD Reveals More Details," Troy, *The Record* (October 12, 2008).
[165] "NYS Approves $650M for AMD," Troy, *The Record* (December 2, 2008); "State Oks AMD Benefits Transfer," Schenectady, *The Daily Gazette* (December 18, 2008).
[166] "AMD Deal Gets Federal Approval," Albany, *The Times Union* (January 7, 2009).
[167] "AMD Ballot Succeeds on Second Try," Schenectady, *The Daily Gazette* (February 19, 2009).

## 4.4.3 Closing the Deal: The Birth of GlobalFoundries

On February 2, 2009, AMD and ATIC presented officials from the town of Malta with comprehensive details of construction plans for an 883,100 square foot wafer fabrication building, a three-story support building, an administration building, and a central utility building. The structures were to be built according to "green building standards." Anthony Tozzi, Malta's building and planning coordinator, indicated that the town's planning board would consider a "soil disturbance permit" which would allow ground clearing activities to begin; a temporary construction permit; and final site plan approval. Tozzi said that "we intend to expedite the process somewhat, but we also intend to do our due diligence."[168]

The AMD-ATIC deal was closed on March 2, 2009, and the Foundry Company was renamed GlobalFoundries. AMD's former CEO Hector Ruiz became Chairman of the new company. At the unveiling ceremony Ruiz thanked the numerous elected officials who had worked with AMD on the project, commenting that "the cooperation between government and business is what made this possible. We've gone through three governors during the course of building this, and all three have been supportive."[169]

### 4.4.3.1 Reaching a Deal with the Construction Trades

Construction of GlobalFoundries "Fab 2" was expected to involve 1,600–1,800 construction workers. In March 2009, after a meeting between GlobalFoundries and representatives of organized labor ended "without reaching a resolution satisfactory to unions," a labor spokesman, Edward Mallory, stated that "We have grave concerns."[170] Labor's supporters in the state legislature called upon Governor David Paterson to intervene, arguing that because GlobalFoundries was receiving state incentives, Fab 2 was a public works project and should be built by local workers at prevailing union wage levels. GlobalFoundries indicated it would try to make 70% of the project available to unionized workers but "that level of commitment fell short of what [was] sought by the unions."[171] One state senator said that "I do not

---

[168] "AMD Unveils Site Plans," Glens Falls, *The Post-Star* (February 3, 2009).

[169] "GlobalFoundries Newest Name for AMD in Malta," Troy, *The Record* (March 5, 2009).

[170] "Unions Went to Work at Fab," Albany, *The Times Union* (March 19, 2009). Malloy was President of the New York State Building and Construction Trades Council, an umbrella group representing unionized construction workers in New York State.

[171] A GlobalFoundries spokesman said the company expected at least 70% union labor but hoped to avoid entering into a labor agreement. "The goal is to come to a[n] understanding without having to put it in writing." He noted that certain technical aspects of the fab's construction would require bringing in outside help. "70 Percent Union Labor to go Into Factory Construction. GlobalFoundries Says," Glens Falls, *The Post-Star* (March 24, 2009).

want to see people from outside the state at the construction site. I don't want to see license plates from South Carolina or Vermont… We want local people working here, local vendors."[172]

Governor Paterson initially asked GlobalFoundries and labor groups to "work things out on their own," but with the continued absence of an agreement threatening to stall construction of Fab 2, the governor's office took "the lead role in working out a labor solution between GlobalFoundries and the buildings trades."[173] Ultimately, the Paterson administration working closely with GlobalFoundries Mike Russo, who was as a former Union official, managed to broker a positive-sum deal between GlobalFoundries and the unions pursuant to which the parties would enter into a project labor agreement under which "union wages will be paid by all firms chosen to work on the project whether they are union or non-union companies."[174] Among other things, the state "sweetened the pot" for the company, augmenting the $1.2 billion incentives package with an additional $15 million for "unplanned costs." Labor spokesman Malloy commented that "it's a great agreement."[175]

#### 4.4.3.2 Formal Commitment by the Parties

On June 9, 2009, GlobalFoundries delivered a letter to New York State development officials formally committing to build "Fab 2," a wafer fabrication facility, an action which cleared the way for the company to receive the $650 million incentive package from the state.[176] The company closed on the $7.8 million sale of the 223-acre site in Luther Forest the following day. GlobalFoundries CEO Douglas Grose commented that "we're basically starting down the path of spending the money and building the building." Grose underscored the importance of the state incentives: "To be very honest, this levels the playing field in terms of where we locate the plant."[177]

---

[172] "Unions Want Work at Fab," Albany, *The Times Union* (March 19, 2009); "Paterson's Help Sought in Dispute," Albany, *The Times Union* (March 20, 2009).

[173] "Paterson's Help Sought in Dispute," Albany, *The Times Union* (March 20, 2009); "Labor Deal Stalls Chip Fab Factory," Albany, *The Times Union* (May 9, 2009).

[174] The agreement reportedly included a cap on wage increases during the life of the project, with no raises above 3%. Seventeen percent of the project would be reserved for companies not engaging in collective bargaining. Up to 7% of the project could go to specialty contractors when union labor could not perform specific tasks. "Deal Includes Union Pay," Albany, *The Times Union* (June 3, 2009).

[175] "Deal Includes Union Pay," Albany, *The Times Union* (June 3, 2009).

[176] GlobalFoundries was also slated to receive about $550 million in Empire Zone tax incentives and infrastructure support. "Work to Begin on Chip Plant," Schenectady, *The Daily Gazette* (June 10, 2009).

[177] "Work to Begin on Chip Plant," Schenectady, *The Daily Gazette* (June 10, 2009).

### 4.4.3.3 Local Development Agreements

Conclusion of development agreements with Malta and Stillwater was necessary to secure the local clearances necessary for construction operations to begin. In 2008, before AMD had made a final commitment to the Luther Forest fab project, the Malta Town Board began drawing up lists of local improvement projects with which the company might assist.[178] AMD executive Ward Tisdale indicated that the company was supportive and that "it's part of our corporate values that we contribute to the communities where we operate."[179] In December 2008, the Malta Town Board disclosed a tentative "development agreement" pursuant to which the company would make phased payments to the town totaling about $4 million:

- $1 million would be given to the town at the time of groundbreaking for the fab to be used to develop ball fields on 34 acres of land owned by the town adjacent to the technology campus.
- $750 thousand would be paid into a newly created trust fund when the shell of the fab building was complete, which was expected in 2010.
- Two payments of $1.125 million each would be made into the trust fund when the first semiconductor was produced (expected in 2011) and when the plant reached full production (2012).[180]

AMD pledged a $1 million "host community payment" to the town of Stillwater.[181] The company also contributed $100 thousand to both Malta and Stillwater to fund town master planning studies to assess the impacts of the project.[182] GlobalFoundries subsequently assumed AMD's commitments.[183]

---

[178] Malta's Town Board hoped that AMD would pay half the cost of a community center expansion project that was already under way and half the cost of a central fire state that could serve both Malta and Round Lake. "Town Raps AMD Plant Wish List," Glen Falls, *The Post-Star* (March 20, 2008).

[179] Tisdale indicated that AMD's local charitable contribution in areas in which it operated was weighted two-thirds in favor of education and the other third on the basis of need. AMD also responded positively to a proposal that it would make "a concerted effort to hire locally and use local vendors." "Town Raps AMD Plant Wish List," Glen Falls, *The Post-Star* (March 20, 2008).

[180] The agreement provided with respect to the trust fund that 90% of the interest on the fund was to be paid out in the form of grants, to be made available to projects driven by the town governments, private organizations, nonprofits, and other entities at the discretion of the board of trustees of the fund. The five-member board would control the remaining interest generated by the fund. Two trustees would be appointed by AMD, two by the town, and the fifth by the four other trustees. To the extent that a third party made a "substantial contribution" to the fund, the trustees could expand the membership of the board to enable participation by the new donor. "Malta, AMD Discuss Benefits," Saratoga Springs, *The Saratogian* (December 7, 2008).

[181] "Malta, AMD Discuss Benefits," Saratoga Springs, *The Saratogian* (December 7, 2008); "$4M AMD Community Gift Detailed," Schenectady, *The Daily Gazette* (December 3, 2008).

[182] "AMD to Submit Plans After Holidays," Schenectady, *The Daily Gazette* (December 23, 2008).

[183] "In Brief," Albany, *The Times Union* (June 9, 2009).

### 4.4.3.4 Groundbreaking

In March 2009, GlobalFoundries was granted a soil disturbance permit from the Town of Malta, opening the way for cutting trees and clearing the site at Luther Forest.[184] Approval was conditioned on the company signing off on the Development Agreement with the town (which awaited GlobalFoundries' final purchase of the land) and finding a project manager for Malta to oversee the work and serve as an ombudsman for local residents.[185] Those hurdles were cleared in June 2009, and the site clearance began with workers and machines provided by the Delaney Group, a construction firm based in Mayfield, New York.[186] A formal groundbreaking was held in July.[187]

## Bibliography[188]

AMD, Intel Race to the Bottom. (2006, June 28). *Forbes*.
AMD's New Investors Make Big Bet on Chip Plants. (2008, October 8). Ocala, *Star Banner*.
AMD's Q4 Retreats Under Competition—ATI By Adds Debt, Intel's New Chip 'Inflicting Pain'. (2007, January 4). *San Jose Mercury News*.
Cross Subsidy in NY. (2007, January/February). *IEEE Power and Energy Magazine*.
Derbyshire, K. (2002, June). Building a Fab—It's All About Tradeoffs. *Semiconductor Magazine*.
Fabless Future: Struggling AMD Spin-off Factories. (2008, October 7). *Associated Press*.
Foundry Sales Defy IC Decline. (2015, October 22). *EETimes*.
Future Horizons. (2012, February 16). *Smart 2010/062: Benefits and Measures to Setup 450mm Semiconductor Prototyping and Keep Semiconductor Manufacturing in Europe*. Luxembourg: Office of Official Publications.
GlobalFoundries. (2011). *Reaping the Benefit of the 450mm Transition*. Semicon West.
GlobalFoundries. (2010). Gold Shovel Project of the Year. (2010, July). *Area Development*.
How Semiconductors Are Made. *Intersil*. http://rel.intersil.com/docs/lexicon/manufacture.html.
Income Surges for AMD—But Threat of Price War With Intel Raises Concern. (2006, July 21). *San Jose Mercury News*.
*Luther Forest Technology Campus GEIS: Statement of Findings*. (2004, June 14). Draft adopted by Stillwater Town Board.
Malta and Stillwater Town Boards Approve PDD for Luther Forest Technology Campus. (2008, June 3). *New York Real Estate Journal*.
Malta Zones Research Site Industrial. (1999, December 28).

---

[184] "Chip Maker Gets OK to Clear Land," Schenectady, *The Daily Gazette* (March 12, 2009).

[185] "Fab Land Deal Nears," Albany, *The Times Union* (April 1, 2009).

[186] "Earth Movers Rumble Onto Site," Albany, *The Times Union* (June 16, 2009).

[187] "More Chips Coming to Saratoga County," Troy, *The Record* (July 25, 2009).

[188] As noted in the front matter of this book, the study also drew on interviews carried out by the authors and numerous articles from *The Times Union* (Albany), *The Daily Gazette* (Schenectady), the *Albany Business Review* (Albany), *The Post-Star* (Glens Falls), *The Record* (Troy), *The Saratogian* (Saratoga Springs), *The Buffalo News* (Buffalo), The *Observer-Dispatch* (Utica), The *Daily Messenger* (Canandaigua), and the *Post-Standard* (Syracuse). These are not individually included in the bibliography.

Munsey, J. S. (2006, May). Project Case Study: High Tech Land Development. *Civil & Structural Engineer.*

National Research Council. (2013). *Best Practices In State and Regional Innovation Initiatives: Competing in the 21st Century.* C. W. Wessner (Ed.) Washington, DC: The National Academies Press.

New York's Big Subsidies Bolster Upstate's Winning Bid for AMD's $3.2 Billion 300-MM Fab. (2006, June 10). *Site Selection.*

The Peridot Capitalist. (2007, January 27. AMD, Intel Price War Revisited. http://www.Peridotcapital.com/2007/01/amd-intel-price-war-revisited.html.

Ruiz, H. (2013). *Slingshot: AMD's Fight to Free an Industry from the Ruthless Grip of Intel.* Austin, Texas: Greenleaf Book Group Press.

Samsung to Invest Additional $9.2 billion in its $14.44 billion fab. (2015, April 15). *KitGuru.*

Schneier, E. V., Murtaugh, J. B., and Pole, A. (2010). *New York Politics: A Tale of Two States.* Armonk and London: M.E. Sharpe.

Sematech. (1991). *Annual Report.* Austin: Sematech.

Semiconductor Industry Association. Policy Priorities: Tax. http://www.semiconductors.org/issues/tax/tax.

Semiconductor Industry Association. (2003, October). *China's Emerging Semiconductor Industry: The Impact of China's Preferential Valve Added Tax as Current Investment Trends.*

Taiwanese IT Pioneers: Robert H.C. Tsao. (2011, February 17). Recorded Interview. Computer History Museum.

Town of Malta Regulations. Chapter 167, Article VII, § 167–26.

Town of Stillwater Code. Art. XI § 211–162A(i)(a).

TSMC to Spend $10 Billion Building 450mm Wafer Factory. (2012, June 12). *Reuters.*

Tucker, F. M. (2008, Fall). The Rise of Tech Valley. *Economic Development Journal.*

The U.S. $4B Project that Got Away from Singapore. (2009, March 27). *The Business Times.*

# Chapter 5
# The Infrastructure Buildout: A Detailed Look

**Abstract** The transportation, water, and electric power infrastructure necessary to support semiconductor manufacturing in Saratoga County did not exist at the time the corporate predecessor of GlobalFoundries committed to build a wafer fabrication plant at a local site. A wide-ranging effort to secure regulatory approvals and build new infrastructure was required. Although this process encountered delays and setbacks, by the time the fab was built and became operational, the necessary infrastructure was in place.

When Advanced Micro Devices (AMD) announced in 2006 that it would build a chip fab at the Luther Forest Technology Center, the site itself was "more woods than 21$^{st}$ Century industrial park." The infrastructure for semiconductor manufacturing was not in place. However, following the zoning approvals by the town boards of Malta and Stillwater in 2004, intensive planning and design work had been under way for the water and sewer systems, utilities, and roads that a new wafer fabrication facility would require. Local officials predicted that the necessary infrastructure would be put in place within 2 years, roughly the length of time needed to build a new chip fab plant. Malta Supervisor Paul Sausville commented in June 2006 that "in the grand scheme of things, all these things are not huge compared to a multibillion-dollar construction effort on the part of a computer chip manufacturer."[1] As of June 2006—

- A comprehensive plan to deliver the requisite volumes of water to the site, involving construction of a purification plant on the Hudson River and 28 miles of pipeline was in the final stages of environmental review.
- Saratoga County was designing the Round Lake Bypass, a 1.6-mile road from I-87 Exit 11 around Round Lake village, construction of which prior to a chip fab starting operations was a condition of Malta's zoning approval.

---

[1] "Infrastructure Required Before Plan Can be Built," Schenectady, *The Daily Gazette* (June 21, 2006).

- The line route for two new 115-kilovolt power lines to serve the chip fab had been mapped and surveyed.[2]

Many of the local regulatory approvals needed for infrastructure expansion in LFTC itself and in its immediate vicinity had already been secured through the pre-permitting exercise, although in some cases the Planned Development District (PDD) legislation would require amendment.

The buildout of the infrastructure needed to support a semiconductor fab in Luther Forest required the collaboration of a number of state, regional and local development authorities, the Board of Supervisors of Saratoga County, a number of town boards, the Saratoga Water Authority, two electrical utilities, a Saratoga County environmental advocacy organization, numerous private entities and individuals, and state political leaders. The process was not smooth or free of acrimony. Some landowners objected to and resisted construction of pipelines, power lines, and roads crossing their property. But at the end of the day, when the GlobalFoundries fab was ready to start operations at the end of 2011, the infrastructure needed to support it was in place.

Observers of the process credit the fact that most of the key players sought the same basic outcome—semiconductor manufacturing in the region and the high-tech jobs associated with it. An added factor underlying the infrastructure success was the skill and efficiency of local engineering and construction firms, which generally completed projects on time and within budget once the necessary approvals were secured. Finally, at a crucial juncture the state provided funding to keep the process on track and moving forward.

## 5.1 State Infrastructure Funding

The State of New York's agreement with AMD included a commitment to assist the infrastructure buildout for the new fab with financial support. State money flowed into the infrastructure projections through numerous channels, most significantly Empire State Development (ESD). State financial support for LFTC infrastructure is shown in Table 5.1.

## 5.2 The Federal Dimension

Although most of the funding and regulatory approval necessary for the creation of infrastructure for the AMD plant involved the actions of state and local authorities, the federal government played a role as well, reflecting the engagement of the

---

[2] "Officials Put Focus on Luther Forest Infrastructure," Albany, *The Times Union* (June 24, 2006); "Infrastructure Required Before Plant Can Be Built," Schenectady, *The Daily Gazette* (June 21, 2006).

**Table 5.1** State of New York Financial Support for Luther Forest Technology Center Infrastructure

| Need | Cost (millions of dollars) | Financing source | Amount (millions of dollars) |
| --- | --- | --- | --- |
| Electric transmission line | 40.9 | ESD grant | 30.3 |
| Cold Spring Road Corridor | 14.1 | ESD grant | 8.1 |
| National gas line-application | 1.0 | ESD grant | 3.7 |
| Natural gas line-completion | 8.0 | Land sale account | 2.9 |
| Interest and project management fees | 3.5 | DASNY grants | 4.7 |
| | | DOE grant | 1.5 |
| | | Other funding | 16.1 |
| Second water source—Consultant | | DASNY grant | 0.3 |

Note: *ESD* Empire State Development; *DASNY* Dormitory Authority of the State of New York; *DOE* Department of Education
Source: Empire State Department Corporation, *LFTCEDC—Luther Forest Infrastructure Capital II—NYSEDP and Update City-by-City (x043, x044)* (March 25, 2010)

New York congressional delegation. Senator Charles Schumer was a frequent public advocate of the project.[3] Then-Congresswoman Kirsten Gillibrand and her office played a lead facilitative role in the infrastructure effort, working with Senate Majority Leader Bruno's office and convening federal, state, and local authorities and others involved in regulatory approvals and in filling funding gaps.[4] Gillibrand successfully pursued federal funding through earmarks, including $1,477,500 secured in 2010 for the upgrading of power transmission lines serving the Luther Forest Technology Campus.[5]

## 5.3 Collateral Benefits for Communities

The infrastructure upgrades required to support GlobalFoundries' manufacturing operations generated numerous large and small benefits for surrounding residents and communities, including upgraded and expanded water supply systems, improved electricity delivery and reliability, expanded waste water treatment, and new roads and trail systems. In a number of cases, the imminent arrival of the chip fab provided the political and financial impetus for local improvements which residents had been seeking for years or decades without success. The Chairman of the

---

[3] "Schumer Offers Assurance of Luther Forest Fab," Troy, *The Record* (March 25, 2008); "Area's Tech Future Launched," Schenectady, *The Daily Gazette* (July 25, 2009); "Schumer: Plant Overdue," Albany, *The Times Union* (January 10, 2012).

[4] "Water Supply Transfer Moves Forward, County Expected to Get Permit," Schenectady, *The Daily Gazette* (January 19, 2008); "Locals View Gillibrand as Upstate Watchdog," Saratoga Springs, *The Saratogian* (January 24, 2009).

[5] "Federal Earmarks Set for Luther Forest, Glenn Falls Civic Center, Among Others," Glenn Falls, *The Post-Star* (December 18, 2007).

Saratoga County Board of Supervisors, Thomas Wood, observed in 2012 that "the public, as well as the company, benefit from the infrastructure improvements."[6]

## 5.4 Luther Forest Technology Campus Financial Concerns

As infrastructure work proceeded, the effort was hampered by the mounting financial difficulties of the Luther Forest Technology Campus Economic Development Corp. (LFTCEDC) which was expected to provide much of the infrastructure needed with the Luther Forest Campus itself. LFTCEDC had expected to raise the necessary funds through sale of additional parcels of land within the campus to other manufacturers locating near GlobalFoundries, but a number of factors combined to frustrate this objective. The Town of Malta had zoned the campus for "nanotechnology" manufacturing, which precluded other forms of industrial investment. The end of the Empire Zone program in mid-2010 eliminated one of the principal attractive features of the campus from the perspective of potential investors.[7] Compounding the problem, the town of Malta prohibited tax breaks for businesses in Luther Forest "to protect local taxpayers."[8] Absent revenue from property sales, LFTCEDC was dependent on periodic infusions of public money to continue its infrastructure work.

By 2010, GlobalFoundries was expressing concern publicly over infrastructure delays in Luther Forest. Travis Bullard, a spokesman for the company, said in October 2010 that "We have very serious concerns about meeting our timelines to having the fab up on time. It's been an issue for a long time."[9] He said that the company had worked closely with all of the parties and "doesn't care who owns the surrounding industrial park or builds the infrastructure, as long as the necessary improvements are made."[10]

LFTCEDC had received loans from the state for infrastructure improvements during the era in which Joseph Bruno was the state Senate Majority Leader, with a verbal understanding that at some future point the loans would be converted to grants or written off.[11] Following Bruno's retirement, Empire State Development threatened to foreclose on loans that had been made to LFTCEDC, giving ESD control of the Luther Forest campus.[12] The prospect of state takeover reportedly dismayed local

---

[6] "GlobalFoundries Chip Fab Plant Fosters a Ripple Effect Felt Far and Wide," Saratoga Springs, *The Saratogian* (July 24, 2012).

[7] Interview with Dennis Brobston, president, Saratoga Economic Development Corporation (October 28, 2015).

[8] "One Troubled Tech Park in Malta," Schenectady, *The Daily Gazette* (April 6, 2014).

[9] "GlobalFoundries Unveils New Tech Lab as Cloud of Uncertainty About State's Threat to take Over Luther Forest Loans," Saratoga Springs, *The Saratogian* (October 27, 2010).

[10] "Tech Park Ownership at Issue," Schenectady, *The Daily Gazette* (October 26, 2010).

[11] Michael Relyea, the head of LFTCEDC, said that the state had loaned $12 million through ESD

5.5 Water 137

officials in Saratoga County. LFTCEDC pointed out that part of its difficulty in proceeding with infrastructure work was attributable to the fact that ESD had been slow to reimburse Luther Forest for about $15 million in infrastructure work.[13]

## 5.5 Water

"The toughest piece of infrastructure to put together [for the AMD fab was] the water supply."[14] AMD's wafer fabrication plant in Luther Forest would require two to three million gallons of water per day to clean silicon wafers after they were etched with chemicals. When the AMD deal with New York was concluded in 2006, a water system capable of delivering such volumes to Luther Forest did not exist, but Saratoga County was committed to having adequate supplies of public water available to AMD at its site by the early fall of 2008.[15] The Capital region enjoyed virtually unlimited water resources, including the Hudson River and lakes and aquifers close to the site. However, the area's fragmented political jurisdictions had a checkered history with respect to the establishment of an integrated countywide water purification and distribution system. In 2005, on the eve of AMD's selection of the Luther Forest site, Saratoga County was still served by a "bewildering array of public and private water systems."[16] According to the New York State Health Department, at that time Saratoga County was served by 111 different water systems relying on surface reservoirs and wells.[17]

### 5.5.1 An Early Attempt at Consolidation

In 1990 New York enacted legislation creating the Saratoga County Water Authority, tasked with consolidating municipal water service for as many as 13 municipalities in the county under one central authority.[18] The concept underlying the authority

---

and that it was the hope of LFTCEDC that "the debt would be forgiven." See "Luther Forest Chief Rejects State Takeover of the 1,414-Acre Park Where GlobalFoundries is Building its Chip Fab," Saratoga Springs, *The Saratogian* (October 29, 2010).

[12] "Default, Demands and Dismay," Albany, *The Times Union* (October 29, 2010).

[13] "Waiting on a $15 M Promise," Albany, *The Times Union* (December 18, 2010).

[14] "Officials Put Focus on Luther Forest Infrastructure—Various Projects to Move Past Planning Stages in Wake of AMD Deal," Albany, *The Times Union* (June 24, 2006).

[15] "Water Plan Rolling Thanks to Chip Fab—System to be up and Running in Fall '08," Schenectady, *The Daily Gazette* (June 23, 2006).

[16] "Large Water Source Sought for Decades," Schenectady, *The Daily Gazette* (March 6, 2005).

[17] "Large Water Source Sought for Decades," Schenectady, *The Daily Gazette* (March 6, 2005).

[18] "Cuomo Gets Bill of Water Authority," Albany, *The Times Union* (June 3, 1990). The Water Authority was created by the legislature at the request of the Saratoga County Board of Supervisors. "Politics of Water Unclear on Costs," Albany, *The Times Union* (October 2, 2002).

was to create a system that would draw water from the upper Hudson River north of Glens Falls, purify it at a new plant to be built in Moreau, and deliver it to central and southern Saratoga County through a newly built pipeline.[19] However, a number of municipalities balked at joining the newly constituted Water Authority.[20] Soon after it was formed, the Water Authority launched an ill-fated legal effort to take control of a private Malta-based water company serving the Luther Forest area, Alexander Mackay's Saratoga Water Services, Inc., which at the time was the largest water distribution entity in the county.[21] Mackay challenged the Water Authority on constitutional and procedural grounds "at every turn,"[22] and although the Water Authority prevailed in every legal forum, it ran out of money in the process. The legal battle lasted seven years, and in 1998 the Water Authority, suffering from indebtedness and a virtually complete lack of operating revenues, abandoned its effort. In the settlement reached with Mackay, Authority Chairman David Wallingford gave his word that the Water Authority would not re-initiate any legal action against Mackay or his company as long as he remained Chairman.[23]

## 5.5.2 Revival of the Hudson River Plan

In 2002, the Saratoga County Board of Supervisors began studying a water plan similar to that originally envisioned when the Saratoga County Water Authority was created, which would bring water from the upper Hudson River down to Saratoga Springs and the rest of the county.[24] The Water Authority itself, however, was by this time "dead in the water," with virtually no operating assets and debts of roughly $275,000.[25] Its seven members, originally appointed by the County Board of Supervisors in 1998 for one- and two-year terms, had not been reappointed, and the Water Authority's chairman commented in 2005 that "we're all lame ducks

---

[19] "Politics of Water Unclear on Costs," Albany, *The Times Union* (October 2, 2002). Plans for the project were developed by Clough Harbour & Associates in 1990 and 1995 under the auspices of the Saratoga County Water Authority. "River Water is Favored by Officials—Study Offers an Alternative to Saratoga Lake," Schenectady, *The Daily Gazette* (June 26, 2002).

[20] "Grandin Blames County for Water Vote," Albany, *The Times Union* (August 30, 1990); "Water Authority Deadline Postponed Until January," Albany, *The Times Union* (December 6, 1990).

[21] "Private Water System Sought by Authority," Albany, *The Times Union* (December 20, 1991).

[22] "Effort to Seize Water System Ends—Malta Vote Leaves County Authority too Broke to Continue," Schenectady, *The Daily Gazette* (December 10, 1998).

[23] "Authority Drops Water Takeover," Albany, *The Times Union* (December 10, 1998); "Effort to Seize Water System Ends—Malta Vote Leaves County Authority too Broke to Continue," Schenectady, *The Daily Gazette* (December 10, 1998).

[24] Part of the impetus for this plan was to forestall any effort to develop nearby Saratoga Lake as a water source. Residents of Saratoga Springs and surrounding communities feared that they would not be able to use the lake for recreation if it were a primary source of drinking water. "Politics of Water Unclear on Costs," Albany, *The Times Union* (October 2, 2002).

[25] "Water Authority to Sell Sole Asset," Albany, *The Times Union* (June 26, 2005).

5.5 Water

appointed to positions on the county Water Authority. We serve at their pleasure."[26]

The prospect that a semiconductor fab might locate in Luther Forest, providing major demand to augment that of local communities, changed the equation, making the Hudson River project appear more feasible as well as necessary. In May 2006, the Board of Supervisors voted to provide a $15 million interest-free loan to a "revitalized" Saratoga County Water Authority, which would be tasked with creating a system to deliver water from the upper Hudson River to central and southern areas of the county. The new Water Authority was to take over the project when the county's Water Committee completed the project design and environmental reviews.[27] Seven new Water Authority board members were named by the county in October 2006. The reconstituted Water Authority was to have the power to borrow up to $30 million against future water sales revenue, "allowing the county to finance the project without having to increase property taxes."[28]

The wafer fab appeared to make the project economically viable.[29] The county's financial planners indicated that in order to break even, the new water system needed to sell a minimum of 3.65 million gallons a day at a price of $2.05 per thousand gallons. The county had secured tentative commitments from three towns along the potential route for a total of about 2.2 million gallons per day.[30] Assuming all three towns participated in the plan, AMD's fab, with a forecast demand of two to three million of gallons per day, would push the system well over the 3.65 million break-even minimum. The Chairman of the county's Water Committee, Ray Callanan, commented that the AMD fab "makes this whole project feasible from the financial point of view."[31]

---

[26] "Water Authority to Sell Sole Asset," Albany, *The Times Union* (June 26, 2005).

[27] The loan used 50% of the county's budget surplus. It was to be paid back over a period of 32 years. "County Loan to Fund Water System—$15 million Earmarked to Guarantee Supply for Computer Chip Plant," Albany, *The Times Union* (May 18, 2006).

[28] "Water Authority Adds Some New Members—Board of Supervisors Taps 7 to Aid River Plan," Schenectady, *The Daily Gazette* (October 18, 2006).

[29] John Lawler, who subsequently became chairman of the Saratoga County Water Authority (SCWA), likened the launch of the water project to "a perfect storm," citing the confluence of strong residential growth, Majority Leader Bruno's efforts to attract a chip fab, and the prospects for an AMD fab, which "made perfect timing for the SCWA to begin work on the plant." See "Meeting Focuses on Water System," Saratoga Springs, *The Saratogian* (July 25, 2008).

[30] The town of Ballston committed to 400,000 gallons per day, Wilton committed to about 300,000 gallons (later upped to 500,000) per day, and Clifton Park committed to 1.5 million gallons per day. Clifton Park's commitment was contingent on its securing four of the seven seats on a yet-to-be-created water authority. Saratoga Springs did not commit to the plan, and its representatives took the position that AMD's fab could be supported at lower cost by running a pipe from the Hudson River in Stillwater to Luther Forest. "Tech Park News Helps Water Plan," Saratoga Springs, *The Saratogian* (June 22, 2006).

[31] "Official: Chip Fab Plan Makes Water Project More Urgent," Schenectady, *The Daily Gazette* (July 7, 2006).

### 5.5.2.1 Accelerating Timetable

Callanan stressed the urgency of moving ahead with the water plan "a lot faster ... than we have in the past," with completion required six months earlier than planned. Immediate hurdles included securing a wetlands disturbance permit from the Army Corps of Engineers, preparation of an environmental impact statement, and the securing of easements over 184 private plots of land along the planned pipeline route. While work proceeded on these tasks, a number of alternative competing proposals to supply water to the AMD fab materialized.[32] In addition, Alexander Mackay, who had successfully withstood efforts by the original Water Authority to seize his company, filed a number of legal challenges to the new effort.[33] An attorney for Saratoga County commented in 2008 that "just the existence" of Mackay's lawsuits had been "damaging the project."[34]

## 5.5.3 Reconstitution of the Water Authority

In March 2007, Saratoga County approved a $250,000 loan from the county's general fund to enable the reconstituted Water Authority to start up operations, hire an executive director and a lawyer, and rent office space.[35] In August, the new Water Authority named William Simcoe, a professional engineer who was serving as the assistant water commissioner of the city of Albany, to be executive director.[36]

---

[32] "Official: Water Project More Likely With Chip Plant Coming," Schenectady, *The Daily Gazette* (July 7, 2006). In August 2006, the mayor of Albany was reportedly considering a pipeline plan pursuant to which a pipeline could be built at Albany's Loudonville Reservoir and run up either the I-87 median or along Route 9, terminating at the AMD plant. The Chairman of the Saratoga County Board of Supervisors commented that "these types of conversations should have started years ago if they were interested. It is too late at this point. They are way, way behind." See "County Leader Douses Pipe Plan," Albany, *The Times Union* (August 5, 2006). Clifton Park officials reportedly were discussing a plan to supply the new fab by buying comparatively pure water not requiring extensive treatment from the city of Schenectady, which enjoyed access to abundant water from the local Great Flats Aquifer. "Water Plan Could Include AMD," Schenectady, *The Daily Gazette* (December 21, 2006); In 2007, the president of Saratoga Water Services, a private water company in Malta, said that his company could supply the AMD fab more quickly and at lower costs by using two high-yield wells in Stillwater. "River Water Project Opponent Pitches Alternative Plan," Schenectady, *The Daily Gazette* (April 17, 2007).

[33] Mackay's first lawsuit reportedly complained that relevant officials had not complied with New York's Environmental Quality Review Act with respect to the water project. A second action challenged the permit issued by New York's Department of Environmental Conservation enabling the project to proceed. Animosity between Mackay and local officials reportedly dated back nearly two decades. In the 1990s, the Town of Malta had tried for seven years to take over Mackay's water company. "New Plan for Tech Campus Water," Albany, *The Times Union* (April 17, 2007).

[34] "Saratoga County Water Pipeline Lawsuit Dismissed," *The Post-Star* (April 22, 2008).

[35] The chairman of the Water Authority, John Lawler, commenting on the need for the funding, said that "right now we don't have the money to buy a pencil." See "Water Authority Given Loan—County Supervisors Approve $250G to Cover Startup Costs," Schenectady, *The Daily Gazette* (March 21, 2007).

[36] "Water Authority Gets its First Director," Schenectady, *The Daily Gazette* (August 22, 2007).

In April 2007, the SEDC signed a contract with the Saratoga County Water Authority to buy 2.45 million gallons of water a day to supply tenants of the Luther Forest Technology Campus, including AMD. Taken together with signed contracts with Wilton and Ballston for a total of 875,000 gallons, the Water Authority held signed contracts for 3.3 million gallons of daily water sales—enough, according to proponents, to justify construction of the new system. "The money will be there," said a spokesperson for Senate Majority Leader Bruno.[37] New York's Environmental Facilities Corp. extended a grant of $11.25 million to support the new water project, augmenting the $10 million state grant previously announced in 2005.[38] The state Dormitory Authority announced another $10 million grant for the project in May 2007. SEDC indicated it would loan the project $10 million, and the Water Authority planned to issue bonds to finance the remaining $24 million in construction costs.

### 5.5.4 Construction Begins

The Army Corps of Engineers issued a permit for the county water project in May 2007.[39] The county, which had solicited construction bids for the project in November and December 2006, awarded eight contracts within two weeks of receiving clearance from the Army. The county authorized the contractors to begin work.[40] A groundbreaking ceremony was held on June 2, 2007 featuring Senate Majority Leader Bruno operating a backhoe at the work site, commenting that "I'm glad I'm where I am and can help direct resources to this region." The water line would be the first piece of infrastructure actually to be built in anticipation of the AMD fab.[41]

### 5.5.5 Further Legal Obstacles

Although construction activity on the water project began in August 2007, legal and regulatory issues remained. The state Department of Environmental Conservation (DEC) permit for the project named the Saratoga County Board of Supervisors as

---

[37] "Stake Oks $10 M for Water Plan," Schenectady, *The Daily Gazette* (May 11, 2007).

[38] "County Authority Signs Deal to Sell Water to Tech Campus," Schenectady, *The Daily Gazette* (April 18, 2007).

[39] The Army Corps attached conditions to its permit requiring the county to offset the water projects' effects on wetlands through two county-funded mitigation projects, creating of a new wetlands area in Wilton and removal of log and debris jam from Kayaderosseras Creek, opening the creek to boaters from Saratoga Lake. "Logjam to be Cleared from Kayaderosseras," Schenectady, *The Daily Gazette* (August 15, 2007).

[40] "$67 M Water Plan Advances—Saratoga County Supervisors Approve Deals with 8 Contractors," Albany, *The Times Union* (May 19, 2007).

[41] "Water Milestone Marked—Backhoe Ceremony Celebrates County System," Schenectady, *The Daily Gazette* (June 2, 2007).

the party authorized to perform the work, but the work was being performed by the newly created Water Authority, and in order to finalize state grants and the sale of bonds to finance the project, the permit needed to be transferred to the name of the Water Authority, an action at which DEC initially balked.[42] In addition, the Water Authority had not been able to secure easements from all of the residential and commercial property owners along the new water system's route. The Canadian Pacific Railroad was seeking conditions for use of its right-of-way.[43] The Water Authority indicated that if necessary, "they'll take land for the water line by eminent domain."[44]

An immediate concern was a dispute between Saratoga County Water Authority and the state Environmental Facilities Corporation over the terms of the EFC's $11.5 million grant for the water project. EFC wanted the Water Authority to sign a letter stating that it would prevail in any current or future litigation, a guarantee that the Water Authority declined to provide. In explaining its position, spokesman for EFC acknowledged that the lawsuits filed against the Water Authority by Alexander Mackay had "given us pause." The EFC's deferral of its grant pending resolution of the dispute in turn held up the planned $37 million bond issue because in order to proceed with the bond transaction, the Water Authority was required to be "in possession of all anticipated grant money." The impasse reportedly was sufficiently serious to jeopardize the prospects for the construction of the AMD fab.[45] Following negotiations in which EFC's spokesperson acknowledged that "everybody in the state realizes how important this project is," the dispute was resolved through an agreement in which EFC demonstrated "willingness to change [its proposed] language."[46]

---

[42] DEC granted the county supervisors a permit, but when the Water Authority asked for the permit to be transferred, given that name of the actual plans had changed with the transfer of authority over the project, DEC indicated that "the agency needed to know more about the impact of the project." The issue was resolved when DEC issued the permit to the Water Authority in March 2008. Receipt of the permit put the Water Authority "in a better position to issue bonds and collect the rest of the money the water authority needs to build the project." See "Saratoga Gets Key Water Permit," Albany, *The Times Union* (March 8, 2008).

[43] "Saratoga County's Water Authority has Yet to Get Easements for Fewer Than 16 of the More Than 140 Residential Property Owners Along the County Water System's 28-mile Route," Saratoga Springs, *The Saratogian* (October 24, 2007).

[44] "Water Milestone Marked—Backhoe Ceremony Celebrates County System," Schenectady, *The Daily Gazette* (June 2, 2007); "Saratoga County's Water Authority has Yet to Get Easements for Fewer Than 16 of the More Than 140 Residential Property Owners Along the County Water System's 28-mile Route," Saratoga Springs, *The Saratogian* (October 24, 2007).

[45] "Grant Interpretation Jeopardizes AMD," Saratoga Springs, *The Saratogian* (March 26, 2008); "Water Worry for AMD," Troy, *The Record* (March 26, 2008); "Feud Over Water Project—Saratoga County Agency Upset by State's Demands for Funding," Albany, *The Times Union* (March 25, 2008).

[46] "Pipeline Plans Move Forward," Saratoga Springs, *The Saratogian* (April 21, 2008); "Water Worry for AMD," Troy, *The Record* (March 26, 2008).

In March 2008, the Water Authority began eminent domain proceedings against 40 properties with respect to which the owners had refused easement and offers of compensation, involving about six miles of the project's route.[47] In April, the state Supreme Court upheld the Water Authority's ability to conduct construction activities without interruption on properties with respect to which the owners had refused easement.[48]

The litigation initiated against the Water Authority and other county and state organizations by Alexander Mackay was resolved in 2008. The New York Supreme Court dismissed an effort by Mackay to annul the permit issued for the water project by the state Department of Environmental Conservation and to block the Water Authority's awarding of construction contracts.[49] Concurrently, Mackay and the Water Authority entered into a settlement agreement providing that Mackay's private water company would be the back-up water company for the Luther Forest complex in the event that the Water Authority's water deliveries were interrupted. Mackay agreed to drop a case against the Water Authority that was still pending before a state appellate court and the Water Authority agreed not to "unreasonably" try to block any future expansion plans of Mackay's company.[50]

## 5.5.6 Construction and Operation

In contrast to the effort to secure the legal and regulatory clearances for Saratoga's Hudson River water project, the actual construction work proceeded relatively smoothly. The first construction work began in August 2007 at the site of the water treatment plant in Moreau, and by October contractors were working to install pipe along the entire 28-mile route between Moreau and Luther Forest.[51] In May 2008, with over 10 miles of pipe laid, the Water Authority reported that construction was on time and under budget.[52] The target completion date of December 2009 was

---

[47] "Water Authority Files for Eminent Domain," Saratoga Springs, *The Saratogian* (March 31, 2008). The easements averaged 10–30 ft in width and required owners to move houses, sheds, and other structures and give up further use of the land involved once the pipeline was built. "Water Board Ups the Pressure," Albany, *The Times Union* (March 29, 2008).

[48] "Court Decision Clears Way for Water Pipeline," Troy, *The Record* (April 21, 2008).

[49] "Saratoga County Water Pipeline Lawsuit Dismissed," *The Post-Star* (April 22, 2008).

[50] "Water Project Overcomes Last Legal Obstacle," Schenectady, *The Daily Gazette* (April 26, 2008).

[51] "Logjam to be Cleared from Kayaderosseras," Schenectady, *The Daily Gazette* (August 15, 2007); "Saratoga County: Water Authority Has Yet to Get Easements for Fewer than 16 of the More Than 140 Residential Property Owners Along the County System's 28-mile Route," Saratoga Springs, *The Saratogian* (October 24, 2007).

[52] "$67 M Water Project on Time, Under budget," Saratoga Springs, *The Saratogian* (May 22, 2008).

missed due to a series of minor "snags"[53]; however, the system was up and running in February 2010, and "actually delivering water to a customer," the town of Wilton, which had contracted for 300,000 gallons of water per day.[54] Connections between the new water system and the AMD/GlobalFoundries' fab, by then under construction, were established in October 2010.[55]

The new water system created what was apparently an abundant new water source for Saratoga County communities and potential industrial users. The purification plant at Moreau was designed to produce 14 million gallons of water per day, far more water than immediately required by the chip fab and participating communities. The plant utilized a state-of-the-art membrane micro-filtering system that removed all materials down to 0.1 μm, or 1 ten-millionth of a meter, a screen sufficiently fine to filter out bacteria as well as particulates. A Water Authority official commented that "in terms of membrane filtration plants, this is probably the largest in New York State."[56] GlobalFoundries' disclosure in 2010 that it was considering expansion of Fab 8 from the originally planned 210,000 square feet to 300,000 square feet raised the prospect that fab's water needs would increase from the 3.1 million gallons of water per day to roughly 4.6 million gallons per day. Saratoga County Water Authority officials commented that "we can handle it," reflecting the fact that the new system could bring as much as ten million gallons per day of treated water from the upper Hudson to Luther Forest.[57]

In addition to the main water system construction, the Water Authority built a 100-foot high steel water tank in a corner of the Luther Forest Technology Campus. The tank holds five million gallons of water, or roughly one day's supply, at the site. The Luther Forest Technology Campus Economic Development Corp. funded the $4.5 million cost of the tank, the need for which became evident from discussions with AMD. "If there were an interruption, they need a certain amount of water to shut down their process," said LFTCEDC executive director Mike Relyea.[58] Construction of the tank was completed in August 2010.[59]

---

[53] Problems included "computer glitches," "glitches with controls," missing equipment, and use of an extension cord instead of permanent wiring. "Water Plant Opening Awaits Tweaks, Tests," Schenectady, *The Daily Gazette* (January 29, 2010); "Saratoga County, N.Y. Waterline Work Continues, with Some Snags," Glenn Falls, *The Post-Star* (January 16, 2010); "Construction Snags Slow Work on Water Project," Albany, *The Times Union* (January 30, 2010).

[54] "Water Flows in County's $67 M System," Schenectady, *The Daily Gazette* (February 24, 2010).

[55] "Work Coming Along Inside GlobalFoundries," Schenectady, *The Daily Gazette* (October 22, 2010).

[56] "Town Taps into Water System," Schenectady, *The Daily Gazette* (June 3, 2010).

[57] "Bigger Chip Fab Called No Problem," Albany, *The Times Union* (May 4, 2010).

[58] "Developer to Pay $2.5 M for Luther Forest Water Tank," Schenectady, *The Daily Gazette* (June 24, 2008).

[59] "Tech Park 'Open for Business,'" Albany, *The Times Union* (March 12, 2009).

## 5.5.7 Operational Challenges for the New System

In 2011, tests by the state Health Department found potentially unhealthy levels of haloacetic acids and other disinfection byproducts in water treated by the Water Authority, a problem attributed to buildup of organic material in the system following flooding in the Hudson River.[60] The Moreau treatment plant, which was built "without an extensive ability to filter out organic materials," was part of the problem, coupled with the amount of time the Water Authority's treated water sat in holding tanks or water pipes before use, allowing more time for chemical reactions creating haloacetic acids.[61] GlobalFoundries' planned 2012 startup which would result in use of much more water—increasing flow through the system—was seen as an eventual solution. In the interim, the Water Authority brought the contaminant level into compliance by adjusting its water treatment patterns and periodically flushing its pipeline.[62]

A second issue was raised by the state Department of Health in a 2012 report which warned that the Water Authority might not be able to produce enough water to meet peak demand. The report stated that when the Moreau purification plant's membrane filters were clogged by particles in raw intake water from the Hudson River, or when workers were cleaning the filters, the plant's capacity plummeted from 14 million gallons of water a day to 2.3 million gallons per day—not enough to serve the daily needs of GlobalFoundries' fab, much less the drinking water requirements of the county. The Water Authority indicated it was looking at options to improve filtration, including replacement of membranes and construction of a second filtration system. A Water Authority spokesperson commented that the system was new, and that bugs were thus inevitable:

> *It's no secret we were building the airplane as we were flying it. It hasn't been easy, but everyone comes together to work hard and make the right decisions. We need to focus on what we've done. We're operating. We're providing water. We'll get there, we'll reach perfection.*[63]

## 5.5.8 Water Supply Agreement with GlobalFoundries

When the Water Authority's new system became operational in 2010, GlobalFoundries was constructing its fab and did not need three to four million gallons per day that would be required when its manufacturing operations began in

---

[60] Haloacetic acids are linked to cancer and other health problems by the EPA. They are created from the chemical reaction of chlorine with organic materials in water such as leaves and algae. The EPA sets a maximum acceptable level of 60 parts per billion for drinking water. The county Water Authority's treated water registered 70 ppb. "Unhealthy Chemicals in County's Water System," Schenectady, *The Daily Gazette* (August 26, 2011).

[61] "Unhealthy Chemicals in County's Water System," Schenectady, *The Daily Gazette* (August 26, 2011).

[62] "Wilton Still Not Buying County Water Even Though it Has Improved," Saratoga Springs, *The Saratogian* (October 6, 2011).

[63] "Heads Up on Water Supply," Albany, *The Times Union* (February 5, 2012).

2012. In 2010 GlobalFoundries bought 300,000–400,000 gallons per day from the Water Authority to support testing of piping, pumps, and utilities as the construction project proceeded, based on 30-day contracts.[64] Negotiations on a long-term supply agreement were protracted and lasted through the end of 2011, during which time GlobalFoundries continued to rely on 30-day supply contracts. In September, two members of the Water Authority voted against renewal of another 30-day agreement, "citing frustration that a long-term deal hadn't been reached."[65]

In 2012, the Water Authority and GlobalFoundries reached agreement on a 10-year water supply contract pursuant to which the company would pay $2.75 per thousand gallons in 2012 and $2.50 per thousand gallons in 2013. Thereafter, the company would pay $2.75 per thousand gallons when its average daily volume was below four million gallons a day and $2.50 when average daily use was greater than four million gallons. At the time the deal was reached, the company was using about two million gallons a day as it installed and tested manufacturing equipment but expected increased consumption with the startup of manufacturing operations.[66] The prices to which GlobalFoundries agreed were substantially higher than the Water Authority's wholesale rate of $2.05 per thousand gallons rate, which the company had been paying in 2010 and through most of 2011. The prices also exceeded the rate of $2.08 per thousand gallons agreed with the Water Authority's municipal customers for 2012.[67]

### 5.5.9 GlobalFoundries Seeks a Second Water Source

The state's 2012 warning that the Water Authority's delivery capability could sometimes fall dramatically below the volumes needed by users underscored the need for reliable back-up water sources. Virtually from the moment it committed to building a fab in Luther Forest, GlobalFoundries expressed concern over the need to establish a second source for water supply for the fab in case the Water Authority's water

---

[64] "GlobalFoundries Water Deal in Works," Schenectady, *The Daily Gazette* (December 14, 2010); "Chip Fab Water Plan Wanted," Albany, *The Times Union* (October 15, 2011). The Luther Forest Technology Campus Economic Development Corporation had originally planned to set up a transportation company that would buy water from the county Water Authority and sell it to GlobalFoundries. However, given the magnitude of GlobalFoundries' needs and its sensitivity to water availability, it was decided that the company and the Water Authority should deal with each other directly. This arrangement required clearance from legal counsel because of a prohibition on retail sales of water by the Water Authority. "One Plant Issue Set, One to Go," Schenectady, *The Daily Gazette* (March 4, 2011).

[65] "County Water Authority, GlobalFoundries Reach Tentative Agreement on Water Terms," Saratoga Springs, *The Saratogian* (December 15, 2011).

[66] "Water Authority Reaches Deal with GlobalFoundries," Schenectady, *The Daily Gazette* (January 14, 2012).

[67] "Water Authority Reaches Deal with GlobalFoundries," Schenectady, *The Daily Gazette* (January 14, 2012).

deliveries were interrupted for whatever reason. In fact, as part of the original Grant Disbursement Agreement with New York State ESD, GlobalFoundries was guaranteed a fully redundant second water source as any possibility of an interruption in water supply to a semiconductor fab could be catastrophic. During the economic downturn that occurred during fab construction, ESD asked for relief from the requirement to construct "a fully redundant second water source" because of the cost. ESD undertook a study to evaluate other options that could provide an adequate back-up supply and determined that the Mackay source together with additional holding tanks could meet the need. It was also determined that the proposed solution was in line with the settlement between Mackay and the Water Authority as the Mackay source would be used as a back-up source. Discussions between GlobalFoundries and Luther Forest officials began to stall and threatened the development of adequate water supplies as well as power infrastructure within LFTC. While GlobalFoundries' employees "were not convinced that a groundwater source [such as Mackay's] could provide [adequate water] for the company's computer chip factory,"[68] it was adequate as a back-up source. A problem arose when it was explained that maintaining the (Mackay) back-up source would require that it would operate at a reduced but continuous basis so as to maintain a certain level of turbidity in the line and a consistent water chemistry in the event it needed to be turned on in the event of a failure of the main system. The Water Authority insisted that the litigation settlement with Mackay provided that his company was to be the second source, which meant it could not run at any level on an ongoing basis, and as such it would be considered as a primary source. It would not collaborate in efforts to explore alternatives.[69]

Frustrated, GlobalFoundries sent a letter to Empire State Development Chairman Dennis Mullen in October 2010 expressing concerns over the unresolved back-up water issue and other infrastructure concerns. The ESD, understanding that LFTC's inability to provide needed infrastructure was threatening the project, responded by initiating steps to take over the Luther Forest Technology Campus, a move which reportedly "outraged local officials who have contracts, understandings, and relationships with the existing LFTC management."[70] ESD also retained Clough Harbour & Associates, the engineering firm that had designed the original Hudson River water plan, to assess the alternatives available to provide a reliable second source of water to GlobalFoundries.[71] Ironically, although the Water Authority was "disappointed and surprised" to learn of the Clough Harbour study and opposed its preparation, the study identified a solution involving Mackay's Saratoga Water

---

[68] A second issue was the composition of secondary source water, which needed to be "consistent with or similar to" water being delivered to the chip fab from the Moreau plant on the upper Hudson. "Water Fears Precipitated Seizure," Albany, *The Times Union* (November 5, 2010).

[69] "Saratoga County Water Authority Says it Won't Be the Sole Water Provider for GlobalFoundries' Chip Plant by Malta," Saratoga Springs, *The Saratogian* (March 2, 2011).

[70] "Firm Studies Chip Plant Water Sources," Schenectady, *The Daily Gazette* (December 5, 2010).

[71] "Firm Studies Chip Plant Water Sources," Schenectady, *The Daily Gazette* (December 5, 2010); "Water Fears Precipitated Seizure," Albany, *The Times Union* (November 5, 2010).

Services—a scenario backed by the Water Authority—as "the best solution at least in the near term."[72] In 2014, GlobalFoundries submitted applications with Malta and Stillwater to build two 5-million gallon water tanks at its Fab 8 site which would be supplied with water by Saratoga Water Services. A GlobalFoundries executive commented that the arrangement represented the "long-planned, long-contemplated answer to the redundant water supply required here in support of our Fab 8 semiconductor manufacturing process." The town boards approved construction of the two tanks in January 2015.[73]

Milestones in the effort to provide an adequate water supply to the Luther Forest manufacturing site are summarized in Box 5.1.

---

**Box 5.1: Milestones—Water for Luther Forest Semiconductor Fab**

| Year | Event |
|---|---|
| 1990 | Saratoga County Water Authority created |
| 1998 | Water Authority settles litigation with Alexander Mackay's Saratoga Water Services |
| 2006 | Advanced Micro Devices chip fab plan announced<br>Revitalized Saratoga County Water Authority launched |
| 2007 | Legal challenge by Mackay to new Water Authority launched<br>Saratoga Economic Development Corporation contracts with Water Authority for water and supply chip fab<br>Funds raised for Glens Falls-Luther Forest water line<br>Army Corps of Engineers issues permit<br>Groundbreaking for construction held |
| 2008 | Water Authority initiates eminent domain proceedings against 40 landowners<br>Mackay litigation settled |
| 2010 | New water line connected to fab site in Luther Forest |
| 2012 | Water supply agreement between GlobalFoundries and Water Authority |
| 2015 | Malta and Stillwater town boards approve GlobalFoundries contribution of two 5-million gallon water tanks in Luther Forest Technology Campus |

---

## 5.6 Sewers

In 2006, following the disclosure that AMD would build a chip fab at the Luther Forest site, the Saratoga County Sewer Commissioners announced plans to expand the capacity of the county sewage treatment plant from 21.3 million gallons per day to 50 million gallons per day by July 1, 2009. The additional capacity was "needed to meet the industrial needs of the $3.2 billion Advanced Micro Devices computer chip plant." In addition, SEDC was working with the C.T. Male engineering firm to

---

[72] "Second Fab Water Line Likely to be Late," Albany, *The Times Union* (June 11, 2011).

[73] "Second Water Tap for Chip Fab," Albany, *The Times Union* (December 11, 2014); "Two 5-Million Gallon Water Tanks Get OK," Albany, *The Times Union* (January 9, 2015).

design a 10-million gallon sewer line that would connect the Luther Forest site and the main trunk sewer line for the county in Halfmoon, creating sufficient capacity to support two chip fabs.[74]

Expansion of the Saratoga County sewage treatment plant began in the summer of 2008 with a projected cost of $52 million.[75] Most of the funding for the project was raised by the county through the issuance of debt, to be paid off over time by users of the sewer system through sewer charges.[76] SEDC paid for the construction of the new trunk sewer connecting the Luther Forest site with the county system in Halfmoon. The county succeeded in securing very favorable interest rates on its bonds, with Moody's and S&P citing "the expected economic impact of the computer chip plant as a prime reason why the county bonds are a potentially good investment."[77] The refurbished sewer system featured more efficient state-of-the-art equipment including new ultraviolet and air defusing technology.[78]

In addition to the county's expansion project, the Luther Forest Technology Campus Economic Development Corp. oversaw the construction of a 3.5-mile sewer line—"enormous as private sewer lines go"—with a 10-million gallons-per-day capacity connecting the chip fab site within the county system. Construction began in 2008 and was completed in 2011, at a cost of about $6 million. As of the beginning of 2013, GlobalFoundries was discharging 1.6–1.7 million gallons per day, now 3.4 mgd, of wastewater with the line, still leaving substantial capacity for potential additional industrial tenants in Luther Forest.[79] In 2013, with the LFTCEDC in financial distress, the 3.5-mile connector line was taken over by Saratoga County.[80]

## 5.7 Roads and Trails

The Luther Forest chip fab site was located less than two miles from two major transportation arteries, Interstate 87 and New York Route 9. The transportation challenge was to improve the quality of the road connections between these highways and the proposed fab while minimizing the adverse effects of increased traffic on the affected communities.

---

[74] "Sewer System Needs Capacity—Commissioners Planning for Arrival at Chip Plant at Tech Park," Schenectady, *The Daily Gazette* (July 22, 2006).

[75] "Treatment Plant Expansion Begins in Saratoga County," Schenectady, *The Daily Gazette* (November 24, 2008).

[76] "Chip Fab Plant Helps County Borrow Money," Schenectady, *The Daily Gazette* (July 16, 2009).

[77] "Chip Fab Plant Helps County Borrow Money," Schenectady, *The Daily Gazette* (July 16, 2009).

[78] "Sewer Expansion on Track in Saratoga County," Glens Falls, *The Post-Star* (November 29, 2009).

[79] "Tech Park Sewers to Become County Property," Schenectady, *The Daily Gazette* (January 24, 2013).

[80] "County Sewer District Owns Line Serving Chip Fab Plant," Schenectady, *The Daily Gazette* (March 3, 2013).

## 5.7.1 Round Lake Bypass: Regulatory Hurdles

The village of Round Lake lies between I-87 and New York Route 9 to the east of I-87 Exit 11. "An island of quaintness," it is comprised of numerous examples of Victorian-era architecture and "buggy-width" streets.[81] As the surrounding region developed, traffic flow from Exit 11 to Route 9 through the village increased to the point that by the late 1990s residents were concerned that traffic threatened the bucolic character of the village.[82] The idea of constructing a bypass from Exit 11 which would route traffic around the north side of the village to Route 9 was discussed for many years.[83] However, as a local planning official said of the bypass proposal in 2000, "people say it's a good idea, and then laugh at their palm and say it's never going to happen."[84]

During the pre-permitting town meetings convened in 2003 to discuss the proposed Luther Forest Technology Campus, SEDC presented a proposal for a bypass road around the village of Round Lake. SEDC's Ken Green argued that the bypass road would create more direct access from Exit 11 to the Luther Forest chip fab site while at the same time reducing traffic flows through the village, bringing traffic back to pre-1987 levels. Despite considerable audience skepticism, Round Lake Mayor Dixie Lee Sacks argued that the bypass was a potential way to solve the village's traffic problem, and she emerged as a vocal advocate for the concept.[85] The Malta and Stillwater zoning legislation that emerged from the pre-permitting exercise, approving the 1350-acre Luther Forest site for a chip fab, solidly endorsed the bypass proposal, actually requiring as a condition of rezoning the site that a bypass be built around Round Lake village and be completed before the first chip fab opened.[86]

In 2004, the New York State Department of Transportation (DOT) began a $500,000 study to determine the feasibility of a bypass road.[87] The state DOT study

---

[81] "Round Lake Battles to Retain Victorian Character and Charm," Schenectady, *The Daily Gazette* (January 4, 1998).

[82] "Round Lake Battles to Retain Victorian Character and Charm," Schenectady, *The Daily Gazette* (January 4, 1998). Daily traffic flows through the village in 1987 consisted of about 4000 vehicles. A 2003 Saratoga County study found that the flow would reach 12,000 vehicles by 2025 even if the proposed Luther Forest Technology Park was not built. "Technology Park Road Plan Received With Skepticism," Albany, *The Times Union* (February 5, 2003).

[83] A bypass was discussed in the early 1990s when the village was developing a new land-use plan. "Malta Board to Review Plan for Round Lake Area," Albany, *The Times Union* (September 7, 1993).

[84] "Round Lake Wary of New Master Plan," Schenectady, *The Daily Gazette* (March 17, 2000).

[85] "We have a problem with traffic in the village," she said. "We cannot have any more traffic." See "Technology Park Road Plan Received With Skepticism," Albany, *The Times Union* (February 5, 2003).

[86] "Vote Paves Way for $1.2 M Bypass Design—Construction Could Begin Next Summer," Schenectady, *The Daily Gazette* (March 16, 2005).

[87] "State to Study Bypass for Round Lake," Schenectady, *The Daily Gazette* (March 5, 2004). The money for the study was to be drawn from federal transportation funds reportedly obtained for "unspecified Luther Forest Technology Campus transportation improvements" by US Rep. John E. Sweeney.

pegged the cost of a bypass at $13.5 to $15 million, sketching out two alternative routes, both of which began just to the east of Exit 11 on Curry Road near the Round Lake Firehouse, proceeded north parallel to I-87 for about a mile before veering east to join state Route 9.[88]

In March 2005, the Saratoga County Board of Supervisors authorized a $1.2 million engineering design for a highway bypass around Round Lake pursuant to a "complex arrangement" involving federal highway funds.[89] The county hoped to complete the design of the bypass by September 2006, but the county and its consultants ultimately decided that the project would require an environmental impact statement, primarily because of the project's potential impact on wetlands.[90] Saratoga County notified potentially affected residents that it would take properties along the bypass route by eminent domain if "amicable" negotiations failed.[91]

### 5.7.2 Saratoga PLAN

The proposed route for the bypass required crossing Ballston Creek, the county-owned Zim Smith Trail, and a former trolley line owned by Saratoga's preeminent environmental organization, Saratoga PLAN, which had originally been formed in

---

[88] "Round Lake Bypass's Cost Put at $13.5–$15 M," Schenectady, *The Daily Gazette* (September 30, 2004). Exit 11's exit ramps fed onto Curry Road, which ran roughly perpendicular to I-87 in an east-west direction. At the time of the study traffic exiting I-87 moving eastbound on Curry Road passed directly through the village.

[89] A spokesman for Congressman John Sweeney noted that federal highway bill contained $8.15 million for unspecified Luther Forest-related transportation improvements (these funds were distinct from the $500,000 federal grant arranged by Sweeney in 2004 to enable the state DOT to conduct its feasibility of the bypass). The Saratoga supervisors planned to allocate $1.2 million of the new federal money to the forthcoming reconstruction of a local road in Edinburg. County money from that project would thus be freed up by the federal grant and redirected to the Round Lake bypass engineering work. This arrangement was seen as necessary because the federal government could demand a refund if it paid directly for the bypass engineering work, but the bypass was not actually built. "Vote Paves Way for $1.2 M Bypass Design—Construction Could Begin Next Summer," Schenectady, *The Daily Gazette* (March 16, 2005).

[90] An engineering consultant working with the county commented that "the wetlands impacts are going to have to be mitigated. The wetlands are pretty much all over out there." "Environmental Study is Needed," Schenectady, *The Daily Gazette* (December 23, 2005).

[91] "Nothing Easy About Tech Campus Plans," Schenectady, *The Daily Gazette* (February 4, 2006). The "Zim Smith Trail" was an 8.8-mile unpaved trail running from Ballston Spa to Halfmoon through Clifton Park, Round Lake, Malta, and Ballston along the original right-of-way for the Delaware and Hudson Railroad. Plans to pave the trail for easier use by bicyclists and walkers, thus creating the "backbone" of a countywide bicycle and hiking trail system had been discussed since the 1970s. "Bicycle Trail Effort Revived," Schenectady, *The Daily Gazette* (July 17, 1999).

reaction to the proposed development of the Luther Forest Technology Campus.[92] Saratoga PLAN hired an attorney to fight the county's eminent domain proceedings and criticized the engineering design's proposed treatment of the trail as "essentially defective." However, Saratoga PLAN indicated that it would prefer to "open a dialogue with the county rather than fight," reflecting the fact that the bypass initiative appeared to open possibilities with respect to the county's trail system. Julia Stokes, Chairman of Saratoga PLAN, commented that "if the county proceeds with the bypass, there are some opportunities to do several things with the trail system."[93] In March 2006, Saratoga PLAN and the county announced that they would work together to ensure that trails were developed, not destroyed, by the Round Lake bypass.[94]

### 5.7.3 Round Lake Bypass: Construction

Saratoga County conducted the initial environmental and design work for the bypass, supported by $1.2 million in state funds, but state legislation enacted in 2006 provided for the state takeover of the project in December 2006. The construction work was entirely funded by the state at a cost of about $37 million.[95] The opening of construction bids for the bypass was scheduled for the spring of 2007. However, the project was delayed as the terms of a wetland disturbance permit were negotiated with the Army Corps of Engineers. In January 2007, state officials indicated they expected to have the wetlands permit "soon," and that construction work should start in mid-May.[96] In the end, the permit was not issued until December 2007.[97]

---

[92] Saratoga PLAN ("preserving land and nature") is a 501(c)(3) nonprofit conservation organization dedicated to "preserving the rural character, natural habitats and scenic beauty of Saratoga County ... [by] ... helping communities create plans that balance growth with conservation ... ." http://www.saratogaplan.org/about/. It is funded through a combination of contributions by individuals and businesses, foundation grants, government, and fundraising events. "A Promise to Keep the Land Pristine—Saratoga PLAN Works on its Own and Partners with Others to Preserve Open Spaces," Albany, *The Times Union* (March 25, 2007).

[93] "Group: Bypass Plan Design Defective," Albany, *The Times Union* (March 10, 2006).

[94] "County Trails Part of Bypass Project," Albany, *The Times Union* (March 30, 2006).

[95] New York Department of Transportation, *Round Lake Bypass Project* (Final Environmental Impact Statement P.I.N.1807.01, prepared by MJ Engineering and Land Surveying, PC, 2016); "State Will Provide Bypass Funding. Round is Needed for Luther Forest," Schenectady, *The Daily Gazette* (August 24, 2006); "Tech Park Road Work Could Begin in April," Schenectady, *The Daily Gazette* (February 29, 2008).

[96] "Hurdles Leapt Bypass Work to Start in Spring—Road to Reduce Traffic Through Round Lake," Schenectady, *The Daily Gazette* (January 27, 2007).

[97] "Round Lake Bypass Bids Delayed—Permit Needed Before Construction," Schenectady, *The Daily Gazette* (May 25, 2007); "Town Nears Deal on Luther Forest Roads," Schenectady, *The Daily Gazette* (January 26, 2008).

5.7 Roads and Trails 153

Construction of the Round Lake Bypass was a more complex proposition than its 1.6-mile length suggested. The firm heading up the project, Rifenburg Construction, built the road itself, and two roundabouts at the intersections at Route 9 and Curry Road (off Exit 11) and constructed a 640-foot, four-span automobile bridge, a pedestrian bridge, a large box culvert, a mechanically stabilized earth wall, and retention ponds and wetlands areas. The project required blasting 65-foot wide corridors through shale rock for 1500 feet.[98] The state also set aside 98 preservation acres around the highway, created 5 acres of new wetlands, planted 2800 trees and shrubs, and built new retention ponds to limit runoff water from the paved areas.[99]

Despite engineering challenges, the Round Lake Bypass was completed on time and opened for vehicular traffic in July 2009. Round Lake village would "no longer be a cut-through for cars and trucks headed to the Northway." Round Lake Mayor Sacks commented after the ribbon-cutting ceremony that:

*This is basically a wonderful thing for the village. Traffic has done nothing but increase over the years. We've had this in our master plan for several years and never thought it would be a reality. This was a gift to us [from the state].*[100]

### 5.7.4 Other Road Projects

In addition to the Round Lake Bypass, a number of other road projects were undertaken within and closely adjacent to the Luther Forest campus to ensure access to the planned fab. The federal government allocated $6.52 million to widen and expand intersections around the campus, with work beginning in 2006.[101] At the end of 2005 the federal government allocated $4.5 million to rebuild and pave Cold Springs Road, then a dirt road, which ran north-south immediately east of the campus.[102]

Within the Luther Forest campus, federal and state funding of $37 million was allocated for the construction of 5.5 miles of interior roads, with construction the responsibility of the Town of Malta. This work included an extension of Stonebreak Road, a paved road that ran east from Route 9/67 into the campus itself and beyond to the chip fab site, forming a primary entrance.[103] Malta agreed to handle future bidding for road construction and to accept long-term responsibility for the roads.[104]

---

[98] "Bypass Project Fast Tracked," *Professional Surveyor Magazine* (January 2010).

[99] "Go Ahead, Bypass Round Lake," Albany, *The Times Union* (July 22, 2009).

[100] "Bypass Road Open to Traffic—Path to Spare Village by Serving as Direct Route to Northway," Saratoga Springs, *The Saratogian* (July 22, 2009).

[101] "Federal Money Earmarked for Tech Campus Roadwork," Saratoga Springs, *The Saratogian* (July 29, 2005).

[102] "Luther Area to Get Road Funding—$10 Million Slated for Paving, New Intersections," Schenectady, *The Daily Gazette* (December 20, 2005).

[103] "Luther Forest Road Work Delayed—Town Awaits Decision on State, Federal Environmental Permits," Schenectady, *The Daily Gazette* (May 1, 2007).

[104] "Town Agrees to Accept Tech Park Roads," Schenectady, *The Daily Gazette* (October 4, 2006).

This work was completed in 2010.[105] Finally, between 2006 and 2012, thirteen new roundabouts were constructed in Malta, including five around Northway Exit 12, to alleviate traffic congestion.[106]

### 5.7.5 Upgrading the Trail System

The infrastructural preparation for a chip fab in Luther Forest included the creation and improvement of a network of recreational trails connecting the fab site with trails serving the rest of Saratoga County, enabling commuting by cycling, jogging, and walking. The chip fab project built upon and accelerated a grass roots movement in the Capital Region which, beginning in the 1990s, saw the construction of numerous new trails.[107] The Zim Smith Trail, which would be crossed by the Round Lake Bypass, was an important part of longstanding local efforts to create a trail system unifying Saratoga County.[108] The trail was a largely unpaved 8.8-mile track running from Ballston Spa to Halfmoon through Clifton Park, Round Lake, Malta, and Ballston along the right-of-way of the Delaware and Hudson Railroad.[109] Crucially, it ran under I-87 north of Exit 11, providing a route for pedestrians and cyclists to get across the highway.[110] Plans to pave the entire length of the trail for easier use by cyclists, horses, and walkers—thus creating the backbone of a county-wide bike, riding, and hiking system—had been discussed since the 1970s, but the county's efforts to secure state or federal funding for the estimated $1.5 million cost had failed.[111]

---

[105] "Tech Park Work Nears Completion," Schenectady, *The Daily Gazette* (April 18, 2010).

[106] "Public Rises up About Roundabouts," Schenectady, *The Daily Gazette* (October 25, 2012).

[107] The Capital Region trails varied from dirt paths to paved 6- to 8-feet wide multi-use routes suitable for bicycles and wheelchairs. By the late 1990s, the Capital Region was "crisscrossed" with "hundreds of miles of trails," including 65 miles of Clifton Everywhere Park, 45 miles in and around Saratoga Springs, and shorter trails in Malta, Ballston, and Ballston Spa. "The Trail to Everyone," Albany, *The Times Union* (June 1, 1997).

[108] Zimri Luce Smith was a retired Air Force Colonel who settled in Saratoga Springs in 1976. A member of Saratoga's Design Review Commission, he worked on historic preservation projects in Saratoga Springs and at Saratoga National Historical Park. An advocate for trails, he served on the committee that created the Saratoga County Heritage Trail system by negotiating easements with private landowners. When Smith died in 1994, the committee named the trail for him. "Work on Trail Progresses," Albany, *The Times Union* (August 14, 2007).

[109] Saratoga County took over the D&H right-of-way in the 1960s. "County to Build New Trail," Albany, *The Times Union* (December 17, 2002).

[110] "Count to Build New Trail," Albany, *The Times Union* (December 12, 2002).

[111] "Malta's Paving May be the Start of Inter-Town Bicycle Trail," Schenectady, *The Daily Gazette* (September 8, 1998). The town of Malta paid to pave a 6700-foot stretch of the trail in 1998 between East Line and Ruble Roads. "Bicycle Trail Effort Revived," Schenectady, *The Daily Gazette* (July 17, 1999).

A 2002 federal grant made $686,400 available to transform the Zim Smith Trail into a paved and multipurpose pathway, and Saratoga County contributed another $176,000. In May 2004, when the towns of Malta and Stillwater voted to approve rezoning of an area inside Luther Forest for nanotechnology manufacturing, they codified a requirement that before certificates of occupancy were issued for buildings at the site, the Luther Forest Technology Campus Economic Development Corporation would complete the construction of 7.5 miles of paved pathways and trails inside the campus as well as a link to the Zim Smith Trail.[112] Up to this point, reflecting various delays, the Zim Smith Trail itself had remained "more dream than reality," although county and municipal officials had begun brush clearing activity.[113]

### 5.7.5.1 Completing the Zim Smith Trail

Saratoga PLAN saw the prospect of a chip fab, with the conditions requiring construction of the bypass and paved trails, as an opportunity to realize the Zim Smith Trail vision. PLAN Chairwoman Julia Stokes said in 2006 that:

> This is a real opportunity to take a regional trail system that connects five towns and link it with trails to Stillwater and even National Battlefield. When something happens in Luther Forest, one of the conditions is construction of the Round Lake Bypass, and you really need to have the trail plan ready then. It would be really wrong to lose this opportunity.[114]

The broader infrastructural effort to establish support for the chip fab helped the Zim Smith Trail initiative in a number of ways. Construction of the water line by the Saratoga Water Authority paid for a bridge over Mourning Kill stream that could be utilized by the trail, eliminating a longstanding obstacle. In addition, construction of the water line north toward Moreau created a pathway that could be used to extend the trail in the future.[115] Two road projects, the Round Lake Bypass and the removal of an overpass on Route 67 paid for paving of the trail at the crossing areas.[116] The

---

[112] Town of Stillwater Code, Art. XI § 211–184.

[113] "Zim Smith Recreation Trail Remains Elusive Goal," Schenectady, *The Daily Gazette* (June 10, 2006). "County to Build New Trail," Albany, *The Times Union* (December 17, 2002). In 2003 and 2004, County officials worked with crews of municipal workers and prison inmates to clear the trail, which "was much more overgrown that anyone had anticipated." A conundrum was presented by the need to cross the Mourning Kill stream in Ballston. The idea of moving an old highway bridge to the site turned out to be cost-prohibitive. The U.S. Department of Transportation had delayed release of the federal grant money, saying that "the county still has issues to work out," such as what would happen where the proposed trail encountered CP Rail just south of Ballston Spa. "Saratoga County Plans to Pave Nine-Mile Trail," Schenectady, *The Daily Gazette* (November 6, 2004); "Zim Smith Recreation Trail Still a Dream," Schenectady, *The Daily Gazette* (June 10, 2006).

[114] "Study Seeks Options for Linking Trails—Improvements Will Coincide With Bypass Work," Schenectady, *The Daily Gazette* (March 20, 2006).

[115] "Work on Trail Progresses," Albany, *The Times Union* (August 14, 2007).

[116] "Construction Bids Approved for Trail—County Ready to Move Forward with Zim Smith," Schenectady, *The Daily Gazette* (May 4, 2007).

Route 67 overpass was replaced by the state Department of Transportation with a culvert enabling bicycles and horses to pass through it on separate paths.[117] Saratoga PLAN's Julia Stokes commented that "We were thrilled. DOT has done a great favor for Saratoga County on this."[118]

In 2009 Saratoga County, having expended the original $686,000 federal grant money on construction, received another $1.7 million in federal stimulus money to complete construction of the Zim Smith Trail.[119] However, state Department of Transportation officials concluded that federal stimulus money could not be used for work on private property. This affected the section of trail that crossed CP Rail property pursuant to an arrangement under which the county leased the property from the railroad rather than acquiring it outright. Accordingly, the county agreed to contribute $550,000 of its own funds to complete the segment of trail.[120]

In October 2010, the Zim Smith trail was completed.[121] In 2012 the trail became one of 54 in the United States to be designated a national recreation trail, and the only trail in New York State to bear this designation.[122]

### 5.7.5.2 Connecting the Zim Smith and Luther Forest Trails

The completion of the Zim Smith Trail itself left unresolved the issue of how and when the trail would be connected to the Luther Forest trail system, a precondition set forth in the Malta and Stillwater zoning codes for AMD to begin operations. In December 2006, SEDC requested that the Town Boards of Malta and Stillwater modify the 2004 zoning approvals for the chip fab in Luther Forest to relieve SEDC and its subsidiary, LFTCEDC, from responsibility for establishing a paved connector trail from Luther Forest to the Zim Smith Trail and other connector trails outside the campus before AMD could begin operations. SEDC said that satisfying that requirement "could be delayed by engineering studies and other factors beyond the Campus' control, with the result that AMD's startup could be delayed."[123] With trail advocacy groups objecting to the request, the town boards retained the condition but provided that they might grant an exemption if trail work was delayed.[124]

---

[117] "Overpass Replacement will Help Trail," Schenectady, *The Daily Gazette* (February 20, 2008).

[118] Route 67 Crossing Set to be Replaced—Project to Make Room for Trail," Schenectady, *The Daily Gazette* (February 20, 2008).

[119] "Zim Smith Trail Work to Finish With Fed Aid," Schenectady, *The Daily Gazette* (May 14, 2009).

[120] "Stimulus Won't Fund All Trail Work," Schenectady, *The Daily Gazette* (July 15, 2009).

[121] "At Long Last, Zim Smith Trail is Done," Schenectady, *The Daily Gazette* (October 16, 2010).

[122] "Malta's Zim Smith Named National Recreation Trail," Troy, *The Record* (June 3, 2012).

[123] "Town Closer to Approving Zoning for New Chip Plant," Schenectady, *The Daily Gazette* (December 29, 2006).

[124] "SEDC Seeks Break on Trial Requirement," Schenectady, *The Daily Gazette* (December 14, 2006); "Town Closer to Approving Zoning for New Chip Plant," Schenectady, *The Daily Gazette* (December 29, 2006).

In 2007 Saratoga PLAN received a $100,000 legislative grant, a "member item from Senate Majority Leader Joseph L. Bruno," for planning recreational trail connections in the Ballston Creek Valley, "the first step toward linking the Zim Smith Trail and the trolley line over to the tech park area."[125] The Town of Malta was able to use state transportation grant money left over from building roads in the Luther Forest Technology Campus to finance the remainder of the cost of establishing a 1.3-mile paved connector trail linking the Zim Smith and Luther Forest Trail System.[126] However, the proposed trail ran across a 21-acre parcel of land owned by Clifton Park resident Ronald Wayne Van Patten who refused to negotiate an easement with the town. Malta commenced eminent domain litigation against Van Patten in 2011, and in June 2012 the state Supreme Court validated the Town's eminent domain petition. The ruling gave the Town of Malta the authority to construct the paved connector link, establishing access to the GlobalFoundries site for pedestrian and bicycle commuters.[127]

## 5.8 Electric Power

Semiconductor fabrication plants require an adequate and extremely reliable supply of electric power. AMD submissions to the Town of Malta indicated that its chip fab would need about 40 megawatts of electricity (the average usage of 12,000–35,000 homes), and if three fabs were built at the site, 120 or more megawatts would be needed.[128] Reliability was crucial given that a chip fab at that point in time could not be without power for an average of more than 100 ms or the manufacturer would suffer major financial losses.[129]

---

[125] "Grant Boosts Efforts to Connect Trails," Schenectady, *The Daily Gazette* (April 6, 2007).

[126] The proposed connector trail would meet the Zim Smith Trail near the point at which it passed under I-87, use I-87's eastern embankment to cross Ballston Creek, then run along the former trolley line owned by Saratoga PLAN and under the Round Lake Bypass bridge. From there it would proceed uphill to Route 9 (across land owned by Van Patterson) and on to connect with the Luther Forest Trail system. "Zim Smith, Tech Campus Trail Contract Near," Schenectady, *The Daily Gazette* (August 1, 2012).

[127] "Court to Decide Whether Zim Smith Trail Can be Connected to Recreational Trail Network in Luther Forest Technology Campus," Saratoga Springs, *The Saratogian* (May 23, 2012); "With Eminent Domain, Malta Obtains Key Property that Will Connect Zim Smith Trail to Luther Forest Trail System," Saratoga Springs, *The Saratogian* (June 18, 2012).

[128] "Utility May Add Second Power Line," Schenectady, *The Daily Gazette* (March 25, 2008).

[129] "Chip Fab a Chance to Cash in—AMD Could Generate Big Business for Companies Like National Grid, Others in Region," Albany, *The Times Union* (July 21, 2006). In Austin Texas, Freescale Semiconductor Inc., one of the largest local chip manufacturers, reportedly lost $20,000,000 as a result of four electric power outages between 2002 and 2006. Freescale "stopped short of threatening to leave the city because of the problems" but "the problems caused a major stir in Austin, which depends heavily on the semiconductor industry for jobs and economic development." See "A Grid for the 21st Century," Albany, *The Times Union* (October 22, 2006).

## 5.8.1 New Power Infrastructure: Regulatory Hurdles

SEDC's State Environmental Quality Review Act (SEQR) environmental impact statement submitted to the Malta and Stillwater town boards in 2004 during the pre-permitting process called for creation of an electric power infrastructure to support the construction and operation of a chip fab plant at the Luther Forest site.

- To support the first phase of fab construction, the plan called for establishment of a temporary electric service from National Grid's Malta substation via a 2.5-mile 13.2-kV express distribution circuit which would run along existing power poles and highway right-of-way.
- To support the fab's eventual operation, the plan called for creation of a new 115 kV substation to serve the Luther Forest Technology Campus, including capacitor banks (clusters of energy storage devices), in order to provide adequate voltage performance to 115 kV transmission lines connecting the new substation to National Grid's existing transmission system. The substation would continue to ensure electricity flow to the fab even if the line to the LFTC was lost.
- Two extensions of 115 kV double circuit lines into the campus were proposed: 1) a 2.5-mile extension to National Grid's Malta substation, crossing Route 9, and 2) a 5.9-mile line connecting the campus with New York State Electric & Gas Corp's Mulberry substation in Stillwater. (The electricity itself was to be supplied by National Grid from its system across the Hudson but run over NYSEGC's grid to Luther Forest.) The plan called for creation of 500-foot wide corridors along the routes of the 115 kV lines.[130]

During the pre-permitting review of the Luther Forest proposal by the two Town Boards, SEDC argued that the power lines to be built into the LFTC should be overhead in order to keep costs down and improve reliability. Some residents objected to the anticipated visual appearance of the lines. The issue was resolved in an agreement pursuant to which SEDC would pay the town $1.5 million to allow the overhead lines. SEDC subsequently changed its stance, asking that the downtown segment of the lines go underground because of the high cost associated with acquiring land along the downtown corridor as real estate prices rose. SEDC and Malta agreed that the downtown lines would go through a concrete underground duct bank for 1.1 miles, in return for which the mitigation payment SEDC was to make to the town would be reduced to $975,000.[131]

---

[130] *Luther Forest Technology Campus GEIS: Statement of Findings*, Draft adopted by Stillwater Town Board (June 14, 2004), pp. 9–10. John S. Munsey, "Project Case Study: High Tech Land Development," *Civil and Structural Engineer* (May 2006).

[131] "Luther Forest Tech Park Plans Put Power Lines Underground," Schenectady, *The Daily Gazette* (October 3, 2007).

## 5.8.2 New Power Infrastructure: Construction

Construction of the substation and the new electric transmission lines began in 2009. The 115 kV transmission line connecting LFTC to the Mulberry Substation was energized in September 2010. The line from LFTC to the Malta Substation was energized in October 2010.[132]

In addition to establishing the necessary power connections in the immediate vicinity of Luther Forest, National Grid undertook a very substantial effort to upgrade electrical transmission systems across Upstate New York, the Northeast Region Reinforcement Strategy. These upgrades were based on anticipated regional growth as well as the needs of the AMD/GlobalFoundries' fab or fabs. In 2009, National Grid told the state Public Service Commission that "without improvements to the northeast electrical system, the development of the [Luther Forest] campus could be jeopardized along with the economic benefits to customers in the region."[133] The utility disclosed plans in 2009 to replace 115 kV lines between Ballston and Saratoga Springs with higher-capacity, more efficient 115 kV lines "to improve system reliability and prepare for the future power demands at the GlobalFoundries computer chip plant in Malta."[134]

In addition to the local improvements, National Grid prepared a $66 million project to upgrade the regional power supply by installing a new 33-mile 115 kV line running from the Spier Falls hydropower plant in Moreau south through Saratoga County alongside an existing line to Rotterdam in Schenectady County.[135] The new line would provide "additional reliability and service to approximately 45,000 commercial and residential customers."[136] A GlobalFoundries representative commented that from the company's perspective, this project was "absolutely necessary."[137] In February 2011, the New York Public Service Commission approved National Grid's

---

[132] Empire State Development, *National Grid—Luther Forest Infrastructure Capital II—Upstate City-by-City (x044)* (December 14, 2011).

[133] "Powering Up for the Future," Albany, *The Times Union* (July 2, 2009). National Grid said the planned upgrades were "tied to both general residential and commercial growth and plans for Advanced Micro Devices to build a computer chip factory in Malta that could use as much power as a small city." Utility May Add Second Power Line," Schenectady, *The Daily Gazette* (March 15, 2008).

[134] "Power Line Improvement Set for Reliability, Added Service," Schenectady, *The Daily Gazette* (May 28, 2009).

[135] "Utility Plans 33-Mile Line," Albany, *The Times Union* (February 19, 2010).

[136] National Grid, *Spier Falls to Rotterdam 115 kV Transmission Line Project,* http://www9.nationalgridus.com/transmission/spier_rotterdam.asp.

[137] "Neighbors Unhappy With Plans for New Power Line," Schenectady, *The Daily Gazette* (July 23, 2010). A National Grid manager indicated that Saratoga County had the fastest growing electricity demand anywhere in National Grid's service area, even before GlobalFoundries was operating. He warned that "the existing system performs marginally at times of high demand" and that "problems will become worse unless improvements are made." Ibid.

proposed transmission line from Spier Falls to Rotterdam.[138] Construction began in November 2011 with an 18-month completion horizon.[139]

Since 2007 National Grid has been engaged in "Connect 21," an initiative to invest $3 billion in the Upstate New York electric grid. As part of this effort, in 2015 it opened its $50 million Eastover substation in Speigletown, New York, 15 miles from Malta, to prevent overloading of nearby stations and to increase reliability in the region, including the site of GlobalFoundries' fab. According to a National Grid spokesperson, the new substation featured "some of the latest utility technology," would "increase our ability to detect outages and limit the number of customers out of power," and "adds resiliency to the system, providing for back-up power sources in case of an interruption."[140] Mike Russo, speaking for GlobalFoundries, welcomed the new substation, commenting that "We use a lot of power. We use quality power. We need to have a resilient system."[141] (See Box 5.2 for a discussion of National Grid's support for regional economic development.)

---

**Box 5.2: National Grid Support for Regional Economic Development**
The driving force behind the establishment of the necessary electric power capability was a major private utility, National Grid, which was committed to regional economic development that would increase demand for electricity. In the 1990s, National Grid's corporate predecessor, the Niagara Mohawk Power Corporation, was a private utility serving Upstate New York. It agreed to divest its power generation business in 2000 and became exclusively a provider of power transmission and distribution services. During the 1990s, Niagara Mohawk collaborated with the Saratoga Economic Development Corporation (SEDC), Empire State Development, and the Center for Economic Growth (CEG) on efforts to attract a chip fab to Upstate New York. In 2002, the United Kingdom's National Grid Group, which held a monopoly on power transmission in England and Wales, acquired Niagara Mohawk, which operated thereafter as "Niagara Mohawk, a National Grid company."[142]

---

[138] "Controversial Power Line to Service GlobalFoundries Approved," Glens Falls, *The Post-Star* (February 17, 2011).

[139] "National Grid Begins Construction on 115-kV Line in New York," *Transmission Hub* (November 2, 2011).

[140] National Grid, "New Eastover Sub Station Provider Base for Continued Growth in New York Capital Region," Press Release (June 4, 2015).

[141] "National Grid Power Boost," Albany, *The Times Union* (June 4, 2015).

[142] The National Grid Co. was created in 1990 when the United Kingdom broke up and privatized its state-run electric power sector. National Grid took over the high-voltage transmission system. National Grid Co. went public in 1995 as the National Grid Group. "NIMO, National Grid to Merge at End of Month," Syracuse, *The Post Standard* (January 17, 2002).

> Following the acquisition, Niagara Mohawk/National Grid remained an integral part of the New York State economic development team marketing the state as a destination for high-tech manufacturing investment. Michael King, a National Grid executive, observed in 2006 that "Luther Forest Technology Campus is a one-of-a-kind project that will represent one of the largest electrical loads in New York State."[143] National Grid invested $1 million in promoting the state to the semiconductor industry and contributed $750,000 to SEDC for site preparation at Luther Forest, including engineering, aerial mapping, and environmental and vibration studies.[144] The power industry journal, *Electric Light and Power*, recalled that
> 
> *National Grid and the "NY Loves Nanotech" team marketed the region's assets at industry trade shows like SEMICON West and SEMICON Europe, and the annual Semiconductor Industry Association meeting in Silicon Valley. In addition to hitting the road with its pitch to industry, the team hosted industry leaders at the annual Albany Symposium and Global Business Issues in Semiconductors and Nanotechnology, held at Lake George each September.*[145]

## 5.9 Natural Gas

The GlobalFoundries fab would require an estimated 465,000 cubic feet of natural gas per hour at 20 pounds per square inch. A second fab would more than double the required gas delivery rate to about 994,000 cubic feet per hour. As of 2011, these forecast rates of demand were "well beyond the capacity of National Grid's medium-pressure gas system … serving the GlobalFoundries site." In 2011, National Grid submitted a proposal to the State Public Service Commission to build a 4-mile-long, 12-in. pipeline connecting an existing National Grid line in Ballston to a new gas regulator station to be located in Malta.[146] Preparation of the application was funded by a $1.4 million grant from ESD.[147] Construction of the $10 million line was to be paid for by GlobalFoundries and the State of New York; other National Grid customers would "not have to help foot the bill." National Grid's project manager, Ed

---

[143] John S. Munsey, "Project Case Study: High Tech Land Development," *Civil and Structural Engineer* (May 2006).

[144] "Chip Fab a Chance to Cash in … AMD could Generate Big Business for Companies Like National Grid, Others in Region," Albany, *The Times Union* (July 21, 2006).

[145] "Utilities Can Play a Role in Attracting High-Tech Industry," *Electric Light & Power* (November 1, 2006).

[146] National Grid, *GlobalFoundries: New Gas Transmission Line,* (Article VII Application for a Certificate of Environmental Compatibility and Public Need, August 2011), p. 2.

[147] "State Redirects $1.4 M to National Grid for Luther Forest Work," Glens Falls, *The Post-Star* (December 16, 2010).

Wencis, said that

> The basic driver for this project is to support GlobalFoundries' energy needs. To meet their requirements, we have to build this infrastructure. This will be supplying Global as well as the Luther Forest Technology Campus.[148]

In June 2012 the PSC approved construction of a gas pipeline which would be capable of delivering 994,000 cubic feet of gas per hour to GlobalFoundries.[149] The pipeline and an associated regulator station were constructed in 2013 by Feeney Brothers of Dorchester, Massachusetts, an effort which "presented many challenges as it varied from wet running sand to shale to consolidated rock."[150]

## Bibliography[151]

Bypass Project Fast Tracked. (2010, January). *Professional Surveyor Magazine*.
Empire State Development. (2001, December 14). *National Grid—Luther Forest Infrastructure Capital II—Upstate City-by-City (x044)*.
Empire State Department Corporation. (2010, March 25). *LFTCEDC—Luther Forest Infrastructure Capital II—NYSEDP and Update City-by-City (x043, x044)*.
Luther Forest Technology Campus GEIS: Statement of Findings. (2004, June 14). Draft adopted by Stillwater Town Board.
Munsey, J. S. (2006, May). Project Case Study: High Tech Land Development. *Civil & Structural Engineer*.
National Grid. Spier Falls to Rotterdam 115kV Transmission Line Project. http://www9.national-gridus.com/transmission/spier_rotterdam.asp.
National Grid. (2011, August). GlobalFoundries: New Gas Transmission Line. Article VII Application for a Certificate of Environmental Compatibility and Public Need.
National Grid. (2015, June 4). New Eastover Sub Station Provider Base for Continued Growth in New York Capital Region. Press Release.
National Grid Begins Construction on 115-kV Line in New York. (2011, November 2). *Transmission Hub*.
New York Department of Transportation. *Round Lake Bypass Project*. Final Environmental Impact Statement.
NIMO, National Grid to Merge at End of Month. (2002, January 17). Syracuse, *The Post Standard*.
Town of Stillwater Code. Art. XI § 211–184.
Utilities Can Play a Role in Attracting High-Tech Industry. (2006, November 1). *Electric Light & Power*.

---

[148] "Natural Gas Pipeline to Service GlobalFoundries," Saratoga Springs, *The Saratogian* (January 10, 2012).

[149] "PSC Approves Gas Line for Chip Fab," Albany, *The Times Union* (June 16, 2012).

[150] http://feeneybrothers.com/projects/national-grid-3/.

[151] As noted in the front matter of this book, the study also drew on interviews carried out by the authors and numerous articles from *The Times Union* (Albany), *The Daily Gazette* (Schenectady), the *Albany Business Review* (Albany), *The Post-Star* (Glens Falls), *The Record* (Troy), *The Saratogian* (Saratoga Springs), *The Buffalo News* (Buffalo), The *Observer-Dispatch* (Utica), The *Daily Messenger* (Canandaigua), and the *Post-Standard* (Syracuse). These are not individually included in the bibliography.

# Chapter 6
# The Launch of GlobalFoundries

**Abstract** The construction of the GlobalFoundries wafer fabrication plant in Luther Forest was one of the largest building projects ever undertaken in the United States. The original plans were revised on a number of occasions to expand their scope. Construction was completed on time, notwithstanding numerous challenges, and was characterized by labor peace and the amicable resolution of disputes with local governmental units and residents.

GlobalFoundries estimated that construction of its first fab—now dubbed "Fab 8" rather than "Fab 2"—would require nearly as much human effort as the construction of the 102-story Empire State Building, with the work forecast to take 3 years and require five million man-hours of labor (versus seven million for the construction of the skyscraper). The fab would incorporate enough cement to build a 4-lane highway 11 miles long and would use 75 miles of pipe. Earth-moving of 1.1 million cubic yards of dirt would be required which, if loaded into a string of dump trucks lined up bumper-to-bumper, would stretch from New York City to just short of Montreal, 325 miles.[1] These figures increased as the original plan for the fab expanded. The construction of the fab was noteworthy not only for its scale but also the complexity and sophistication of the engineering challenges involved. In a 2010 interview, Rick Whitney, President of U.S. Operations for M+W Group, which oversaw the construction, summarized what was involved in the following way—

> In consideration of the sensitivity of the manufacturing process, wafer fabrication plants are constructed to the highest quality standards possible. Special structural systems are put in place that limit any potential vibration. The use of chemicals and gases in the production of the wafers requires the air supply, process abatement and ventilation to be very well-tuned. Application-specific fire protection and safety systems must be fail-safe since there is the utmost concern for human safety and protecting the sophisticated manufacturing equipment whose cost far exceeds the cost of building construction.

---

[1] "Dirty Work Begins at Site to Level Miles of Earth," Saratoga Springs, *The Saratogian* (July 31, 2009); "Construction of Chip Plan Seen as Monumental Effort," Schenectady, *The Daily Gazette* (June 16, 2009).

**Table 6.1** GlobalFoundries investment at the Malta site

| Year | Amount of investment (dollars) | | | |
|---|---|---|---|---|
| | Construction | Tooling | Other | Total |
| 2010 | 436,843,709 | 211,703 | 0 | 437,055,412 |
| 2011 | 545,283,316 | 1,738,634,978 | 10,579,117 | 2,294,497,411 |
| 2012 | 160,679,616 | 1,373,292,364 | 65,480,891 | 1,599,452,871 |
| 2013 | 458,716,236 | 484,154,939 | 75,079,382 | 1,017,950,557 |
| 2014 | 1,157,840,835 | 3,745,917,003 | 85,084,824 | 4,988,842,662 |
| 2015 | 618,065,232 | 2,299,898,617 | 89,722,813 | 3,007,686,662 |
| 2016 | 943,588,331 | 303,457,590 | 96,651,394 | 1,343,697,315 |
| 2017 | 169,453,336 | 894,097,628 | 33,664,428 | 1,097,215,391 |
| Total | 4,490,470,611 | 10,839,664,822 | 456,262,849 | 15,786,398,281 |

Source: Business Annual Reports (BAR) submitted by GlobalFoundries to the State of New York

> Layered on top is the need for flexibility to accommodate ever-changing wafer manufacturing technology. During the useful life of a fab, especially one that serves as a foundry, tools are constantly refurbished, upgraded and replaced by new tools to accommodate new product requirements.[2]

Between 2010 and 2017, GlobalFoundries invested $15,786 billion at the Malta site (see Table 6.1). The principal source of these outlays was the Mubadala Investment Company PJSC, the sovereign wealth fund of Abu Dhabi.

## 6.1 The Choice of M+W Group

The engineering firm M+W US Inc., based in Dallas, was chosen to oversee the construction of the fab. M+W US was part of M+W Group (formerly known as M+W Zander), part of Austria's Stumpf Group, a real estate and industrial conglomerate. M+W had previously built semiconductor fabs around the world, including IBM's fab in East Fishkill, the SUNY Albany NanoTech complex, and AMD's two fabs in Dresden which were spun off to GlobalFoundries.[3] In February 2010, M+W US disclosed that it was moving its North American headquarters from Dallas to Watervliet, NY, a move that would bring 250 additional jobs to the state.[4] M+W's Alan Asadoorian, who oversaw the construction of the GlobalFoundries fab, explained the company's decision to relocate:

> Albany is known worldwide as a leader in the (development) of nanotech and nanoscience, as evidenced by the companies who have centered here—it is the center of the universe as it is now in nanotechnology.[5]

---

[2] "GlobalFoundries' Lead Construction Contractor Gives a Progress Report," Schenectady, *The Daily Gazette* (February 21, 2010).

[3] "More Chips on the Table" Albany, *The Times Union* (February 9, 2010).

[4] "On the Heels of High-Tech: M+W Group Moving Headquarters to Watervliet Arsenal," Saratoga Springs, *The Saratogian* (February 10, 2010).

[5] "GlobalFoundries Ripple Effect Felt," Troy, *The Record* (March 10, 2010).

**Table 6.2** Local firms involved in early-stage fab construction

| Firm | Location of headquarters | Task |
|---|---|---|
| Christian Steel | Guilderland, New York | Erect steel for main fab building |
| MLB | Malta, New York | Concrete foundations |
| Delaney Group | Mayfield, New York | Land clearance |
| Jersen Construction | Waterford, New York | Foundations, central utility building |
| BCI Construction Services | Albany, New York | Construction of on-site office building |
| Bonded Concrete | Watervliet, New York | Concrete batch plant |
| Stone Bridge Iron & Steel | Gansevoort, New York | Supply steel for central utility building |

Source: "Chip Fab Rises in Luther Forest," Troy, *The Record* (December 13, 2009)

In mid-2010, GlobalFoundries awarded M+W Group the contract to install over $3 billion in semiconductor manufacturing equipment at the site.[6]

## 6.2 Involvement of Local Firms

By September 2009, 150 acres of land had been cleared at the fab site and concrete-pouring began.[7] A reporter who visited the site in October commented that hundreds of acres of forest had been "bulldozed as flat as you please" and was "bristling with steel pilings, fortified with concrete walls, rumbling with oversize trucks."[8] Much of this work was undertaken by local Capital Region firms and workers (see Table 6.2). A spokesman for M+W said that "with so many qualified workers in the area, there was no reason to import."[9] In addition to the local subcontractors directly engaged in the construction itself, other local businesses provided support, in some cases, through temporary premises established at or near the site.[10]

---

[6] "Tool Award at Chip Factory," Albany, *The Times Daily Union* (July 1, 2010).

[7] "Chip Plant Work Going Well," Schenectady, *The Daily Gazette* (September 2, 2009).

[8] "Factory Site No Longer Just a Forest," Schenectady, *The Daily Gazette* (October 25, 2009).

[9] "Chip Fab Rises in Luther Forest," Troy, *The Record* (December 13, 2009).

[10] United & Taylor Welding Supply, a local supplier of welding, safety, and general construction supplies, opened a store (in a trailer) inside Luther Forest and did a brick business supplying construction firms at the site. Schenectady restauranteur Angelo Mazzone was engaged to cater food to construction workers in a large tent on the site. "Chip Fab Work Striking In Its Scope, Complexity," Schenectady, *The Daily Gazette* (February 21, 2010); "Welding Supply Meets Demand," Saratoga Springs, *The Saratogian* (March 11, 2010).

## 6.3 Expansion of Original Construction Plan

In March 2010, GlobalFoundries applied to the town of Malta to permit expansion of the original manufacturing clean room from 210,000 square feet to 300,000 square feet, adding roughly $2 billion in additional investment, for a total investment of over $6 billion. GlobalFoundries indicated that it foresaw the potential need for additional capacity to serve customers arising out of its merger with Singapore's Chartered Semiconductor, also a foundry. The addition could be built within the original timetable for the fab.[11]

Malta's Planning Board approved GlobalFoundries' proposed expansion in April 2010. GlobalFoundries indicated that there was no real need for an extensive environmental review given the thorough environmental studies conducted in 2004 and 2008. The Planning Board concurred, and no other approvals were required because the expanded fab would remain within the size limits established by the Town's zoning legislation.[12] The Saratoga Water Authority and National Grid indicated that they would have "no problem" supplying the additional water and power required to support the expanded fab.[13]

Then in February 2011, GlobalFoundries disclosed plans for a new 221,000 square foot administrative building next to the chip fab.[14] The new facility, Admin 2, would initially house 450 workers supporting the first fab, but plans the company submitted to the Town of Malta indicated that the building was designed to be "the administrative hub of a second factory known as Module 2."[15] The Malta Planning Board quickly approved the new building, the only approval required before construction could start.[16] Planning and Development Director Tony Tozzi indicated that the first approval reflected the fact that "GlobalFoundries came in with a very well-prepared application and worked day-to-day with town staff and consultants to address all the issues the town would have."[17]

---

[11] "GlobalFoundries Seeking to Expand," Schenectady, *The Daily Gazette* (March 29, 2010).

[12] The expansion plan called for the additional square footage at the manufacturing plant and the adjoining central utility building as well as expansion of the on-site electrical substation. "Board Oks GlobalFoundries Expansion," Schenectady, *The Daily Gazette* (April 21, 2010); "Plan to Enlarge Chip Plan Headed for Fast Approval," Schenectady, *The Daily Gazette* (April 8, 2010).

[13] "Bigger Chip Fab Called No Problem," Albany, *The Times Union* (May 4, 2010).

[14] "Chip Fab Office, Jobs on Way," Albany, *The Times Union* (February 19, 2011).

[15] "GlobalFoundries Builds With Future in Mind," Albany, *The Times Union* (February 24, 2011).

[16] "GlobalFoundries Building Plan Gets OK," Albany, *The Times Union* (March 16, 2011).

[17] "Second Major Office Building OK'd for GlobalFoundries," Schenectady, *The Daily Gazette* (March 17, 2011).

## 6.3 Expansion of Original Construction Plan

> **Box 6.1 Good Relations with the Construction Trades**
>
> A 2014 article in *New York* observed that "tech companies and labor union have never been friends" and that "tech's executive class was opposed to unions from the beginning."[18] Robert Noyce, the co-founder of Intel and one of the inventors of the integrated circuit, once said that "remaining non-union is essential for survival for most of our companies."[19] New York has been described as "the most unionized state in the country."[20] The construction trades are powerful, and a work stoppage involving even one has the potential to shut down a project as the other trades halt work in solidarity.[21] Accordingly, it is noteworthy that the construction of the GlobalFoundries fab, one of the biggest projects in the history of the state, took place without any work stoppages.
>
> GlobalFoundries reached a comprehensive agreement with the construction unions in 2009 which established the foundation for a productive partnership during the construction of the fab. GlobalFoundries was represented in these talks by Michael Russo, the former Capital District Director for then-Congresswoman Kirsten Gillibrand; Russo had previously held leadership posts in organized labor with the glass-molders union.[22] In 2009, GlobalFoundries, general contractor M+W Zander, and an organization representing local construction unions concluded a Project Labor Agreement (PLA), facilitated by Russo, pursuant to which the companies agreed to pay union-scale wages to union and nonunion workers employed at the site, and the unions committed not to strike or "engage in other actions that would impede construction."[23] This deal, which has grown to be the largest private

---

[18] "Silicon Valley, Meet Organized Labor," *New York* (October 7, 2014).

[19] David Bacon, *Organizing Silicon Valley's High Tech Workers,* http://dbacon.igc.org/Unions/04hitec2.htm.

[20] "How Did New York Become the Most Unionized State in the Country?" *The Nation* (September 3, 2014).

[21] "Crane Operators Threaten Strike at Many Construction Sites," *The New York Times* (June 30, 2006).

[22] "GlobalFoundries Appoints Russo to Gov't Relations Team," Schenectady, *The Daily Gazette* (May 12, 2009).

[23] "GlobalFoundries, Unions Strike Labor Deal," Glen Falls, The Post-Star (June 4, 2009).

PLA in history, held up throughout the construction, and other factors helped cement good relations. The company set up a cafeteria for the workers run by Capital Region restaurant magnate Angelo Mazzone, which served inexpensive meals that were highly popular.[24] Russo, who subsequently became GlobalFoundries' Director of Government Affairs, summarized the company's relationship with the building trades in 2013—

> The trades have been very progressive. We've laid our cards on the table and talked about how we can improve training. We're working with them to develop curricula so their workers are ready to work in the fab environment, so they know what a clean space is. It's a totally different animal, building these large fabs. We have to make sure the labor is available when we need it. For a fab, that can mean thousands of workers right away.[25]

## 6.4 Construction of the Building

During the winter of 2009–2010, the steel shell of the manufacturing building was erected. In addition, 150-foot long, 47-ton steel roof trusses and concrete slabs ("waffle tables") that would underlay the manufacturing floor were lifted and installed. Town officials responsible for building code inspections during the construction process indicated they were surprised at how smoothly the project was going. Malta Building and Planning Director Anthony Tozzi told the Town Board that GlobalFoundries had been "very responsive."[26]

By October 2010, 1500 people were working at the chip fab site. Connections to the Saratoga County water and sewer systems were established, and in the 139,000 square foot utility building adjacent to the fab, hundreds of feet of pipe and microfilter for deionization and purification of Saratoga County water were installed. Werner Greyling, M+W's project manager, said that "it's the biggest ultra-pure water system I've ever seen." In the fab, work was proceeding on a clean room the size of three football fields where roughly 600 production tools would be installed. A 15-ton elevator was being built to lift tools from ground level to the second floor, where the clean room was to be located.[27]

---

[24] Gary Moon of Queensbury, general foreman of the pipefitters at the site, said in 2011, "Let me tell you, in 44 years in the business, I've never eaten like this. I've worked all over the place, and wow, you never get food anywhere near this good." See "Never Mind the Lunchbox," Albany, *The Times Union* (February 16, 2011).

[25] National Research Council, Charles W. Wessner (rapporteur), *New York's Nanotechnology Model: Building the Innovation Economy* (Washington, DC: The National Academies Press, 2013), p. 88.

[26] "Chip Fab Work Striking In Its Scope, Complexity," Schenectady, *The Daily Gazette* (February 21, 2010). GlobalFoundries provided Malta with $1.5 million to hire experts to inspect the construction. Malta contracted with two local engineering firms, Chazen Cos. of Troy and Evergreen Engineering of Albany, to assist the town in reviewing site plans and codes and inspecting the plant itself. "GlobalFoundries Plant Generates Demand for Site Inspectors," Albany, *Business Review* (March 18, 2010).

[27] "Work Coming Along Inside GlobalFoundries," Schenectady, *The Daily Gazette* (October 22, 2010).

## 6.5 Tool Installation

Installation of manufacturing systems in the fab began in the summer of 2011, and the company began moving employees into Fab 8 to begin manufacturing tool testing. Gas, acid, and chemical systems used in the manufacturing process were hooked up. Tool installation was an 18-month process, with machines arriving from elsewhere in the United States as well as Japan, the Netherlands, Israel, and Germany. In January 2012, GlobalFoundries indicated that although tool installation was still under way, it had begun processing its first wafers, 32-nm chips for its customer IBM.[28] The company indicated that most of 2012 would be spent testing and adjusting production process to eliminate defects, with full-scale production expected to start thereafter.[29]

GlobalFoundries terminated its contract with M+W for the installation of tools in the fab in mid-2012. By this point hundreds of tools had been put in place and GlobalFoundries' workforce had grown to 1200 people. GlobalFoundries indicated that going forward it could oversee tool installation and subcontractors itself from that point forward. A company spokesman said that the move reflected "the natural evolution of the project. Basically what we're doing is accelerating our ability to perform the hook-up work directly with the subs…. [O]ur team has grown to a point where we can do things on our own. Two years ago, we didn't have any people to do hook-up."[30]

## 6.6 Further Expansion

In December 2012, construction began on the 90,000 square foot expansion of the manufacturing clean room at Fab 8. Turner Construction, a national firm with an Albany office, was chosen as the general contractor. Space for this expansion had been included at the back of the Fab 8 building when the basic shell was being built in 2010–2011, but the internal space had been left empty.[31]

### 6.6.1 Proposed Second Fab

Prior to this, in June 2011, GlobalFoundries presented New York State officials with a proposal for a second wafer fabrication facility at the Luther Forest site, which would be "at least as large as the 300,000 square foot factory now under

---

[28] "Warehouse an Important Stop for Chip Plant Tools," Schenectady, *The Daily Gazette* (April 7, 2012); "Chip Plant's Laborer Set to Move In," Schenectady, *The Daily Gazette* (April 27, 2011); "No Slowdown Planned at Chip Factory," Albany, *The Times Union* (January 14, 2012).

[29] "GloFo to Make New IBM Chip," Schenectady, *The Daily Gazette* (January 10, 2012).

[30] "M+W Out of Job in Malta," Albany, *The Times Union* (March 3, 2012).

[31] "$2.3B in Work Starts at Fab 8," Schenectady, *The Daily Gazette* (December 22, 2012).

construction by the company." The company reportedly sought state incentives on a scale "something comparable to $1.4 billion in cash and tax breaks it received for the initial factory." According to one source Governor Cuomo was "playing hardball" with GlobalFoundries and "refusing the company's heavy demands for more cash."[32] GlobalFoundries indicated that it was looking at various potential sites around the world for its next fab and that it sought incentives from New York because "building a chip fab in New York costs $1 billion more than it does in Asia or other parts of the world."[33] In November 2011, GlobalFoundries disclosed that it would "not expand in Malta without more financial assistance from New York."[34]

### 6.6.2 A New Technology Development Center

The GlobalFoundries expansion proposal also called for construction of a research and development facility at the Luther Forest site.[35] In September 2012, the company submitted a site plan application amendment to the Malta Planning Board for a new facility, a 565,000 square foot Technology Development Center (TDC), housing new laboratories, manufacturing, research and development, and "some back-end components not limited to production."[36] The proposed research center would nearly double GlobalFoundries' workforce to about 3000.[37] The Board approved the proposal in October 2012. Malta town Supervisor Paul Sausville commented that GlobalFoundries "initially promised 1465 jobs, and this is twice that. It's all about jobs, stimulating the economy and creating jobs for our young people so they don't have to leave here."[38] The R&D center did not receive new state incentives, although the entire GlobalFoundries complex benefited from Empire Zone tax credits.[39]

The cost of the new R&D center was estimated at $2 billion.[40] The new center would have facilities to manufacture, analyze, and run prototype processes prior to full production at Fab 8.[41] A company spokesman explained that much of the

---

[32] "Cuomo Balks at Cash for Second Plant," Albany, *The Times Union* (June 4, 2011); "Talks for Added Site Cool," Albany, *The Times Union* (July 1, 2011).
[33] "Where Will the Chips Fall?" Albany, *The Times Union* (July 10, 2011).
[34] "GlobalFoundries Delays Fab Plans," Albany, *The Times Union* (November 12, 2011).
[35] "R&D Center Plan in Malta," Albany, *The Times Union* (July 10, 2011).
[36] "GlobalFoundries Announces it May Build an Additional Semiconductor Facility in Malta," Saratoga Springs, *The Saratogian* (September 25, 2012); "Chip Facility Set to Grow," Albany, *The Times Union* (September 26, 2012).
[37] "Board OKs GloFo Center Plan," Schenectady, *The Daily Gazette* (October 18, 2012).
[38] "Board OKs GloFo Center Plan," Schenectady, *The Daily Gazette* (October 18, 2012).
[39] "CEO Signals 'Fab 8.2,'" Schenectady, *The Daily Gazette* (January 11, 2013).
[40] "GlobalFoundries to Invest $2 Billion in New Malta Research and Development Facility," Saratoga Springs, *The Saratogian* (January 8, 2013).
[41] "Saratoga County Approves $387M Tax Break for GlobalFoundries Projects," Saratoga Springs, *The Saratogian* (March 18, 2013).

research GlobalFoundries conducted to push the boundaries of chip manufacturing was conducted with IBM at SUNY Albany's CNSE. However, moving from "lab to fab" is challenging, and testing new technologies at an operating fab "takes up precious space and equipment intended to make money for the company." The new center was intended to test what the company learned at CNSE and "perfect it before it is moved to the fab."[42] Specific themes for the new R&D center were "likely to include interconnect and packaging for the 3D stacking of chips, advanced mask-making and the use of extreme ultraviolet lithography."[43]

M+W Group was selected by GlobalFoundries as the primary contractor for the Technology Development Center.[44] Construction began in early 2013 on a site the size of two city blocks. The project included laying down a 12-foot thick concrete foundation using 200 truckloads of concrete per day.[45] The construction of the TDC was a project of sufficient magnitude to require other construction to expand Fab 8's gas yard and industrial utilities. By the winter of 2013–2014, 3500 construction workers were employed at the site working under the auspices of an expanded PLA, primarily working on the TDC, as well as 2100 permanent employees at Fab 8.[46]

## 6.7 Initial Operations

GlobalFoundries began test producing limited numbers of 12-inch wafers at the end of 2011, and throughout 2012 it continued to produce chips in small volumes. Fab 8's first customer was IBM, and by the end of 2012 it was serving "multiple customers."[47] Fab 8 initially produced devices using 32-nm design rules but was concurrently "working on perfecting making chips at the 20 nm and 14 nm level, technologies that will fuel the ongoing smart phone revolution and other popular mobile computing devices such as the iPad for the next 5 or 6 years."[48]

GlobalFoundries showed particular interest in technologies which could facilitate the creation of interconnected 3D chip stacks, widely seen as a promising response to the mounting challenges associated with further miniaturization of semiconductor line widths. In April 2012, the company disclosed that later in the year it would begin manufacturing a 20-nm-based chip for smartphones and consumer devices based on "through-silicon vias" (TSVs), vertical holes drilled through silicon and filled with copper to enable 3D stacking of chips, reducing power con-

---

[42] "A GlobalFoundries Growth Spurt," Albany, *The Times Union* (January 9, 2013).

[43] "GlobalFoundries Plans New York R&D Center," *EE Times* (January 9, 2013).

[44] "Firm to Build Research Center," Schenectady, *The Daily Gazette* (March 6, 2013).

[45] "Nano Tech Valley: Construction Keeps Rolling," Saratoga Springs, *The Saratogian* (June 23, 2013).

[46] "GlobalFoundries a True Growth Business," Schenectady, *The Daily Gazette* (February 16, 2014).

[47] "GloFo to Make New IBM Chip," Schenectady, *The Daily Gazette* (January 10, 2012).

[48] "Firm Betting on Chip Plant," Albany, *The Times Union* (December 21, 2012).

**Table 6.3** GlobalFoundries' Malta fab utilizes the company's cutting-edge technology nodes

| GlobalFoundries fab location | Technology nodes | Capacity (wafers/month) |
|---|---|---|
| Malta | 7 nm, 14 nm | 60,000 |
| East Fishkill | 90–22 nm | 20,000 |
| Burlington | 350–9 nm | 40,000 |
| Dresden | 28–12 nm | 80,000 |
| Singapore | 180–4 nm | 161,000 |
| Chengdu (forthcoming) | 180–130 nm | – |

Source: GlobalFoundries (May 2017)

sumption, and increasing memory capacity.[49] In 2014, GlobalFoundries unveiled a technology for reducing the "keep-out zones" (buffers) around TSVs, reducing the cost of using stacked devices.[50]

In April 2014, Samsung announced it would license its FinFET manufacturing technology to GlobalFoundries to enable the company to mass-produce 14-nm 3D devices for smartphones.[51] In mid-2015, GlobalFoundries disclosed that it had decided to "repurpose the new Technology Development Center to focus less on research, than on commercial manufacturing for a major customer." Although none of the companies involved would confirm it, the work was said to involve manufacture of 14-nm devices for Samsung pursuant to which that firm supplied chips for Apple smartphones.[52]

## 6.8 GlobalFoundries Acquisition of IBM Chipmaking Operations

In October 2014, GlobalFoundries and IBM disclosed a deal pursuant to which GlobalFoundries would acquire and operate IBM's semiconductor manufacturing facilities in East Fishkill, New York, and Essex Junction, Vermont. GlobalFoundries was to become IBM's exclusive supplier of 22-, 14- and 10-nm semiconductors for 10 years, thus guaranteeing it a long-term, continuous flow of revenue. In addition, IBM committed to pay GlobalFoundries $1.5 billion. GlobalFoundries committed to offer jobs to substantially all IBM employees at the East Fishkill site, enabling the company to fulfill its original commitment to the State of New York.[53]

---

[49] "Fab 8 to Build Milestone New Chip This Year," Schenectady, *The Daily Gazette* (April 27, 2012).

[50] "GloFo Shows Progress in 3D Stacks," *EE Times* (March 19, 2014).

[51] "Samsung Licenses 3D Chip Manufacturing Tech to GlobalFoundries to Win More Orders," *Reuters* (April 17, 2014).

[52] "GloFo Seeks Boost in Tax Relief," Schenectady, *The Daily Gazette* (July 21, 2015).

[53] "Town Officials Satisfied New Chip Plant Would be Quieter," Schenectady, *The Daily Gazette* (May 2, 2013).

IBM indicated that despite sale of its manufacturing units it would continue to invest $3 billion over 5 years in semiconductor technology research. GlobalFoundries would enjoy "primary access" to the results of this research.[54] GlobalFoundries gained access to IBM's portfolio of 10,000 U.S. patents and additional international patents. In addition, GlobalFoundries acquired know-how in specialized areas of semiconductor technologies with implications for growing markets. As a trade journal noted at the time—

> *IBM has some of the deepest semiconductor design, manufacturing and packaging expertise on the planet—broader and deeper in some areas than Intel—and plans to continue reaping benefits from that technology for at least the next decade. It has expertise in stacked die (both 2.5D and 3D ICs) as well as silicon photonics; microfluidics, silicon on insulator (SOI); silicon germanium (SiGe) and RF SOI*[55] *The deal also gave GlobalFoundries IBM's advanced process technology, including air gap technology, and IBM's advanced packaging technology for 3D ICs.*[56]

The IBM deal also gave GlobalFoundries access to IBM's "very large set of customers for their own technologies."[57] GlobalFoundries took over IBM's application-specific integrated circuit (ASIC) business, enabling it to create and produce custom ASICs and mass-produced ASICs for a wider market than that served by IBM. IBM's ASIC technology included intellectual property and design capability necessary to develop customized ASICs for wired communications access.[58] Charles Janac, president CEO of system-as-a-chip developer Arteris, commented in November 2014 that—

> *With the IBM chip assets, GlobalFoundries should be able to compete for some of the largest foundry deals out there. They become truly global. The challenge here is to make this much larger business work operationally. They have to make IBM Microelectronics profitable and they have to rationalize a complex portfolio of facilities, organizations and processes. Can they service the largest ASIC customers cost effectively and reliably at this new level of business? It's very promising but quite a challenge.*[59]

---

[54] Ed Sperling "An Inside Look at the GlobalFoundries—IBM Deal," *Semiconductor Engineering* (November 20, 2014).

[55] "GloFo to Get 2 IBM Facilities," Schenectady, *The Daily Gazette* (October 21, 2014); "GlobalFoundries: No Layoffs in IBM Chip Units," *Associated Press* (October 20, 2014).

[56] Ed Sperling "An Inside Look at the GlobalFoundries—IBM Deal," *Semiconductor Engineering* (November 20, 2014). RF SOI technology is used to make a range of key radio frequency (RF) devices, including switches and other components in smartphones and tablets. IBM, the world's largest source of RF SOI devices, could not keep pace with demand. GlobalFoundries has additional RF SOI capacity in Singapore that can be brought to bear on the smartphone market. "GF Closes on IBM Chip Business Purchase," *Semiconductor Engineering* (July 1, 2015).

[57] "GF Closes on IBM Chip Business Purchase," *Semiconductor Engineering* (July 1, 2015).

[58] Ed Sperling "An Inside Look at the GlobalFoundries—IBM Deal," *Semiconductor Engineering* (November 20, 2014).

[59] Ed Sperling "An Inside Look at the GlobalFoundries—IBM Deal," *Semiconductor Engineering* (November 20, 2014).

## 6.9 The Leading-Edge Foundry

The combination of GlobalFoundries and IBM's chipmaking technology and know-how has enabled Fab 8 to emerge as the most advanced semiconductor foundry in the world, the "leading and bleeding edge" of high-performance microelectronics manufacturing.[60] Fab 8 utilizes more advanced process technology (currently 14 nm, moving to 7 nm) than any other fab in the company's global operations (see Table 6.3). The company is adding capacity at the current 14-nm node and proceeding with aggressive plans for 7 nm and 5 nm.

GlobalFoundries' principal competitor in the foundry market, TSMC of Taiwan, holds over 50% of the world foundry market, and GlobalFoundries sees pursuit of cutting-edge process technologies as a way of differentiating itself from the market leader.[61]

In February 2017, GlobalFoundries announced that it would increase its manufacturing capacity at sites in Germany, New York, Singapore, and Chengdu, China. The company planned a 20% increase in capacity to produce 14-nm FinFET devices, with the investments to be completed by the end of 2018.[62]

In September 2016, GlobalFoundries senior vice president and manager of Fab 8, Tom Caulfield, announced that the company would invest "billions of dollars" to develop technology for 7-nm chips at its factory in Malta/Stillwater—in effect, skipping the 10-nm node and moving directly from 14 to 7 nm. A team of 700 employees was assigned this task, designed to increase chip performance by 30% and reduce production costs by 30%.[63] In June 2017—less than a year later—GlobalFoundries announced that its "7LP" 7-nm FinFET process was available to customer companies to begin building actual products based on the 7-nm process. While additional investments need to be made to scale production, GlobalFoundries expects that the first 7LP products will launch in mid-2018, with volume increasing rapidly during the latter half of the year.[64]

GlobalFoundries has also made significant strides in developing next generation 5-nm process technology. A "major breakthrough" in 5 nm was announced in June 2017 by a consortium which included SUNY Poly, GlobalFoundries, IBM, and Samsung. The partners had succeeded in fabricating 5-nm transistors using EUV

---

[60] Tom Patton, GlobalFoundries' chief technology officer, said in 2015 that advanced research at Fab 8 leveraged "people moving in from IBM, "Inside GlobalFoundries' Feb 8," *EE Times* (August 18, 2015).

[61] "Inside GlobalFoundries' Fab 8," *EE Times* (August 18, 2015).

[62] "GlobalFoundries Will Boost Capacity in New York, Germany, Singapore and China," *Albany Business Review* (February 10, 2017).

[63] "GlobalFoundries Will Invest Billions of Dollars to Develop Next Generation of Chips," *Albany Business Review* (September 15, 2016).

[64] "GlobalFoundries" Fires up its 7-nm Leading Performance Forges," *The Tech Report* (June 16, 2017). The 7LP process uses optical lithography but EUV is "compatible" if a user requires it. GlobalFoundries indicates that the 7-nm process offers a 40% device performance improvement over the 14-nm process and reduced power requirements of up to 60%. "GlobalFoundries on a Roll With 7 nm and 5 nm Announcements," *Forbes* (June 15, 2017).

lithography and "stacked nanosheets," a process IBM had been exploring for over a decade.[65] This development, featuring creation of a functioning 5-nm test chip, confounded predictions that nothing "under 7 nm would be possible or coming any time soon."[66]

However, in August 2018, GlobalFoundries announced a shift in strategy aimed at maximizing profitability. The company froze developmental work on 7-nm technology with an eye toward targeting high-volume end markets with 14-nm technology. The markets include the Internet of Things, automotive, data centers, and computer networks.[67]

## 6.10 Addressing Local Issues

The startup of construction at the GlobalFoundries site saw the emergence of a number of issues between the company and the surrounding towns and residents. Largely unanticipated during the planning phases, these issues, some of which were contentious, were resolved or moving toward resolution by the time GlobalFoundries began large-scale operations in 2013.

### 6.10.1 Dispute over Property Tax Assessments

GlobalFoundries had concluded a payment-in-lieu-of-taxes (PILOT) agreement with the Saratoga County Industrial Agency pursuant to which instead of paying local property taxes, it would make direct payments to the Ballston Spa Central School District (75%) and the Town of Stillwater and the Stillwater School District (25%) according to an agreed formula based on the assessed value of the property.[68] However, the methodology for assessing the value of that property became the subject of dispute.

In 2010, the Town of Malta proposed to assess the GlobalFoundries fab, then under construction, at a value of $160 million for local tax purposes, or about 10% of the value of all property in the town.[69] GlobalFoundries objected in a filing with the town assessor's office, arguing that the $160 million figure was based on investment expenditures to date, but that the actual market value was closer to $55 million.[70] A town review board rejected GlobalFoundries' plea in July 2010.[71]

---

[65] "Huge Breakthrough at SUNY Poly with 5 nm Chips," Albany *The Times Union* (June 5, 2017).
[66] "GlobalFoundries on a Roll with 7 nm and 5 nm Announcements," *Forbes* (June 13, 2017).
[67] "What's Next for GlobalFoundries After its Pivot," *Albany Business Review* (October 10, 2018).
[68] "Fab 2 Could Benefit Schools," Albany, *The Times Union* (August 4, 2009).
[69] "$160M Chip Plant Assessment Proposal," Schenectady, *The Daily Gazette* (May 1, 2010).
[70] "Chip Maker Fighting Tax Bill," Schenectady, *The Daily Gazette* (May 26, 2010).
[71] "Fab's Tax Bill Firm," Albany, *The Times Union* (July 2, 2010).

GlobalFoundries filed a lawsuit asking a state Supreme Court judge to intervene and set a lower assessment on the fab.[72]

In May 2011, with the litigation still pending, the Malta Assessor set a value of $400 million on the 70-percent completed fab, which would yield tax bills of over $8.4 million, most of which would go to the Ballston Spa School District.[73] GlobalFoundries again objected, taking position that the market value of the fab was only $210 million.[74] As in the prior year, GlobalFoundries filed a lawsuit asking a Supreme Court judge to lower the assessment.[75] The towns faced the prospect that if GlobalFoundries won, "the governments and the school district to which the plant pays taxes would lose out on a combined $6 million in revenue."[76]

In April 2012, following a year of intensive settlement negotiations, GlobalFoundries and five local taxing entities reached a deal on the assessment of the fab for property tax purposes:

- Tax assessments would be based solely on the value of the fab buildings, not the high-cost equipment inside.
- A long-term (47-year) assessment formula was reached based on the work of expert appraisers retained by the parties.
- GlobalFoundries would drop its Supreme Court legal challenges to the 2010 and 2011 assessments, eliminating the risk to the school district of a court-mandated refund.
- The fab would be assessed at a value of $635 million for 2012 (resulting in at least $13.5 million in tax revenue) and would gradually drop, reflecting depreciation, to $125 million after 15 years.
- The building's assessment would remain fixed at $125 million from 2027 to 2059.[77]

The parties professed satisfaction at the settlement. The Chairman of Saratoga County's Law and Finance Committee, Alan Grattidge, said that "I like the idea that this is a long-term settlement, so that this isn't a recurring problem."[78] Stillwater Town Supervisor Edward Kinowski commented that "if we look at it right now for squaring our taxes, it's a great Rx for Stillwater."[79] An attorney for the company said

---

[72] "Suit Filed Over Chip Plan Assessment," Schenectady, *The Daily Gazette* (July 30, 2010).

[73] "GlobalFoundries' Tentative Assessment is $400M," Schenectady, *The Daily Gazette* (May 3, 2011).

[74] "GlobalFoundries Challenging Value Assessment of its Malta Microchip Manufacturing Plant for a Second Time," Saratoga Springs, *The Saratogian* (May 25, 2011).

[75] "Malta Sued in Tax Dispute," Albany, *The Times Union* (August 10, 2011).

[76] "Tentative Deal Reached in GlobalFoundries Legal Battle Over Assessments," Saratoga Springs, *The Saratogian* (February 28, 2012).

[77] "Supervisors Settle on Assessment for GlobalFoundries," Schenectady, *The Daily Gazette* (April 18, 2012).

[78] "Chip Plant Tax Deal Nearly Ready," Schenectady, *The Daily Gazette* (April 5, 2012).

[79] "Supervisors Settle on Assessment for GlobalFoundries," Schenectady, *The Daily Gazette* (April 1, 2012).

6.10 Addressing Local Issues    177

that "this eliminates any future litigation and it allows everyone to move forward with certainty about the amount of revenue to be received." GlobalFoundries spokesperson, Travis Bullard said of the settlement that—

> We've always been prepared to pay our tax obligations based on a fair and reasonable assessment of the Fab 8 property. GlobalFoundries challenged the first two years of property tax assessments because we felt the assessments were not reasonable. But we've worked collaboratively for over a year with the towns of Malta and Stillwater and the Ballston Spa and Stillwater central school districts to reach this approved settlement of agreement.[80]

The settlement benefitted local property owners. In 2012, a local school official observed that—

> [T]he property taxes generated solely by the chip plant's $635 million assessed value will allow the [Ballston Spa Central School District] to actually lower the property tax rates for each of the other property owners this year.... Those tax revenues go to the county and towns.[81]

### 6.10.2 Sales Tax Exemptions

The original package of state incentives offered to AMD by the Saratoga County Industrial Development Agency included a waiver of $27.8 million in sales taxes for the $800 million construction of the buildings at the fab site. There was no arrangement, however, with respect to the manufacturing equipment to be installed, the estimated value of which was as much as $7 billion. In 2010, GlobalFoundries asked the New York Department of Taxation and Finance for a ruling on whether the tools were exempt from sales tax as manufacturing and R&D equipment, but securing a ruling, the company feared, "could take a year."[82] In the interim, GlobalFoundries asked the Saratoga County Industrial Development Agency, which, as a state Industrial Development Agency (IDA), held authority to grant sales tax exemptions, for roughly $100 million in sales tax exemptions so that GlobalFoundries "could place tax-free orders [for equipment] while they wait for answers from the state." The Saratoga County IDA approved $111 million in sales tax exemptions ($12 million for construction costs and $99 million for equipment costs).[83]

---

[80] "Saratoga IDA Approves GlobalFoundries Property Assessment Plan," Glens Falls, *The Post-Star* (May 15, 2012).

[81] "GlobalFoundries Chip Fab Plant Fosters a Ripple Effect Felt Far and Wide," Saratoga Springs, *The Saratogian* (July 24, 2012).

[82] "Chip Maker Seeks New Tax Breaks," Schenectady, *The Daily Gazette* (December 3, 2010).

[83] "Chip Maker Seeks New Tax Breaks," Schenectady, *The Daily Gazette* (December 3, 2010); "IDA Approves Another $111 Million in Benefits for Computer Chip Factory," Glens Falls, *The Post-Star* (December 21, 2010). State law provides that manufacturing equipment is exempt from sales tax, but the determination of which pieces of equipment are exempt are subject to a case-by-case review. "Chip Plant Exemptions Get OK," Schenectady, *The Daily Gazette* (June 14, 2011).

Then in May 2011, with the state's ruling still pending, GlobalFoundries asked the Saratoga County IDA for additional sales tax exemptions of $305 million, based on $5.7 billion in anticipated expenditures on equipment as well as $40 million for construction of the new administration building. The company pointed out that "most manufacturers in the county get such exemptions from the IDA if they add capacity."[84] In June 2011, the IDA approved sales tax exemptions valued at $405 million, most of which was an exemption from estimated sales taxes on $5.7 billion in semiconductor manufacturing equipment.[85] In March 2013, the IDA approved another roughly $387 million in sales tax exemptions in connection with GlobalFoundries' proposed Technology Development Center and Fab 8.2.[86]

### 6.10.3 Noise Abatement

The site for Fab 8 was located about a quarter mile through woods from the nearest homes in the Luther Forest housing development. GlobalFoundries committed to adhere to noise limits set forth in the town zoning approvals, and a company spokesman said "hopefully people around here won't even know that a 1.3-million-square-foot factory is being built."[87] To contain noise at the site, M+W oversaw construction of a 40-foot berm.[88]

Noise from the operation of the fab itself was not expected to be a problem when the building was being designed, so the construction did not utilize acoustic absorption materials.[89] However, in July 2011, with construction underway, residents of Malta and Stillwater complained about a "humming noise" coming from the fab site which seemed to "carry through the woods." The company indicated that although the noise level was lower than required by the town's zoning rules, it had built a temporary noise-containment unit behind its power supply unit (made of bales of hay and steel crates) and had hired Vibration Engineering Consultants of Santa Cruz, CA, to study possible additional measures.[90] The company believed the noise problem emanated from its continuous power supply (CPS) units, which provide

---

[84] "GlobalFoundries Seeks Tax Breaks," Albany, *The Times Union* (May 5, 2011).

[85] "Chip Plant Exemptions Get OK," Schenectady, *The Daily Gazette* (June 14, 2011).

[86] "Saratoga County IDA Approves $387M Tax Break for GlobalFoundries Projects," Saratoga Springs, *The Saratogian* (March 18, 2013).

[87] "Plant Builders Vow to Adhere to Noise Limits," Schenectady, *The Daily Gazette* (June 19, 2009). The zoning approval required construction noise at the GlobalFoundries property line not to exceed 55 dB by day and 45 dB by night. Normal conversation occurs around 60 dB. Ibid.

[88] "Dirty Work Begins at Site. Miles of Earth to Level Facility Site," Saratoga Springs, *The Saratogian* (July 31, 2009).

[89] "Town Officials Satisfied New Chip Plant Would be Quieter," Schenectady, *The Daily Gazette* (May 2, 2013).

[90] "Hum Bothers GloFo Neighbors," Schenectady, *The Daily Gazette* (July 26, 2011).

backup emergency power, and disclosed plans to construct baffles and partitions around the CPS units to muffle the sound.[91] The baffles, which were installed in December, reduced noise but did not eliminate it.[92]

In January 2012, GlobalFoundries met with local residents and admitted that its plan to muffle sound from the CPS units had "largely failed," and that the next step was to address noise emanating from the roof of the building, where nine spinning flywheels were based.[93] In March, the company disclosed plans to install customized silencers on exhaust stacks on the top of the main utility building to be operational by June.[94] David James, GlobalFoundries Fab 8 facilities manager, commented that "we're trying to do what we can, but I can't make it crickets and birds anymore."[95]

In June 2012 meetings with residents, GlobalFoundries' most recent measures received mixed rather than negative reactions. One resident said that he barely heard the sound and that "it mostly is gone as far as I am concerned."[96] Another said that he noticed an improvement, but could still hear some noise: "It's generally around a C sharp, it reminds me of a noisy pool pump." The company indicated it would conduct a 30-day test to analyze the effects of the hum-quieting measures, noting that preliminary tests showed that "the noise varies from house to house and that cloud cover, humidity and foliage all affect the sound."[97] It is worth noting that while the sound emanating from the site was below the mandated noise thresholds, GlobalFoundries unilaterally undertook additional measures to abate as much sound as possible, at substantial additional cost. For example, GlobalFoundries sprayed cellulose insulation inside the utility building to reduce noise reverberations inside the building.[98] Collectively, the remediation measures taken were sufficient to convince town officials that any new fab built at the site would not encounter similar noise issues.

---

[91] "GlobalFoundries in Malta Unveils Long-Term Plan for Lowering Noise Levels, Talks About Ways to Alleviate Traffic Congestion," Saratoga Springs, *The Saratogian* (August 30, 2011).

[92] "Silencer Should Reduce Chip Fab Noise by June," Schenectady, *The Daily Gazette* (March 13, 2012).

[93] "Malta Chip Plant's Noise Making Neighbors Mad," Glens Falls, *The Post-Star* (January 11, 2012).

[94] "Silencers Should Reduce Chip Fab Noise by June," Schenectady, *The Daily Gazette* (March 13, 2012).

[95] "Electric Noise Hope to be Reduced," Schenectady, *The Daily Gazette* (March 29, 2012).

[96] "GlobalFoundries Says it has Found a Way to Minimize Noise Emitted From its Plant," Saratoga Springs, *The Saratogian* (June 28, 2012).

[97] "The 'Hum' Goes On For Some," Albany, *The Times Union* (June 29, 2012).

[98] "Town Officials Satisfied New Chip Plant Would be Quieter," Schenectady, *The Daily Gazette* May 2, 2013).

# Bibliography[99]

Bacon, D. *Organizing Silicon Valley's High Tech Workers*. http://dbacon.igc.org/Unions/04hitec2.htm.

Crane Operators Threaten Strike at Many Construction Sites. (2006, June 30). *The New York Times*.

GF Closes on IBM Chip Business Purchase. (2015, July 1). *Semiconductor Engineering*.

GlobalFoundries. Business Annual Reports (BAR) submitted to the State of New York. n.d.

GlobalFoundries on a Roll With 7 nm and 5 nm Announcements. (2017a, June 15). *Forbes*.

GlobalFoundries Fires up its 7-nm Leading Performance Forges. (2017b, June 16). *The Tech Report*.

GlobalFoundries: No Layoffs in IBM Chip Units. (2014, October 20). *Associated Press*.

GlobalFoundries Plans New York R&D Center. (2013, January 9). *EE Times*.

GlobalFoundries Plant Generates Demand for Site Inspectors. (2010, March 18). Albany, *Business Review*.

GloFo Shows Progress in 3D Stacks. (2014, March 19). *EE Times*.

How Did New York Become the Most Unionized State in the Country? (2014, September 3). *The Nation*.

Inside GlobalFoundries' Fab 8. (2015, August 18). *EE Times*.

National Research Council. (2013). *New York's Nanotechnology Model: Building the Innovation Economy*. C. W. Wessner (rapporteur). Washington, DC: The National Academies Press.

Samsung Licenses 3D Chip Manufacturing Tech to GlobalFoundries to Win More Orders. (2014, April 17). *Reuters*.

Silicon Valley, Meet Organized Labor. (2014, October 7). *New York*.

Sperling, E. (2014, November 20). An Inside Look at the GlobalFoundries—IBM Deal. *Semiconductor Engineering*.

---

[99] As noted in the front matter of this book, the study also drew on interviews carried out by the authors and numerous articles from *The Times Union* (Albany), *The Daily Gazette* (Schenectady), the *Albany Business Review* (Albany), *The Post-Star* (Glens Falls), *The Record* (Troy), *The Saratogian* (Saratoga Springs), *The Buffalo News* (Buffalo), The *Observer-Dispatch* (Utica), The *Daily Messenger* (Canandaigua), and the *Post-Standard* (Syracuse). These are not individually included in the bibliography.

# Chapter 7
# Economic Impact of New York's Nanotechnology Investments

**Abstract** Despite ongoing skepticism in some quarters, the economic payoffs for the Capital Region from New York's investments in nanotechnology have been substantial, particularly in regard to employment. Indeed, the benefits for the region in terms of jobs, investment, and growth have exceeded all forecasts. The substantial investments required to attract GlobalFoundries to the region have resulted in a great many more jobs than were either anticipated or required. Instead of 1200 jobs, GlobalFoundries actually created over 3500 direct jobs at the Luther Forest site, while preserving some 2000 jobs at IBM's former operation in East FishKill. GlobalFoundries' presence reflects roughly $17 billion in private and public investments in facilities and equipment. Moreover, the state and private investments in CNSE created another 4000 jobs within CNSE and its industrial partners in Albany, although this number has recently declined to closer to 3400.

Direct employment gains of over 9000 jobs have been complemented by large numbers of indirect jobs, that is, those within the GlobalFoundries supply chain. In an unanticipated development, construction jobs have ranged as high as 3500 at some points, and hundreds of construction workers are still active at the GlobalFoundries site in Malta/Stillwater. The high salaries associated with high-tech employment have also had major ramifications for the growth of the regional economy, thereby creating thousands of induced jobs in sectors as diverse as hotels, restaurants, banking, and retail sales. Depending on the multipliers used, the indirect and induced jobs range from 20,000 to nearly 50,000 with the higher numbers more accurate. Total direct, indirect, induced, and construction jobs attributable to nanotechnology are in the 60,000–80,000 range. In short, the dynamic effects of the initial investments have resulted in massive private-sector investment, thousands of high-quality, high-tech related jobs, while also providing major reputational gains for the region.

For two decades New York's investments in nanotechnology have raised questions about their actual economic effects. Forecasts of new jobs and broader economic ripple effects have frequently met with skepticism or a wait-and-see attitude. A common observation has been that a state incentive package of $1 billion for a semiconductor plant employing 1000 people cost the taxpayers $1 million per job.[1] Despite the scale and ongoing nature of the investments, some have expressed concern that these new private-sector investments will prove transient.[2] Some question how the presence of local high-tech research and manufacturing facilities means anything of consequence for the surrounding region. As one individual interviewed for this study expressed it, people see "a bunch of guys in lab coats behind walls, doing secret stuff" and the larger community "doesn't see the benefit."[3]

Fortunately, the continued growth and scope of investments are now putting these longstanding questions to rest, and observers note that the term "Tech Valley" has stopped being regarded by some as a joke. The fact that the nation as a whole was beginning to take note of GlobalFoundries' investments was underscored in 2012 when ABC's "Made in America with Diane Sawyer" reported that the employment and regional development impact already exceeded expectations even though production was not yet in full swing.[4] The Albany *Times Union* commented in 2013 that nanotechnology was having a transformational effect on the Capital Region, which it characterized as "a sixth age in the ongoing reinvention of the region's economy." Conceding that the changes in many cases represent "a revolution not visible to the naked eye," the *Times Union* cited a Brookings study that ranked the Capital Region 18th out of all metro areas in the United States in terms of patents per million residents. "[T]he high tech revolution is here. It is real, and it is accelerating. It is high time, it would seem, to retire the 'Smallbany' parochial put-downs."[5]

New York's nanotechnology initiatives have fostered not only research-related jobs and economic activity but high-tech manufacturing on a significant and growing scale. Jack Kelly, who was part of the SEDC team that pursued chip fab investors in the early- and mid-2000s, commented in 2013 that the level of investment by GlobalFoundries had already far surpassed expectations for the Luther Forest site and that it had happened far more quickly than had been anticipated:

> *If nothing else ever happens at Luther Forest, it's still the greatest economic development success story to date for New York state and possibly the nation.*[6]

---

[1] "Chip Plan Cost is $1M for Each Job," Schenectady, *The Daily Gazette* (June 22, 2006); "Chip Factory Bonanza is Prophesied," Schenectady, *The Daily Gazette* (April 8, 2008).

[2] The fact that ESD saw it necessary to secure a contractual commitment from AMD/GlobalFoundries for a sustained level of minimum employment is indicative of the wary perspective of state leaders. The agreement which Advanced Micro Devices entered into with the State of New York in 2006 provided that AMD would invest in a wafer fabrication plant at the Luther Forest site and create 1205 jobs at the site and maintain them for 7 years. "New York's Big Subsidies Bolster Upstates' Winning Bid for AMD's $3.2 Billion 300 mm Fab," *Site Selection* (July 10, 2006).

[3] Interview with Albany-based business executive (September 16, 2015).

[4] "Made in America: Global Companies Expand in US Towns," *ABC News* (April 30, 2012).

[5] "Region Thrives Amid High-Tech Revolution," Albany, *The Times Union* (March 19, 2013).

[6] Kelly's remarks predated most of GlobalFoundries' massive investments in a new Technology Development Center. "Has 'Tech Valley' Peaked?" Albany, *The Times Union* (November 10, 2013).

**Box 7.1 New York Counties Comprising Tech Valley**

| Lower Hudson | Capital District |
|---|---|
| Putnam | Albany |
| Rockland | Rensselaer |
| Westchester | Montgomery |
| | Columbia |
| **Mid-Hudson** | Saratoga |
| Dutchess | Schenectady |
| Orange | Warren |
| Sullivan | Schoharie |
| Ulster | Washington |

Table 7.1 Annual average unemployment rates

| Year | Capital Region (%) | New York State (%) |
|---|---|---|
| 2016 | 4.0 | 4.8 |
| 2015 | 4.6 | 5.3 |
| 2014 | 5.2 | 6.3 |
| 2013 | 6.5 | 7.7 |
| 2012 | 7.5 | 8.5 |
| 2011 | 7.3 | 8.3 |
| 2010 | 7.5 | 8.6 |
| 2009 | 7.0 | 8.3 |
| 2008 | 5.0 | 5.4 |
| 2007 | 4.0 | 4.6 |
| 2006 | 3.9 | 4.5 |
| 2005 | 5.0 | 5.0 |

Source: New York State Department of Labor

## 7.1 Employment Effects

For the average citizen, the most important measure of economic performance is employment—jobs created, jobs preserved, and jobs supported indirectly through expenditures by firms in the form of payroll, taxes and fees paid, and payment for locally procured goods and services. Unemployment in the Capital Region has been dropping since 2013 and is now lower than it was at the onset of the recession in 2008. The region consistently outperforms the state as a whole (as shown in Table 7.1) as well as the national average of 4.7%.

As the Albany *Times Union* reported in January 2016, the jobs picture is bright: "in 2015 Capital Region unemployment fell, private-sector jobs hit a record high, construction surged and people rediscovered the area's downtowns."[7] Job growth in

---

[7] "Capital Region Grows Even Bigger, Stronger," Albany, *The Times Union* (January 3, 2016).

manufacturing in the computer and electronics sector has been particularly dramatic. James Ross, a state Department of Labor analyst based in the Capital Region, said that the region, together with Long Island, had the lowest unemployment rate in the state, which he attributed to the growth in manufacturing jobs, which hit the highest level since 2006: "That's counter to what the historical trends were. We were in a slow steady drop in manufacturing jobs."[8] Ross indicated that the Capital Region's resurgence in manufacturing employment was "largely due to GlobalFoundries in Saratoga County, General Electric in a few places in the region, digital imaging in Rensselaer and the Regeneron Pharmaceuticals announcement that it will be building more facilities." He observed that "higher volume at these companies will spur job growth at their support companies."[9]

In a 2014 presentation on the upstate economy, William C. Dudley, President and CEO of the Federal Reserve Bank of New York, made the following comments—

*And there's some especially good news to report from the Capital Region's manufacturing sector: jobs are growing, and growing quite strongly—a testament to the success of the area's burgeoning high tech sector. Since the recovery began, the area has added over 4,000 manufacturing jobs, many of which are tied to computers and electronics. That's an increase of nearly 25 percent. This pace of job growth is four times as great as the national pace. In fact, there are now more people employed in the Capital Region's manufacturing sector than before the recession. Not many places can say that, and this is certainly not true for the nation as a whole.*[10]

Nanotechnology's jobs impact was particularly evident in Saratoga County. Construction of GlobalFoundries fab in Malta/Stillwater began at the end of 2009, and by the end of 2012 the fab was producing semiconductors for multiple customers. By then Saratoga County was experiencing the most rapid employment gain of any county in the state, with new jobs concentrated in the manufacturing and hospitality sectors.[11] In mid-2014 Saratoga County had the lowest unemployment rate in the entire Capital Region—4.5%—compared with the statewide average of 6.6%.[12] In January 2016 the county's unemployment rate of 4.4% was better than the average for the region (4.6%) and the state (5.3%).[13]

---

[8] "Rossi Region Holding its Own in Job Market," Schenectady, *The Daily Gazette* (February 28, 2016).

[9] "Jobless Rate in Region Improving," Albany, *The Times Union* (December 23, 2015).

[10] William C. Dudley, "The National and Regional Economy," Remarks at RPI, Troy, New York (October 7, 2014).

[11] "Saratoga County Sees the Highest Employment Gain in New York State," Saratoga Springs, *The Saratogian* (February 27, 2013).

[12] "Region's Unemployment Rate Continues to Drop," Albany, *The Times Union* (July 22, 2014).

[13] New York State Department of Labor.

## 7.1.1 How Accurate Were the Forecasts Compared to the Results?

New York's investments in university research infrastructure in nanotechnology were based on studies of how successful high-technology clusters evolved in places like Silicon Valley, Austin, and North Carolina's Research Triangle. Forecasts drawing on these precedents concluded that investments in university-based research would eventually foster not only research jobs but significant numbers of local high-tech manufacturing jobs and associated indirect jobs. Because the state is undertaking a new generation of nanotechnology-based economic development initiatives, which will be based on similar studies, it is useful to recall what was forecast a decade ago and to benchmark these projections against actual experience. In fact, the forecasts, which were widely considered to be overly optimistic at the time, vastly understated the positive economic impact the state's investment in nanotechnology would actually have in the region.

In 2008, two respected consultancies prepared forecasts of the economic effects, including employment effects, of a 300 mm semiconductor fabrication plant in the region. Their projections of 5000–8000 new jobs were controversial at the time, and the value of the state's investments was questioned. Ironically, while most of the studies' critics charged that the forecasts were too rosy, those forecasts have in fact proven to have been far too conservative. Still, in the interviews conducted for this study between 2015 and 2017, it became evident that many local leaders still thought of the employment impact of nanotechnology in terms of the 2008 studies, and some continued to disparage those estimates as overly optimistic. That perspective is not only wrong but wrong by an order of magnitude. The number of direct jobs alone attributable to nanotechnology actually exceeds the 2008 estimates for jobs of all types (direct, indirect, and induced). (See Box 7.2.)

---

**Box 7.2 Employment Multipliers**

Employment multipliers are used to estimate the employment impact of a company or industry within a region. They measure the number of "direct," "indirect," and "induced" jobs attributable to the company or industry.

- *Direct jobs* refer to individuals employed by the company and sometimes include individuals working full time on company premises who are employees of vendors serving the company.
- *Indirect jobs* refer to individuals employed by supply chain, logistics, and service providers serving the company.
- *Induced jobs* are jobs attributable to direct and indirect employees spending money in the region (e.g., healthcare providers, hospitality industries, retail, home building, and sales).

> In calculating employment effects, it is normal practice to take the number of direct jobs, for which precise data is usually available from employers, and apply a multiplier to estimate the number of indirect and induced jobs the company or industry has fostered. Thus a multiplier of 4.4 for particular industry would mean that for every direct job, 4.4 other jobs are created (indirect and induced).

### 7.1.1.1 The Semico Study

In 2008 National Grid, CEG, and Mohawk Valley Edge, an economic development organization operating in Oneida County,[14] engaged the Phoenix-based consultancy Semico Research Corporation to prepare a forecast of the employment impact of a hypothetical 300 mm wafer fabrication plant in Upstate New York. CEG President and CEO Mike Tucker said that—

> *With the investment the state has made in this industry and the public concerns about whether it was worth it, we thought we needed to be able to demonstrate from an independent source the region is well poised for the future and additional investment will only help to ensure a return on what has already been spent.*[15]

Semico forecast that if a 300 mm fab were built in Tech Valley or the Mohawk Valley, 1160 direct jobs and 435 on-site support jobs would be created by the second year of operations. Semico also forecast 1500 direct construction jobs while the fab was being built and another 2550 indirect jobs associated with the construction. Semico forecast 2414 indirect jobs associated with operation of the fab or 1.52 indirect jobs for each on-site job.[16] Total employment forecast by Semico was 5514 new jobs. Semico President Jim Feldhan commented that—

> *[S]ome critics of large government incentives spent to attract companies to build new facilities usually don't factor in the creation of other businesses and jobs that spring up to support the lured business. If you had $1 billion in investment to build a chip fab that employed 1,000 people, that might seem like $1 million a job, but fortunately the semiconductor fab doesn't work in a vacuum. Operating a semiconductor fab takes a tremendous amount of support industries.*[17]

Feldhan observed that a new fab would require computer sales and maintenance services, warehousing, chemical disposal, private security, and a specialized business to "clean" suits worn by fab workers: "[A Fab] isn't going to want to send those to California every week for cleaning. Some local entrepreneur is probably going to start that business."[18] Semico forecast that the average fab worker would earn

---

[14] "Semiconductor Industry Study to be Released," Albany, *The Times Union* (April 1, 2008).
[15] "AMD Incentives Called Good Investment," Schenectady, *The Daily Gazette* (April 2, 2008).
[16] Semico Research Corporation, *Upstate New York: Assessing the Economic Impact of Attracting Semiconductor Industry* (March 2008).
[17] "AMD Incentives Called Good Investment," Schenectady, *The Daily Gazette* (April 2, 2008).
[18] "AMD Incentives Called Good Investment," Schenectady, *The Daily Gazette* (April 2, 2008).

$40,000 a year, with engineers getting $70,000 and managers $110,000. It predicted that 70% of the fab's workforce would be recruited locally but that 70% of the managers would be drawn from outside the region.[19]

The Semico study was criticized and even derided in some quarters in the state. A spokesperson for the organization New Yorkers for Fiscal Fairness argued that the study should not have included construction jobs in the job creation totals because of their temporary nature.[20] An opinion piece in the Schenectady *Daily Gazette* commented that—

> By the time the economic impact consultants get done with their multipliers, we will benefit from millions, tens of millions, hundreds of millions in new tax revenues and billions in new "economic activity" as they call it. The employees at the factory will spend the money they earn, and other people, like dentists and plumbers, who receive the money, will spend it again, and so on and so on, until pretty soon we're all flush within a 50-mile radius, much like the two families in the old joke who prospered by taking in each other's laundry.[21]

A few observers ventured the opinion—little noticed at the time—that Semico's job forecasts actually understated the potential employment impact of a new semiconductor fab on the region. Gary Dyal, the Executive Director of CVD Equipment Corporation, a Long Island-based maker of chemical vapor deposition equipment for the semiconductor industry, which operated a plant in Saugerties, New York, said that he doubted the number forecast by Semico because the ongoing nanotechnology research activity at the Albany NanoCollege was going to "enable many new markets for semiconductor chips." He said, "I think their estimates are very conservative. I think the growth will be much more explosive."[22]

### 7.1.1.2 The Ehrlich Study

In 2008, AMD sponsored a study of the regional economic impact of a new chip fab in conjunction with the spinoff of its manufacturing operations, which would create the entity ultimately named GlobalFoundries.[23] The study was prepared by a local consultancy headed by Everett M. Ehrlich, an eminent economist and statistician.[24] The Ehrlich study concluded that the AMD fab would exceed the Semico estimate of 1160 jobs for a hypothetical fab and would employ 1465 people by the end of

---

[19] "Study Touts Impact of Chip Fab," Albany, *The Times Union* (April 2, 2008).

[20] "A Future Filled With Promise," Albany, *The Times Union* (April 20, 2008).

[21] Carl Strock, "AMD Not Looking Like Big Winner," Schenectady, *The Daily Gazette* (April 10, 2008).

[22] "AMD Incentives Called Good Investment," Schenectady, *The Daily Gazette* (April 2, 2008).

[23] "Numbers to Leave Us All Breathless," Albany, *The Times Union* (October 17, 2008).

[24] Everett M. Ehrlich, *Manufacturing, Competitiveness, and Technological Leadership in the Semiconductor Industry* (2008). Ehrlich served in the Clinton Administration as Under Secretary of Commerce for Economic Affairs. As the chief executive of the country's statistical system, he led the first comprehensive strategic review of U.S. economic statistics in four decades. He supervised the redesign of the 2000 census.

2014. Like Semico, Ehrlich noted that wafer fabrication operations require the on-site presence of large numbers of individuals employed by other companies providing various support services and functions. He estimated that the AMD fab would have 550 individuals employed by vendors working on-site by the end of 2014. He forecast 1600 construction jobs during the construction of the fab. Using the multiplier of 2.25 indirect jobs for each on-site job—the metric being used at the time by Empire State Development—he forecast that the fab would also support 4500 indirect jobs, for a total of 8115 jobs. He commented after the release of the study that—

> If you use the multipliers that Empire State Development uses, there are about 5,000 more jobs, and that's a total of 6,500. This area will be a viable player when it comes to locating facilities anywhere in the world. When you create a payroll of $290 million a year, it has to go somewhere.[25]

But Joe Dalton, President of the Saratoga County Chamber of Commerce, warned that the forecast economic benefits were difficult for the average person to grasp, cautioning "Joe Six Pack doesn't believe in multipliers."[26]

### 7.1.1.3 Not Considered in the Forecasts: IBM's East Fishkill Operations and Albany NanoTech

The Ehrlich and Semico forecasts did not make any estimates with respect to the economic effects associated with IBM's then-existing semiconductor manufacturing operations at East Fishkill. However, the state's investments in nanotechnology were undertaken not only in the hope of attracting a new manufacturer to the region but also in order to retain a local semiconductor manufacturing presence by IBM. It was known that the company might either invest in upgrading that operation to the 300 mm technology level or, alternatively, wind down the existing 200 mm operation; either prospect would have significant implications for the mid-Hudson counties of Tech Valley, Dutchess, Orange, and Putnam counties. The Semico and Ehrlich studies also did not examine the employment impact of Albany NanoTech at its own site on Fuller Road, which in 2016 supported more workers—4000—than either of the Global/Foundries fabs at Malta/Stillwater and East Fishkill. The employment effects at these two locations were relevant to the future of Tech Valley but were outside the scope of the Semico and Ehrlich studies.

The Semico and Ehrlich studies were wrong, not because they overstated the employment impact of nanotechnology—as they were criticized for doing at the time—but because they understated it by substantial margins. Direct employment, which is readily verifiable through actual headcount, was projected at 1595 by Semico and at 1465 by Ehrlich for a hypothetical wafer fabrication plant. GlobalFoundries had more than twice that many direct employees on site in 2015, a

---

[25] "Consultant Sees Likely Spinoffs from AMD Days Ahead in Region," Schenectady, *The Daily Gazette* (November 19, 2008).

[26] "Region Now Major High-Tech Player," Albany, *The Times Union* (November 19, 2008).

total of 3538. When direct employment at the NanoCollege and GlobalFoundries' East Fishkill plant is added, the regional nanotechnology direct employment total of 9623 exceeds the original forecasts by over 600%. The Semico and Ehrlich estimates of on-site construction employment were likewise far too pessimistic with respect to both headcount and duration of employment.

The Semico and Ehrlich studies also underestimated the number and duration of construction jobs for the building of the fab. Semico forecast 1500 direct construction jobs and Ehrlich 1600 for under 2 years duration. In fact, at the peak of construction activity, the fab accounted for roughly 3500 construction jobs, many of them nearly double the duration originally forecast. These jobs paid union-scale wages, reflecting the unique Project Labor Agreement (PLA) negotiated by GlobalFoundries and the construction trades, the largest PLA in the history of the United States.[27] There have been no strikes or work stoppages at GlobalFoundries.

## 7.1.2 The Direct Employment Situation in 2016

In 2016, 8 years after the Semico and Ehrlich studies, data is available with respect to actual on-site employment at Tech Valley's three nanotechnology hubs—the GlobalFoundries fabs in Malta/Stillwater and East Fishkill and the CNSE complex in Albany. The available data show on-site employment of nearly 10,000 workers at the three sites, thousands of construction jobs, and inferentially, tens of thousands of indirect jobs. The actual number of indirect jobs attributable to the state's investments in nanotechnology is more difficult to calculate with precision, but there is evidence of substantial regional economic activity and new employment that would not exist without the presence of GlobalFoundries at its two sites and Albany Nanotech. The new jobs have resulted from the provision of supporting goods and services for large-scale, sustained construction activity; the development of local supply chain activity; and the ripple effects felt in the housing, hospitality, retail, health care, and other support services sectors.

### 7.1.2.1 On-Site Employment

Table 7.2 shows actual on-site employment at the three principal hubs of semiconductor-related research and production in the 16 counties comprising New York's Tech Valley. The figures include direct employees and contractors who are employed by other companies and work full time at each site.

---

[27] National Research Council, Charles W. Wessner (rapporteur), *New York's Nanotechnology Model: Building the Innovation Economy* (Washington, DC: The National Academies Press, 2013), p. 88. As noted, GlobalFoundries and the unions benefited from the contribution of Mike Russo, who played a key role in reaching the labor agreement that overcame a major potential stumbling block to the project's operation.

**Table 7.2** Semiconductor-related on-site employment in Tech Valley

| Institution | Location | Number of on-site workers | Year data point derived |
|---|---|---|---|
| College of Nanoscale Science and Engineering | Albany | 4000+[a] | 2016 |
| GlobalFoundries | Malta/Stillwater | 3538[b] | 2015 |
| GlobalFoundries | East Fishkill | 2085[c] | 2015 |
| Total | | 9623 | |

[a]College of Nanoscale Science and Engineering, Quick Facts, *http://www.sunycnse.com/AboutUs/QuickFacts.aspx*. Note: Headcount was 4261 for 2015–2016 but fell to 3391 in 2016–2017. For these purposes, a figure of 4000 has been used
[b]GlobalFoundries statistics. Totals as of December 2015
[c]"GlobalFoundries: E. Fishkill Site Key to Overall Strategy," *Poughkeepsie Journal* (November 12, 2015). Figure based on reported total of 1780 workers in 2015 plus assumed contract workers based as same ratio as for the GlobalFoundries Malta/Stillwater site (1780 + 305 = 2085)

**Table 7.3** Employment impact of GlobalFoundries Luther Forest fab forecast/actual

| | Number of jobs | | | |
|---|---|---|---|---|
| | Commitment made to Empire State Development | Ehrlich Study Projection (2008) | Semico Study Projection (2008) | Actual (2015) |
| Direct jobs | 1200 | 1465 | 1160 | 3023 |
| On-site support jobs | – | 550 | 435 | 515 |
| Indirect/induced jobs | – | 4500 | 2419 | 17,300[a] |
| Construction jobs | – | 1600 (for 2 years) | 1500 (for 2 years) | 900/1100 (est. average over 5 years) |
| Construction "multiplier" jobs | – | 2700 (for 2 years) | 2550 (for 2 years) | 1512/1870[b] (est. average over 5 years) |

[a]Based on Semiconductor Industry Association multiplier of 4.89 indirect jobs for each direct job
[b]Based on Ehrlich/Semico multiplier of 1.7

In 2008 Semico estimated that the average worker in a hypothetical semiconductor fab would earn about $40,000 a year. In 2016 GlobalFoundries employees earned an average of $92,733 a year. About 61% of GlobalFoundries' employees are natives of New York State, and 93% were residents of the United States when hired.[28]

---

[28] GlobalFoundries.

### 7.1.2.2 GlobalFoundries' Malta/Stillwater Operations

Employment data from GlobalFoundries' operations in Malta/Stillwater demonstrate that the direct job estimates made in 2008 by Ehrlich and Semico for a new 300 mm fab have been exceeded by over 100%—the fab directly employing 3023 workers at the end of 2015 with another 515 on-site jobs attributable to vendors. As shown in Table 7.3, construction employment has been far more extensive and sustained than forecast in 2008.

It should be noted that in August 2018 GlobalFoundries announced that it would halt work on the development of 7 nm technology and that it would downsize its workforce in November to reflect that change in strategy. It filed a notice with the New York Department of Labor to the effect that it would cut 424 positions at the Malta wafer fabrication site and 31 research positions at SUNY Poly.[29]

### 7.1.2.3 The IBM/GlobalFoundries Site at East Fishkill

In 2015 IBM announced that it would no longer manufacture microelectronics devices and would cease operations in its Microelectronics Division. GlobalFoundries was the only company interested in acquiring the Division's assets and in maintaining operations in what were then IBM fabs in East Fishkill, New York, and in Burlington, Vermont. The acquisition by GlobalFoundries not only saved most of the jobs at these sites but strengthened the foundation of what is now referred to as the "Northeast Technology Corridor."

IBM operated semiconductor manufacturing facilities in East Fishkill, in New York's Dutchess County, from the 1960s through 2015, when it transferred the facilities to GlobalFoundries.[30] GlobalFoundries reported that as of November 2015 it employed 1780 workers at the site, not counting on-site contract workers from other companies.[31] With respect to on-site workers employed by other companies, to the extent that GlobalFoundries' East Fishkill site supports approximately the same ratio of such employees as Luther Forest, East Fishkill would have another 305 contract workers, for a combined total of 2085 workers on-site.

This study documents the fact that IBM was on a path toward disinvestment in New York State in the early 1990s, cutting 6000 jobs and preparing to end the production of semiconductors at East Fishkill.[32] The company's stance changed to one

---

[29] "GlobalFoundries Job Cuts Total 455," Schenectady, *The Daily Gazette* (August 31, 2018).

[30] In the 1960s, IBM's East Fishkill facility produced more semiconductor material than all of Silicon Valley, although all of its output was for its own internal consumption. "How Green is our Hudson Valley for Technology?" Albany, *The Times Union* (November 6, 1998).

[31] GlobalFoundries was reportedly hiring new employees, including individuals to provide tool maintenance, a function that had previously been outsourced by IBM. "GlobalFoundries: E. Fishkill Site Key to Overall Strategy," *Poughkeepsie Journal* (November 12, 2015).

[32] "State Offers $40M in Loans in Effort to Help IBM Grow," Albany, *The Times Union* (January 30, 1994).

**Table 7.4** Employment at the Albany NanoTech complex

| Year | Number of industry employees | Number of paid graduate students | Number of engineers, scientists, administrators | Others | Total |
|---|---|---|---|---|---|
| 2010–2011 | 1625 | 164 | 513 | 0 | 2302 |
| 2011–2012 | 2042 | 184 | 623 | 25 | 2874 |
| 2012–2013 | 2047 | 189 | 663 | 50 | 2949 |
| 2013–2014 | 2934 | 199 | 642 | 75 | 3850 |
| 2014–2015 | 3400 | 153 | 705 | 100 | 4358 |
| 2015–2016 | 3209 | 132 | 877 | 43 | 4261 |
| 2016–2017 | 2747 | 87 | 557 | 0 | 3391 |

Source: SUNY Poly

of the renewed investments in and after 1995 as New York invested in nanotechnology and particularly in the research facilities at SUNY Albany. IBM committed to build a 300 mm fab at East Fishkill in 1997.[33] The jobs that were created at that site, and continue to exist today, are directly attributable to the state's investments in nanotechnology. The fact that GlobalFoundries was present in New York and able to assume ownership of the East Fishkill site—preserving the jobs at that location—is likewise attributable to the state's nanotech investments.

The jobs associated with IBM's former operation therefore should be counted in any assessment of the employment effects of the state's investments in nanotechnology.[34] IBM's decision in 2000 to build a $2.5 billion 300 mm semiconductor fab at East Fishkill was substantially attributable to the proximity of the rapidly expanding nanotechnology research facilities at SUNY Albany.[35] The values of New York's investments at the East Fishkill site are acknowledged by local leaders. As noted in the Albany *Times Union*—

---

[33] "IBM Plant a Likely Magnet," Albany, *The Times Union* (October 11, 2000).

[34] The State of New York has periodically made substantial investments in the IBM site to ensure that the company continued to upgrade its facilities there as semiconductor technology advanced rather than move production to another region. See "Odds Long, Rewards Great to Get Microchip Plan in Area," *Buffalo News* (May 17, 1998).

[35] "IBM Plant a Likely Magnet," Albany, *The Times Union* (October 11, 2000). In 2000, IBM CEO Lou Gerstner said that the decision to build the new fab in East Fishkill reflected "IBM's ongoing relationship with the Center for Advanced Thin Film Technology, which is currently the only university-based research facility for 300 mm technology." Gerstner said that "even more important that the state's tax and regulatory reform … were the nearby universities such as the University at Albany and Rensselaer Polytechnic Institute in Troy that have really made the kinds of investments in training that we need." "Plant's Benefit to Area Touted—IBM's Facility May Spark Local Opportunities," Schenectady, *The Daily Gazette* (October 11, 2000).

7.1 Employment Effects

> When you ask officials in Dutchess County whether New York's investments in IBM Corp.'s computer chip fab expansion in East Fishkill were worth the price, they don't skip a beat. "They are the highest-paying jobs outside of the finance industry," said Anne Conroy, President of the Dutchess County Economic Development Corp. It is wealth creating. It is impossible to [overvalue] the value to the economy.[36]

#### 7.1.2.4 Employment at the Albany NanoTech Complex

In 2014–2015, 4358 workers were employed at the Albany NanoTech Complex on Fuller Road. These include employees of the College of Nanoscale Science and Engineering and the SUNY-affiliated management companies overseeing the site as well as on-site employees of companies engaged in research projects at the complex. Because the NanoTech Complex is a research site with manufacturing limited to small-volume production for research purposes, it is unlikely that it supports as many supply chain jobs as a commercial semiconductor fab. However, the NanoTech Complex clearly supports a significant number of off-site supply chain and induced jobs. The on-site headcount fell to 4261 in 2015–2016 and to 3391 in 2016–2017. (See Table 7.4.)

#### 7.1.2.5 Other Nanotechnology-Related Direct Employment

A significant number of nanotechnology-related jobs in Tech Valley exist because firms involved in the sector have relocated to be near what is now increasingly recognized as one of the leading regional centers of nanotechnology in the world. Foremost among these firms is M+W Group, perhaps the world's principal designer and builder of semiconductor fabs and other high-technology manufacturing facilities, which moved its US headquarters from Texas to New York in 2010. In 2015, M+W Group reportedly had 2000 employees active in New York State.[37] These workers are not part of GlobalFoundries or the Albany NanoTech supply chains and therefore are not included in estimates of indirect "supply chain" employment. Arguably, they should be added to the direct employment figures for nanotechnology noted above.

### *7.1.3 Estimating Multiplier Employment Effects in the Semiconductor Industry*

Although "Joe Six Pack" may not believe in multipliers, virtually all economic studies of the impact of manufacturing on jobs in a region accept the notion that in addition to individuals directly employed by a manufacturer, additional jobs in the area

---

[36] "A Future Filled With Promise," Albany, *The Times Union* (April 20, 2008).
[37] "No State Subsidy for Zen Tenant," Albany, *The Times Union* (March 28, 2015).

are attributable to the manufacturers' presence. In some studies these jobs are lumped together in one category commonly called "indirect jobs." In other studies the term "indirect jobs" is used more narrowly and applied only to employment by firms that directly supply goods or services to the manufacturer—that is, "supply chain" jobs. Jobs attributable to local firms providing goods and services to the employees of the manufacturer and its suppliers (such as retail sales, health care, and laundry) are sometimes termed "induced jobs."[38]

The 2008 Ehrlich study estimating the employment impact of an AMD fab categorized a forecast 1465 direct jobs and 550 related fab-service jobs as "direct." It used a multiplier derived from ESD to calculate that an additional 2.25 "indirect" jobs would be created for each new direct job, or about 4500 jobs.[39] Semico's 2008 forecast envisioned 435 "outside fab support" jobs and 2419 "support industry and local business jobs."[40] The 2.25 multiplier used by Ehrlich to estimate the indirect employment effects of GlobalFoundries was derived from standard metrics employed by ESD across industry sectors as a rough tool for assessing the economic impact of an employer. The 1.52 multiplier used by Semico was even more conservative than the ESD number. Semico's president said that his "company was extremely conservative with its economic assumptions, which helps with the credibility of the results."[41]

The Semiconductor Industry Association uses a multiplier which is substantially higher than the 2.25 indirect-jobs-per-on-site job used by ESD and Ehrlich and the 1.52 multiplier used by Semico. SIA's Director of Industry Statistics and Economic Policy, Falan Yinug, estimates that every job in the semiconductor manufacturing sector accounts for 4.89 indirect jobs. SIA states that—

> The employment multiplier for the U.S. semiconductor industry is relatively high compared to multipliers in other industries. This means that the U.S. semiconductor industry has an outsized positive effect on job creation in other sectors compared to many other industries.[42]

---

[38] Elizabeth Scott and Howard Wial, *Multiplying Jobs: How Manufacturing Contributes to Employment Growth in Chicago and the Nation*, (Chicago: University of Illinois at Chicago, May 2013), p. 4.

[39] Everett M. Ehrlich, *Manufacturing, Competitiveness, and Technological Leadership in the Semiconductor Industry* (2008), p. 4.

[40] Semico Research Corporation, *Economic Impact of the Semiconductor Industry on Upstate New York*, (February 2008), p. 61.

[41] "A Future Filled With Promise," Albany, *The Times Union* (April 20, 2008).

[42] Semiconductor Industry Association, *U.S. Semiconductor Industry Employment* (January 2015). Travis Bullard, a GlobalFoundries spokesperson, would use higher multipliers than SIA. He estimated in 2012 that for every job inside the fab, "there are four to five support jobs outside, such as workers who take care of the clean-room clothing." In addition to such supply chain-related jobs "are maybe an additional five to six indirect jobs, such as openings in new restaurants. Many of these workers need new apartments, grocery stores, and day-care centers." "Can American Manufacturing Really Be Cornerstone of Economic Revival?" *The Christian Science Monitor* (February 8, 2012).

## 7.1 Employment Effects

Application of SIA's multiplier to GlobalFoundries' total of 3538 on-site jobs would yield a total of 17,300 indirect jobs. Application of the same multiplier at GlobalFoundries' East Fishkill site produces another 10,196 jobs.[43]

A growing body of academic work supports use of a higher indirect jobs multiplier for the semiconductor industry than those used by Semico and Ehrlich in 2008. These analyses differentiate between "tradable" or "regional export" industries (usually manufacturers), on the one hand, and "nontradable" or local services industries, on the other hand. Firms in tradable industries produce products which are "exported" and sold elsewhere. Nontradable industries, which account for the vast majority of jobs, provide local services such as healthcare, retail sales, dry cleaning, and dentistry.[44] Enrico Moretti, a Professor of Economics at the University of California at Berkeley, argues that innovation-sector jobs in the traded sector generate a more powerful effect in the localities where they operate—

> My research, based on an analysis of 11 million American workers in 320 metropolitan areas, shows that for each new high tech job in a metropolitan area, five additional local jobs are created outside of high tech in the long run. In essence, in Silicon Valley, high-tech jobs are the *cause* of local prosperity, and the doctors, lawyers, roofers and yoga teachers are the *effect*.[45]

Moretti offers several explanations for the extremely high multiplier effects exerted by innovative tradable industries like the semiconductor industry. Average wages are far higher than the regional averages, resulting in more disposable income to spend locally. High-tech companies require many local business services, including information technology support, graphic design, business consultants, and specialized legal and security services. Finally, high-tech manufacturers support particularly dense clusters of supply chain firms which themselves tend to pay higher average salaries and demand specialized services.[46]

Even if one accepts the more conservative multipliers used by the Ehrlich and Semico studies, the substantially higher number of actual on-site jobs in GlobalFoundries' operations than the number forecast in 2008 is indicative that the indirect job estimates from the 2008 studies were too low. Using Ehrlich's and the ESD's multiplier of 2.25 indirect jobs for each on-site job, GlobalFoundries

---

[43] The European Semiconductor Industry Association uses a multiplier virtually identical to that employed by SIA, calculating that the 200,000 direct jobs attributable to the industry in Europe in 2011 supported 1,000,000 indirect jobs.

[44] Elizabeth Scott and Howard Wial, *Multiplying Jobs: How Manufacturing Contributes to Employment Growth in Chicago and the Nation* (Chicago: University of Illinois at Chicago, May 2013).

[45] Enrico Moretti, *The New Geography of Jobs* (Boston and New York: Mariner Books, 2013), p. 60, original emphasis.

[46] Enrico Moretti, *The New Geography of Jobs* (Boston and New York: Mariner Books, 2013), p. 62.

currently supports 7960 indirect jobs in the region. Using Semico's lower multiplier of 1.52 indirect jobs for each on-site job, GlobalFoundries supports 5378 indirect jobs, still a higher figure than forecast by either Ehrlich or Semico in the 2008 studies.

## 7.1.4 The Sustained Impact of Construction Employment

Construction jobs are episodic as phases of construction begin, reach a peak of activity, and wind down. Simply adding the number of construction jobs at a particular point in time to direct or indirect jobs associated with regular operation of a fab is an apples-and-oranges comparison, and construction jobs arguably are best considered as a separate employment category. Since the year 2000 construction jobs have been associated with major new semiconductor facilities at GlobalFoundries in Saratoga County, the NanoCollege in Albany, and IBM's site in East Fishkill. In addition, building the infrastructure for the GlobalFoundries fab has given rise to large numbers of construction jobs for projects such as the Round Lake bypass and the new water line from Moreau to Saratoga Springs.

### 7.1.4.1 GlobalFoundries Construction Employment

In 2008 skeptics who questioned the economic impact of a new chip fab in Upstate New York argued that construction jobs should not be considered because "they are temporary, lasting less than 2 years."[47] However, the skeptics did not envision the sheer scale and duration of the construction activity that would actually occur. In mid-2013 *The Saratogian* reported that "GlobalFoundries has been in a constant state of construction since July 2009."[48] At the time, 1200 workers were active at the site with the total building toward 3000 by year's end. In the winter of 2014–2015, roughly 3500 construction workers were employed on-site, mainly working on the Technology Development Center.[49]

In 2015, long after completion of the original fab, 900–1100 construction workers were active at the GlobalFoundries site. Behind the site was "basically a mobile city," comprised of dozens of temporary offices housing companies providing plumbing, pipe-fitting, electrical work, and construction goods and services.[50] In

---

[47] "A Future Filled With Promise," Albany, *The Times Union* (April 20, 2008).

[48] "Nano Tech Valley: Construction Keeps Rolling," Saratoga Springs, *The Saratogian* (June 23, 2013).

[49] "GlobalFoundries a True Growth Business," Schenectady, *The Daily Gazette* (February 16, 2014). In 2014 so many works were employed at the site that they were asked to use a separate entrance from Cold Springs Road in Stillwater to avoid traffic congestion at the factory's main gate. Ibid.

[50] "GlobalFoundries a True Growth Business," Schenectady, *The Daily Gazette* (February 16, 2014).

## 7.1 Employment Effects

2015, anticipated new construction at the site prompted the Malta Planning Board to review designation of 16 acres for a new and expanded "trailer city" near the site—

> There are more than 80 trailers at GlobalFoundries today, but some projections say there could be as many as 200 trailers... Application documents indicate that contractors foresee remaining at the Fab 8 site for another five to eight years.[51]

Ehrlich's 2008 study forecast that construction of the AMD fab would result in 1600 direct jobs and an additional 2700 indirect jobs through economic multiplier effects.[52] The 2008 Semico study, based on data from other 300 mm fab construction projects, forecast that construction of a hypothetical upstate fab would give rise to 1500 direct jobs as well as 2550 indirect jobs based on a multiplier of 1.7.[53]

Neither the Ehrlich nor the Semico studies envisioned that the initial construction of a 300 mm fab would be augmented, as actually occurred, by new construction projects including expansion of the original design and addition of new structures These included a second 210,000 square foot administrative office building, about the size of three Walmart Supercenters, to house the workforce and the 565,000 square foot Technology Development Center, which entered the construction phase in 2013. As a result, rather than directly employing around 1500 construction workers for a temporary project of under 2 years' duration, between 2009 and early 2016 the work at the Luther Forest site created about 20,000 temporary construction jobs, and total construction man-hours are estimated at 10 million.[54] During the 2014–2015 period, GlobalFoundries' average daily "burn rate" (expenditure) for construction was $4.5 million per day.[55]

The construction jobs created at the Luther Forest site in and after 2009 coincided with the deepest recession the United States had experienced since the Great Depression. The construction industry was particularly hard hit.[56] The construction work provided by the GlobalFoundries fab enabled local construction firms, their workers, and the region to ride out the recession and avoid the worst effects experienced in other regions. Bob Fortune, Vice President at BCI Construction, a contractor working on the GlobalFoundries fab, indicated that without that work, his firm would have had to lay off workers—

---

[51] "Towns Weigh Relocating Fab 8 'Trailer City,'" Schenectady, *The Daily Gazette* (April 21, 2015).

[52] Everett M. Ehrlich, *Manufacturing, Competitiveness, and Technological Leadership in the Semiconductor Industry* (2008), p. 3.

[53] Semico Research Corporation, *Upstate New York: Assessing the Economic Impact of Attracting Semiconductor Industry*, March 2008, p. 53.

[54] Semico Research Corporation, *Upstate New York: Assessing the Economic Impact of Attracting Semiconductor Industry*, March 2008, p. 53.

[55] GlobalFoundries statistics.

[56] "Building Permits in Serious Decline—Construction Firms Must Adapt or Go Out of Business," *Fort Wayne News Sentinel* (January 14, 2008); "Pawning Tools of the Trade: Construction Workers in Hock, Others Sell Gold," Riverside, *The Press-Enterprise* (March 9, 2008); "Construction Recovery will be Slow," *The Grand Rapids Press* (December 9, 2009).

**Table 7.5** GlobalFoundries expenditures to New York-based vendors

| Firm | New York home base | Type of work | Amount billed to GlobalFoundries (millions of dollars) |
|---|---|---|---|
| Turner Construction | Albany | General contractor | 64.9 |
| Stone Bridge | Gansevoort | Structural steel | 43.3 |
| FPI Mechanical | Cohoes | Mechanical | 42.8 |
| Danforth Co. | Victor | HVAC/plumbing | 30.6 |
| Air Liquide | Feura Bush | Process chemicals | 30.4 |
| Western International | Fishkill | Mechanical/piping | 29.2 |
| Total Facilities Solutions | Watervliet | Electrical | 28.0 |
| Piller USA | Middletown | Power systems | 22.3 |
| Graybar Electric | Albany | Site lighting | 18.9 |
| T. Lemme Mechanical | Albany | Pipework | 13.4 |
| LeChase | Schenectady | Site work | 9.3 |
| George J. Martin & Sons | Rensselaer | Electrical | 9.3 |
| DLC Electric | Troy | Electrical | 9.1 |
| Selby & Smith | Albany | Material supplier | 8.7 |
| Sano-Rubin | Albany | Interview work | 8.7 |
| James H. Malloy | Loudonville | Site work | 7.6 |
| MLB Construction | Ballston Spa | Concrete | 7.0 |
| CG Power | Albany | Distribution/power transformers | 6.9 |
| AJS Masonry | Clifton Park | Masonry | 6.6 |
| SRI Fire Sprinkler | Albany | Fire sprinklers | 6.0 |
| Jersen Construction | Waterford | Concrete | 5.4 |
| J. Keller & Sons | Castleton-on-Hudson | Sewer work | 4.9 |
| WW Patenaude Sons | Mechanicville | Painting | 4.0 |
| DiGesare Mechanical | Schenectady | Plumbing | 3.8 |
| Trane | Latham | Chillers | 3.7 |
| TEC Protective Coatings | Waterford | Painting | 3.0 |
| Northeast Air | Albany | HVAC | 2.4 |
| Monaghan & Laughlin | Hudson Falls | Roofing | 2.4 |
| J.W. Stevens | Albany | Boilers | 2.1 |
| Otis Elevator | Albany | Elevators | 1.5 |
| Johnson Controls | Albany | Air handling units | 1.4 |
| Emerick Associates | Cohoes | Pumping | 1.2 |

(continued)

7.1 Employment Effects

**Table 7.5** (continued)

| Firm | New York home base | Type of work | Amount billed to GlobalFoundries (millions of dollars) |
|---|---|---|---|
| Kelly Bros. | Ballston Spa | Doors, frames, hardware | 0.9 |
| Mid-Hudson | Feura Bush | Exhaust fans | 0.7 |
| R.F. Peck | Albany | Exhaust fans | 0.6 |
| FCS Group | Lynbrook | Painting | 0.4 |
| B.R. Johnson | Syracuse | Overhead doors | 0.3 |
| Thermal Environmental Sales | Clifton Park | Centrifugal separators | 0.3 |
| Mechanical Testing | Waterford | Air/water balancing | 0.2 |
| DS Specialties | Mooers | Material supplier | 0.1 |
| Buckley Associates | Albany | Exhaust fans | 0.1 |

Source: "Complete Chip Plant Construction Gives Big Boost to GS Subcontractors," *Albany Business Review* (August 17, 2015)

*This has helped tremendously. If we didn't have this we'd be like a lot of other parts of the state or country, just desperately looking for work. We're fortunate to have this.*[57]

In a 2015 report filed with the Saratoga County Industrial Development Authority, GlobalFoundries itemized the amounts it had been billed by the construction contractor, Turner Construction, and subcontractors working on the 90,000 square foot expansion of Fab 8, which began at the end of 2012. The figures, shown in Table 7.5, convey the magnitude of expenditures by GlobalFoundries benefitting New York-based vendors.

### 7.1.4.2 Indirect Construction Employment

In 2016, the Schenectady *Daily Gazette* commented on indirect employment resulting from construction activity at the GlobalFoundries site, cautioning that "good numbers" on such jobs "aren't always easy to come by." However, it observed that—

*When construction started in 2009, temporary construction workers rented hotel rooms, houses and apartments across the Capital Region, many of them for jobs that lasted as long as five years. Millions of dollars were spent on rent alone, but any estimate would be only a guess.*[58]

A Schenectady-based restaurateur, Angelo Mazzone, contracted to provide food services for the construction workers and M+W Group administrators at the site,

---

[57] "Construction: GlobalFoundries Work Comes at the Right Time for Contractors," *Albany Business Review* (March 29, 2010).
[58] "GloFo By the Numbers: A Look at the Specs Behind the Massive Semiconductor Operation," Schenectady, *The Daily Gazette* (February 28, 2016).

hired 12 employees to staff an on-site cafeteria.[59] As construction began, local "restaurants in particular [were] noticing the impact."[60]

The Ehrlich and Semico studies forecast that for each direct construction job, construction of the fab would support 1.7 indirect jobs. Based on this 1.7 multiplier and the 900–1100 direct construction jobs attributable to the GlobalFoundries site, construction activity gave rise to another 1512–1870 indirect jobs. Because construction has been continuous, the Ehrlich/Semico estimates of indirect jobs based on an under-2 years construction timetable substantially understate the indirect employment impact of GlobalFoundries' construction projects. Factoring in the peak direct employment at key phases in the construction would also result in substantially higher figures for actual indirect employment.

### 7.1.5 Supply Chain Employment in Detail

The GlobalFoundries fabs at Malta/Stillwater and East Fishkill rely on long and complex supply chains for the equipment, materials, and services they require to sustain operations. Although many of the firms comprising the supply chain are located outside of New York, a number of their employees are present in the state and some suppliers have established a significant local presence—

- **Sumitomo (SHI) Cryogenics of America** (SCAI). In 2014 SCAI, a manufacturer of cryogenic equipment, moved into a 1600 square foot facility in Malta "near its main customer, GlobalFoundries." SCAI systems create near-vacuum conditions inside machines etching electronic circuits onto silicon wafers.[61]
- **Janitronics**. Janitronics is an Albany-based provider of specialized cleaning services which is responsible for many cleaning functions at the GlobalFoundries fab. On a typical day, the company has over 180 Janitronics employees on the site who work in the clean rooms to remove contaminants. In all (after accounting for managers), GlobalFoundries engages 200 Janitronics personnel on a full time basis. "Global is key to our business." The company hires high school graduates and virtually all of its recent hires grew up in the region.[62]
- **Beard Integrated Systems**. Beard is a Texas-based provider of process piping

---

[59] "Local Caterer Chosen to Feed Workers Building Chip Plant," Schenectady, *The Daily Gazette* (December 18, 2009).

[60] Since GlobalFoundries construction began and its operations came online, Mazzone Hospitality has leveraged its cafeteria services at GlobalFoundries to serve New York State offices in Albany, several new restaurants, and a growing catering business. This is a vivid example of regional business growth and induced job creation. "County Sales Tax Revenue Up Again," Schenectady, *The Daily Gazette* (September 21, 2010).

[61] "Cryogenics Group Opens Facility in Malta, is Air Pump Supplier for GlobalFoundries," *Saratoga Business Journal* (January 8, 2014).

[62] Interview with Jim Harris, president, Janitronics Facility Services, Albany, New York (November 29, 2016).

with a long history of specialized work for the semiconductor industry. In 2015 it moved into offices in Malta near the GlobalFoundries fab where it will work for the company on future fab expansion projects, bulk specialty gases, lateral systems, ultrapure water, and smaller projects.[63]

- **PeroxyChem**. GlobalFoundries consumes three 20,000 L truckloads of hydrogen peroxide per day, requiring an 1800-mile journey from a chemical plant operated by PeroxyChem in Texas. In 2015, ground was broken on a $30 million PeroxyChem hydrogen peroxide purification plant in Saratoga Springs, establishing a local source of supply for GlobalFoundries. The plant employs 15 workers with an annual payroll estimated at $670,000.[64] Fourteen of these workers were local hires.[65]
- **Axcelis Technologies**. In 2013 Axcelis, a Massachusetts-based maker of capital equipment for the semiconductor industry, leased 3000 square feet of space at the Clifton Park Flex Park at Northway Exit 10 "to support GlobalFoundries Fab 8.1 at the Luther Forest Technology Campus."[66]
- **Edwards Vacuum**. In 2014 Edwards Vacuum, a US-based supplier of clean room equipment to the semiconductor industry, opened what it characterized as a "world class facility" in 3000 square feet of space at the Great Oaks Office Park in Guilderland. The new site enabled the company to be "close to clients at the SUNY College of Nanoscale Science and Engineering and GlobalFoundries' Fab 8 computer chip factory in Malta."[67]
- **Mazzone Hospitality**. Mazzone Hospitality, based in Clifton Park, operates a catering business, restaurants, and on-site cafeterias in a number of industrial sites in the Capital Region. The cafeterias, operated by Mazzone's Prime Business Dining division, include the Global Cafe at GlobalFoundries. The company employs 1100 people in the Capital Region. Mazzone employees' on-site presence at GlobalFoundries has grown from 12 when it began operations in 2011 to 50–60 workers at present serving GlobalFoundries around the clock. Virtually all of Mazzone's personnel grew up in the region. The company's hires are usually high school graduates who are trained onsite for 40–50 h.[68]

Supply chain employment is expected to grow as GlobalFoundries' suppliers establish permanent local offices and facilities.

Out-of-state suppliers usually rely on logistics firms operating in the state to deliver equipment and materials to the GlobalFoundries fabs or to nearby ware-

---

[63] "Beard Integrated Systems Moves Into Malta Offices Working With GlobalFoundries Plant," *Saratoga Business Journal* (September 9, 2015).

[64] "GloFo Contractor to Build Hydrogen Peroxide Plant," Schenectady, *The Daily Gazette* (May 14, 2015).

[65] Telephone interview with Stephanie Montary, Peroxy Chem (January 16, 2017).

[66] "Axcelis Technologies Lands in Clifton Park Aided by National Grid Infrastructure Grant," *Saratoga Business Journal* (June 5, 2013). [Fab 8 was re-designated 8.1.]

[67] "Local Office Spaces Blend Amenities, Innovation," Albany, *The Times Union* (March 12, 2014).

[68] Interview with Angelo Mazzone, owner, Mazzone Hospitality, Clifton Park, New York (November 29, 2016).

houses. These include freight forwarding companies, expediters, and specialized transport companies. Most equipment and materials are shipped via air freight to JFK Airport in New York City and trucked the 200 miles to GlobalFoundries' fab in Malta/Stillwater. Some inbound shipments arrive in oceanborne containers and may arrive by rail via a Norfolk Southern intermodal container freight facility in Mechanicville which can handle shipments of bottled gas.[69] Some shipments are temperature-sensitive and require special trucks and storage. James Carey, an executive with Clancy Moving Systems, based in Putnam County, who handled some of the specialized trucking, commented in 2012 that "This [servicing GlobalFoundries] has really brought us up into the area."[70]

The large freight forwarders serving GlobalFoundries have set up local warehouses. Panalpina, a global logistics firm based in Switzerland, opened a 73,000 square foot warehouse facility in Clifton Park in 2011. The warehouse features climate and humidity controls for the storage of sensitive tools, as well as high ceilings and enlarged doors to accommodate outsized equipment. "GlobalFoundries' coming is really what brought us to the region," said Matt Brockway, Panalpina's strategic accounts manager, in 2012.[71] Suppliers of semiconductor manufacturing equipment have also established warehouses to house spare parts near the GlobalFoundries fab in Malta/Stillwater. Applied Materials operates a warehouse in Albany. Tokyo Electron and ASML operate warehouses near the fab. These storage facilities typically employ light industrial workers with a "few full blown engineers."[72]

In 2017 Albany-based Arnoff Moving and Storage established an $11.6 million global logistics hub in Malta, New York, adjacent to the Luther Forest Technology Campus. Arnoff renovated a 72,000 square foot former auto supply factory, adding a 25,000 square foot warehouse. Arnoff moved 50 jobs from Albany to Malta and created 40 new positions, looking to fill an additional 15 openings.[73]

In addition to employment by supply chain companies to sustain their operations, these firms frequently require construction projects to establish and expand their local presence. As the supply chain develops, it is fostering ongoing construction activity and related employment.

---

[69] "From Pigs to Nanochips," *The Journal of Commerce* (August 19, 2013).

[70] "Warehouse an Important Stop for Chip Plant Tools," Schenectady, *The Daily Gazette* (April 7, 2012).

[71] "Firm Aids at Fab Needs," Albany *The Times Union* (March 20, 2012); "Warehouse an Important Stop for Chip Plant Tools," Schenectady, *The Daily Gazette* (April 7, 2012).

[72] Interview with Jason VanBuren, GlobalFoundries (January 27, 2016).

[73] Arnoff Moving & Storage, "Arnoff Company Opens to $11.6 Million Logistics Hub in Saratoga County," *www.arnoff.com/blog* (June 17, 2017).

## 7.1.6 "Induced" Employment in Depth

The term "induced" employment is used to refer to jobs created to provide goods and services for a manufacturers' workforce which are not jobs at companies which comprise the manufacturers' supply chain. Thus, in Saratoga Springs, "local firms have sprung up to meet the needs of the administration and staff of GlobalFoundries, including hotels, housing, restaurants and shops, as well as expanded medical facilities and diverse community services."[74] Induced jobs reflect expenditures by workers directly employed by a manufacturer and its supply chain firms, visitors to the manufacturer from outside the region, and outlays by the manufacturer in the form of local taxes, fees, and charitable donations.

In 2015 GlobalFoundries paid its full-time employees at the Luther Forest site an annual average salary of $91,235, or more than twice the $40,000 estimated by Semico in 2008. Its gross payroll for 2015 was about $350 million, a substantial amount of which was undoubtedly spent by GlobalFoundries employees on goods and service in the surrounding region. To the extent that employees at the former IBM site at East Fishkill are compensated at a comparable level as those at Malta/Stillwater, the annual payroll at East Fishkill adds another $163 million. Use of Semico's 2008 average salary of $40,000 with respect to the individuals employed at the Albany NanoComplex produces another $160 million.

In total, workers at these three sites dispose of annual income totaling between one-half and three-quarters of a billion dollars. These sums are being spent on housing, cars, other goods and services, and local tax payments in the region. Visitors to the three sites from outside the region account for additional expenditures, particularly in the hospitality sector. Manifestations of such spending are increasingly observable across the region.

### 7.1.6.1 Housing

The housing industry employs local workers in real estate development, sales, rental, relocation, appraisal, lending, and property management businesses as well as construction workers when buildings are built or renovated. As GlobalFoundries ramped up its production operations, the surge in new residents associated with the fab increased demand for housing.

Much of the initial impact of GlobalFoundries on the regional housing market was in rentals, and a number of area developers launched building projects to expand the stock of rental housing. In 2013 Saratoga County Chamber of Commerce President Todd Shimkus said that "There are at least a half-dozen rental housing projects that have gone into Saratoga County from Clifton Park north to Malta and Saratoga Springs directly related to people relocating to the area to work for

---

[74] "Nano Tech Valley: A Pebble in the Pond—GlobalFoundries' Effect on the Community Around It," Saratoga Springs, *The Saratogian* (June 23, 2013).

**Table 7.6** Recent data on capital region home sales

| County | Closed sales | | Median sale price | |
|---|---|---|---|---|
| | 2015 quantity | Increase or decrease since 2014 | 2015 price (thousands of dollars) | Increase or decrease since 2014 (%) |
| Albany | 2437 | 12 | 200 | 3 |
| Schoharie | 248 | 33 | 115 | 6 |
| Montgomery | 259 | 30 | 99 | – |
| Schenectady | 1491 | 17 | 152 | – |
| Saratoga | 2370 | 7 | 253 | 5 |

Source: Greater Capital Association of Realtors

GlobalFoundries."[75] In mid-2015, a relocation director at Berkshire Hathaway Blake observed that in the Capital Region inquiries about single-family rentals were up 20% in the first half of 2015 compared with the same period in 2014. She said that the 20% was indicative of "a significant increase of interest" and that it was "driven in part by hiring at GlobalFoundries in Malta and IBM at SUNY Polytechnic Institute in Albany."[76]

Developers based in Florida and western New York have disclosed plans for construction of 512 upscale rental units on a site in Malta about 3 miles from the GlobalFoundries fab. Phase I of this project, "Grande Ville at Park Place," envisions construction of 292 one-, two-, and three-bedroom units in 18 buildings. The developers indicate they will acquire a 5500 square foot nearby community center, enlarge it to 9000 square feet, and renovate it for use by tenants. The developers "were attracted to the region by the growth of the nanotechnology industry."[77]

Sales of houses have also increased, most notably in Saratoga County but also in Warren, Washington, Albany, and Rensselaer Counties.[78] In January 2016 Laura Burns, the CEO of the Greater Capital Association of Realtors, cited newly released data showing an improving year-over-year picture in home sales, commenting that "Job-wise, there's a lot of investment in technology firms such as GlobalFoundries, Albany's Convention Center and all of the activity in Schenectady with the coming casino." In all, the Capital Region's housing market was characterized as the best since the 2008 recession.[79] (See Table 7.6.)

The advent of GlobalFoundries has also been noted as a factor underlying new

---

[75] Shimkus cited as examples developer Sonny Bonacio's Market Center in Saratoga Springs and Ellsworth Commons in Malta. The Market Center project created 124 rental apartments and 31,000 square feet of retail space. Ellsworth Commons was a $53 million project which built 22 townhouses, 310 lofts and apartments, and 73,000 square feet of retail space. "GlobalFoundries Expansion Could Spark New Building Boom," Glens Falls, *The Post-Star* (January 27, 2013).

[76] "One Family Rentals Fly Off the Lists," Schenectady, *The Daily Gazette* (July 6, 2015).

[77] "500 Housing Units in Malta are Planned by Developers from Florida, Western NY," *Saratoga Business Journal* (May 7, 2015).

[78] Between November 2011 and November 2012, sale of homes in Saratoga County increased by 17.5%. "GlobalFoundries Expansion Could Spark New Building Boom," Glens Falls, *The Post-Star* (January 27, 2013).

[79] "Housing Market Robust," Schenectady, *The Daily Gazette* (January 23, 2016).

## 7.1 Employment Effects

**Table 7.7** Building permit data

| Town | Number of building permits issued | Number of units | Cost (millions of dollars) |
|---|---|---|---|
| Halfmoon | 143 | 183 | 36.6 |
| Colonie | 150 | 266 | 49.6 |

Note: Table includes data from January to November 2014

home construction in the region around Malta/Stillwater. Two towns, Halfmoon in Saratoga County and Colonie in northern Albany County, led all Capital Region municipalities for the first 11 months of 2014 in issuing building permits (see Table 7.7). Halfmoon Planning Director Richard Harris commented on these figures in early 2015, saying that "builders have told me this anecdotally, but much of the residential development that's happening here is a result of the people coming in from out of the area to work in Malta. It's the growth in Malta that's doing it." Harris observed that with GlobalFoundries having created 3000 permanent jobs and planning additional hiring, "new residential construction is expected to continue in the communities surrounding Malta as these employees look for homes nearby."[80] In 2013, an official of Halfmoon's building department said he realized how good GlobalFoundries was for home construction in his town after a line of houses in one development had filled up, noting—

> Six right in a row. One right next to another, all in one development. They were all GlobalFoundries employees of some sort.... I talk to developers and builders, asking where this person comes from or that person, and they all say GlobalFoundries.[81]

### 7.1.6.2 Hospitality Industry

The construction of the GlobalFoundries fab and subsequent startup of operations resulted in a robust growth in demand for nearby temporary lodging, ranging from one-night lodging to extended-stay arrangements involving hybrid "part-hotel, part-condo" arrangements.[82] Each of these facilities employs workers from surrounding communities. Todd Shimkus, President of the Saratoga County Chamber of Commerce, said that sales and service people associated with the technology industry need a room "for a week, 2 weeks, a month," while making sales or working on projects.[83] In 2013 Todd Garofalo, President of the Saratoga Convention and Visitors Bureau, said that "GlobalFoundries has been bringing employees from around the world to Saratoga for training or temporary assignments at Fab 8. They are sometimes staying in local hotels for weeks… or using commercial lodging while they

---

[80] "Construction Boom in Suburbs," Schenectady, *The Daily Gazette* (January 19, 2015).

[81] "GloFo Keeping Builders Busy," Schenectady, *The Daily Gazette* (January 3, 2013).

[82] "Hotel Occupancy Rates up Sharply for City, County," Schenectady, *The Daily Gazette* (April 20, 2012).

[83] "Pavilion Grand Suites Takes Shape," Schenectady, *The Daily Gazette* (June 16, 2013).

learn the area and shop for a house."[84]

As of 2013 Saratoga Springs had 1645 hotel rooms, with another 1000 available elsewhere in Saratoga County.[85] By mid-2014, "a new wave of hotel construction" was "on the drawing board or under construction along the Northway corridor from Colonie to Saratoga Springs."[86] The ten new projects involved nearly 1000 additional rooms. While these projects typically involved investments in the $10–12 million range, some were much more expensive, such as the expansion of the Rip Van Dam Hotel in Saratoga Springs, an investment of $45 million. Shimkus observed in 2014 that—

> The makeup of the consumer renting hotel rooms in Saratoga County has changed. We have a significantly higher percentage of corporate travel than ever before.

The hotel construction boom has not been limited to Saratoga Springs. At the end of 2015, a new 107-room Hilton House 2 Suites was under construction in Malta itself which "will offer extended-stay lodging likely geared to visitors to the GlobalFoundries computer chip plant, which has out-of-town employees and plant vendors visiting for several weeks at a time."[87] In Clifton Park, a "hotel zone" has materialized around the downtown area near Northway Exit 9 which offers ready access to Albany, Schenectady, Saratoga Springs, "or the growing number of high-tech sector jobs in Malta." As of early 2016, the Clifton Park hotel zone had 710 rooms—triple the number in 2000—and was projected to grow to 1000 rooms by 2017 "if projects now under construction or on the drawing board become reality."[88]

In a February 2016 feature on the hotel construction boom, the Schenectady *Daily Gazette*, which a decade before had disparaged the prospective impact of a chip fab on the region, observed that—

> The arrival of GlobalFoundries in Malta has meant a lot to the hospitality industry, as have the billions of dollars being invested by government and private industry at the SUNY Polytechnic Institute Campus in Albany. The $12 billion GlobalFoundries semiconductor plant alone, with its 3,000 employees, has had a significant impact on lodging needs since construction started in 2009, since construction workers, suppliers and employees involved in training sometimes come to the area for weeks at a time. Also people relocating to the area for new jobs sometimes book extended stays while deciding where to live.[89]

One of the hotel industry professionals interviewed for the *Daily Gazette* story was Deborah Charbonneau, General Manager of Homewood Suites, the newest hotel in Clifton Park. Charbonneau said that she had personally seen the changes

---

[84] "Outlook 2013: GlobalFoundries Becoming an Economic Powerhouse," Schenectady, *The Daily Gazette* (February 17, 2013).

[85] "Hotel Fever in Spa City," Albany, *The Times Union* (May 14, 2013); "Four New Hotels to Grace Saratoga Springs Skyline," Saratoga Springs, *The Saratogian* (May 4, 2013).

[86] "Growing Region Tells Hotel Story," Albany, *The Times Union* (May 11, 2014).

[87] "Work Under Way on Hotel, Restaurant," Schenectady, *The Daily Gazette* (December 1, 2015).

[88] "Exit 9 'Hotel Zone' Matures, and is Still Growing: Town May Reach 1000 Rooms," Schenectady, *The Daily Gazette* (February 29, 2016).

[89] "Exit 9—'Hotel Zone' Matures, and is Still Growing: Town May Reach 1000 Rooms," Schenectady, *The Daily Gazette* (February 29, 2016).

GlobalFoundries has brought to the southern part of Saratoga County. She moved to the area from California in 2008 and got a job at the then-new Hyatt Place Hotel at Exit 12 in Malta, subsequently moving to the Hilton Gardens in Clifton Park Center and finally to the General Manager position at the Homewood Suites. She said "I've seen the town kind of change, I've built my career on it."[90]

Numerous new restaurants have opened in the region, with GlobalFoundries and CNSE often cited as factors.[91] In the 12-month period between June 2011 and June 2012, restaurants and bars created 658 new jobs in Saratoga County. Angelo Mazzone, whose Mazzone Hospitality had operated the food tent at the GlobalFoundries construction site, invested $250,000 to open GlobalCafe, an on-premises cafeteria serving the GlobalFoundries workforce.[92] In 2012 Albany-based BBL Hospitality noted that downtown Malta was "underserved" with casual-dining restaurants in light of the development of the GlobalFoundries fab, and announced it would open a Recovery Sports Grill on Route 9, with 30–35 full-time employees.[93] A month later the owners of Wheatfields Restaurants, known for premises-made pastas and other Italian specialties, announced they would open a Wheatfields in Malta, with one owner commenting, "I live in Malta, and I see all this growth going on. I think Malta can support several high-quality restaurants."[94]

### 7.1.6.3 Retail

Nanotechnology-related operations in the Capital Region have drawn significant new investments in retail stores throughout the region. In 2013 the General Manager of Crossgates Mall in Albany, Joseph Castaldo, noted that Lord & Taylor and Designer Michael Kors would open stores at the mall in 2014 and that Crate & Barrel would open up a "pop-up" store during the Christmas season, one of only four nationwide. Castaldo observed that the region had grown wealthier and that the technology sector's impact should not be overlooked—

> When we do leasing for any tenant, whether Lord & Taylor or Dave & Buster's…the boom in the nanotech college, GlobalFoundries, the Global 450 consortium, all of that has been a selling point. It changes your whole demographic profile.[95]

Ted Potrikus, Senior Vice President of the Retail Council of New York State,

---

[90] "Exit 9—'Hotel Zone' Matures, and is Still Growing: Town May Reach 1000 Rooms," Schenectady, *The Daily Gazette* (February 29, 2016).

[91] "Saratoga County Sees the Highest Employment Gain in New York State," Saratoga Springs, *The Saratogian* (February 27, 2013).

[92] "Thanks to GlobalFoundries, Malta is Fertile for Foodies," *Albany Business Review* (September 14, 2012).

[93] "Growing Recovery Grill to Sport New Location," Schenectady, *The Daily Gazette* (July 3, 2012).

[94] "Wheatfields Operators to Open Eatery in Ellsworth Commons," Schenectady, *The Daily Gazette* (August 23, 2012).

[95] "Tech Boom Luring Stores to Region," Albany, *The Times Union* (August 16, 2013).

**Table 7.8** Tech Valley semiconductor R&D and manufacturing jobs impact

|  | Direct employment (2015)[a] | Estimated indirect employment (2015) based on— | | |
|---|---|---|---|---|
|  |  | ESD multiplier (2.25) | SIA multiplier (4.89) | Moretti multiplier (5.00) |
| GF Malta/Stillwater | 3538 | 7960 | 17,300 | 17,690 |
| GF E. Fishkill | 2085 | 4691 | 10,196 | 10,425 |
| CNSE Albany | 4000 | 9492 | 19,560 | 20,000 |
| Totals | 9623 | 22,143 | 47,056 | 48,115 |

[a]Includes direct employees and full-time on-site employees of vendors

cited the proximity of the NanoCollege as the reason for retailers' interest. "A company like Lord & Taylor wouldn't make the investment it's going to make—this isn't a kiosk opening for the holiday season—(until) they've done the math and take a look at the population and the area."[96]

New small retail outlets have also opened near the nanotechnology hubs, frequently occupying retail space in new mixed-use housing and retail complexes. In 2013 Delmar Opticians, based in Delmar, New York, opened a second office in Ellsworth Commons in Malta, a mixed commercial/residential development, "drawn by the excitement created by the GlobalFoundries manufacturing plant in Malta."[97]

#### 7.1.6.4 Healthcare

The thousands of workers at the Capital Region's nanotechnology hubs have increased the demand for healthcare services. When GlobalFoundries began work as the Luther Forest fab, its general contractor, M+W Group, contracted with Saratoga Hospital to establish and staff an on-site clinic to support the construction workforce.[98] In September 2010, Saratoga Hospital and Albany Medical Center announced a partnership to provide outpatient care, including emergency services, at a center to be established as the first facility at the Saratoga Medical Park in Malta, a 140-acre site owned by Saratoga Hospital near the GlobalFoundries fab. SEDC spokesperson Dennis Brobston welcomed the initiative, stating that—

---

[96] "Tech Boom Luring Stores to Region," Albany, *The Times Union* (August 16, 2013). The President of Albany-based real estate firm, the Harvard Group, commented that "The tech boom has brought in people who didn't live here before. We now have a number of jobs that are higher-paid." Ibid.

[97] Thomas Hughes, Jr., owner of Delmar, said that he had first come to Malta 18 months previously to visit a friend, and "I couldn't believe what was going on. The area was really booming." See "Development in Malta Prompts Delmar Optician to Open Saratoga County Office," *Saratoga Business Journal* (August 6, 2013).

[98] "Saratoga Hospital Staffing Clinic at Luther Forest Site," Schenectady, *The Daily Gazette* (November 16, 2010).

7.1 Employment Effects

**Table 7.9** Direct plus indirect/induced employment impact

| Multiplier | Total direct plus indirect/induced based on multiplier |
|---|---|
| Empire State Development (2.25) | 31,766 |
| Semiconductor Industry Association (4.89) | 56,679 |
| Moretti (5.0) | 73,738 |

> *I applaud Saratoga Hospital and Albany Medical Center for their foresight to forge this partnership and invest at Exit 12, Saratoga County. GlobalFoundries' investment at the Luther Forest Technology Campus will attract thousands of high-tech jobs within Fab 8 and throughout the surrounding area. This workforce will expect a world class quality of life, which includes quality healthcare. Saratoga Hospital and Albany Med will be in the center of it all.*[99]

The new facility, Malta Med Emergent Care, opened in 2013, featuring a $17.3 million building, sophisticated equipment including MRI, CT, X-ray and ultrasound, and offices for doctors from the collaborating hospitals. Malta Med represented the first phase of a long-term developmental plan, with the next phase to involve long-term care and medical office space.[100]

### 7.1.7 Summarizing Employment Effects

The direct employment associated with semiconductor research and manufacturing operations in Tech Valley totaled over 9000 jobs in 2015. While indirect/induced jobs cannot be calculated with precision, they substantially exceed direct employment even utilizing the multiplier used by ESD for all manufacturing—2.5 induced jobs for each direct job—which is almost certainly too conservative given the high income levels and spending power enjoyed by research and manufacturing employees in the semiconductor industry. Table 7.8 depicts estimated indirect employment in Tech Valley based on 2015 direct employment totals utilizing multipliers developed by ESD for all manufacturing, by the Semiconductor Industry Association for semiconductor manufacturing, and by Professor Enrico Moretti of the University of California at Berkeley for the "innovation sector." As the table shows, the SIA and Moretti multipliers would suggest that the advent of semiconductor research and manufacturing in Tech Valley has resulted in creation of additional 45,000–60,000 indirect/induced jobs in the region. The ESD multiplier, based on all manufacturing industries, produces a figure of 20,000 additional indirect/induced jobs. While it is conceivable that the SIA and Moretti multipliers yield a figure that is too high, the

---

[99] Albany Medical Center, "Albany Med and Saratoga Hospital Announce Joint Venture," Press Release, September 15, 2010.
[100] "Malta Med Emergent Care Facility Opening at Northway Exit 12," Saratoga Springs, *The Saratogian* (May 31, 2013).

**Table 7.10** GlobalFoundries local tax payments and fees, 2011–2016

| | Amount (dollars) | |
|---|---|---|
| | Malta/Ballston Spa SD | Stillwater |
| Town taxes | 359,417 | 3,072,327 |
| School taxes (PILOT) | 49,736,680 | 13,624,369 |
| Development fees | 4,769,368 | 1,729,283 |
| Fire department | 2,611,346 | 14,995 |
| Library | 403,999 | – |
| Foundations | 732,851 | 341,350 |
| Total | 58,710,991 | 18,440,936 |

Source: GlobalFoundries

**Table 7.11** Additional GlobalFoundries taxes and fees

| Year | Amount of Saratoga County Taxes paid (dollars) | Payments to LFTCEDC (dollars) |
|---|---|---|
| 2011 | 365,350 | – |
| 2012 | 1,477,451 | 43,602 |
| 2013 | 1,532,107 | 46,806 |
| 2014 | 1,563,328 | 44,625 |
| 2015 | 1,576,342 | 46,689 |
| 2016 | 1,498,037 | 26,171 |
| Total | 8,012,615 | 207,892 |

Source: GlobalFoundries

number based on the ESD multiplier is probably far too low, a fact that is reinforced by what can be gleaned from anecdotal empirical information regarding supply chain and induced employment in the region.

As shown in Table 7.9, when the direct and indirect/induced job totals are combined, the result, depending on the multiplier used, ranges from 31,766 (almost certainly too low) to 56,679–73,738 (defensible based on recent work by the semiconductor industry and academia). To the extent that the Capital Region's perception of the impact of nanotechnology is still defined by the 2008 forecasts, it is simply out of step with reality.

## 7.2 Other Regional Expenditures by Nanotechnology Firms

In addition to spending on payroll, construction projects, and procurement of goods and services from supply chain firms, GlobalFoundries, CNSE, and other nanotechnology firms and organizations make other expenditures that affect the local economy. These include payment of local taxes and fees, payments to utilities, and charitable donations.

**Table 7.12** Total grants made by the GlobalFoundries-Town of Malta Foundation and the GlobalFoundries Stillwater Foundation through 2016

| Year | Value GlobalFoundries-Town of Malta Foundation grants (thousands of dollars) | GlobalFoundries Stillwater Foundation grants (thousands of dollars) | Number of recipient organizations |
|---|---|---|---|
| 2011 | 37.5 | – | 13 |
| 2012 | 60.0 | – | 16 |
| 2013 | 164.0 | – | 38 |
| 2014 | 172.9 | 242[a] | 70 |
| 2015 | 151.9 | 49.8 | 56 |
| 2016 | 146.6 | 49.5 | 51 |
| Total | 732.9 | 341.4 | |

[a]GlobalFoundries. Includes grants made in 2013

## 7.2.1 Local Taxes and Fees

Between 2011 and 2016, GlobalFoundries paid over $77 million in taxes, fees, and payments to foundations to the towns of Malta and Stillwater, which had a cumulative population of 23,052 in the 2010 census.[101] (See Table 7.10.) In 2015 Moody's Investor Service upgraded the Town of Malta's credit rating from Aa2 to Aa1, "a reflection of GlobalFoundries' impact on the local economy and the town's continued financial health." Aa1 is the second highest rating that Moody's gives municipalities and the upgrade "could mean lower interest rates when Malta borrows money in the future."[102]

The school taxes paid by GlobalFoundries to the Ballston Spa Central School District, where the fab is located, have been sufficiently large to enable property tax reduction for other property owners in the district.[103] For 2015–2016 Malta's school budget tax levy was $46,716,018, of which GlobalFoundries was expected to pay about 19.7%. Stillwater's 2015–2016 school budget tax levy was $9,348,840, of which GlobalFoundries was expected to pay about 27%.[104] These GlobalFoundries' payment totals do not include property taxes paid by its employees, and the company has also funded technology-related education programs, such as a Tech Valley robotics competition.[105] The Albany *Times Union* reported in 2014 that—

> It may be hard to measure the direct economic impact that GlobalFoundries and other high-tech companies are having on the region. But … the most visible impact is on local school

---

[101] "GlobalFoundries Filling Local Coffers," Albany, *The Times Union* (August 12, 2014).

[102] "A. First," Schenectady, *The Daily Gazette* (August 3, 2015).

[103] "GlobalFoundries Chip Fab Plant Fosters a Ripple Effect Felt Far and Wide," Saratoga Springs, *The Saratogian* (July 24, 2012).

[104] GlobalFoundries.

[105] GlobalFoundries Share Success With Surrounding Community," Saratoga Springs, *The Saratogian* (November 13, 2014); "GlobalFoundries to Work With Schools on High Tech Training," Schenectady, *The Daily Gazette* (August 29, 2013).

**Table 7.13** Examples of grants made by the GlobalFoundries-Town of Malta Foundation

| Grant value | Recipient | Activity supported |
|---|---|---|
| 10,000 | Ballston Spa School District | Expand robotics team program |
| 20,000 | Malta-Stillwater Emergency Medical Services | Acquisition of mechanical cardio-pulmonary resuscitation device |
| 10,000 | Rebuilding Together Saratoga County | Building materials for home repairs for those in need |
| 7000 | Malta citizen preparedness program | Trains citizens for volunteer work in emergencies |
| 11,750 | Saratoga Bridges | Kitchen renovations for community-based home for those with developmental disabilities |
| 21,800 | Community Foundation for Greater Capital Region | Engineering education program run by Ballston Spa School District |

> districts. During the 2013-14 school year, local districts received $11.1 million in tax payments from GlobalFoundries, nearly all of the $13 million in payments the company made last year as a result of its deal with local municipalities.... The money had a major impact, accounting for 11 percent of the total budgets of both districts [Ballston Spa and Stillwater] at a time when most districts are scrambling to meet budget shortfalls.[106]

As shown in Table 7.11, GlobalFoundries also pays taxes to Saratoga County and makes annual payments to the Luther Forest Technology Campus Economic Development Corporation (LFTCEDC) to contribute to roadway and common area maintenance within Luther Forest.

## 7.2.2 GlobalFoundries Charitable Local Expenditures

Pursuant to development agreements with the Towns of Malta and Stillwater, GlobalFoundries has created two non-profit foundations. The foundations have supported nearly 200 community-based organizations with grants totaling nearly $1 million as of early 2016 as shown in Table 7.12.

The foundation boards include a mix of GlobalFoundries representatives and local residents. In 2014–2015 the GlobalFoundries-Town of Malta Foundation made a diverse array of awards.[107] Examples of grants made are shown in Table 7.13.

In 2009 GlobalFoundries contributed $1 million to the Town of Malta to cover part of the cost of building youth ball fields in Luther Forest.[108] The landscaping and architecture firm LA Group of Saratoga Springs was awarded a $110,000 contract

---

[106] "Taxes Bolster Schools," Albany, *The Times Union* (September 13, 2014).

[107] "Area Schools Fare Well in GloFo Foundation Grants," Schenectady, *The Daily Gazette* (December 18, 2015); GloFo Foundation Awards $172,915 in Grants," Schenectady, *The Daily Gazette* (January 2, 2105).

[108] "Town Gets $1.1 Million from Tech Park," Schenectady, *The Daily Gazette* (July 29, 2009).

to design the fields.[109] The Town of Malta augmented GlobalFoundries' $1 million with funds from its recreation budget. The sports complex opened in 2013 featuring two softball fields and two soccer fields. The fields proved sufficiently popular that the town decided in 2014 to add a large concession building with bathrooms, a smaller concession stand near the most distant fields, and sheltered decks for players' use in rainy conditions.[110] In addition, Brown's Beach, which had been a public beach and lake access on Saratoga Lake dating back to the 1800s, had been closed to the public in recent years and was slated for private development. Through an agreement with the Town of Stillwater in which GlobalFoundries would pay $3 million at the time of the ground breaking of a second fab to be applied toward the purchase of the beach with a guarantee that it would remain a public beach in perpetuity, the town was able to purchase the beach and return what had been a favorite local destination for more than a century back to public use.

## 7.3 Conclusion

As outlined in this chapter, the regional and state efforts to build out this unique research and manufacturing cluster have resulted in a major success. Indeed in many ways it has significantly exceeded the original expectations, not to mention the state's investment requirements. Specifically—

- **Direct employment** attributable to nanotechnology research and manufacturing totals nearly 10,000, and at GlobalFoundries average compensation of full-time employees is over $92,000 per year.
- **Indirect and induced employment** exceeds 20,000 jobs and may total nearly 50,000 jobs.
- **Construction employment** has vastly exceeded 2008 forecasts, which foresaw 1500–1600 jobs for 2 years. The GlobalFoundries Malta/Stillwater site has involved as many as 3500 jobs at peak and still employs hundreds of construction workers 8 years after ground was broken for the fab. Most of these jobs feature union-scale compensation, reflecting a seminal agreement reached in 2009 between GlobalFoundries and the local building trades. There have been no strikes.
- **Total employment** attributable to New York's investments in nanotechnology clearly approaches 60,000–80,000 direct, indirect, and induced jobs in addition to the very substantial jobs related to construction.

Much of New York's success is due to the combination of effective and committed state, university, and private sector leadership over a sustained period. Their efforts were successful not just because of continuity of leadership but also because

---

[109] "Firm to Design Tech Campus Ballfield," Schenectady, *The Daily Gazette* (September 10, 2010).

[110] "Town to Add Athletic Complex," Schenectady, *The Daily Gazette* (February 7, 2014).

of a willingness to make available internationally competitive bids through instruments such as the Empire State Development Corporation. The result has been significantly enhanced levels of economic activity, and more importantly, the generation of a much larger number of high-quality, high-paying jobs—both at GlobalFoundries and at CNSE—than was originally anticipated. The sustained effort has also had an enormously positive impact on the region's national and global reputation, where it is now seen as an outstanding center of advanced research and manufacturing.

## Bibliography[111]

500 Housing Units in Malta are Planned by Developers from Florida, Western NY. (2015, May 7). *Saratoga Business Journal*.
Albany Medical Center. (2010, September 15). Albany Med and Saratoga Hospital Announce Joint Venture. Press Release.
Arnoff Moving & Storage. (2017, June 17). Arnoff Company Opens to $11.6 Million Logistics Hub in Saratoga County. http://www.arnoff.com/blog.
Axcelis Technologies Lands in Clifton Park Aided by National Grid Infrastructure Grant. (2013, June 5). *Saratoga Business Journal*.
Beard Integrated Systems Moves Into Malta Offices Working With GlobalFoundries Plant. (2015, September 9). *Saratoga Business Journal*.
Building Permits in Serious Decline—Construction Firms Must Adapt or Go Out of Business. (2008 January 14). *Fort Wayne News Sentinel*.
Can American Manufacturing Really Be Cornerstone of Economic Revival? (2012, February 8). *The Christian Science Monitor*.
College of Nanoscale Science and Engineering. Quick Facts. http://www.sunycnse.com/AboutUs/QuickFacts.aspx.
Construction Recovery will be Slow. (2009, December 9). *The Grand Rapids Press*.
Cryogenics Group Opens Facility in Malta, is Air Pump Supplier for GlobalFoundries. (2014, January 8). *Saratoga Business Journal*.
Development in Malta Prompts Delmar Optician to Open Saratoga County Office. (2013, August 6). *Saratoga Business Journal*.
Dudley, W. C. (2014, October 7). The National and Regional Economy. Remarks at RPI, Troy, New York.
Ehrlich, E. M. (2008). *Manufacturing, Competitiveness, and Technological Leadership in the Semiconductor Industry*.
From Pigs to Nanochips. (2013, August 19). *The Journal of Commerce*. http://www.techhistoryworks.com/silicon-valley-history/2016/9/21/the-father-of-silicon-valley.
GlobalFoundries: E. Fishkill Site Key to Overall Strategy. (2015, November 12). *Poughkeepsie Journal*.
Made in America: Global Companies Expand in US Towns. (2012, April 30). *ABC News*.
Moretti, E. (2013). *The New Geography of Jobs*. Boston and New York: Mariner Books.

---

[111] As noted in the front matter of this book, the study also drew on interviews carried out by the authors and numerous articles from *The Times Union* (Albany), *The Daily Gazette* (Schenectady), the *Albany Business Review* (Albany), *The Post-Star* (Glens Falls), *The Record* (Troy), *The Saratogian* (Saratoga Springs), *The Buffalo News* (Buffalo), The *Observer-Dispatch* (Utica), The *Daily Messenger* (Canandaigua), and the *Post-Standard* (Syracuse). These are not individually included in the bibliography.

National Research Council. (2013). *New York's Nanotechnology Model: Building the Innovation Economy*. C. W. Wessner (rapporteur). Washington, DC: The National Academies Press.

New York's Big Subsidies Bolster Upstate's Winning Bid for AMD's $3.2 Billion 300-MM Fab. (2006, June 10). *Site Selection*.

Pawning Tools of the Trade: Construction Workers in Hock, Others Sell Gold. (2008, March 9). Riverside, *The Press-Enterprise*.

Scott, E., and Wial, H. (2013, May). *Multiplying Jobs: How Manufacturing Contributes to Employment Growth in Chicago and the Nation*. Chicago: University of Illinois at Chicago.

Semico Research Corporation. (2008a, February). *Economic Impact of the Semiconductor Industry on Upstate New York*.

Semico Research Corporation. (2008b, March). *Upstate New York: Assessing the Economic Impact of Attracting Semiconductor Industry*.

# Chapter 8
# Educating and Training a High-Tech Workforce

**Abstract** The Capital Region was able to attract major inward investments by high-technology companies largely because its educational institutions ensured the availability of skilled and educated manpower. However, the growth of the region's technology-intensive industries has exceeded forecasts, and tech firms are warning of a "skills gap" (e.g., a major shortfall in available workers with the requisite knowledge and skill sets). Across the region, educational institutions are scrambling to respond with new investments, programs, and initiatives. This effort is Tech Valley's single most important, and complex, challenge.

In 2009 the Semiconductor Industry Association surveyed its members to determine what factors were given the most weight in evaluating potential sites for location of new facilities. The survey revealed that by far the most important consideration was the availability of a well-educated local workforce capable of supporting semiconductor design and manufacturing.[1] Awareness of that fact led the architects of Tech Valley to emphasize the region's strong educational institutions in their outreach efforts to semiconductor manufacturers in and after 1998.[2] Semiconductor executives visiting the region were given tours of RPI, SUNY Albany, and other

---

[1] Semiconductor Industry Association, *Maintaining America's Competitive Edge: Government Policies Affecting Semiconductor Industry R&D and Manufacturing Activity* (March 2009).

[2] Brian McMahon, Executive Director of the New York State Economic Development Council, recalls that when Capital Region economic developers went to Semicon West and similar gatherings of semiconductor industry leaders, SUNY Albany "sent representatives who could explain what, technologically, was going on" at the NanoCollege: interview, Albany, New York (October 28, 2015). See also "Area Colleges Getting into the Tech Valley Game," Schenectady, *The Daily Gazette* (February 22, 2004). AMD/GlobalFoundries decision to establish the nation's first semiconductor foundry near Albany was based, in substantial part, on its reasoning that "[T]he nearby community colleges could help provide workers with some of the skills it needed, and nearby universities such as Rensselaer Polytechnic Institute and the University at Albany could help supply some of the higher-level talent." See "Can American Manufacturing Really Be the Cornerstone of Economic Revival?" *Christian Science Monitor* (February 8, 2012). The CEO of a major Capital Region engineering company commented in 2016 that a large percentage of his company's employees come out of New York universities—"higher education here is incredible, [the schools are] still putting out a product, doing what they are supposed to do." Interview in Albany, New York (September 16, 2016).

**Table 8.1** Highlights of Albany Times Union Survey of Local Investments by Colleges and Universities (2016)

| Institution | Projects | Value of investments (millions of dollars) |
|---|---|---|
| SUNY Polytechnic | NanoFab Xtension, ZEN building (renewables), NanoFab East and North | 891 |
| SUNY Albany | ETC (Emerging Technology and Entrepreneurship) complex, new engineering school renovations, data center, Mohawk Tower | 800 |
| Rensselaer Polytechnic | Center for Biotechnology, Experimental Media and Performing Arts Center, East Campus Athletic Village | 400 |
| Union College | Renovations, Peter Irving Wold Science Building | 127 |
| HVCC | Science Center, TEC-SMART, Haas Center for Advanced Manufacturing Skills | 116 |
| Sienna College | New residence and dining halls, Stewart's Advanced Instrumentation and Technology Center, technology upgrades | 111 |
| Skidmore College | Student residences, dining hall, Arthur Zankel Music Center | 113 |
| College of Saint Rose | Residence hall, Massry Center for the Arts, Thelma P. Lally School of Education | 90 |

Source: "College Spending Works to Meet Student Needs," Albany *The Times Union* (November 27, 2016)

educational institutions. Years before any semiconductor manufacturers committed to locate in the region, Senate Majority Leader Joseph L. Bruno "adopted" Hudson Valley Community College (HVCC) in Troy, steering state grant money to HVCC to enable training programs in semiconductor manufacturing and in alternative and renewable energy.[3]

Public and private investments in higher education in New York have been substantial and sustained. A 2016 informal survey of Capital Region colleges and universities revealed a sweeping array of capital expansion initiatives under way or recently completed across the region, particularly at institutions with strong engineering programs such as RPI, Union College, and SUNY Polytechnic (see Table 8.1).

## 8.1 The Skills Gap

Paradoxically against this background, unmet human resource needs, particularly in technology-intensive occupations, are constraining further regional economic growth. At the end of 2016, *The Daily Gazette* of Schenectady reported on a chronic shortage of workers with the skill sets necessary for high-tech manufacturing:

---

[3] Joseph L. Bruno, *Keep Swinging: A Memoir of Politics and Justice* (Brentwood, TN: Post Hill Press, 2016), pp. 137–138. Bruno reflected that "I didn't know how this would play out, yet of one thing I was certain. Unlike my father, these young men and women would retire with more than a 10 dollar-a-week pension." *Ibid*.

## 8.1 The Skills Gap

*Hard though it may be to believe, there are companies—solid manufacturing companies with good reputations that offer jobs with benefits—that are begging for people to trade their time for a paycheck. GlobalFoundries—the tech giant that has changed the region's economic landscape—has around 100 technician jobs it can't fill, and the hiring sign outside the Quad/Graphics printing plant in Saratoga Springs seems to be permanent. [A local economic development official said] "I hear over and over again that we can't find the workers we need."*[4]

Part of the problem in 2016 was simply Saratoga County's unemployment rate of 3.6%—meaning that "anyone half-talented already has a job" (see Footnote 4). But another issue is a longstanding skills deficit in the regional work force. Although local educational institutions, trade unions, and other public and private training organizations in the Capital Region have been working for decades to address this challenge, the increase in demand within the region for skilled workers continues to outpace the supply.[5] The human resources firm Linium Recruiting, which conducts quarterly surveys of Capital Region company personnel executives to compile a Hiring Index, found in its first survey of 2017 that 82% of the companies surveyed were finding it "challenging" to recruit "highly skilled" technology workers, up from 70% in the prior survey in 2016.[6] Only 11% of the companies surveyed thought the situation was improving.[7]

The regional skills challenge is broader than the needs of the semiconductor industry. In 2015, a McKinsey & Company study of the Capital Region's economy reported that over 50% of employers across a range of industries were experiencing a shortage of workers with the necessary skills, which they regarded as a major barrier to their further expansion within the region. The report also cited points of weakness in the region's generally strong educational system, including declining enrollment in institutions of higher learning, declining high school graduation rates, and disparities in graduation rates across race and income lines.[8] Carolyn Curtis, Academic Vice President of Hudson Valley Community College, observes that many regional high school graduates are not "college ready."[9]

---

[4] "Getting to Work Closing the 'Skills Gap,'" Schenectady, *The Daily Gazette* (September 24, 2016).

[5] The region has been grappling with the skills issue for over two decades. In the 1990s manufacturers in the Hudson Valley region complained about the lack of relevant skills of locally-recruited workers. In 1997, MiCRUS, a joint venture of IBM and Cirrus Logic that operated an 8-in. semiconductor wafer fabrication plant in East Fishkill, New York, observed that in its workforce of 900, "most of the newly hired workers have no experience with semiconductor production." A production manager at Blasch Precision Ceramics in Menands observed that while "area colleges have supplied engineers and other white collar workers . . . some manufacturing workers came to the company needing better math skills, greater aptitude reading blueprints, and more savvy with computer applications such as spreadsheets." "State Targets Jobs of Future," Albany, *The Times Union* (January 5, 1997).

[6] "Finding Hiring-Tech Workers Difficult," Albany, *The Times Union* (January 18, 2017).

[7] "High Tech Hiring Becomes High Stress in the Capital Region," Albany, *The Times Union* (January 17, 2017).

[8] McKinsey & Company, *Capital 20.20: Advancing the Region through Focused Investment* (McKinsey & Company, 2015); "Jobs Skills Gap Drives Search for Solutions," Albany, *Times Union PLUS* (August 6, 2016).

[9] Interview, Troy, New York (June 7, 2016).

High-tech industries' demand for skilled workers is volatile. Companies not only staff up on a crash basis to cope with surging demand, but retrench sharply in downturns.[10] A natural tension exists between the often-mercurial industry demand for more workers with specialized skills and the degree to which the education and training pipeline can or should be adjusted, expanded, and/or reoriented to respond to such needs. The continuing migration of high-tech firms to the Capital Region from elsewhere suggests that New York remains competitive for now with other regions in this respect, but that cannot be taken for granted in the future. Darren Suarez, Director of Government Affairs at the Business Council of New York, said in 2013 that the State Department of Labor had projected a 135% increase in science, technology, education, and math (STEM)-related computer electronics manufacturing jobs in the Albany area between 2008 and 2018, "driven by growth in this sector," and that "our economic future will be determined by our ability to educate those individuals."[11]

## 8.2 Institutional Challenges

While tech industry manpower needs are subject to wide and sudden variations, the educational pipeline is by nature long and difficult to modify, a mismatch between demand and supply elements that has been exacerbated by regional planners' inability to foresee future manpower needs with any degree of precision.[12] As noted in Chap. 7, the actual manpower requirements of the GlobalFoundries wafer fabrication facility, as well as those of the associated construction workforce, greatly exceeded forecasts made in 2008, and other new high-tech companies, and projects are present in the region that were not envisioned a decade ago. For example, GlobalFoundries is currently running at around a 9% turnover rate. That means over 300 new hires a year are required just to stay even with current operation needs. Uncertainty with respect to manpower will persist given the fact that the existing

---

[10] The US computer industry cut nearly 60,000 jobs in 2014. In 2015 the Chairman of MIT's physics department "lamented that an entire generation had been told that this was a great national emergency, that we needed scientists," at the time. "Now they are on the street and they feel cheated." "Is there a US Engineering Shortage? It Depends Who You Ask," *Power Electronics* (August 19, 2015).

[11] National Research Council, Charles W. Wessner (rapporteur), *New York's Nanotechnology Model: Building the Innovation Economy* (Washington, DC: The National Academies Press, 2013), p. 63.

[12] Given the long timeframe associated with education and apprenticeship, even very successful programs cannot have an impact in the market in the short run. For example, the highly acclaimed and widely studied Pathways in Technology Early College High School (P-TECH) program combining academic and workplace learning took in its first students in 2012 but will not produce its first graduates until 2018.

## 8.2 Institutional Challenges

methodology and data necessary to track and forecast high-tech manpower needs are not adequate for the task.[13]

But even if forecasting tools were available to identity with accuracy the future workforce demands facing the region, "as technology evolves, the skills that STEM employers seek are changing faster than our education system has adapted to meet them," as a 2017 study of STEM education in New York State observed.[14] Stanley Litow, IBM's Vice President of Corporate Citizenship and Corporate Affairs commented in 2017 that—

> *The academic decision-making change process doesn't operate in the same way as other sectors.... The university president may see what is needed, but the process takes a long time. [New York education policy makers] need to think very seriously about how to incorporate the changing requirements of the labor force into curriculum more quickly. For example, someone who got a computer science degree five years ago didn't need digital or design skills.*[15]

SUNY, which enrolls more students in the Capital Region than all other regional institutions of higher learning combined, has long been criticized for over-regulation which impedes introduction of new curriculum in response to technological change.[16] Within the SUNY system, introduction of new programs and modifications of existing academic programs require a multi-stage approval process involving reviews and sign off by SUNY itself, the State Education Department, the Board of Regents, and the Governor (see Fig. 8.1). An institution seeking the new or modified program must submit a program proposal with appropriate supporting

---

[13] Laura I. Schultz of SUNY Polytechnic and a number of her academic colleagues observed in 2015 that "evidence on skills gaps and likely needs with respect to the regional economy is limited. [Existing projections and assessments] lack the detail necessary to guide the development and/or expansion of degree or training programs geared to nano-related industry." Laura I. Schultz, et. al., "Workforce Development in a Targeted, Multisector Economic Strategy: The Case of New York University's College of Nanoscale Science and Engineering," in Carl Van Horn, Tammy Edwards, and Todd Green (eds.) *Transforming US Workforce Development Policies for the 21st Century* (Kalamazoo: W.E. Upjohn Institute for Employment Research, 2015), pp. 343–344. Estimating the "demand and employment in STEM fields is difficult because there is no single accepted definition for a STEM job. Estimates of the number of STEM jobs range from 5 million to 19 million, according to the National Science Foundation, depending as what is included. Many are technical jobs that don't require even a bachelor's degree." "Is There a US Engineering Shortage? It Depends on Who You Ask," *Power Electronics* (August 19, 2015).

[14] Allison Armour-Garb, *Bridging the STEM skills Gap: Employer/Educator Collaboration in New York* (Public Policy Initiative of New York State, Inc., January 2017), p. 5. Hudson Valley Community College spent several years developing a degree program in semiconductor manufacturing in anticipation of the particular requirements of Advanced Micro Devices, but when AMD transferred its manufacturing operations to the corporate entity which eventually became GlobalFoundries, the new management wanted different skillsets and HVCC had to redesign its curriculum. Interview with HVCC president, Drew Matonak, Troy, New York (June 8, 2016).

[15] Allison Armour-Garb, *Bridging the STEM skills Gap: Employer/Educator Collaboration in New York* (Public Policy Initiative of New York State, Inc., January 2017), p. 19

[16] John W. Kalas, "SUNY Strides into the National Research Stage," in John B. Clark, W. Bruce Leslie, and Kenneth P. O'Brien (eds.), *SUNY at Sixty: The Promise of the State University of New York* (Albany: SUNY Press, 2010), p. 161.

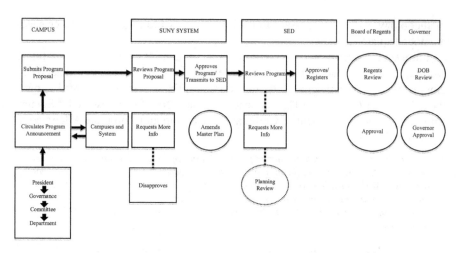

**Fig. 8.1** SUNY program development and review process. (Source: State University of New York, *Guide to Academic Program Planning* (October 2013))

documentation, a process that educators interviewed for this study analogized to a grant application, and some proposals "get stuck somewhere in the approval process."[17] Carolyn Curtis, academic vice president of Hudson Valley Community College, observes that "State Ed is a big stumbling block." New programs take too long to approve and the college cannot enroll students in a non-approved program.[18] The ability of institutions of higher education to introduce new curriculum is vital to a region's competitiveness (see Box 8.1).

> **Box 8.1 Curriculum and Innovation**
> The ability of US universities to quickly introduce new scientific and technological curriculum was a principal factor underlying eventual US leadership in the information industries. "American dominance of the computer software industry was overwhelmingly due to the remarkable speed with which its university faculties were able to develop and to introduce an entirely new academic curriculum in computer science" between 1959 and 1965.[19] In semiconductors, which constitute the core of all information technology

---

[17] The entire program approval process is detailed in State University of New York, *Guide to Academic Program Planning* (October 2013).

[18] Interview, Troy, New York (June 7, 2016).

[19] Nathan Rosenburg, "America's Entrepreneurial Universities," in David M. Hastfed, *The Emergence of Entrepreneurship Policy: Governance, Start-ups and Growth in the US Knowledge Economy* (Cambridge: Cambridge University Press, 2003).

> hardware, Stanford's Department of Electrical Engineering introduced a course devoted to integrated circuit design and fabrication soon after these devices began becoming available in 1961, and Stanford augmented its faculty with industry engineers at the cutting edge of the emerging technology.[20] New York institutions must prove nimble in a similar manner and historically, at least, have demonstrated the ability to do so. The electric power industry is usually dated as emerging in 1882 when Thomas Edison's Pearl Street Station in New York City became operational. In 1883 Cornell University introduced a course in electrical engineering and awarded its first doctorate in the subject in 1885.[21]

This chapter surveys numerous innovative and promising educational programs under way in the Capital Region in response to the skills gap, but most of these are pilot or model initiatives involving relatively small numbers of students and/or workers. Even the best cannot necessarily be scaled up quickly without risking erosion of quality standards.[22] Drew Matonak, President of HVCC, indicates that with sufficient funding he could double the size of the college's highly successful Advanced Manufacturing Program (AMP) but that he could not triple it over the near term.[23] But as the Public Policy Institute of New York states in its 2017 study of the STEM skills gap—

*If New York wants to maximize the economic future of its students and businesses, creating model programs is not enough.*[24]

During her tenure as SUNY Chancellor, Nancy Zimpher viewed SUNY's sheer size as "an immense opportunity to take ideas to scale" but that scale-up was not an easy proposition. "I think scale is a big challenge in our society. We have great

---

[20] Stanford's Provost, Frederick Terman, recruited the most talented engineers from Silicon Valley companies and made them "adjust professors" at Stanford to teach students and faculty about consent trends in the semiconductor industry. Nathan Rosenburg, "America's Entrepreneurial Universities," in David M. Hastfed, *The Emergence of Entrepreneurship Policy: Governance, Start-ups and Growth in the US Knowledge Economy* (Cambridge: Cambridge University Press, 2003).

[21] Nathan Rosenburg and Richard R. Nelson, "American Universities and Technical Advance in Industry," *Research Policy* (1994).

[22] Thus in 2016 Emily Reilly, Director of Human Resources at GlobalFoundries, praised one of Hudson Valley Community College's semiconductor-related associates degree programs but pointed out that the program graduates 25 individuals, whereas in the single year of 2015 her company hired 300 technicians. Moreover, her company had to compete with its own vendor companies to hire some of the 25 HVCC graduates. "The technician pipeline is a constant challenge," she indicates. Interview, Malta, New York (January 27, 2016).

[23] Interview, Troy, New York (June 8, 2016). On the problem of scaling up institutions without loss of excellence, see Robert I. Sutton, "Scaling: The Problem of More," *Harvard Business Review* (October 3, 2013).

[24] Allison Armour-Garb, *Bridging the STEM skills Gap: Employer/Educator Collaboration in New York* (Public Policy Initiative of New York State, Inc., January 2017), p. 5.

experiments, we have great examples of one-off practices, but we haven't been able to scale our work in education and science and science technology...."[25]

Finally, at each educational level, many graduates, in some cases a majority, leave the region after graduation, so regardless of the knowledge and skills levels they have attained in the course of their education, they are not available to local employers.[26] Over half of RPI students leave the area after graduation.[27] RPI students who are skilled at coding "usually leave Troy for jobs in Silicon Valley, Seattle, New York City or Boston." RPI Professor Mukkai Krishnamoorthy, the Director of RPI's Center for Open Source Software, said in 2016 that "most of them go to the West Coast to Google, Amazon, IBM. Very few stay here. Those that do start their own companies."[28] The pattern is not universal, however, particularly with respect to individuals who grew up in New York State prior to their entry into college or an apprenticeship program. CNSE admits a high percentage of native New Yorkers in its nanotechnology programs and "way more" of them stay in the region after graduating than is the case with respect to other engineering programs in the region.[29]

## 8.3 Workforce Requirements of High-Tech Manufacturing

High-tech manufacturing requires a far different kind of workforce than the traditional mid-twentieth-century assembly line laborers who performed repeated, relatively unskilled tasks under the close supervision of low-level managers.[30] Technology-intensive manufacturing requires individuals who can function well on a team, solve problems, exercise leadership, pay close attention to detail and who possess an array of specialized skills, including proficiency in information technology and mechanized tools. Semiconductor companies establishing a presence in the Capital Region found that significant numbers of individuals with the necessary personal qualities and character were present in the region, including displaced

---

[25] "A Conversation with Nancy Zimpher," Syracuse, *The Post-Standard* (August 7, 2011).

[26] Ray Rudolph, CEO of the engineering group CHA Companies, speaks of a regional "brain drain," observing that millennials want "urban living" and are drawn to large metropolitan center outside the region. Interview, Albany, New York (September 16, 2015).

[27] Interview with Jonathan Dordick, vice president for research, Rensselaer Polytechnic Institute, Troy, New York (September 17, 2015).

[28] "No Coders, No Tech Valley: Decoding Albany's High-Tech Future," *Albany Business Review* (June 3, 2016).

[29] Interview with Professor Laura Schultz of the College of Nanoscale Science and Engineering, Albany, New York (November 30, 2016).

[30] In the twentieth century, many factories in what is now sometimes called the rust belt utilized assembly line techniques which were a legacy of the theories of Frederick Winslow Taylor and the production methods utilized by Henry Ford. These systems minimized the importance of worker judgment and skill. See generally David Montgomery, *The Fall of the House of Labor: The Workplace, the State, and American Labor Activism 1865–1925* (Cambridge: Cambridge University Press, 1989), pp. 229 and 251.

workers from traditional manufacturing industries, military veterans, and graduates from local schools. However, in many cases they lacked necessary math and computer-related skills, and in most cases required additional training before they could function in a semiconductor manufacturing environment.[31]

### 8.3.1 Semiconductor Manufacturing: Education and Training Requirements

The workforce at the GlobalFoundries manufacturing facility in Malta/Stillwater is broken down roughly as 10% senior managers, 30% engineers, and 60% technicians and operators.[32] In addition, the GlobalFoundries employee workforce is augmented by permanent on-site employees seconded by supply chain companies and, at any given time, skilled construction workers in numbers ranging from a few hundred to several thousand. Thus, taken as a whole, the GlobalFoundries workforce encompasses an extraordinarily broad range of educational backgrounds and specialized skills.

The senior echelons of semiconductor manufacturing firms are dominated by individuals with 4-year college degrees, many of these advanced degrees. GlobalFoundries recruits doctoral graduates for the company's research and development needs who are "fluent" in the field of nanotechnology and who typically make a presentation on their dissertation as part of their job application process. "These are individuals who have the new knowledge and research expertise that we need to have as part of a company that relies on innovation to stay competitive." Individuals with master's degrees "perform tasks that require strong knowledge of semiconductors" and are tasked with "testing the success of the company's products and ensuring that they meet the needs of the company paying for them." Supervisory personnel who oversee technicians and direct operations in the clean rooms and who bear some responsibility for the manufacturing process tend to have 4-year bachelor's degrees (see Footnote 32). Individuals responsible for managerial, marketing public communications, finance, legal, and human resources functions have a mix of bachelor's and advanced degrees. (See Table 8.2.)

Members of GlobalFoundries' engineering group, comprising nearly a third of the workforce, are expected to have the minimum of a bachelor's degree in a

---

[31] "Ex-plant Workers Find New Jobs in Help-Hungry High Tech Sector," *Albany Business Review* (June 2, 2008). Emily Reilly, GlobalFoundries' director of human resources, says that "the military is a huge source of hires for us." Interview, Malta, New York (January 27, 2016). Travis Bullard, a spokesman for GlobalFoundries, said in 2015 that veterans had skills and training in areas such as preventive maintenance, troubleshooting, and equipment maintenance and that the "professionalism of the military is very cohesive with the manufacturing environment . . . . [A]nybody who's getting out of the military soon, we'd love to hear from them." See "HR Director Provides Look at GloFo Hiring Process," Schenectady, *The Daily Gazette* (February 26, 2015).

[32] "Talking About Fab 8's Work Force." Albany, *The Times Union* (September 9, 2012).

Table 8.2 Higher education backgrounds relevant to GlobalFoundries workforce needs

| | Intern | Associate's degrees | Bachelor's degrees | Master's degrees | Doctorate |
|---|---|---|---|---|---|
| Manufacturing | x | x | x | x | x |
| Engineering | x | x | x | x | x |
| Supply Chain and Procurement | x | | x | x | |
| Research and Development | x | | x | x | x |
| Design | x | | x | x | x |
| Sales and Marketing | | | x | x | x |
| Information Technology | x | x | x | x | |
| Human Resources | x | | x | x | |
| Finance | x | | x | x | |
| Legal | | x | x | | x |
| Communications and Public Relations | x | | x | x | |

Source: GlobalFoundries (2017)

relevant engineering discipline (electrical, mechanical, and chemical) or an advanced degree in a relevant field of science, such as materials science, physics, or mathematics. As shown in Box 8.2, engineers at GlobalFoundries are responsible for an extremely broad range of functions relevant to semiconductor design, manufacturing, testing, and applications.

---

**Box 8.2 Categories of Engineering Expertise Relevant to GlobalFoundries Luther Forest Operations**

| | |
|---|---|
| • Process Controls | • Fab Automation |
| • Process Integration | • Implant |
| • Back End of Line Integration | • Industrial and Operations Research |
| • Front End of Line Integration | • IT Systems |
| • CFM Process | • Memory |
| • CVD/PVD | • Metrology Process |
| • Device | • Modeling |
| • Diffusion Equipment | • Photolithography |
| • Failure Analysis Product | • Product Test |
| • Electroplating Process | • Safety and Environmental |
| • Equipment | • Senior Device R&D |
| • Development | • Quality Assurance |
| • Technology Development | • Technology and Integration |
| • Fab Systems Application | • Yield Enhancement |

Source: GlobalFoundries (2017)

Nearly two-thirds of the GlobalFoundries workforce is comprised of technicians and operators who actually work in the clean rooms and run the machines and facilities and technicians who operate the Fab's utilities, heating, and cooling and other infrastructure.[33] (See Box 8.3.)

---

**Box 8.3 Technician and Operator Positions at GlobalFoundries Luther Forest Facility**

| Facilities | Manufacturing Services |
|---|---|
| • HVAC | • Factory Automation |
| • Instrumentation and Controls | • Warehouse Operations |
| • Electrical | • Fab Systems Set-up |
| • Chemical | • Factory Information Control Systems |
| • Ultrapure Water | *EHS/Security* |
| • Waste Water Treatment | • Security |
| *Engineering* | • Environmental |
| • Process Technician | • Safety |
| • Maintenance Technicians | *Operations* |
| • Process Integration Technicians | • Wafer Fab Technicians |
| • Analytical Lab Technicians | • Wafer Fab Operators |

Source: GlobalFoundries (2017)

---

Technician and operator jobs do not necessarily call for individuals with 4-year college degrees.[34] Many GlobalFoundries help-wanted postings for high-skill technician positions require only a high school diploma but are highly demanding with respect to personal qualities such as the ability to function on a team, attention to detail, motivation, and physical stamina, as well as specialized knowledge and skills (see Box 8.4). Pedro Gonzalez, former head of staffing for GlobalFoundries Fab 8, said in 2012 with respect to such jobs that "this is where the community colleges come in. That's where the sustainable workforce for this business is."[35] Indeed for

---

[33] In 2009 Michael Fancher, CNSE's vice president for business development and economic outreach said that 2500 people were working in the Nanocomplex, and it was not "just high-tech specialists who were benefitting…. There's a role for tradesmen as well. The labs and clean rooms need to be fitted with specialized fixtures and pipes and trained individuals are needed to do that work. [This creates] an opportunity for professionals like plumbers and electricians to develop new skills for the new market." See "Nano Center on Job Magnet in Albany," *Utica Observer-Dispatch* (July 19, 2009).

[34] Clean room jobs entail 12-h shifts wearing a "bunny suit" and facilities workers often handle hazardous chemicals. Mistakes in operating the manufacturing equipment or contamination of waters can cost a company "millions of dollars in a matter of seconds." See "Fab Job in a Bunny Suit: Educators See New World of Opportunity at Fab 8 Factory," Albany, *The Times Union* (June 26, 2011).

[35] "Who's Hiring? Many Hope It's GlobalFoundries," Glens Falls, *The Post Star* (January 15, 2012).

semiconductor makers, prior industry experience or relevant training at a community college is seen as more valuable than 4-year degrees. Thus about half of the students enrolled in Hudson Valley Community College's 2-year semiconductor manufacturing programs in 2014 already held 4-year bachelor's degrees. An HVCC Faculty member observed that

> We've always thought that a four-year degree is the answer to getting a good-paying job. Now, they are experiencing, it's not always the right answer.[36]

---

**Box 8.4 Demanding Work: GlobalFoundries Job Posting for Process Technicians for Thin Films (2013)**

**Job Summary:**
**GLOBALFOUNDRIES** Fab 8 is seeking highly skilled and motivated technicians to become part of our state of the art 300 mm factory in Malta, New York. These positions will be required to sustain and run the factory floor with primary responsibility of resolving equipment and process issues. The Manufacturing Technician 1 position is a shift position (working alternating weeks of three and four 12-h shifts per week). **GLOBALFOUNDRIES** is staffing for four shifts, with 50% of the positions assigned to the night shift.

**Specific Responsibilities:**
- Process wafers using defined procedures
- Operate metrology/inspection equipment and interpret results
- Review Statistical Process Control charts for process quality and react to out of control conditions including defect troubleshooting
- Perform visual inspections (quality check)
- Recover from process and tool interruptions
- Use standard software application (MS Office) and specific programs (SAP, MES, ASPECT) creating reports and documenting procedures
- Identify and address potential areas for improvement and optimize tool availability, cycle time, utilization, and cost
- Perform engineering experiments by following instructions
- Complete all required reporting and documentation
- Understand and follow all health, safety, and environmental procedures and requirements
- Actively participate in continuous improvement processes and in the Emergency Response Team
- Train new team members

---

[36] Penny Hill, "Preparing Middle-Class Workers for Middle-Skill Jobs," *Marketplace* (April 25, 2014).

**Required Qualifications:**
- High school diploma
- Fluent in English Language—written & verbal
- Able to perform shift work on a 12 h per day shift schedule
- Able to work in a cleanroom environment per semi-conductor protocol/requirements
- Able to wear all required clean room protective clothing and equipment for normal 12 h per day shift (excluding breaks)
- Ability to perform work in a standing position for majority of 12 h shift (excluding breaks)
- Able to lift a minimum of 30 pounds on a periodic basis throughout the shift (preventive maintenance) and/or process issues
- Demonstrated technical skills and knowledge of semiconductor processing and process equipment
- Demonstrated ability to operate computer and system interface programs to ensure appropriate computing and analysis of production information
- Demonstrated ability to follow detailed instructions and procedures to complete tasks and required documentation
- Demonstrated work performance in an environment requiring high level of attention to detail and timeliness
- Ability to handle multiple tasks simultaneously and prioritize activities
- Strong team player with ability to work well within a global team

**Desired Qualifications:**
- Working knowledge of Statistical Process Control methodologies and systems
- Equipment or process maintenance experience in semiconductor manufacturing
- Familiarity with lean processes and activities and Kaizen teams

GlobalFoundries operates an extremely rigorous training program for new hires for technician and operator positions, and it takes roughly 6 months for them to "get up to speed." The program includes classes and assignment to an individual trainer and an experienced "buddy." The company defines what skills the trainees must learn and prepares extremely detailed checklists of operations they are able to perform properly before they are certified. Experienced workers partner with them, observe their performance, and ensure they are fully capable of operating in a real manufacturing environment.[37] This training program is in essence an internal apprenticeship program, one which the company has offered as the basis of a certified advanced manufacturing apprenticeship program in cooperation with institutions on a state and

---

[37] Interview with Emily Reilly, GlobalFoundries' head of human resources, Malta, New York (January 27, 2016).

national level. GlobalFoundries also partners with Hudson Valley Community College, which trains workers for the company, a process which is funded by the company.[38]

GlobalFoundries' ongoing need for thousands of technicians and operators is consistent with the original vision of the architects of New York's Tech Valley, who believed that high-tech manufacturing could at least partially offset the impact of the decline of traditional manufacturing in the region. Former Senate Majority Leader Joseph L. Bruno recalls in his memoirs that in his aspirations for Tech Valley, a significant role would exist for middle-skill workers who grew up in the Capital Region and would remain there if good job opportunities were available—

*When most people think of high tech, they imagine rich brainiacs, visionary entrepreneurs, and software and hardware engineers. However, the companies that we hoped would come would require technical workers—the blue-collar men and women of the future, if you will. Muscle wouldn't be of any use to them. Training, that was the key to a decent-paying career.*[39]

"Blue collar men and women of the future"—former Senate Majority Leader Joseph L. Bruno.

---

[38] Interview with Carolyn Curtis, academic vice president, Hudson Valley Community College, Troy, New York (June 7, 2016).

[39] Joseph L. Bruno, *Keep Swinging: A Memoir of Politics and Justice* (Brentwood, TN: Post Hill Press, 2016), p. 137.

## 8.3.2 In-migration of Workers: Advantages and Limitations

In order to remain globally competitive, leading-edge firms in the semiconductor industry chain must be able to recruit the best talent on a global basis, an imperative which underlies the US semiconductor industry's opposition to immigration and residency restrictions that limit its ability to recruit and retain foreign talent.[40] In 2012 GlobalFoundries reported that of its workforce in Malta/Stillwater, then about 1100 employees, roughly half came from outside the region from 30 different countries, and half were local hires, with a company commitment to hire locally when possible. The head of recruiting for GlobalFoundries Fab 8 said that

> *Implicit in our name is "global." We cannot be competitive without having an international workforce. The niche we're doing in semiconductor is such a narrow skill set that it just didn't exist in this region.*[41]

At the same time, importing talent has limitations, beginning with relocation and settlement costs that are not incurred with respect to local hires. More importantly, if relocating workers are not happy with their new environment, they are likely eventually to leave. As a result of such factors, semiconductor manufacturing, design, and supply chain firms locate their operations in venues where there is an actual and/or potential pool of local talent, reflecting the existence of a strong local education and training infrastructure.[42]

GlobalFoundries has an equal focus on developing external education and workforce development programs intended to improve the education system so that it can meet the demands of high-tech businesses and advanced manufacturers as well as identify the best and brightest talent coming out of the university systems and from around the world. In an interview, Mike Russo, who is the GlobalFoundries executive responsible for external education and workforce development initiatives in the United States, said—

> *We have an obligation to our company and the regions in which we operate to do what we can to improve the (education) system so that we can ensure the talent we need exists where we do business, and citizens, regardless of their background, can benefit from rewarding careers. In the U.S. we have and will continue to lead efforts to develop programs that can be scaled and sustained throughout the state and nation to provide a pathway to those careers.*

The arrival of large semiconductor enterprises in the Capital Region began with Sematech's phased move from Austin to Albany, followed by other semiconductor-related firms such as Tokyo Electron, M&W Zander, and ultimately GlobalFoundries. Initially these firms brought in hundreds of personnel from outside the region who

---

[40] Between 2009 and early 2012, GlobalFoundries Fab 8 had obtained 214 H-1B visas for foreign employees, but the visas only allowed the workers to remain in the United States temporarily. Each visa costs a company an estimated $3600, with decisions taking months. "Visa Spike a Chip Hike," Albany, *The Times Union* (April 10, 2012).

[41] "Who's Hiring? Many Hope It's GlobalFoundries," Glens Falls, *The Post-Star* (January 15, 2012).

[42] Semiconductor Industry Association, *Maintaining America's Competitive Edge* (2009).

possessed requisite skillsets. By 2008 over 1600 scientists, researchers, and other staff were working at Sematech's 450,000 square-foot complex in Albany, many of them former residents of Austin, Texas. The influx of high-tech professionals from outside the region was so extensive that a number of local businesses and development organizations developed specialties in enabling re-locating high-tech employees to make smooth transitions into the Capital Region.[43] GlobalFoundries established its own relocation operation, offering immigration services, assimilation presentation materials, language lesson benefits, assistance with schools, and employee family day activities to assist in the formation of community networks.[44]

Ultimately, however, it is questionable whether a large high-tech manufacturing operation that must recruit most of its workers from other countries and regions can remain viable over the long run, particularly with respect to the line technicians, operators, and facilities managers that comprise the bulk of the work force.[45] GlobalFoundries expects that it will eventually recruit most of its workforce from the Capital Region or adjacent regions in the Northeastern United States—assuming the skills gap is surmounted.

An overview of the regional education and training infrastructure reveals a wide-ranging effort by institutions and individuals at all levels to overcome the skills deficit. Major tech companies like IBM, GE, GlobalFoundries, and M&W Zander are engaged in various educational initiatives in collaboration with regional schools and colleges. The Capital Region is blessed with a rich and diverse array of institutions of higher learning, and during the past decade virtually all of these institutions have established new curricula, degree programs, and work-study initiatives to engage high-tech firms and to enable students to transition successfully into jobs with those firms.[46] These efforts are being reinforced by improvements in the region's K-12

---

[43] In 2008 Bob Blackman, a vice president at Realty USA, said that his job was to help professionals relocating to Albany from Austin or other parts of the world to find jobs at International Sematech. He said, "I work with companies to help their employees to bring their families in. When people come from an area that's very different from our area, like Austin, it's not just selling them a house. They want to know about school systems and all of that." See "Sematech Deal Brings Business, High-Tech Jobs," Schenectady, *The Daily Gazette* (February 14, 2008). The Center for Economic Growth operates a division called Talent Connect (formerly Tech Valley Connect) that facilitates integration of immigrants into the regions by assisting with housing and providing information about schools and healthcare, geographical orientation, and a newsletter explaining local customs.

[44] "Helping Newcomers Assimilate," Albany, *The Times Union* (March 30, 2016).

[45] "If GlobalFoundries hires someone from Arizona or Texas, which have established semiconductor industries, it's on the hook for moving expenses. And some of those hires don't stay, for any number of reasons, including the frosty upstate weather." See "Preparing Middle-Class Workers for Middle-Skill Jobs," *Market place* (April 25, 2014).

[46] In 2011 a group of four professional site-selectors, who specialized in finding new locations for business clients, toured the Capital Region. One of them, Lee Higgins from Austin, Texas, said that he was "amazed at the number and concentration of higher learning institutions producing the talent needed to work at places such as GlobalFoundries' new $4.6 billion semiconductor plant in Malta . . . . That is a huge plus and one of the main reasons I believe there's going to be the ability to generate skilled labor." See "After Touring Capital Region, Analysts Say Tech Valley Shows Potential for Growth," Saratoga Springs, *The Saratogian* (August 5, 2011).

schools, including greater emphasis on STEM education and transitioning students to postgraduation education and training opportunities.

## 8.4 Four-Year Institutions of Higher Learning

The Hudson Valley is the site of the United States' first successful introduction of applied technology and science at the collegiate level, a tradition that has been maintained down to the present day in the founding educational institutions of the region.[47] The United States Military Academy at West Point, the first technical and engineering college in the country, introduced in the 1820s a curriculum centered on civil engineering based on a teaching method featuring small class size, self-study, and daily homework, which the Academy retains today.[48] Rensselaer Polytechnic Institute (RPI), the oldest degree-conferring engineering school in the English-speaking world, was founded in 1824; its early catalog stated that among its objectives was "to cause the student to commence with practical applications of science, as he will better understand elementary principles after he has become acquainted with the end and object of them."[49] Union College, founded in Schenectady in 1795, established a scientific curriculum in 1828 and a civil engineering program in 1845, becoming "the one traditional liberal arts school in the first half of the Nineteenth Century to make a thoroughly uncompromising and effective place for applied science in the course of study." As one observer noted—

> *The designers and builders of the country's canals and railroads were overwhelmingly graduates of the military academy at West Point, R.P.I. and Union College. . . . These institutions also were changing the civil engineer in the United States from a product of training on the job to a professional formally instructed in an educational institution.*[50]

---

[47] Frederick Rudolph, *Curriculum: A History of the American Undergraduate Course of Study Since 1636* (San Francisco: Josey-Bass, 1977), p. 63.

[48] During the Revolutionary War, the Continental army was dependent on foreign émigrés for engineering skills, and the establishment of the US Military Academy was in substantial part intended to address this need. Colonel Sylvanius Thayer, superintendent of West Point from 1817 to 1833, made civil engineering the foundation of the Academy's curriculum. During the nineteenth century, West Point graduates were primarily responsible for building the nations' first railroad lines and modern bridges, roads, and harbors. United States Military Academy at West Point, "A Brief History of West Point," <http://www.usma.edu/wphistory/sitepages/home.aspx>.

[49] Rensselaer Polytechnic Institute, *Catalog* (1828). Stephen van Rensselaer, the principal founding patron of RPI, expressed his vision for the institution, shared by co-founder Amos Eaton, in a letter that became part of RPI's charter. The school was founded "for the purpose of instructing persons, who may choose to apply themselves, in the application of science to the common purposes of life . . . . I am inclined to believe that competent instructors may be produced in the school at Troy, who will be highly useful to the community in the diffusion of a very useful kind of knowledge, with its application at the business of living." Letter from Stephen van Rensselaer to the Reverend Samuel Blatchford, November 5, 1824, reproduced in part in Thomas Phelan, D. Michael Ross, and Carl A. Westerdahl, *Rensselaer: Where Imagination Achieves the Impossible* (Albany: Mount Ida Press, 1995), p. 30.

[50] Frederick Rudolph, *Curriculum: A History of the American Undergraduate Course of Study Since 1636* (San Francisco: Josey Bass, 1977), p. 63, citing Daniel H. Calhoun, *The American Civil Engineer: Origins and Conflict* (Cambridge, MA: Technology Press, 1960).

**Table 8.3** Students enrolled in STEM majors at Capital Region colleges and universities

| Institutions | STEM enrollment |
|---|---|
| Rensselaer Polytechnic Institute | 4541 |
| SUNY Albany | 2391 |
| Albany Medical College | 840 |
| Union College | 793 |
| SUNY Empire State College | 699 |
| Siena College | 588 |
| SUNY Polytechnic Institute | 321 |
| The Sage College | 220 |
| The College of Saint Rose | 193 |
| Union Graduate College | 181 |
| Skidmore College | 149 |

Source: Upstate Revitalization Initiative *Capital 20.20* (2015) based on US Department of Education Statistics

While venerable institutions like RPI and Union College and the rapidly emerging College of Nanoscale Science and Engineering (CNSE) dominate assessments of the Capital Region's resources in higher education, the region has over 20 colleges and universities. The other institutions have developed niches based on their traditional strengths and emerging industry needs and by focusing primarily on ancillary occupations, and they have avoided "head-on competition with the research powerhouses at CNSE and RPI."[51] Local colleges like Siena College and the College of Saint Rose have developed reputations for educating students employable by local technology companies "ranging from high level technical people to interns."[52] Mark Sullivan, president at the College of Saint Rose, observed in 1998 that local liberal arts schools like his own institution were assets in the effort to attract high-tech business to the region, part of a "rich stew of higher education institutions that offer virtually anything that economic development specialists or corporate relocation specialists look at when they want to locate their plants.... We don't produce wafers for silicon plants, but we produce trained personnel to build those facilities."[53] Table 8.3 shows recent data on the number of students enrolled in STEM majors at Capital Region colleges and universities.

---

[51] "Area Colleges Getting Into 'Tech Valley Game,'" Schenectady, *The Daily Gazette* (February 22, 2014).

[52] "Not Yet 'Tech Valley' but Getting Closer," Albany, *The Times Union* (March 7, 1999); "Area Software Firms Seek New Talent," Schenectady, *The Daily Gazette* (November 19, 1999).

[53] "Doesn't Take Degree to See Colleges' Business Assets," Albany, *The Times Union* (February 22, 1998). A 2011 profile of a number of the region's 4-year institutions observed that "each has turned out leaders of national prominence, men and women who are now at the top of their field in the areas of medicine, law, science, politics, literature and business. To many of them ... the time they spent at these area institutions was just what they needed to succeed in life on the big stage." See "From Graduates to Greatness," Schenectady, *The Daily Gazette* (June 19, 2011).

Finally, New York State has numerous excellent colleges and universities which are located outside of the Capital Region but which are major sources of graduates entering employment in Tech Valley. GlobalFoundries has a list of "key schools," which are priority targets for its recruiting, all of them located within the state: RPI, Rochester Institute of Technology, Cornell, SUNY Polytechnic, SUNY Albany, SUNY Buffalo, and Clarkson University. Clarkson, in Potsdam, NY, operates the 20-year-old Center for Advanced Materials Processing (CAMP) under the direction of Professor S. V. Babu, a leader in the field of chemical-mechanical planarization, a critical enabling technology for the fabrication of semiconductor logic and memory devices.[54] Not coincidentally, Clarkson is a "huge" source of recruits for GlobalFoundries.[55] Monroe Community College was favorably mentioned in interviews conducted for this study for doing a "nice job in Rochester" and being "strongly tied to local industry."[56]

## 8.4.1 Rensselaer Polytechnic Institute

Sometimes characterized as "MIT on the Hudson," RPI "trained the shock troops for the American industrial revolution, and more than its share of generals" in the nineteenth century, with RPI graduates responsible for mapping the West, building the transcontinental railroad, constructing the first Ferris Wheel and the Brooklyn Bridge, and staffing emerging technology-intensive companies like GE, Westinghouse, and Standard Oil.[57] As of the end of 2016 RPI graduates and faculty included 84 members of the National Academy of Engineering, 17 members of the National Academy of Sciences, and 8 members of the Institute of Medicine.[58] In the Troy area, where RPI is located, the school fostered an innovation-based industrialization which has been characterized as the nineteenth century's version of Silicon Valley.[59]

---

[54] "Babu Leads Clarkson's CAMP to Prominence," Massena, *Daily Courier-Observer* (June 23, 2015).

[55] Interview with Emily Reilly, Malta, New York (January 27, 2016).

[56] Interview with Ray Rudolph, chairman, CHA Companies, Albany, New York (September 16, 2015).

[57] Stuart W. Leslie, "Regional Disadvantage: Replicating Silicon Valley in New York's Capital Region," *Technology and Culture* (2001), pp. 238–239.

[58] "Rensselaer Polytechnic Institute Faculty Lauded," *RPI News* (December 7, 2016). The Institute of Medicine is now known as the National Academy of Medicine.

[59] P. Thomas Carroll, a former RPI faculty member, pointed out in a 1999 address that "Troy manufacturers were among the first in the nation to realize that, once the canals and railroads provided cheap transportation costs to a geographically widespread market economy, and only after that happened, it would make good business sense for a small number of centralized operations in a single city to manufacture, at various times and in various plants, 75,000 stoves a year, a million horseshoes and a quarter-million Arrow shirts a week, and a million detachable collars and cuffs a day. Invented in the 1820s by Hannah Lord Montague, the detachable shirt collar proved to be one of the many adaptations to modernity, akin to our adoption of the microwave oven that Troy

Throughout the twentieth century RPI maintained a single-minded focus on providing the best-quality undergraduate engineering training while competing schools of engineering built up major graduate engineering programs and, beginning in World War II, secured massive federal research funding.[60] During the war, MIT received $117 million through military research contracts and Caltech $83 million, while RPI garnered only $200,000. RPI's eclipse by other educational institutions was paralleled by the erosion of Troy's manufacturing economy, which saw the closure or out-migration of apparel, steel, and other manufacturing firms. After World War II the local economy of Troy collapsed so precipitously that the city's architecture remained intact as "one of the most perfectly preserved nineteenth Century downtowns in the United States."[61]

George Low, who served as RPI's president from 1976 until 1984, is widely credited with reversing the school's decline, encouraging new investments in the college by government, industry, and alumni. RPI university-industry research centers established under Low "were quite successful, especially considering that the university did not have the national stature of Stanford or MIT."[62] In a 1986 survey of 20 US colleges achieving academic excellence, the authors reported that "while RPI was considered an engineering school with high expenses and low faculty salaries in the 1970s, today the college has quality students and excellent technical programs."[63] After Low's death his RPI research centers continued to "draw contributions from industry and remain[ed] resilient during the 1980s" (see Footnote 62). However, Low was followed by a succession of RPI presidents characterized as "interim caretakers, without strong agendas."[64]

---

invented for those struggling to make every day urban life function smoothly." P. Thomas Carroll, "Designing Modern America in the Silicon Valley of the Nineteenth Century," *RPI Magazine* (Spring 1999). RPI graduates founded Troy-based Gurley Precision Instruments in 1845 to manufacture high-quality surveying instruments, and the company continues to produce precision instruments down to the present day. Sanford Cluett, an RPI graduate, joined Troy-based Cluett, Peabody & Co., the maker of Arrow shirts, where he developed sanforization, the process for preshrinking woven fabrics. Stuart W. Leslie, "Regional Disadvantage: Replicating Silicon Valley in New York's Capital Region," *Technology and Culture* (2001), pp. 239–240; "Business: Shirt Tale," *Time* (February 21, 1938).

[60] In the postwar era, RPI "concentrated on its undergraduate student programs, becoming one of the best training grounds in engineering. And in the past 50 years, those within the university acknowledge, this emphasis pulled RPI away from the research realm. The school has subsequently watched the money, the staff and the reputation associated with research go elsewhere." Schenectady, *The Daily Gazette* (April 2, 2000).

[61] Stuart W. Leslie, "Regional Disadvantage: Replicating Silicon Valley in New York's Capital Region," *Technology and Culture* (2001), p. 240; "A Walk into 19th Century Troy, NY," *Upstate Earth* (June 27, 2012).

[62] Elizabeth Popp Berman, *Creating the Market University: How Academic Science Became an Economic Engine* (Princeton and Oxford: Princeton University Press, 2012), p. 130.

[63] "Study Finds RPI an Inspiring Role Model," Albany, *The Times Union* (August 5, 1986).

[64] "RPI Will Devote More Attention to Research—Locally, University at Albany Has Stolen Much of the Spotlight," Schenectady, *The Daily Gazette* (April 2, 2000).

## 8.4 Four-Year Institutions of Higher Learning

In 2000 the RPI Board of Trustees chose Shirley Ann Jackson, who held a doctorate in physics from MIT and who was former chairperson of the Nuclear Regulatory Commission, to become the president of RPI, giving her a mandate to propel the institution into the ranks of the top 10 research universities in the United States. Jackson moved quickly, developing a strategic blueprint, the "Rensselaer Plan," securing a $360 million anonymous gift, expanding the faculty and launching new research centers.[65] Jackson's goal for RPI was to attain national and international recognition in several specific fields, notably information technology and biotechnology, rather than "being pretty good in a whole bunch of fields" (see Footnote 64). Her stated goals were to grow the faculty by 100 and to double the number of doctoral degrees awarded from 125 to 250 between 2001 and 2011.[66]

Criticized by some faculty members for her top-down management style, Jackson has retained the solid support of the Board of Trustees, which renewed her contract for a second 10-year period in 2010 and in 2014 declared that she had "more than surpassed the expectations of the Board."[67] She proved to be the driving force behind the creation of RPI's Center for Computational Innovation (CCI) and the emergence of RPI as a major force in supercomputing, where it operates the most powerful supercomputer at any private university in the United States. Jackson has in many respects achieved her original goals, carrying forward the vision of George Low.

RPI remains a leader with respect to undergraduate education of engineers and scientists. A GlobalFoundries spokesman noted in 2013 that more GlobalFoundries engineers, technicians, and new college graduates come from RPI than from any other institution.[68] RPI also operates some very well-established and -recognized (worldwide) graduate research programs. Six RPI graduate departments in the School of Engineering now rank in the top 30 nationally. Under Jackson's tenure, the RPI research enterprise has grown from $37 million per year to well over $100 million at present.

---

[65] "A Visionary for RPI President: Shirley Ann Jackson is Charging Hard—And Fielding Flak—As She Pursues an Ambitious Agenda for the School," Albany, *The Times Union* (July 14, 2002).

[66] "Research Initiatives Take Center Stage in RPI Plans" Schenectady, *The Daily Gazette* (February 25, 2001).

[67] "RPI prepares to celebrate Dr. Jackson's 15th Anniversary as President," Troy, *The Record* (July 25, 2014). Under Jackson's tenure, applications for admissions to RPI more than tripled between 1999 and 2014, with SAT scores for entering freshmen up 104 points during the same period. Under Jackson, RPI hired 440 new faculty including 302 new tenure track and tenured individuals. Sponsored research grew from $37 million in 1999 to $100 million in 2014. Under Jackson new undergraduate and graduate programs were launched in Science, Engineering, Architecture, Humanities, Arts, and Social Sciences, as well as in the Lally School of Management, and launched new interdisciplinary degree program, Information Technology and Web Science. Ibid.

[68] National Research Council, Charles W. Wessner (rapporteur), *New York's Nanotechnology Model: Building the Innovation Economy* (Washington, DC: The National Academies Press, 2013), p. 68.

### 8.4.1.1 RPI Research Centers

As noted in Chap. 2, during his tenure as RPI's president, George Low's strategy for revitalizing the institution emphasized the creation of research centers, an incubator, and a technology park in which students and faculty could collaborate with established technology-intensive companies or, in some cases, start their own enterprises.[69] From their inception the research centers have been interdisciplinary in nature, which has required institutional and cultural changes at RPI over time.[70] Roland W. Schmitt, a GE executive who became president of RPI in 1988, regarded RPI's innovative alliances with public- and private-sector actors, embodied in the research centers, as one of the college's greatest assets, noting that "The centers here at RPI are the national prototype."[71] A 2012 Princeton University study compared the performance of RPI's research centers favorably with those of Caltech and the University of Minnesota, citing RPI's leveraging "the sponsorship of industry with sponsorship of government."[72] A 2006 academic assessment of the RPI research centers opened during or immediately after Low's tenure concluded that—

> Over the past two decades, these efforts have resulted in a host of institutional and programmatic elements in and around RPI, creating a comprehensive innovation milieu. The Troy area possesses a broad knowledge base with a developed infrastructure. Therefore, RPI emphasizes a diversity of technology sectors with strong information systems and manufacturing industry, though practically software and biotechnology are dominant.[73]

- **George M. Low Center for Industrial Innovation (CII)**. CII, which George Low pioneered but which was constructed after his death in 1984, consolidated under one roof various RPI research centers that had previously been "spread out over the RPI campus, on other sites in Troy and Watervliet."[74] CII currently houses numerous thematic research centers and laboratories, most notably the Center for Materials, Devices, and Integrated Systems—dedicated to research themes which include interconnect technologies, power electronic systems, and micro

---

[69] "RPI's Low Left Quite a Legacy," Albany, *The Time Union* (October 6, 2016).

[70] Interdisciplinary research precludes narrowly focused research themes and typically results in multi-authorized publications of research papers. Questions thus arose over individuals' contributions. Initially young faculty in those programs suffered in peer reviews which provided the basis for promotion and tenure. The solutions evolved included encouragement of faculty to also focus on themes within their discipline; less weight being accorded individuals' authorship; and changing other criteria for assessment. C. W. Le Maistre, "Academia Linking with Industry—The RPI Model" *IEEE Xplore* (October 1989), p. 208.

[71] "Senior GE Executive Named 16th President of RPI," Albany, *The Times Union* (January 24, 1988).

[72] Elizabeth Popp Berman, *Creating the Market University: How Academic Science Became an Economic Engine* (Princeton and Oxford: Princeton University Press, 2012), p. 131.

[73] Leonel Corona, Jérôme Doutriaux, and Sarfraz A. Mian, *Building Knowledge Regions in North America: Emerging Technology Innovation Poles* (Cheltenham, UK, and Northampton, MA: Edward Elgar, 2006), p. 49.

[74] "RPI shows off Research Center," Albany, *The Times Union* (November 19, 1986).

contamination—and the Center for Automation Technologies and Systems (CATS). Dan Walczyk, an RPI faculty member, commented in 2012 with respect to CII's Advanced Manufacturing Laboratory that it taught students not only how to design products but also how to design the systems needed to manufacture those products efficiently and with minimal waste, making it unique in the United States, noting "No one else does it soup to nuts."[75]

- *Center for Biotechnology and Interdisciplinary Studies (CBIS)*. CBIS, one of the most advanced research facilities in the United States, was launched by Shirley Jackson in 2004 to draw on multiple disciplines to produce breakthroughs in medicine and health.[76] Highlighting its interdisciplinary character, CBIS' facility houses no departmental offices, only laboratory space where researchers "can collaborate with people from . . . many disciplines and use the latest equipment and resources."[77] Research themes include molecular biology, regenerative medicine, protein synthesis, bioengineering, and bioinformatics. In its first 10 years, CBIS trained 1000 undergraduates, awarded 200 PhD's, published 2000 peer-reviewed articles, and raised $130 million in external funding.[78]
- *Center for Directed Assembly of Nanostructures*. RPI's Center for Directed Assembly of Nanostructures is an NSF-funded program focusing on new methods for assembling novel functional materials and devices from nanoscale building blocks. Staffed by interdisciplinary teams of researchers and students with backgrounds in nanotechnology science and engineering, the center partners with private firms, international research centers, universities, and innovative K-12 educational programs.
- *Center for Computational Innovations (CCI)*. CCI, originally established in 2006 as the Computational Center for Nanotechnology Innovations (CCNI), is a $100 million partnership between RPI, IBM, and the State of New York centered on one of the world's most powerful supercomputers (see Box 3.3 in Chap. 3). CCNI facilitates research utilizing RPI faculty and students and the Center's high-performance computing for simulation, modeling, and manipulation of big data.[79]
- *Center for Lighting Enabled Systems and Applications (LESA)*. RPI's Center for Lighting Enabled Systems and Applications (LESA) has been established through grants from the National Science Foundation and operates as a consortium including Boston University, the University of New Mexico, and industry

---

[75] "Mission: Make Something," Albany, *The Times Union* (May 3, 2012); "Pushing the Boundaries," Albany, *The Times Union* (April 24, 2011).

[76] "Decade of Growth in RPI Biotech Unit," Albany, *The Times Union* (September 9, 2014).

[77] "RPI Center Celebrates First Decade," Troy, *The Record* (September 11, 2014).

[78] Rensselaer Polytechnic Institute, "Center for Biotechnology & Interdisciplinary Studies," <http://biotech.rpi.edu>.

[79] See summary of remarks of Shirley Anne Jackson in National Research Council, Charles W. Wessner (rapporteur), *New York's Nanotechnology Model: Building the Innovation Economy* (Washington, DC: The National Academies Press, 2013), p. 69.

partners.[80] The center is on the verge of graduating from a decade of NSF support, with that support reflecting an acknowledgment at the federal level of the quality of its work.

### 8.4.1.2 RPI's Incubator and Technology Park

During George Low's tenure, RPI established in 1980 one of the first business incubators in the United States (see Box 2.3 in Chap. 2) and in 1981 launched the Rensselaer Technology Park, which was modeled on the Stanford Industrial Park established by Fred Terman in 1951.[81] Between 1981 and 2010, the RPI incubator housed a total of 250 start-up companies, many of which remained in the Capital Region after leaving the incubator. Jason Edwards, president and CEO of the start-up CORESense, commented in 2013 that "the incubator is a phenomenal program for a start-up company," citing the access accorded to broad networks of financial, marketing, and operational support.[82] Several individuals interviewed for this study indicated that many of the Capital Region's successful start-ups came out of RPI, reflecting that institution's entrepreneurial character. The RPI incubator has produced some of the area's best business success stories, including Evocative Design, Dais Analytic Corporation, Velan Studios, Ecovative, MapInfo Corporation, Vicarious Visions, and 1st Playable Productions.[83] However, in 2010 RPI decided that it would no longer physically house companies at its incubator and that the incubator's existing tenants would need to leave. Dinah Adkins, former president of the National Business Incubation Association, which had always held up the RPI incubator as "an example of best practices for university business incubators," commented that "the closing of the RPI incubator is a travesty. It's very sad. Here was a top program in the world."[84]

A year after closing the RPI incubator's physical site, RPI launched a new business incubation program, Emerging Ventures Ecosystem (EVE), specializing in start-up businesses in the energy and environmental sectors. EVE had a central office in downtown Troy but the start-up companies were at scattered locations in

---

[80] "Smart Lights Are a Bright Idea," Albany, *The Times Union* (April 10, 2012); "Internet of Things Result," Albany, *The Times Union* (May 17, 2015).

[81] "Thinking East, Looking West," Albany, *The Times Union* (March 1, 2009).

[82] "Incubators Get Start-ups Off and Running," Schenectady. *The Daily Gazette* (February 23, 2003).

[83] "Ideas in Action," *RPI Alumni Magazine* (Fall 2006).

[84] Nancy L. Zimpher, "Foreword," in Jason E. Lane and D. Bruce Johnstone (Eds.), *Universities and Colleges as Economic Drivers: Measuring Higher Education's Role in Economic Development* (Albany: SUNY Press, 2012), p. xii.

"Change in Direction for RPI's Incubator," Albany, *The Times Union* (February 10, 2010). RPI's move was reportedly based on a decision to revamp its business incubator program to focus exclusively on energy and environmental start-ups. A university official commented that "the question is how coordinated are all of these operations at Rensselaer that have to do with business start-ups. Our feeling was that we could do better." Ibid.

Troy and elsewhere.[85] The new program provided support for start-ups in the pre-seed, seed, and early-stage phases, with services including a board of advisers, funding, networking, a help desk, and technology showcases.[86] James Spencer, director of New Venture Development at RPI, said in a 2013 interview that in contrast to the old RPI incubator, EVE "will not be low rent," and that he envisioned teams or "cohorts" of entrepreneurs, scientists, and engineers entering the incubator, where it would be determined whether their ideas were viable in the real-world marketplace.[87] While it may be premature to assess the new incubator program, its first director left after less than a year when his contract was not renewed, and some of the individuals interviewed for this study from 2015 to 2017 commented that RPI "used to have a good incubator."[88]

RPI's Technology Park never equaled the spectacular success of Stanford's industrial park, but during the first decade of its existence it grew at a "slow, steady pace," attracting 40 businesses and "putting the Capital District on the high-tech map," enabling RPI to "attract top students and provide them with state-of-the-art training."[89] In the park, small companies moved into multitenant buildings built by RPI and larger companies built their own facilities … [so that] by 2007 the park housed over 70 companies employing 2300 people and had an occupancy rate of 96%.[90] In September 2007 a new data center opened in the park, housing an IBM supercomputer capable of performing 100 trillion arithmetic functions per second, part of RPI's Computational Center for Nanotechnology Innovations. In 2009 the park added an additional 200 acres for the construction of a $165 million GE Healthcare Digital X-Ray Detector Production Facility dedicated to mammography. GE began manufacturing digital mammography detectors at the site in 2010.[91] As of 2016 the park housed over 70 technology-intensive companies.[92]

---

[85] "Legacy of RPI's Incubator Program Continues With EVE," Troy, *The Record* (February 8, 2011).

[86] "RPI Launcher Incubator Program," Troy, *The Record* (February 7, 2011).

[87] "RPI to Revive Business Incubator in Downtown Troy," Troy, *The Record* (December 2, 2013).

[88] "RPI Leader to Fill 2 Jobs," Albany, *The Times Union* (July 13, 2012). Esther Vargas, a certified business incubation manager, took over the directorship of EVE in 2014. She concluded that some incubator applicants were underqualified and needed additional support to enter the program. She initiated a summer accelerator, the Rensselaer Emerging Ventures Ecosystem Accelerator Lab, for entrepreneurs ready to move "beyond the idea stage." She secured some funding from a 3-year $350,000 grant from Empire State Development to RPI which she expected to use for support for program participants' prototyping and to renovate space at the RPI campus to enable collaboration—that is, to restore some aspects of the original incubator by establishing a physical site. "Rensselaer Prepares for First Summer Accelerator for Entrepreneurs," *Albany Business Review* (June 3, 2015).

[89] "Times Changes: Technology Park RPI Project Grows Slowly," Albany, *The Times Union* (October 28, 1990).

[90] "Tech Park Aims to Build on Success," Albany, *The Times Union* (June 3, 2007).

[91] "Tech Park Developers Retain Optimistic Outlook," *Albany Business Review* (January 7, 2008); "New Product Lines Help GE Workforce Grow," Troy, *The Record* (September 8, 2013).

[92] "RPI's Low Left Quite a Legacy," Albany, *The Times Union* (October 6, 2016).

### 8.4.2 State University of New York (SUNY)

SUNY's institutional presence in the Capital Region includes SUNY Albany, SUNY Polytechnic, Empire State College, and five community colleges, a number of them very highly regarded.[93] These institutions form part of a statewide network of over 60 campuses. In 2012 SUNY enrolled over 467,000 students, employed 88,000 faculty, and counted over 3 million alumni, and 93% of New York's residents lived within 15 miles of a SUNY campus.[94] Not inappropriately, SUNY's 2010 strategic plan to address economic distress in the wake of the 2008 recession was named "The Power of SUNY," and that power has been repeatedly demonstrated through large-scale investments in research infrastructure that underlie the emergence of Tech Valley.

"SUNY's human capital development role in New York is immense," and it is the primary provider of higher education within the Capital Region. In 2008–2009 1.6 million SUNY alumni lived in the state and in some regions SUNY grads comprised two thirds or more of all college graduates.[95] In the Capital Region, a 2011 survey by the Rockefeller Institute of Government found that in 2008–2009, 166,000 SUNY alumni lived in the region including 103,000 who graduated from SUNY institutions located within the region. Of the 44,200 students enrolled in SUNY institutions in the Capital Region in 2008–2009, 68% grew up in the region.[96]

Nancy Zimpher became Chancellor of SUNY in 2009 in the middle of one of the worst recessions in the past half century. After months of deliberation, she launched an effort to substantially increase SUNY's role as an economic engine, citing the example of the land-grant colleges established pursuant to the Morrill Act of 1862.[97] SUNY established five technology-transfer hubs at research colleges in the SUNY system to improve tech transfer to businesses.[98] Zimpher directed SUNY's efforts at four technology areas where the educational system could have a positive impact on local job growth—nanotechnology, healthcare, high performance computing, and green energy.[99] In 2012, the president of SUNY Adirondack said in an interview that

---

[93] "Area Community Colleges Gain High–Tech Focus," Schenectady, *The Daily Gazette* (September 22, 2009).

[94] Nancy L. Zimpher, "Forward," in Jason E. Lane and D. Bruce Johnstone (Eds.), *Universities and Colleges as Economic Drivers: Measuring Higher Education's Role in Economic Development* (Albany: SUNY Press, 2012), p. xii.

[95] Thomas Gais and David Wright, "The Diversity of University Economic Development Activities and Issues of Impact Measurement," in Jason E. Lane and D. Bruce Johnstone (Eds.), *Universities and Colleges as Economic Drivers: Measuring Higher Education's Role in Economic Development* (Albany: SUNY Press, 2012), p. 36.

[96] Rockefeller Institute of Government of the University at Albany, and the University at Buffalo Regional Institute, *How SUNY Matters: Economic Impacts of the State University of New York* (State University of New York, June 2011).

[97] See summary of remarks of Nancy Zimpher in National Research Council, Charles W. Wessner (rapporteur), *New York's Nanotechnology Model: Building the Innovation Economy* (Washington, DC: The National Academies Press, 2013), pp. 104–105.

[98] "SUNY Chancellor Outlines Goals," Glens Falls, *The Post-Star* (September 30, 2010).

[99] "SUNY Chancellor Speaks at FMCC," Schenectady, *The Daily Gazette* (January 8, 2011).

Zimpher had "really pushed a model to get us to think as a system. . . . When Governor Cuomo talks about SUNY he talks to Nancy Zimpher. She really has helped put SUNY on the forefront and tied us to economic development."[100]

SUNY has developed an array of programs to promote "applied learning" throughout New York's education system.[101] This effort includes work-based activities (work–study, internships, co-ops, and clinical placements), community-based learning, and discovery-based activities such as research and entrepreneurship. One initiative, SUNY Works, is designed expressly to align higher education with labor market needs and engages companies such as GlobalFoundries, IBM, and GE (see discussion earlier in this chapter). A 2015 study of SUNY Works conducted by the Rockefeller Institute of Government concluded that—

> *SUNY Works is unique: There is no other state or system in the US that has advanced a work-based learning initiative on a scale and across the breadth of types of study programs and institutions encompassed by SUNY.*[102]

In September 2017 Kristina Johnson succeeded Nancy Zimpher as Chancellor of the SUNY system. Johnson, who served as federal Under Secretary of Energy under President Obama, holds 118 patents and recently co-founded a clean energy company building and operating hydropower plants, which grew from 1 to 19 plants during her tenure. She has made it clear that she will maintain SUNY's mission of growing the state's economy. "[W]hat we can think about when we think about economic engines is how we can use these individual campuses to become the innovators, whether it's in the arts or in food science, or in the tech sector, or in other areas, energy or energy efficiency."[103]

## *8.4.3  University at Albany: State University of New York (SUNY Albany)*

SUNY Albany enrolls over 17,000 students in nine schools and colleges offering 50 undergraduate majors and 138 graduate programs. It operates through campuses located in Albany, Rensselaer, and Guilderland. Based on size of enrollment, it is the

---

[100] "SUNY Adirondack Getting a Lesson in Growth," Glens Falls, *The Post-Star* (February 18, 2012).

[101] "Applied learning is the application of previously learned theory whereby students develop skills and knowledge from direct experiences outside a traditional classroom setting." Alan Wagner, Ruirui Sun, Katie Zuber, and Patricia Strach, *Applied Work-Based Learning at the State University of New York: Situating SUNY Works and Studying Effects*. (Albany: The Nelson A. Rockefeller Institute of Government, May 2015), p. 1.

[102] Alan Wagner, Ruirui Sun, Katie Zuber, and Patricia Strach, *Applied Work-Based Learning at the State University of New York: Situating SUNY Works and Studying Effects*. (Albany: The Nelson A. Rockefeller Institute of Government, May 2015).

[103] "Dr. Kristina Johnson Begins Tenure as SUNY Chancellor," Albany, *The Times Union* (September 6, 2017).

largest institution of higher learning in the Capital Region. However, as the *Albany Business Review* observed in 2013, at SUNY Albany "most of the focus over the years has been on the college of Nanoscale Science and Engineering and its colorful leader, Alain Kaloyeros…meanwhile, the University at Albany, which is older, employs more people and graduates far more students, has been overshadowed."[104]

The separation of CNSE from SUNY Albany was criticized by former SUNY Albany President Karen Hitchcock in 2013, who pointed out that the institution had invested heavily in the NanoCollege "to enable it to grow and make the entire university and the region stronger."[105] Mathematics professor and former Chair of the SUNY Albany Senate Michael Range offered a different perspective in a 2013 interview, acknowledging that some SUNY Albany faculty "have been feeling that UAlbany [SUNY Albany] did not really benefit much" from CNSE and that "UAlbany's budget has been cut quite a bit while the nanocollege prospered," expressing his view that "UAlbany would be better off on its own." He said that "people wrongfully assume that UAlbany enjoys the same prosperity as the nanoscale college, and separating the two will paint a clearer picture of Albany's finances."[106] SUNY Albany President Robert Jones, who took office in 2013, shared that perspective, supporting the separation because "the nanocollege's independence will give UAlbany faculty and students more opportunities to compete for research and academic resources."[107]

In fact, the separation of CNSE from SUNY Albany in September 2014 was the occasion for a difficult but ultimately productive appraisal of the challenges facing SUNY Albany, which had been accumulating for years and which were taken up by incoming president Jones. Enrollment was declining, which Jones characterized as a "significant financial threat."[108] In 2010 SUNY Albany cut programs in French, Russian, Italian, classics, and theater, all of which were experiencing declining enrollment. Funding for graduate students was reduced.[109] Jones said in 2013 that on SUNY Albany's campuses, the signs of aging were all too apparent—"windows continue to leak, our heating and cooling continues to fail and lecture halls need to be modernized."[110]

---

[104] "Beyond Nano, UAlbany has grown," *Albany Business Review* (August 2, 2013).

[105] "Q&A with Former UAlbany President in Nanocollege Split," *Albany Business Review* (July 24, 2013). One SUNY Albany donor and member of the board of University at Albany Foundation complained in 2013 that "UAlbany [SUNY Albany] sacrificed much for the success of CNSE. A lion's share of tuition dollars and the limited taxpayer and SUNY dollars available to UAlbany has been diverted from mainstream university programs and poured into CNSE. Of the tuition and state dollars made available to UAlbany each year, $20 million is lopped off and directed to CNSE. As a result [UAlbany] rankings suffered, particularly from the axing of programs and a debilitating student to facility ratio." See "SUNY's Risky Play for Power," Albany, *The Times Union* (April 3, 2013).

[106] "Will Kaloyeros Break away from UAlbany?" *Albany Business Review* (March 22, 2013).

[107] "And He can make a Loss Seem like a Win," Albany, *The Times Union* (October 27, 2013).

[108] "Loss of Students Cited as a Threat," Albany, *The Times Union* (April 2, 2013).

[109] "Beyond Nano, UAlbany Has Grown," *Albany Business Review* (August 2, 2013).

[110] "How to Make Breaking UP Easier to Do" Albany, *The Times Union* (July 18, 2013).

Stating that the college needed to reverse the "cumulative effects of insufficient investments over the past decade—investments in people, in our physical plant and in the high-need academic programs," Jones oversaw a number of new renovation and expansion initiatives.[111] In October 2015 he unveiled a plan to "boost academic offerings, enhance the student experience and grow [the college's] financial resources over the next 5 years to reach [an enrollment of] 20,000 students."[112] A centerpiece of the effort was a plan to offer "a public college offering for engineering, which is important, because engineering schools can be expensive."[113] In addition, a new College of Emergency Preparedness, Homeland Security, and Cybersecurity would be established, the first such institution in the nation.[114] As these and other projects moved forward, two state legislators based in the Capital Region observed that "UAlbany [SUNY Albany] is regaining its position as one of the state's and the nation's premier public research institutions."[115]

Jones characterized the creation of the new engineering college as a "game changer."[116] The program expands the College of Computing and Information established in 2005 to create the College of Engineering and Applied Sciences at its downtown Albany campus. The new college will be housed in a renovated Albany high school building which, when completed, will be capable of supporting 1000 students and 180 faculty researchers at the undergraduate and graduate levels. The new engineering college has benefitted from broad bipartisan support in the state legislature as well as advocacy by business leaders from the Capital Region and local officials and residents.[117] Kim Boyer, the former head of RPI's Department of Electrical, Computer and Systems Engineering has been recruited as dean,[118] and Boyer, who is recruiting faculty for the new school, said in 2016 that "offering a strong engineering program at an affordable public university is crucial for the Capital Region."[119]

In February 2016 Governor Cuomo announced that SUNY Albany would construct a 236,000 square-foot Emerging Technology and Entrepreneurship Complex (ETEC) which will "couple cutting edge research with economic development initiatives" to "spur the transfer of ideas and new technologies to commercial

---

[111] "Loss of Students Cited as a Threat," Albany, *The Times Union* (April 2, 2015).

[112] "20,000 Students by '20 the New Goal," Albany, *The Times Union* (October 28, 2015).

[113] "Albany Looks to Engineering, Cybersecurity to Grow Enrollment," Glens Falls, *The Post-Star* (November 27, 2015).

[114] "Albany Looks to Engineering, Cybersecurity to Grow Enrollment," Glens Falls, *The Post-Star* (November 27, 2015); "Nation's First Security College Creates Student Opportunities," *Long Island Examiner* (January 27, 2015).

[115] "Building Toward UAlbany's Future," Albany, *The Times Union* (March 29, 2016).

[116] "Engineering College Set," Albany, *The Times Union* (February 24, 2016).

[117] "Building Powered UAlbany's Future," Albany, *The Times Union* (March 29, 2016).

[118] "UAlbany Pursues 'Game Changer,' Officials Seek $20M to Turn Old School Into College of Engineering," Schenectady, *The Daily Gazette* (February 24, 2016).

[119] "Engineering College Set," Albany, *The Times Union* (February 24, 2016). One student who enrolled in the new college said that she chose it over other engineering colleges because "all the other offers were unaffordable." Ibid.

enterprises." Funded by the state with a budget of $184 million, the complex will house the College of Emergency Preparedness, Homeland Security, and Cybersecurity; the New York State Mesonet; and university facilities supporting start-ups, technology transfer, and small businesses.[120]

SUNY Albany has moved to strengthen its ties with community colleges in the region to facilitate smooth transitions for graduates of 2-year programs into SUNY Albany 4-year degree programs. Although transfer agreements with a number of community colleges have been in place for years, the recent growth at SUNY Albany has made revisions necessary. In February 2017 SUNY Albany and Hudson Valley Community College finalized a transfer agreement that enables a "seamless" transfer path for 34 courses of study including sciences, computer science, and any program at the newly established College of Engineering and Applied Science and the College of Emergency Preparedness, Homeland Security, and Cybersecurity. SUNY Albany is reportedly working on similar transfer agreements with SUNY Adirondack, Fulton-Montgomery Community College, and Dutchess Community College in Poughkeepsie.[121]

#### 8.4.3.1 College of Nanoscale Science and Engineering

The College of Nanoscale Science and Engineering (CNSE) has been the centerpiece of the effort to create Tech Valley, but CNSE's industrial partnerships have tended to overshadow its significant educational role, which involves not only its own curriculum and students but a wide range of collaborations with other educational institutions within the region. CNSE offers bachelor's, master's, and doctoral degrees in nanotechnology and nanoscience. The degree programs are closely linked to CNSE's research and development activities with industry partners, with most advanced degree students participating in that work and some remaining as post-docs. CNSE awarded its first PhD and master's degrees in 2004 and its first bachelor's degrees in 2013. CNSE's tracking data indicates that roughly one third of its undergraduates move on to positions in the nanotechnology industry in the Capital Region and that over half of the advanced degree graduates transition to jobs in New York State, most of them in the nanotechnology field.[122]

In September 2014, CNSE was transferred to the State University of New York at Utica-Rome (SUNY IT), merging with it to form a new entity, SUNY Polytechnic Institute (SUNY Poly). CNSE was reorganized into two Albany-based colleges, the

---

[120] Interview with James Dias, vice president for research, SUNY Albany (September 16, 2015); Office of the Governor, "Governor Cuomo Announces New SUNY Emerging Technology and Entrepreneurship Complex at Harriman Campus," Press Release (February 5, 2016).

[121] "Smooth Transition for Students," Albany, *The Times Union* (February 25, 2017).

[122] Laura I. Schultz, et al., "Workforce Development in a Targeted, Multisector Economic Strategy: The Case of New York University's College of Nanoscale Science and Engineering," in Carl Van Horn, Tammy Edwards, and Todd Green (eds.) *Transforming US Workforce Development Policies for the 21st Century* (Kalamazoo: W.E. Upjohn Institute for Employment Research, 2015), p. 345.

**Table 8.4** CNSE enrollment forecast through 2020

|  | Actual | Forecast | | | | | |
| --- | --- | --- | --- | --- | --- | --- | --- |
| CNSE Student Category | 2014 | 2015 | 2016 | 2017 | 2018 | 2019 | 2020 |
| Legacy SUNY Albany | 345 | 295 | 220 | 100 | 25 | 20 | – |
| SUNY Poly | – | 56 | 144 | 233 | 322 | 411 | 500 |
| Total enrollment | 345 | 351 | 364 | 333 | 347 | 431 | 500 |

Source: SUNY Polytechnic Institute, *Strategic Plan* (June 2016)

**Table 8.5** CNSE Albany site students served, 2015

|  | Number of students | | |
| --- | --- | --- | --- |
| Area of concentration | Undergraduate | Masters | PhD |
| Nanoscale science | 41 | 1 | 36 |
| Nanoscale engineering | 131 | 21 | 71 |
| Undeclared major | 17 | – | – |
| Total | 199 | 22 | 107 |

Source: SUNY Polytechnic Institute, *Strategic Plan* (June 2016)

College of Nanoscale Science (CNS) and the College of Nanoscale Engineering and Technology Innovation (CNETI). CNSE students who were matriculating at the time of the reorganization were given the option of retaining their SUNY Albany affiliation while completing their course of study at CNSE or of transferring to, and receiving degrees from, SUNY Poly. Thus in the 2015–2019 timeframe, the students enrolled at CNSE are comprised of a mix of "legacy" SUNY Albany students and SUNY Poly students, with the latter forecast to account for the entire student body by the fall of 2020. (See Table 8.4.)

As shown in Table 8.5, the SUNY Poly student body is dominated by individuals oriented toward nanoscale engineering. The separation of CNSE from SUNY Albany appears to have had a negative effect on enrollment, possibly as a result of loss of student access to SUNY Albany infrastructure. Incoming undergraduate classes which at their peak numbered about 75 students have dropped to 25–30. Most of the incoming freshmen have been coming to CNSE "since middle school," reflecting the college's numerous outreach programs engaging local K-12 schools. CNSE also gets a "significant number of transfers from nearby community colleges."[123]

CNSE's draw for undergraduate and graduate students has always been the opportunity to work alongside and build relationships with the high-tech companies engaged in research at the college.[124] At the NanoCollege facility at Fuller Road,

---

[123] Interview with CNSE professor Laura Schultz, Albany, New York (November 30, 2016).

[124] One student enrolling in CNSE wrote in 2015 that "I was attracted to this school mainly because of the super-small class sizes (no lecture halls and actual professor–student interaction), industry networks (companies are actually on-site!) and the truly unparalleled technical resources available . . . . They also did a great job during the Open Houses. I distinctly remember one of the speakers saying this place is not for passive knowledge sponger but for people that grab their learning by the throat . . . . CNSE was a no-brainer for me." See "Anyone Else Attending CNSE?," blog post on *College Confidential* (August 15, 2015).

company personnel outnumber CNSE faculty and students by a wide margin.[125] Graduate students can enter into paid internships with industrial partners during their course of study, "complementing the resources they have available for research" and enhancing their prospects for "future employment opportunities at these companies."[126]

### 8.4.3.2 SUNY Empire State College

SUNY Empire State College operates at 35 locations throughout the state, including, in the Capital Region, sites in Albany, Schenectady, Saratoga Springs, Latham, Queensbury, and Johnstown. The college offers associate's, bachelor's, and master's programs and features courses of study that can be tailored to meet individual goals. Most of its student body is comprised of working adults pursuing additional education. The average age of undergraduates is 35 and of master's degree candidates is 40. It offers bachelor's degree programs in manufacturing, computer systems, and information systems and master's degree programs in emerging technology, business management, and innovation management/technology transfer.[127]

## 8.4.4 Union College

Union College in Schenectady has a long tradition of producing leaders in politics, business, and science, including (as of 2003), 1 US president (Chester Arthur), 15 US senators, 91 members of the House of Representatives, 13 governors, and a half-dozen cabinet members, including Lincoln's Secretary of State, William Seward.[128] Its graduates include R. Gordon Gould, the inventor of the laser; Samuel Blumberg, a Nobel Laureate in medicine; Ted Berger, developer of the bionic brain; Richard Templeton, former CEO of Texas Instruments; and many senior executives in high-tech companies.[129] In 1845 the school became the first small college in the country to introduce an engineering program on an equal basis with liberal arts, beginning what

---

[125] "University at Albany Offers the World's First Nanoscale Undergraduate Degree," Glens Falls, *The Star Post* (February 24, 2010).

[126] Unni Pillai, *SUNY College of Nanoscale Science and Engineering* (Monograph, March 2015); "Moving up, But Not Out," Albany, *The Times Union* (May 15, 2011). Ben Backes, who graduated from CNSE in 2011 with a master's degree in nanoscale engineering, immediately took a job as a characterization engineer at IBM's semiconductor fab in East Fishkill, said that "I sent out two resumes and I got two phone calls back. I felt like a pretty attractive candidate." Ibid.

[127] "Colleges Strive for Workplace Link," Albany, *The Times Union* (January 27, 2013).

[128] "Union Alumna Follows Tradition as Schwarzenegger's Chief of Staff," Albany, *The Times Union* (December 3, 2003).

[129] "From Graduates to Greatness," Schenectady, *The Daily Gazette* (June 19, 2011); "Union Graduate Students Hear of Taxes, Dreams," Schenectady, *The Daily Gazette* (June 10, 2012).

would become Union's hallmark "balanced college concept," a curriculum exposing students to both science/engineering and the liberal arts, which has proven highly effective in educating future leaders.[130] By the mid-nineteenth century, over 30 Union College graduates had become college presidents.[131] Within the hard sciences and engineering fields, Union has been offering interdisciplinary studies in science and technology fields, such as nanotechnology, biochemistry, and environmental technology, for 15 years. The interdisciplinary curriculum makes Union College's bioengineering and nanotechnology graduates more attractive to employers, according to Professor Steven Rice, co-director of the bioengineering program.[132]

Union College has a particularly close and mutually beneficial relationship with IBM. IBM's John E. Kelly III, one of the principal architects of Tech Valley, graduated from the college in 1976. A 2003 survey indicated that IBM employed about 400 Union College graduates, including Stephen Mills (class of 1973) who at the time controlled IBM's entire software operation, and Robert Moffat (class of 1978), then a senior vice president in charge of IBM's Integrated Supply Chain. A Union College official, Charlie Carey, observed that "none of these executives (Kelly, Mills, and Moffat) [was an] engineering grad. That's something we like to tout, and I think that's our special niche—that our graduates will make good managers and executives." He indicated that the college's interdisciplinary approach allowed students to train in a science while concurrently gaining exposure to "the larger ideas and issues a manager will be expected to confront."[133]

For its part, IBM has made substantial investments in Union College. In 2002 the company made a gift of $1 million for technical support to improve the college's engineering program.[134] In 2011 IBM disclosed that it was donating to the college the "intelligent cluster," a "supercomputer unmatched on any liberal arts campus nationwide, if not worldwide."[135] The donation, valued at over $1 million, featured 88 servers with 1056 individual processors capable of 9.5 trillion operations per second.[136] In a 2014 visit to Union College, John Kelly observed that "the technology now, with things like the supercomputer, has advanced so much that the students

---

[130] "Balanced College Concept," in Wayne Somers (ed.), *Encyclopedia of Union College History* (Schenectady: Union College Press, 2003), pp. 83–88.

[131] Frederick Rudolph, *Curriculum: A History of the American Undergraduate Course of Study Since 1636* (San Francisco: Josey Bass, 1977), p. 87.

[132] Rice points out that the majority of Union graduates interested in careers in medicine graduate with working knowledge of biomedical devices, applications, and prosthetics. As a result, companies can hire one Union graduate possessing multiple skills rather than two or three people. "Students have a much broader base of experience than most engineers have." See "Union College Offers Interdisciplinary Mix of Studies," Schenectady, *The Daily Gazette* (February 17, 2013).

[133] "Union College Grads Thrive in Careers at IBM," Schenectady, *The Daily Gazette* (February 23, 2003).

[134] "IBM Invests in Future with Union Gift," Albany, *The Times Union* (March 9, 2002).

[135] "Super Donation Goes to Union," Albany, *The Times Union* (May 22, 2011).

[136] "Super Donation Goes to Union," Albany, *The Times Union* (May 22, 2011); "Supercomputer Lends Status to Union's Research Efforts," Schenectady, *The Daily Gazette* (May 30, 2011).

are getting access to world-class equipment and world-class facilities." Union College President Stephen Ainlay said that the supercomputer was enabling students to "work with the human genome, but we couldn't do it if it weren't for that technology," and the computer was enabling research into cures for diseases like Alzheimer's and research in 3D printing and robotics.[137]

### 8.4.5 Siena College

Siena College is an independent Roman Catholic liberal arts college in Loudonville, which is in Albany County. In 2004 David Smith, a Siena spokesman, observed that 60% of Siena's graduates remained in the region within a 90-mile radius, positioning the college to provide alumni contacts with local businesses and recruit a workforce for those businesses, noting—

> When you think Tech Valley, you think high-tech science. But there's another level with all these people in accounting, human resources, marketing, business managers. That's where we see our niche. If you look at all the people who are movers and shakers in this, they are predominantly Siena grads.[138]

Siena has a long history of educating entrepreneurs who have started businesses in many industries, including nanotechnology, software, clean energy, and information technology.[139]

In 2010, Siena College entered into a partnership with CNSE to allow qualified students studying computer science, biology, mathematics, biochemistry, or physics to take undergraduate and laboratory courses at CNSE during their junior year. These courses would count toward their graduation from Siena but also give them an edge for admission into CNSE's master's and doctoral programs. These students would also be eligible for semester-long research projects and summer internships.[140]

Siena College's Stewart's Advanced Instrumentation and Technology Center (SAInT), which opened in 2015, is a multidisciplinary instrumentation center supporting student research in physics, astronomy, biochemistry, biology, chemistry, and environmental science. Students and faculty recently drawn to Siena cite as decisional factors the SAInTs' advanced precision instruments, including numerous nuclear magnetic resonance spectroscopy machines, and one new faculty member commented that "schools three times this size aren't going to have this level of instrumentation. Students enrolling in Siena's science programs can start doing research in their freshman year."[141]

---

[137] "IBM Exec Touts Region, Alma Mater," Schenectady, *The Daily Gazette* (September 13, 2014).

[138] "Area Colleges Getting Into 'Tech Valley Game,'" Schenectady, *The Daily Gazette* (February 22, 2004).

[139] "Siena's Stack Center Unveils Wall of Success," Troy, *The Record* (April 27, 2015).

[140] "Schools Changing to Meet Chip Plant's Needs," Schenectady, *The Daily Gazette* (December 20, 2010).

[141] "Colleges Raise the Stakes," Albany, *The Times Union* (November 27, 2016).

### 8.4.6 Skidmore College

Skidmore College in Saratoga Springs is one of the most highly regarded liberal arts colleges in the United States.[142] Unusually for a liberal arts institution, it offers many innovative business courses and programs, and in 2016 it was ranked 9th nationally in a comparison of small college business programs.[143] One popular course, MB107, the "cornerstone course of the Management and Business Department," uses the case method based on real-world scenarios at assigned companies which students study using a variety of theoretical and analytic tools. Students write several case study analyses and at the end of the semester, working in teams of four, must make an "executive presentation" to a panel of real-world business executives for evaluation.[144] Skidmore students can earn a bachelor's degree from the college as well as an MBA through collaborative programs with Clarkson University or the Clarkson University Capital Region Campus.

In 2013 Skidmore entered into a partnership called New York Executive Clean Energy Leadership (NY EXCEL) to assist business executives in starting up ventures in renewable energy and clean technology. Conceptualized by Catherine Hill, a Skidmore professor of business administration, the project was supported by $400,000 from the New York State Energy Research and Development Authority (NYSERDA). Partners included Brookhaven National Laboratory, the Pace Energy and Climate Center, the New York Battery and Energy Storage Consortium, and the Syracuse Center of Excellence. Participants in NY EXCEL visit New York clean tech sites, and each is required to develop in "Capstone Project," a business plan for a New York State clean tech venture to be pitched to investors at a final workshop. Mentors from the relevant clean tech sectors are assigned to each participant.[145]

### 8.4.7 The College of Saint Rose

The College of Saint Rose in Albany is a liberal arts institution founded by the Sisters of Saint Joseph of Carondelet in 1920. The college plays an outsized role in training the Capital Region's educational workforce. President C. Wayne Williams commented in 2003 that "one-third of the practicing educators in the Capital Region have one degree or more from the College of Saint Rose. We're probably the most popular choice for local educators pursuing their graduate degrees."[146] Joseph

---

[142] US News and World Report's 2017 rankings of National Liberal Arts Colleges ranked Skidmore 38th out of a total of 239 Institutions.

[143] "30 Great Small College Business Degree Programs 2016," *Online Accounting Degree Programs* (July 2016).

[144] Skidmore College, "MB107," <http://www.skidmore.edu/management_business/mb107/index.php>.

[145] "Skidmore, NYSERDA Launch Program for Execs," Troy, *The Record* (November 19, 2013).

[146] "Education Engine Helps Power the Capital Region," Albany, *The Times Union* (February 23, 2003).

Dragone, the former Superintendent of the Ballston Spa School District who proved instrumental in developing programs to orient K-12 students toward careers in innovation and high technology, holds master's and bachelor's degrees from the college.[147]

## 8.5 Community Colleges in the Capital Region

The expansion of high-technology manufacturing in the Capital Region fostered a much greater demand for skilled technicians than for engineers and scientists with 4-year or graduate degrees. Mike Tucker, former president of CEG, commented in 2010 that although it was not widely recognized, 65% of the jobs in a semiconductor manufacturing facility do not require a 4-year college degree but can be performed by individuals with the right 2-year college degrees.[148] This reality has posed a continuing challenge to the region's community colleges, which have moved to establish and refine curricula and training facilities to meet the burgeoning demand for skilled high-tech workers.

The recession which began in 2008 "ushered in a heyday for the nation's community colleges," reflecting the fact that "the job market was bone dry" and higher education was a "place to ride out the storm and gain new skills community colleges, where cheap tuition has always been the draw, saw enrollment increase sharply as a result."[149] As was the case nationwide, surging enrollment placed pressure on Capital Region community colleges to expand capacity and diversify course offerings. In 2009 Schenectady County Community College reported that enrollment had increased by 56% over the past decade, with the largest growth seen in full-time students. A spokesperson for Hudson Valley Community College observed in 2010 that "we are landlocked. This campus was built for 6000 students at capacity. We are now at 13,500 students." Space was so tight at HVCC that the college leased 30,000 square feet of space in nearby Rensselaer Technology Park and moved some of its classes and offices there. Adirondack Community College reported in 2010 that its enrollment had never been higher: "Across the board, we're completely at capacity."[150] New York State was well positioned for the surge in the more cost-effective, real-time delivery of demand-driven education needed for a higher level of entry-level tech jobs that could be offered through community colleges. The State University of New York was and is a nation-leading public institution of its type and size, with half of its 64 campuses being community colleges. It

---

[147] "School District Picks Leader—Ballston Spa Board Hires Albany Administrator Joseph Dragone," Albany, *The Times Union* (May 10, 2008).

[148] "HVCC Opens Tech Training Center," Schenectady, *The Daily Gazette* (February 7, 2010).

[149] "Economic Upturn a Test," Albany, *The Times Union* (October 8, 2016).

[150] "New Degree of Popularity at Community Colleges," Albany, *The Times Union* (January 10, 2010).

offers a unique construct with Senior Vice Chancellor for Community Colleges and the Education Pipeline, Johanna Duncan-Poitier, reporting directly to the SUNY Chancellor and ensuring a concerted, integrated approach in leveraging the state's public community colleges to meet the needs of employers.

### 8.5.1 Hudson Valley Community College

Hudson Valley Community College was founded in Troy, New York, in 1953 under the supervision of SUNY to respond to post-World War II educational needs and the closing of a local veterans' vocational school. HVCC's initial curriculum was largely technical but over time it added courses in science, business, and the liberal arts.[151] HVCC developed course offerings in telecommunications, electrical systems, and computer technology which subsequently provided a solid base for new programs in microelectronics that were developed in and after the late 1990s. The HVCC curriculum was of sufficient caliber that pursuant to a 1997 agreement with RPI, HVCC associate degree credits could count toward a bachelor's degree at RPI.[152] A 2004 assessment of HVCC's evolution commented that "just 15 years ago [HVCC] was viewed by many as a sort of transitional school for students in flux. Today . . . the school has a reputation for offering essential technical training for students and current workers alike."[153] In a 2016 interview, the CEO of a major Capital Region engineering firm said of HVCC that "They do well. They are reacting to what's going on. They support high tech [and are] quick to react to a deficit of trades people."[154]

In 1998 Douglas Baldrey, HVCC's assistant dean of the School of Engineering and Industrial Technology, disclosed that as part of the state's effort to reach out to semiconductor manufacturers, the college was introducing degree and certificate programs "tailored specifically to semiconductor manufacturing technology." HVCC would offer a 2-year associate's degree in electrical engineering with a specialization in semiconductor manufacturing, targeting high school seniors and manufacturing workers who had been laid off. A certificate program would be offered to individuals with existing technology-related associate's degrees who were interested in pursuing work opportunities in the semiconductor industry. HVCC students would enjoy direct access to semiconductor clean rooms, manufacturing equipment, and know-how through tie-ins with RPI's Center for Advanced Interconnect Science and Technology and SUNY Albany's Center for Advanced Thin Film Technology.[155]

---

[151] Hudson Valley Community College, *2015-16 College Catalog*, p. 4.

[152] "Community Colleges Grow With Distinction," Albany, *Knickerbocker News* (May 11, 1987).

[153] "Efforts Under Way in Region to Prepare Employees for Leaner, High Tech Economy," Schenectady, *The Daily Gazette* (March 13, 2004).

[154] Interview, Albany, New York (September 16, 2016).

[155] "HVCC to Lead Training," *Albany Business Review* (September 14, 1998).

Chip Fab '98, New York's first initiative to attract semiconductor manufacturing to the state included a state-sponsored worker-training program in semiconductor manufacturing involving seven community colleges. At the time, Sematech was urging community colleges nationwide to implement semiconductor industry-approved curricula in semiconductor manufacturing, and in June 1998 Sematech approved HVCC's program, following an assessment of its labs, facilities, equipment, and faculty. Community colleges in Dutchess, Sullivan, Orange, and Ulster counties formed a cooperative consolidating Sematech-recommended courses offered at each school (see Footnote 155). However, the collapse of plans to draw a semiconductor manufacturing plant to North Greenbush in 1999 (see Chap. 4) led HVCC to postpone implementation of its semiconductor programs.[156]

The semiconductor project was revived in 2003 when HVCC disclosed it was considering a new program in nanotechnology which would train semiconductor manufacturing technicians as part of the school's Electrical Engineering Technology Program. The new curriculum was intended to augment rather than duplicate offerings at RPI and SUNY Albany. The HVCC program would "focus more on hands-on, applied nanotechnology and less-theoretical facets. . . . HVCC will train the technicians who work in the clean rooms, and the other schools will train the researchers and scientists." Jeff Foley, an HVCC spokesman, said that "HVCC is demonstrating that we respond quickly to workforce needs in the Capital Region . . . about 70% of the workforce in the technology industries coming to the area will need an associate's degree and we anticipate that HVCC will be the institution that prepares many of those workers for rewarding careers" (see Footnote 156).

#### 8.5.1.1 Semiconductor Degree Program

In 2005 HVCC launched a 2-year associate degree program, "Electrical Technology: Semiconductor Manufacturing Technology." The program enrolled 40 students for its first year of courses, which required study in the college's Electrical Engineering Technology Program. Interested students were eligible to pursue a second year of semiconductor manufacturing-specific courses.[157] The five second-year students who became the first class to enroll in the second-year program took first-semester courses such as Semiconductor and Nanotechnology Fabrication Processes, Vacuum and Thin Film Technology, Semiconductor Metrology and Process Control, and Electro-Mechanical Devices and Systems. During the second semester, students gained hands-on experience in the clean rooms at CNSE.[158]

---

[156] "HVCC Planning Nanotech Program," Schenectady, *The Daily Gazette* (January 18, 2003).
[157] "Job Seekers Hope to Clean Up at Nanotech Fair," Albany, *The Times Union* (April 19, 2006).
[158] "HVCC Students Start Specializing in Semiconductor Manufacturing," *Albany Business Review* (September 18, 2006).

Pursuant to an agreement between HVCC and CNSE, HVCC students could take courses at CNSE at the going rate of tuition for HVCC, with the students incurring no extra cost.[159] Several students in the program's first 2-year class were interviewed in 2007 and indicated that the knowledge and skills taught in HVCC's semiconductor course were directly relevant to work they were doing in the GE Global Research Center in Niskayuna.[160] With the start-up of construction of GlobalFoundries Fab 2 in 2009, the company began building a working relationship with HVCC and other nearby community colleges. A company spokesperson said in August 2009 that "we've been meeting with the folks at HVCC, tweaking the curriculum so they can best develop it to meet our needs."[161] HVCC's TEC-SMART campus (described below), adjacent to the GlobalFoundries Fab, "included an area designed to mimic the ultra-clean conditions at the chip factory."[162]

"HVCC's semiconductor program is geared specifically toward meeting the needs of GlobalFoundries."[163] As HVCC President Drew Matonak recalls, following the initial launch of its semiconductor initiative, HVCC dispatched two faculty members to Dresden, Germany, to study the manufacturing operations of AMD, which subsequently transferred its Dresden operations to GlobalFoundries. "What we learned by working with the folks in Dresden was that we had some skill gaps between our program designed for the global workforce needs in our area and the specific workforce needs of GlobalFoundries. We brought that back and our school of engineering and industrial technology developed a specific gap certificate that ensures that students at the point of graduation are well matched with the specific workforce needs at GlobalFoundries."[164]

The transition in ownership of AMD's manufacturing operations from AMD to GlobalFoundries entailed modifications to HVCC's semiconductor training program. AMD had sought entry-level workers with a higher level of semiconductor-specific skills than GlobalFoundries required. GlobalFoundries wanted people with good "foundation" and problem-solving skills whom the company would then train on the specific technologies being applied in its manufacturing facility. HVCC

---

[159] "HVCC Students Gain Skills for AMD Jobs," Schenectady, *The Daily Gazette* (December 15, 2006).

[160] Christopher Perlee, a 29-year-old senior in the HVCC program, also worked at the GE facility. He observed that "the stuff we learn here, I deal with every day. I work with vacuum systems all the time." Hamad Jahangar, a 23-year-old semiconductor student at HVCC, who also worked in GE's lithography lab, commented that "everything we study here, we use there. It helps you a lot." See "Learning Lessons in Growth—Building Capacity to Host More Technology—Related Industries Requires Educational Effort," Albany, *The Times Union* (November 25, 2007).

[161] "Chip Plant Jobs Draw Wide Interest," Schenectady, *The Daily Gazette* (August 16, 2009).

[162] "Getting TEC-SMART," Glens Falls, *The Post-Star* (January 22, 2010).

[163] "Students, Young, Old Prepare for a New High-Tech Job Field," Glens Falls, *The Post-Star* (October 28, 2010).

[164] National Research Council, Charles W. Wessner (rapporteur), *New York's Nanotechnology Model: Building the Innovation Economy* (Washington, DC: The National Academies Press, 2013), Proceedings of Day 1.

melded its semiconductor program into a new one called mechatronics, a 62-credit course of study combining multiple disciplines—computer engineering, microelectronics, mechanical engineering, blueprint schematic reading, electro-fluid power systems, sensors, power distributions, and control engineering. Some HVCC mechatronics graduates go directly to jobs at GlobalFoundries, while others transfer to 4-year mechatronics programs.[165]

HVCC received $95,864 in federal Department of Labor grants announced by Senator Charles Schumer in 2012. An HVCC spokesman said that the funds would be used to expand two of the school's semiconductor manufacturing technology programs—a 25-credit semiconductor technology certificate program offered by HVCC's School of Engineering and Industrial Technologies and HVCC's associate's degree program in electrical technology/semiconductor manufacturing.[166]

### 8.5.1.2 TEC-SMART

In September 2007, Senate Majority Leader Joseph L. Bruno announced a $13 million investment by the state to create an HVCC extension campus in Luther Forest, which would be called the Technology and Education Center for Semiconductor Manufacturing and Alternative Renewable Technologies (TEC-SMART). TEC-SMART was a collaboration between HVCC and the New York State Energy Research and Development Corporation to train and educate workers for the semiconductor manufacturing and alternative energy technology industries.[167] It was envisioned that a new 43,000 square-foot building would be built at the Saratoga Technology of Energy Park (STEP), a Luther Forest site owned by NYSERDA, housing classrooms and labs capable of training 600–800 technicians over the coming decade.[168] The TEC-SMART site was "separated by little more than a row of trees" from the site of GlobalFoundries Fab 2.0.[169]

The TEC-SMART site became operational in early 2010, with 250 students enrolled in courses for the spring (see Footnote 169). Fred Strnisa, a PhD professor teaching TEC-SMART's semiconductor manufacturing class in 2012, said that many of the enrollees were in their 30s and seeking to train for a second career. These students learned "the principles of computer chip manufacturing, working in a mock clean room in the kind of 'bunny suits' worn in real clean rooms, where purified materials need protection from human contact." Penny Hill, the associate

---

[165] Interview with Drew Matonak, president of Hudson Valley Community College (June 8, 2016); "HVCC offers New Engineering Degree in Mechatronics," Troy, *The Record* (March 8, 2015).

[166] "Community Colleges Get $15M for High Tech Training," Schenectady, *The Daily Gazette* (September 20, 2012).

[167] "HVCC and NYSERDA," *TEC-SMART* (project proposal, 2007).

[168] "Bruno: $13M Training Center 'Is the Future,'" Glens Falls, *The Post-Star* (September 13, 2007).

[169] "Getting TEC-SMART," Glens Falls, *The Post-Star* (January 22, 2010).

dean running TEC-SMART, points out that the facility also trains students "to maintain mechanical systems like the air and water-handling systems" in a semiconductor fabrication plant and qualifies students for "jobs at companies that service and supply GlobalFoundries and other large facilities."[170] In September 2012, Hill reported that all 12 participants in that year's semiconductor manufacturing program at TEC-SMART had at least two job offers by March and that many had three by the time they graduated later in the spring. She said "right now we are at capacity. We have a waiting list."[171]

By 2013, HVCC students in the semiconductor manufacturing program were spending most of their first year at the main HVCC campus in Troy learning essential base subjects such as algebra, trigonometry, basic calculus, electrical concepts, and how microcomputers operate. The second year was primarily spent at the TEC-SMART facility focusing on learning semiconductor-specific skills, some of which were taught in the on-site clean room with miniature versions of manufacturing equipment simulating the operation of a 300 mm fab. From 2010 to 2013 the program graduated about 12 students per year. Of these, "any graduate of the program who wants a job has a job," said Phil White, Dean of HVCC's School of Engineering and Industrial Technologies. "GE Global Research tries to hire every student they can."[172]

### 8.5.1.3 Advanced Manufacturing Program

HVCC's Advanced Manufacturing Program trains students in advanced machining processes used to make aerospace parts, power generating equipment, defense equipment, and electronics. All students in the program participate in a year-long senior capstone project in which students work in groups "to manufacture and assemble complex working models to test their precision planning, machining, and assembly skills." Local high-tech companies are so eager to hire graduates of this program that some will hire students before they begin their degree and arrange to pay their entire tuition, if they keep grades high enough.[173] The pre-graduation job placement rate for this program is about 95%, with a number of students in the programs working at part-time jobs during their course of study which lead to full-time employment upon graduation.[174]

In 2015 HVCC received from the Gene Haas Foundation a $1 million gift which is partially enabling construction of a $14 million, two-story addition to the program's existing facility at Lang Hall in Troy to be named the Gene Haas Technology

---

[170] "Outlook 2012: Riding the Nano Wave," Schenectady, *The Daily Gazette* (February 19, 2012).

[171] "The Key to High Tech Jobs," Albany, *The Times Union* (September 11, 2012).

[172] "Nano Tech Valley: a Learning Environment," Troy, *The Record* (June 23, 2013).

[173] Hudson Valley Community College, *2015-16 College Catalog*, p. 104; Interview with Penny Hill, Hudson Valley Community College, Malta, New York (January 27, 2016).

[174] Hudson Valley Community College, *2015-16 College Catalog*, p. 104.

Center. The center will house "the latest machine tools, equipment and labs for metrology, CAD/CAM, metallurgy, electronic controls, machining, assembly and grinding."[175] The expansion will enable HVCC to double its current enrollment of 115 in the 2-year program.[176] A professor in the program commented in 2016 that given growing demand in the region for trained machinists, "We are having to turn people away from the program at this point." With starting salaries averaging $40,000–$50,000, he estimated that salaries from technicians graduating from the center could total $272 million.[177]

## 8.5.2 Schenectady County Community College

Schenectady County Community College (SCCC), a SUNY-affiliated Community College, has attained renown for its programs in culinary arts, music, and hospitality management and, in the past decade, has made a substantial commitment to nanotechnology manufacturing.[178] In 2000, SCCC's associate dean of academic affairs indicated that the college was moving away from the traditional expectation that community colleges function as a stepping stone to 4-year institutions. Instead SCCC was focusing increasingly on short-term training of students for entry-level jobs, including the expansion of curriculum in computers and electronics.[179]

### 8.5.2.1 Nanoscale Materials Program

In 2006, Schenectady-based SuperPower Inc., a major producer of superconducting wire, received a major order for wire and declared its need for qualified technicians "yesterday." In response, in partnership with Union College and SuperPower, SCCC initiated a new associate's degree program in Nanoscale Materials Technology that included courses in materials, vacuum science, and engineering.[180] Ruth McEvoy, who chaired SCCC's Department of Math, Science, and Technology, commented that "this is really a unique program. I have not been able to find anything similar to

---

[175] "Hiring for Cybersecurity?" Albany, *The Times Union* (September 22, 2015). The new building is expected to be ready for occupancy in the summer of 2018. "Haas Center Work Progresses at HVCC," Albany, *The Times Union* (May 24, 2016).

[176] "Hudson Valley Community College will Double Spots Available in Advanced Manufacturing Program," *Albany Business Review* (September 22, 2015).

[177] "$14M Training Center in Works," Albany, *The Times Union* (March 18, 2016).

[178] A 1987 survey found that SCCC graduates earned better-than-average to exceptional starting salaries relative to graduates of 20 other East Coast 2-year colleges. "Grads Exceed Average Earnings," Albany, *Knickerbocker News* (March 12, 1987).

[179] "Colleges Respond to the Job Market," Schenectady, *The Daily Gazette* (February 27, 2000).

[180] "SCCC Dreams Big: $24M Science Facility," Schenectady, *The Daily Gazette* (October 17, 2012).

it in the United States. They'll be qualified to work not only at SuperPower but at any company that requires broad-based skills." SCCC announced plans to buttress this initiative with the $1 million purchase of atomic force and regular optical microscopes, computers, and software, as with the upgrading of two labs for use in electronics, physics, and vacuum science.[181] The $1 million used by SCCC was obtained by a grant from the State of New York. The program added two SuperPower scientists to the faculty, serving as assistant professors, one of whom pointed out that the skillsets being taught would also be applicable to semiconductor manufacturing.[182]

By 2009, SCCC's 2-year Nanoscale Materials Technology Program was producing its first graduates and enrollment had increased from 12 to 22 full- and 7 part-time students. SCCC was reportedly working with GE to create a program to train technicians to work in its advanced battery project for which the company was constructing a $100 million plant in Schenectady.[183] By 2010 SCCC had 42 full-time and 12 part-time students enrolled in its Nanoscale Materials Technology Program. In addition, it launched an associate's degree program in alternative energy and a 1-year certificate program in storage battery technology, a response to GE's disclosure of plans build a $100 million battery manufacturing plant in the region.[184]

SCCC received $436,288 in grant funds from a federal Department of Labor package announced by Senator Schumer in 2012. The school indicated the funds would be used to create a nanotechnology program specifically designed for returning veterans and unemployed workers seeking retraining.[185] In 2010, incoming SCCC President Quintin Bullock said that the college had major plans to "expand the scope of its nanotechnology program to train workers for semiconductor manufacturers like GlobalFoundries Inc." Among other things, Bullock sought to forge closer ties with CNSE and to tailor SCCC's degree program "to provide more of what GlobalFoundries will need from its clean room workers." SCCC was also developing a new renewable energy program to provide training in fuel cell, solar, wind, and battery storage technologies, all of which were relevant to employers in the Capital Region.[186] In 2012 the *Albany Business Review* reported that "the skills taught in Schenectady County Community College's nanoscale materials technology are in such demand that most of the students in that program have jobs before they graduate," mostly at GlobalFoundries and CNSE.[187]

---

[181] "SCCC Plans Tech Degree—Super Power Has Need in Nanoscale Field," Schenectady, *The Daily Gazette* (April 25, 2006).

[182] "Partnership Wired for Future—Schools, Manufacturer Team Up as Worker Training Program," Albany, *The Times Union* (August 15, 2006).

[183] "Area Community Colleges Gain High-Tech Focus," Schenectady, *The Daily Gazette* (September 22, 2009).

[184] "Schools Changing to Meet the Chip Plant's Needs," Schenectady, *The Daisy Gazette* (December 20, 2010).

[185] "Community Colleges Get $15M for High Tech Training," Schenectady, *The Daily Gazette* (September 20, 2012).

[186] "SCCC to Grow Nanotech Program," Albany, *The Times Union* (April 20, 2010).

[187] "As Tech Sector Scrambles to Find Talent, Students with Skills Cash In," *Albany Business Review* (September 14, 2012).

#### 8.5.2.2 Smart Scholars Early College Program

In 2010 SCCC entered into a collaboration with Schenectady High School for a program aimed at 400 high school students who would take college-level courses at SCCC during their junior and senior years of high school, concurrently earning a high school diploma and college credits. The program was initially funded by a $447,500 grant from the Bill and Melinda Gates Foundation and subsequently received a $100,000 grant from the Schenectady Foundation and additional funding from the New York State Education Department. This program included a special STEM-focused curriculum.[188]

### 8.5.3 SUNY Adirondack

SUNY Adirondack, formerly Adirondack Community College (ACC), is a public college with campuses in Queensbury and Wilton, New York. SUNY Adirondack offers 2-year associates' degrees in computer science, engineering science, business, and math and science.[189] In 2011 ACC broke ground on a 32,000 square foot, $7 million facility in Wilton to enable the school to double its student enrollment, to add lab space, and to expand its course offerings. The project was prompted in substantial part by "demand for courses that will prepare students to work at GlobalFoundries and other tech related firms moving into the area."[190] In a 2014 interview, SUNY Adirondack President Kristin Duffy said that the school was "planning to roll out new programs with a focus on advanced manufacturing and electrical technology to capitalize on the growth of GlobalFoundries and the area's technology industry."[191]

In 2015 SUNY Adirondack secured nearly $10 million in state funds to create the Adirondack Regional Workforce Readiness Center in Queensbury, which will house "hands-on learning and workforce training initiatives including labs that simulate healthcare settings for aspiring nurses and a center for businesses to connect with prospective employees."[192] In addition, the college began construction in 2016 of a $17 million NSTEM building (nursing, science, technology, engineering, and math), cofunded by the state and by Warren and Washington Counties. Duffy said the new facilities would allow the college to expand capacity and "help the local business community."[193]

---

[188] "Early College Effort Gets $100K," Schenectady, *The Daily Gazette* (April 30, 2012).

[189] SUNY Adirondack, "Academics," <http://www.sunyacc.edu/academics>.

[190] "SUNY Adirondack Breaks Ground on New Facility at Wilton Campus," Saratoga Springs, *The Saratogian* (October 14, 2011).

[191] "College Seeks to Reinvest Itself, Offer New Programs," Glens Falls, *The Post-Star* (February 16, 2014).

[192] "$10M Headed to SUNY Adirondack for Workforce Center," Albany, *The Times Union* (October 8, 2015).

[193] "SUNY Adirondack Breaks Ground on New Buildings," Glens Falls, *The Post-Star* (October 27, 2016).

## 8.5.4 *Fulton-Montgomery Community College*

Fulton-Montgomery Community College (FMCC) is a 2600-student SUNY community college located in Johnstown, New York (between Albany and Utica). The only institution of higher learning in its two-county sponsorship area, it has traditionally offered liberal arts and career education programs and more recently has added technology-related programs and course offerings.[194] FMCC's Center for Engineering and Technology features atomic microscopes, a nanotechnology clean room, and robotics labs.[195] GlobalFoundries personnel "take part in the college's training programs on a regular basis."[196] FMCC has developed an apprenticeship program in conjunction with regional school districts to use its research infrastructure to introduce high school students to potential high-technology careers. (See program description in section on "HMF BOCES-FMCC Apprenticeships" in this chapter.)

## 8.6 Hybrid Institutions and Initiatives

Within the past decade, a number of hybrid institutional arrangements have emerged in New York which involve collaboration between educational institutions and local companies, combining K-12 and higher education curricula with practical training applicable in the workplace. These programs frequently target young people from modest or disadvantaged backgrounds and/or displaced workers in an effort to enable them to find jobs in local high-technology industries. They are differentiated from traditional early college programs by the active participation of local companies.

### 8.6.1 *Pathways in Technology Early College High School*

The Pathways in Technology Early College High School (P-TECH) began in 2011 as a collaboration at Brooklyn's Paul Robeson High School between IBM, the City University of New York, and the city Education Department to create a combined high school and community college curriculum reinforced with teaching and mentoring in workplace skills. P-TECH established a rigorous 6-year program culminating in both a high school diploma and an associates' degree in applied computer science. Each student was paired with an industry mentor, and in the case of IBM,

---

[194] Research Foundation of SUNY, *SUNY's Impact on New York's Congressional District 21* (2006).

[195] "FMCC-HFM BOCES Collaboration Creates Career Pathway," *Targeted News Service* (December 24, 2011).

[196] "FMCC Expects Tech Demand to Force Growth," Schenectady, *The Daily Gazette* (November 16, 2013).

students were invited to the IBM semiconductor fab in East Fishkill to observe the manufacturing process. Companies also helped train the program's teachers and provided a full-time industry liason person to help develop the curriculum.[197] Graduates were to be "first in line" for jobs at IBM.

P-TECH had a promising beginning, and, in 2013, students from the program's inaugural class, by then in 10th grade, were already taking and passing college-level courses and overachieving on preliminary SAT tests. The program's apparent initial success prompted Chicago to replicate it, drew the endorsement of the US Secretary of Education Arne Duncan (who endorsed P-TECH as a blueprint for similar programs across the United States), and was even singled out for praise by President Obama in his 2013 State of the Union address.[198]

In early 2013 Governor Andrew Cuomo announced a state program designed to "clone" the Brooklyn P-TECH model at multiple sites around New York State.[199] Sixteen projects were launched that year, with ten more added in 2014.[200] The program, funded initially by the state at $28 million, will enable participating students, entering as ninth graders, to graduate 6 years later with associates' degrees at no cost to them.[201] In the Capital Region, three P-TECH programs were launched, in Ballston Spa, Hudson Falls, and Troy (see section on "K-12 Education" in this chapter). As of early 2018 it is too soon for the new P-TECH programs to have produced graduates, but early data on the first entering class showed that 97% of students had passed at least one Regents exam, 91% had passed two Regents exams, and 85% had earned college credits.[202]

The future of P-TECH in New York remains uncertain. The program is expensive: by the 2019–2020 school year, the last of the first class' period of matriculation, P-TECH will have cost the state $42 million. At a 2016 Board of Regents meeting, Board Member Lester Young reportedly asked Education Department staff if they could gather any research and perform a cost-benefit analysis before the board committed to further funds requests. Donna Watson, assistant superintendent for curriculum and instruction in Troy, commented in 2017 that—

> In some ways it would almost be better not to enroll another cohort if you're not sure the funding will last, because you don't want to promise kids something and then take it away halfway through. It's a very expensive program. It's a college degree for every student. That's a significant investment. There's no doubt about it. (see Footnote 202)

---

[197] See summary of the remarks of Darren Suarez of the Business Council of New York in National Research Council, Charles W. Wessner (rapporteur), *New York's Nanotechnology Model: Building the Innovation Economy* (Washington, DC: The National Academies Press, 2013), p. 64.

[198] "A Jobs Crisis? No It's a skills Crisis," *New York Daily News* (January 16, 2013).

[199] "Andy Plans Class Clone," *New York Daily News* (February 27, 2013).

[200] "Cuomo Touts High Tech High Schools," Albany, *The Times Union* (January 22, 2014).

[201] "Free 2-year Degree May Give an Edge," Albany, *The Times Union* (May 22, 2014).

[202] "Early College Plan Taking off," Albany, *The Times Union* (January 22, 2017).

## 8.6.2 SUNY Works

SUNY Works, part of SUNY's broader effort to promote applied learning, uses cooperative education and internships to enable the development of workplace skills and work experience, focusing in particular on "adult non-traditional students."[203] Supported by grants from the Lumina Foundation and the Carnegie Corporation of New York, the program partners some SUNY campuses with local companies which provide unpaid and some paid internships relevant to a student's course of study. During her tenure, SUNY Chancellor Nancy Zimpher personally lobbied CEOs of New York's large companies to participate in the program and secured commitments from IBM, GE, GlobalFoundries, Motorola, and Chevron.[204]

## 8.6.3 Capital South Campus Center

The Capital South Campus Center (CSCC) in Albany is an education and training center which has been characterized as a "hybrid between higher education, workforce development training and employment assistance."[205] In 2011 the city of Albany won a $5 million federal grant from the Department of Housing and Urban Development to create an educational and training center bringing college-level coursework to Albany's South End, part of a broader effort to reverse the neighborhood's deterioration.[206] The development of the center was spearheaded by the Trinity Alliance, a 100-year-old charitable organization based in Albany, in collaboration with the College of Nanoscale Science and Engineering.[207] The Trinity–CNSE alliance focused on nanotechnology themes relevant to "smart cities," with CNSE holding courses and training programs at the new campus, providing nanotech-based "opportunities that have historically bypassed our inner cities."[208] By the time CSCC opened its doors to students in 2014, a number of other area colleges had committed to provide support.[209]

---

[203] State University of New York, "SUNY Works Campus Partnerships," <https://www.suny.edu/suny-works/partnerships>.

[204] "SUNY Chancellor Promises More Internships in 2014," *Associated Press Newswire* (January 14, 2014).

[205] "Demand Growing at Albany Career Training Center," *Albany Business Review* (July 21, 2015).

[206] "$5M Grant Benefits South End," Albany, *The Times Union* (July 13, 2011).

[207] "Trinity at 100: Still Serving Neediest," Albany, *The Times Union* (March 31, 2012).

[208] "Nano Going Downtown," Albany, *The Times Union* (July 17, 2012).

[209] HVCC pledged to host basic 100 level courses at CSCC in every major field of study. CNSE staffed CSCC's Advanced Training and Information Networking (ATTAIN) computer laboratory. Schenectady County Community College offered 1 year certificate programs at CSCC in a number of disciplines, including inventory control and warehouse management. The Sage Colleges provided mentors and tutors. "Trinity Alliance Partners with Area Colleges, Prepares to Open its Capital South Campus Center," Troy, *The Record* (April 3, 2014).

A survey of CSCC's operations during the second half of 2015 found that it was providing adult education and training for significant numbers of people, many of them displaced from other jobs or otherwise disadvantaged. During the last 6 months of 2015, 426 people completed career enhancement courses and/or developed employment or education plans, of whom 161 had earned a certificate, enrolled in college, or found a job by early 2016. CSCC is constructing a nanotechnology clean room with a $500,000 grant which would be used for training lab technicians in conjunction with a partnership with SUNY Polytechnic.[210]

## 8.7  K-12 Education

The K-12 education system in New York State outperforms those of most other states, ranking ninth in the United States in the 2017 performance evaluation by *Education Week*.[211] Within the state, while the Capital Region's K-12 public schools vary in quality and performance metrics, taken as a whole they consistently outperform state averages based on metrics used by the *Albany Business Review*'s "Albany Schools Report," which compares schools' test scores in math, English, science, and social studies and graduation rates and postgraduation plans.[212] New York spends more per K-12 public school student than any other state in the United States except neighboring Vermont and nearly double the national average of $11,709 for the 2014–2015 school year.[213] In the 2015–2016 school year, 61 of the Capital Region's 90 school districts spent over $20,000 or more per pupil, or over 70% more than the national average, and 10 districts spent $30,000, greatly exceeding the national average.[214] As shown in Table 8.6, five Albany area high schools were named to Newsweek's 2016 list of the top 500 in the country.[215]

But weaknesses in the region's K-12 educational infrastructure are apparent in areas such as STEM education and preparation of students for life after graduation. That fact was underscored by a 2013 study by the Capital Region consultancy Camoin Associates, which examined HVCC's semiconductor degree programs. Camoin identified an apparent paradox. The programs required only a modest financial investment by students (tuition then was $3980 per year) and offered a major payoff for graduates in the form virtually assured employment by local high-tech firms at starting salaries of $35,000–$45,000 with excellent prospects for advancement. Yet the

---

[210] "Celebrating a Year of Success in the South End," Albany, *The Times Union* (March 9, 2016).

[211] "Quality Counts 2017: State Report Cards Map," *Education Week* <http://www.edweek.org>.

[212] "How Albany-Area Schools Compare to the State Average on Test Scores," *Albany Business Review* (June 28, 2016).

[213] NEA Research, *Rankings & Estimates: Rankings of the States 2015 and Estimates of School Statistics 2016*, (National Education Association, May 2016) Table H-11.

[214] "Most Districts Cross $20,000 Spending Line," *Albany Business Review* (June 24, 2016).

[215] "5 Albany—area High Schools Make Newsweek's Top 500 in US List," *Albany Business Review* (August 16, 2016). The Newsweek rankings are based on college acceptance and enrollment, SAT

8.7 K-12 Education

**Table 8.6** Albany-area High Schools in Newsweek's 2016 List of Top 500 US High Schools

| School | Location | National Rank |
|---|---|---|
| Shenendehowa High | Clifton Park | 246 |
| Niskayuna High | Niskayuna | 270 |
| Shaker High | Latham | 328 |
| Saratoga Springs High | Saratoga Springs | 363 |
| Ballston Spa High | Ballston Spa | 379 |

size of the graduating classes were small, reflecting both the fact that the programs were undersubscribed to begin with and that many students were dropping out before graduation, primarily during the transition period between the first and second year.[216]

The high attrition rate in HVCC semiconductor programs was attributable to the "rigorous" math and science workload.[217] The fact that students were finding the curriculum too challenging reflected a longstanding structural problem in the New York K-12 school system, the failure of the curriculum in math and science to prepare a substantial proportion of high school graduates for college-level study. A GlobalFoundries spokesman complained in 2013 that with respect to STEM (science, technology, engineering, and math) education, "we're really floundering here...."[218]

Undersubscription in HVCC's semiconductor programs was linked by then Ballston Spa Superintendent of Schools Joseph Dragone to another larger problem, the lack of awareness by K-12 students of good opportunities available in the region and the inadequate vertical integration of educational programs between elementary, secondary, and post-secondary institutions.[219] Camoin Associates noted that "most of the students find out about them later in life, not directly from high schools,

---

and ACT participation and performance, dual enrollment programs, guidance and counseling resources, and AP and IB participation and performance. "Ballston Spa Makes 'Top High Schools,'" *The Ballston Journal* (August 22, 2016).

[216] Camoin Associates, "The Curious Case of GlobalFoundries and Its Workforce: Setting the Stage" (September 5, 2013). <https://www.camoinassociates.com/curious-case-globalfoundries-and-its-workforce-setting-stage>.

[217] "Nano Tech Valley: A Learning Environment," Troy, *The Record* (June 23, 2013). The Camoin Associates study observed that the issue is not unique to HVCC or its semiconductor programs. "[M]any community college degree programs around the country . . . have intense math, science and technology courses with high entrance requirements. Relatively high turnover occurs as students explore and seek at other degrees or careers." Camoin Associates, "The Curious Case of GlobalFoundries and its Workforce: Setting the Stage" (September 5, 2013). <https://www.camoinassociates.com/curious-case-globalfoundries-and-its-workforce-setting-stage>.

[218] "Replanting the STEM Common Core Should Help Students Master Necessary Skills," *Watertown Daily Times* (October 30, 2013). In 2013 Clarkson University indicated that "a significant number of incoming freshmen from New York schools need remedial help to pass freshman calculus." Ibid.

[219] Camoin Associates, "The Curious Case of GlobalFoundries and Its Workforce: Ballston Spa Central School District" (September 5, 2013). <https://www.camoinassociates.com/curious-case-globalfoundries-and-its-workforce-ballston-spa-central-school-district>.

**Table 8.7** BOCES serving the Capital Region

| BOCES | Counties | Examples of training activities |
|---|---|---|
| Capital Region | Albany, Schenectady | • CTE programs (machining, welding, information technology) |
| WSWHE | Saratoga, Warren, Washington | • Computer skills<br>• Electrical and plant maintenance<br>• Welding, wiring, HVAC |
| Questar III | Rensselaer, Greene, Columbia | • CTE programs (information technology, manufacturing, logistics) |
| HFM | Fulton, Montgomery, Hamilton | • CTE programs (engineering technology, computer and IT technology, construction technology) |
| ONC | Otsego, Delaware, Schoharie, Greene | • Robotics, mechatronics |

with the average age of the students in the program being over 30."[220] Dragone (who resigned as superintendent in 2017 to assume a regional leadership role as senior executive in the Capital Region BOCES) responded by helping to establish programs that create career "pathways" for high school students that expose them to post-secondary academic and work environments.[221] (See section on "Clean Technologies and Sustainable Industries" in this chapter.)

## 8.7.1 BOCES in Tech Valley

In New York State, pursuant to legislation enacted in 1948, Boards of Cooperative Educational Services (BOCES) can be established by two or more school districts to share educational services initiatives, an institution originally intended to help rural and/or poor school districts that might not otherwise be able to sustain certain programs to do so through collective action. BOCES originally focused on remedial education and providing vocational training for students who would enter the work force immediately after high school, and as Table 8.7 indicates, in the Capital Region they continue to offer such training programs, including courses relevant to manufacturing often under the rubric of Career and Technical Education (CTE) classes. However, BOCES are also supporting educational initiatives that open up career paths for K-12 students that lead through 2- and 4-year institutions of higher learning into careers in technology-intensive companies.[222]

---

[220] Camoin Associates, "The Curious Case of GlobalFoundries and Its Workforce: Setting the Stage" (September 5, 2013). <https://www.camoinassociates.com/curious-case-globalfoundries-and-its-workforce-setting-stage>.

[221] Interview with Joseph Dragone, Ballston Spa, New York (September 16, 2015).

[222] See generally Rockefeller Institute of Government. *The Supervisory District of Albany, Schoharie, Schenectady and Saratoga Counties: A Study of Potential Educational Reorganization in the Capital Region* (Prepared for New York State Department of Education, July 2007).

### 8.7.1.1 Tech Valley High School

In 2003 two BOCES in the Capital Region, the Questar III BOCES and Capital Region BOCES, comprising 47 school districts, began advocating the establishment of a high school in the region "focusing student attention on the technical professions of the future," an idea which became a priority for Senate Majority Leader Joseph L. Bruno.[223] Enabling legislation for the creation of the 400-student Tech Valley High School (TVHS) was passed by the legislature and signed into law by Governor Pataki in 2005, with an initial infusion of $1.1 million in state funding.[224] The school was launched in September 2007 in temporary quarters at MapInfo Corporation offices in Rensselaer Technology Park, with an entering class of 40 freshmen. The curriculum was designed to meet all the New York State Regents' learning requirements; to focus on math, science, and technology; and to provide students with hands-on, project-based experience relevant to high-tech career opportunities.[225] There were no classrooms: only learning and work spaces. The foreign language offered was Chinese.[226]

Local businesses formed the Business Alliance for Tech Valley High School to provide students with an ongoing connection between the high school and the "real world of work, the businesses and colleges of Tech Valley." The Alliance was co-chaired by an IBM executive, Kevin Leyden, and by Amy Johnson, president of Albany-based Capstone, Inc.[227] GE executives participated in the Alliance's ongoing efforts and in 2009 the company gave $70,000 to TVHS to "help it forge additional ties with the business community."[228]

In 2009 TVHS opened operations in a new 20,000 square foot facility located at SUNY Albany's East Campus (see Footnote 228). In 2013 the school moved to a 22,000 square foot facility on the campus of CNSE on Fuller Road.[229] The move gave TVHS students access to CNSE auditoriums, laboratories, and common spaces and brought them "into close proximity to the research and development

---

[223] "BOCES Has Plans for Tech Valley High," Troy, *The Record* (June 7, 2005); "Child Protection, Hi-Tech School on Majority Leader's Priority List," Troy, *The Record* (June 10, 2005).

[224] "Pataki Signature Puts Tech High on Course," Troy, *The Record* (November 11, 2005).

[225] "Tech Valley High Hires Four New Teachers," Troy, *The Record* (April 4, 2007); "Tech Valley High Boots Up with Celebration," Troy, *The Record* (September 13, 2007).

[226] "For 40 kids, As Adventure Begins Thursday—Tech Valley High Will Be a Very Different Setting that Lets Students Help Lead Education," Albany, *The Times Union* (September 3, 2007).

[227] "Tech Valley High Names Business Co-Chairs," Troy, *The Record* (August 29, 2007). Johnson said in 2008 the school's curriculum was "entirely based on projects in which students apply elements of the New York State curriculum to real world problems and tasks." "We are trying to help the high school with communication, technology and workflow. The kinds of things you world think are commonplace in industry, and are givens, but education really hasn't been exposed to those things." See "Business Alliances Boosts Tech Valley High," Schenectady, *The Daily Gazette* (January 9, 2008).

[228] "Tech Valley High School's New Beginning," Albany, *The Times Union* (October 2, 2009).

[229] "Tech Valley High School Moving to College of Nanoscale Science and Engineering Campus," Saratoga Springs, *The Saratogian* (February 13, 2013).

being conducted on the campus."[230] In 2016, CNSE and TVHS instituted the University in the High School program, giving TVHS teachers adjunct instructor status at SUNY Poly to teach SUNY Poly-approved courses in nanoscale science, for which TVHS students can earn college credits.[231]

In 2012 TVHS enrolled 125 high school students and was operating at about 75% of capacity. Tuition at the school that year was $12,000 per student. School districts participating in the program received partial reimbursement from the state, which provided BOCES funding to each district at a level that varied from district to district. The cost burden caused some districts to drop out of the program. The Questar III BOCES, by contrast, accounted for 85 of the school's students, more than all other districts combined, perhaps because of the manner in which it handled its billing.[232]

### 8.7.1.2 HMF BOCES-FMCC Apprenticeships

Fulton-Montgomery Community College is located in a sparsely populated region in which three county school districts—Hamilton, Fulton, and Montgomery—established HMF BOCES, which collaborates with FMCC in worker training.[233] In 2013 HMF BOCES organized an apprenticeship program in conjunction with FMCC and local businesses to reach students showing signs of academic difficulty in eighth grade to develop workplace skills in a work–study environment.[234]

In 2011, pursuant to a National Science Foundation grant of $625,000, students in HFM BOCES' Engineering Technology program could earn college credits at FMCC as they completed a 2-year curriculum "steeped in nanotechnology and semiconductor manufacturing." This program, which gave high school students "hands-on access to the clean room, electron and atomic microscopes, robotics equipment and other high-tech tools is a circumstance few school districts could

---

[230] "Tech Valley High School Partners with College of Nanoscale Science and Engineering," Troy, *The Record* (February 14, 2013).

[231] "Tech Valley Students Earn College Credits at SUNY Poly," Albany, *The Times Union* (April 6, 2016).

[232] Questar III BOCES included the costs associated with TVHS with those of its career and technical education programs and charged its constituent districts based on a 5-year average of how many services the districts used. This approach enabled districts to send students to TVHS by making it easier to plan for the associated expenses. "Tech Valley: Work in Progress," Schenectady, *The Daily Gazette* (February 19, 2012).

[233] In New York, pursuant to legislation enacted in 1948, Boards of Cooperative Educational Services (BOCES) may be established under two or more school districts seeking to share educational services initiatives. It was originally aimed at rural and/or poor districts that might otherwise not be able to sustain certain educational initiatives through their own resources.

[234] This program was "aimed at providing students with workplace skills to address difficulties business leader have expressed about new workers. Many say they get entry-level worker who don't seem to know how to work or behave in a workplace." See "BOCES to Organize Workplace Training," Schenectady, *The Daily Gazette* (March 28, 2013).

even dream about for their students."[235] High school students completing the program earn a semester's worth of credits at FMCC in algebra, physics, engineering, semiconductor and fiber optic technology.[236] By the end of 2013, 200 of FMCC's 2800 students were majoring in science and engineering fields, and FMCC had forged relationships with GlobalFoundries and CNSE which "elevate[d] learning at FMCC to the point where students like Chris Renda, 39, are moving right into jobs." Renda, who had begun work as a telecommunications technician at age 19, having topped out in that career, was able to retool at FMCC and secure a job at GlobalFoundries in Malta/Stillwater, "a success be attributes to training at FMCC."[237]

### 8.7.1.3 Tech Valley STEMsmart Alliance

In 2012 four BOCES based in the Capital Region formed the Tech Valley STEMsmart Alliance with the support of the Center for Economic Growth. The initiative was designed to enhance STEM education by connecting and scaling up existing but diffused STEM-related programs, partnerships, and curricula. The alliance joined the SUNY Empire State STEM Learning Network.[238]

## 8.7.2 Troy Riverfront P-TECH

In 2013 the Enlarged School District of Troy received a commitment of $2.8 million from the state to establish a P-TECH program in collaboration with Questor III BOCES.[239] Other local partners included Hudson Valley Community College, Simmons Machine Tool Group, GE Healthcare, Regeneron Pharmaceuticals, and the Center for Economic Growth. The first class of 34 freshman began classes in the fall of 2014.[240]

---

[235] The characterization was by Mark Tanner, an HFM BOCES curriculum specialist. The HFM BOCES engineering technology program was directed at the time by Edward Lataka, a veteran engineer with over 30 years of field experience. Curriculum was developed by a team of local high school math and science teachers and FMCC professors and reviewed by an advisory committee comprised of local Tech Valley industry and education leaders. The NSF grant was made pursuant to the Technological Education Industry, Partnership (TEPP) program. "FMCC-HFM BOCES Collaboration Creates Career Pathway," *Targeted News Service* (December 24, 2011).

[236] "BOCES program Aims to Teach Engineering Early," Schenectady, *The Daily Gazette* (January 6, 2012).

[237] Renda commented that "without the education, I had [at FMCC] I would have known nothing about what I was getting myself into. When you walk into a facility like that you can actually speak the language they are speaking." "FMCC Expects, Tech Demand Force Growth," Schenectady, *The Daily Gazette* (November 16, 2013).

[238] "BOCES, Businesses Team up to Advance Education," Troy, *The Record* (April 23, 2012).

[239] "Enlarged School District of Troy to receive $2.8 million NYS P-TECH Grant," Troy, *The Record* (September 6, 2013).

[240] "State Ed Chief Touts Local High-School Tech Program," Saratoga Springs, *The Saratogian* (September 4, 2014).

### 8.7.3 Clean Technologies and Sustainable Industries

In January 2011 the Ballston Spa Central School District launched the Clean Technologies and Sustainable Industries (CTSI) initiative in collaboration with HVCC and NYSERDA. The program established a curriculum of coursework concentrating in clean technologies and sustainable industries enabling high school students to earn up to 20 college credits.[241] Students in the program spend a half day in high school and a half day on the TEC-SMART Campus. Students are accorded extensive opportunities to interact with collaborating businesses, utilizing guest speakers, telepresence systems, and online collaboration tools and to participate in field experience under the guidance of mentors employed by the businesses.[242] The program was boosted by a $167,394 Smart Scholars Early College grant from the New York State Education Department.[243] Ballston Spa Superintendent of Schools Joseph Dragone said that "the idea is to replicate the program throughout the Capital Region in years to come."[244] By 2015, 200 students and 21 school districts in the Capital Region were participating in CTSI.[245]

> **Box 8.5 Clean Technologies and Sustainable Industries Pathways to Higher Education: Curriculum**
>
> The mission of the Clean Technologies and Sustainable Industries initiative's Early College High School is to develop and offer pathways to higher education that lead to careers in STEM fields based on rigorous academic coursework and a collaborative approach to learning. The program offers pathways in four areas:

---

[241] Coursework includes nanotechnology, nanoeconomics, photovoltaic systems 2D AutoCAD, wind opened, and environmental technologies. "College Credit Offered in Clean Tech for Ballston Spa and Saratoga Springs High School Students," Saratoga Springs, *The Saratogian* (April 24, 2011).

[242] Students visit manufacturing sites, which Dragone characterizes as "incredibly important." All students have mentors. Interview with Joseph Dragone, Ballston Spa, New York (September 16, 2015).

[243] "Ballston Spa School District Receives $167,394 Grant to Support New 'Clean Technologies & Sustainable Industries' Program," Saratoga Springs, *The Saratogian* (May 13, 2011). In 2014 NYSCRDA contributed a $200,000 grant to the program. "School Clean Tech Program Gets $200k," Saratoga Springs, *The Saratogian* (January 21, 2014).

[244] "Blowin' in the wind: Jobs," Albany, *The Times Union* (October 19, 2011).

[245] "Business Offers Scholarships for Early College Students," Saratoga Springs, *The Saratogian* (April 22, 2015).

| Clean Energy | Computer Science and Information Systems |
|---|---|
| • Introduction to wind energy<br>• Residential construction wiring<br>• Photovoltaic theory and design<br>• Safety and labor relations<br>• College English<br>• College math<br>• Sociology | • Introduction to computer and information science<br>• Business computing and analytics development<br>• Programming and logic 1<br>• Programming and logic 1—data structures<br>• Informative systems analysis and designs<br>• Database management systems<br>• College English<br>• College math<br>• Sociology |
| Leadership, Innovation, and Entrepreneurship | Mechatronics |
| • Introduction to entrepreneurship<br>• Entrepreneurship process<br>• Principles of marketing<br>• Business communications<br>• Organization and management<br>• College English<br>• College math<br>• Psychology<br>• Computer concepts and applications | • Electricity 1<br>• Semiconductor and nanotechnology overview<br>• C/C++ for technologies<br>• College English<br>• College math<br>• Sociology<br>• Introduction to philosophy |

Source: Ballston Spa Central School District, <http://www.bscsd.org>

CTSI has been very well received in the region from its inception. In its second year of operation (2012–2013), it was expanded to extend access to the program for schools throughout the Capital Region, including the metropolitan areas of Albany, Troy, and Schenectady. Of the 21 students who graduated in the pilot year of 2011–2012, 15 went on to HVCC, four entered 4-year degree programs, and two enlisted in the military. By 2014, 43 students from 12 local high schools were graduating.[246] CTSI received P-TECH designation from the State in 2013. In December 2015, P21, a national organization that lobbies for 21st Century Learning, designated the CTSI as an "Exemplar School," an award that involves rigorous on-site evaluation and interviews.[247]

---

[246] "Clean Technologies and Sustainable Industries Early College High School (ECHS) Program Graduates 43," *The Saratogian* (June 10, 2014).
[247] "School Recognized for Exemplar Learning," Saratoga Springs, *The Saratogian* (December 2, 2015).

### 8.7.4 CNSE NanoHigh Initiative

In 2006 CNSE initiated NanoHigh, a pilot program to develop and implement innovative science and engineering programs at Albany High School. CNSE committed $400,000 to the program to cover the cost of research and activities and $100,000 for fellowships, scholarships, and internships, and by 2008, 33 students were enrolled in the program.[248] Students received classroom instruction at the high school in subjects which included nanoscale patterning and fabrication, principles of self-assembly, nano-biological applications, and fuel cell exploration, and then were able to engage in laboratory exercises at the CNSE facility on Fuller Road.[249] Through this initiative, Albany High School became the first public school in the United States to offer specialized, on-campus courses on nanoscience and nanotechnology.[250] By 2013 the number of NanoHigh graduates had exceeded 100 students.[251]

## 8.8 Training the High-Tech Construction Workforce

Construction of semiconductor fabrication plants and other high-technology manufacturing sites requires a wide range of special skills, including the ability to install massive, high-precision tools, the ability to meet high standards of cleanliness, the creation of infrastructure to handle a variety of exotic and sometimes hazardous materials and gases, and precision welding. Skilled construction workers are needed not only when a new factory is being built but also when it is up and running, to replace and repair existing systems and structures. In January 2017, when GlobalFoundries had no expansion projects under way, there were nevertheless several hundred construction workers present at its manufacturing site in Luther Forest engaged in various projects.[252]

Construction unions represent only about 28% of the construction workforce in the Capital Region, but according the Jeff Stark, president of the Capital Region Building and Construction Trades Council, union members perform about 85% of the large public and private construction jobs in the region. Contractors for projects involving construction of high-tech manufacturing facilities favor union labor because of the assurance that workers hired out of union halls will have the necessary skill sets and professionalism.[253] As Stark noted, "Our clients can call us today

---

[248] "Albany High Students Get a Look at Nanotech Careers," Albany, *The Times Union* (October 25, 2006).
[249] "Nano Science Brought to Life for High School Students," Troy, *The Record* (June 14, 2008).
[250] "Nano Tech Moves to the Head of the Class," Albany, *The Times Union* (December 11, 2008).
[251] "Students Finish NanoHigh Class," Albany, *The Times Union* (May 25, 2013).
[252] Interview with Jeff Stark, Menands, New York (January 26, 2017).
[253] One contractor providing specialized piping to the GlobalFoundries fab in Malta/Stillwater notes that it is union-affiliated because of the high quality of the personnel. The "union hall picks

and get 40 skilled workers." The unions can also draw quickly on their membership in other parts of the United States to respond to surges in need for workers with specific skills (such as building semiconductor fabs).[254]

While union–management relations are sometimes stereotyped as adversarial, according to Stark, the construction trades' relationship with GlobalFoundries is not adversarial and is characterized by frequent meetings with management where union leaders emphasize problem-solving. For example, during the tool-installation phase of construction of the fab, the world-class standard for fab construction held that a maximum of 14 tools could be installed per week. The unions brainstormed with management over red tape problems and various bottlenecks and devised a way to install 18–20 tools per week, beating the world-class standard (see Footnote 254).

### *8.8.1 Center for Construction Trades Training*

In 2004, M&W Zander, a construction firm specializing in semiconductor plants, entered into collaboration with CNSE and the state to create a training center at the Watervliet Arsenal to "help create a base of workers ready to build and to maintain the semiconductor plants many are hoping will call New York State—and the Capital Region—home." Zander committed $2 million to the new Center for Construction Trades Training (C2T), the state committed another $1.95 million, and the Arsenal Business & Technology Partnership, a nonprofit development group, committed $1.95 million. The US Army would contribute $1.4 million to renovate the building in which Zander would operate. CNSE designed and delivered the curriculum "and provided access to its industrial scale facilities for real world experience."[255] The rationale for the center was the reality that building semiconductor facilities "takes skills the average construction worker doesn't have." The center was expected to train 120 trade union members a year, importing skills such as semiconductor clean room construction.[256] Many of the workers who built the GlobalFoundries wafer fabrication plant in Malta/Stillwater were trained in the center.[257]

---

the guys" the company needs. They are certified and well qualified. The union provides contractors with bios, relevant experience, qualifications, and certifications. The career pathways for these individuals involve apprenticeships featuring two nights of instruction per week for 5 years. Interview, Menands, New York (January 26, 2017).

[254] Interview with Jeff Stark, Menands, New York (January 26, 2017).

[255] Laura I. Schultz, et al., "Workforce Development in a Targeted, Multisector Economic Strategy," in Carl Van Horn, Tammy Edwards and Todd Green, *US Workforce Development Policies for the 21st Century* (Kalamazoo: W.E. Upjohn Institute for Employment Research, 2015).

[256] "New Deal in Works at Arsenal," Albany, *The Times Union* (February 10, 2004).

[257] "Moving on Up (State)," Albany, *The Times Union* (August 7, 2011).

## 8.8.2   Trade Union Training Centers

Capital Region construction trade unions operate their own training facilities for apprenticeship programs, which include training in information technology and on-site simulations of the work environments apprentices can expect to encounter on the job.

- In 2012 the United Association of Plumbers and steam fitters initiated construction of a training center in Glens Falls which would provide training relevant to the work the union's members were performing at GlobalFoundries, including a clean room booth that has the same specifications as the work areas in the Luther Forest fab. The center was funded by $2 million from Local 773 members through payroll deductions and a $300,000 grant from the national union.[258]
- In 2014, the International Union of Painters and Allied Trades' District Council 9, based in Albany, bought a 20,000 square foot warehouse in Menands which the union has transformed into the new home of the Finishing Trades Institute of New York, comprising a training center and offices. The center features simulated work environments, computer training, and classrooms.

## Bibliography[259]

30 Great Small College Business Degree Programs 2016. (2016, July). *Online Accounting Degree Programs*.
Andy Plans Class Clone. (2013, February 27). *New York Daily News*.
Armour-Garb, A. (2017, January). *Bridging the STEM skills Gap: Employer / Educator Collaboration in New York*. Public Policy Initiative of New York State, Inc.
Babu Leads Clarkson's CAMP to Prominence. (2015, June 23). Massena, *Daily Courier-Observer*.
Balanced College Concept. (2003). In W. Somers (Ed.), *Encyclopedia of Union College History* (pp. 83-88). Schenectady: Union College Press.
Ballston Spa Makes 'Top High Schools.' (2016, August 22). *The Ballston Journal*.
Ballston Spa Central School District. http://www.bscsd.org.
Berman, E. P. (2012). *Creating the Market University: How Academic Science Became an Economic Engine* Princeton and Oxford: Princeton University Press.
Bruno, J. L. (2016). *Keep Swinging: A Memoir of Politics and Justice*. Franklin, TN: Post Hill Press.
Business: Shirt Tale. (1938, February 21). *Time*.

---

[258] "Ground is Broken on Long-Awaited Training Facility," Glens Full, *The Post-Star* (June 15, 2012).

[259] As noted in the front matter of this book, the study also drew on interviews carried out by the authors and numerous articles from *The Times Union* (Albany), *The Daily Gazette* (Schenectady), the *Albany Business Review* (Albany), *The Post-Star* (Glens Falls), *The Record* (Troy), *The Saratogian* (Saratoga Springs), *The Buffalo News* (Buffalo), The *Observer-Dispatch* (Utica), The *Daily Messenger* (Canandaigua), and the *Post-Standard* (Syracuse). These are not individually included in the bibliography.

# Bibliography

Calhoun, D. H. (1960). *The American Civil Engineer: Origins and Conflict.* Cambridge, MA: Technology Press.

Camoin Associates. (2013, September 5). The Curious Case of GlobalFoundries and its Workforce: Setting the Stage. https://www.camoinassociates.com/curious-case-globalfoundries-and-its-workforce-setting-stage.

Carroll, P. T. (1999, Spring). Designing Modern America in the Silicon Valley of the Nineteenth Century. *RPI Magazine.*

Community Colleges Grow With Distinction. (1987, May 11). Albany, *Knickerbocker News.*

Corona, L., Doutriaux, J., and Mian, S. A. (2006). *Building Knowledge Regions in North America: Emerging Technology Innovation Poles.* Cheltenham, UK, and Northampton, MA: Edward Elgar.

FMCC-HFM BOCES Collaboration Creates Career Pathway. (2011, December 24). *Targeted News Service.*

Gais, T., and Wright, D. (2012). The Diversity of University Economic Development Activities and Issues of Impact Measurement. In J. E. Lane and D. B. Johnstone (Eds.), *Universities and Colleges as Economic Drivers: Measuring Higher Education's Role in Economic Development* (p. 36). Albany: SUNY Press.

Grads Exceed Average Earnings. (1987, March 12). Albany, *Knickerbocker News.*

Hastfed, D. M. (2003). *The Emergence of Entrepreneurship Policy: Governance, Start-ups and Growth in the US Knowledge Economy.* Cambridge: Cambridge University Press.

Hill, P. (2014, April 25). Preparing Middle-Class Workers for Middle-Skill Jobs. *Marketplace.*

Ideas in Action. (2006, Fall). *RPI Alumni Magazine.*

Is There a US Engineering Shortage? It Depends on Who You Ask. (2015, August 19). *Power Electronics.*

A Jobs Crisis? No It's a skills Crisis. (2013, January 16). *New York Daily News.*

Kalas, J. W. (2010). "SUNY Strides into the National Research Stage," In J. B. Clark, W. B. Leslie, and K. P. O'Brien, (Eds.), *SUNY at Sixty: The Promise of the State University of New York* (p. 161). Albany: SUNY Press.

Le Maistre, C. W. (1989, October). Academia Linking with Industry—The RPI Model. *IEEE Xplore.*

McKinsey & Company. (2015). *Capital 20.20: Advancing the Region through Focused Investment.* McKinsey & Company.

Montgomery, D. (1989). *The Fall of the House of Labor: The Workplace, the State, and American Labor Activism 1865-1925.* Cambridge: Cambridge University Press.

National Research Council. (2013). *New York's Nanotechnology Model: Building the Innovation Economy.* C. W. Wessner (rapporteur). Washington, DC: The National Academies Press.

Nation's First Security College Creates Student Opportunities. (2015, January 27). *Long Island Examiner.*

NEA Research. (2016, May). *Rankings & Estimates: Rankings of the States 2015 and Estimates of School Statistics 2016.* National Education Association. Table H-11.

Office of the Governor. (2016, February 5). Governor Cuomo Announces New SUNY Emerging Technology and Entrepreneurship Complex at Harriman Campus. Press Release.

Phelan, T., Ross, D. M., and Westerdahl, C. A. (1995). *Rensselaer: Where Imagination Achieves the Impossible* Albany: Mount Ida Press.

Pillai, U. (2015, March). SUNY College of Nanoscale Science and Engineering. Monograph.

Preparing Middle-Class Workers for Middle-Skill Jobs. (2014, April 25.) *Market place.*

Quality Counts 2017: State Report Cards Map. (2017). *Education Week.* http://www.edweek.org.

Rensselaer Polytechnic Institute. Center for Biotechnology & Interdisciplinary Studies. http://biotech.rpi.edu.

Rensselaer Polytechnic Institute Faculty Lauded. (2016, December 7). *RPI News.*

Replanting the STEM Common Core Should Help Students Master Necessary Skills. (2013, October 30). *Watertown Daily Times.*

Research Foundation of SUNY. (2006). *SUNY's Impact on New York's Congressional District 21.*

Rockefeller Institute of Government. (2007, July). *The Supervisory District of Albany, Schoharie, Schenectady and Saratoga Counties: A Study of Potential Educational Reorganization in the Capital Region.* Prepared for New York State Department of Education.

Rockefeller Institute of Government of the University at Albany, and the University at Buffalo Regional Institute. (2011, June). *How SUNY Matters: Economic Impacts of the State University of New York.* State University of New York.

Rosenberg, N. (2003). America's Entrepreneurial Universities. In D. M. Hurt (Ed.), *The Emergence of Entrepreneurship Policy; Governance and Growth in the US Knowledge Economy.* Cambridge: Cambridge University Press.

Rosenberg, N., and Nelson, R. R. (1994). American Universities and Technical Advance in Industry. *Research Policy* 23:326.

Rudolph, F. (1977). *Curriculum: A History of the American Undergraduate Course of Study Since 1636.* San Francisco: Josey Bass.

Schultz, L., Wagner, A., Gerace, A., Gais, T., Lane, J. and Monteil, L. (2015). Workforce Development in a Targeted, Multisector Economic Strategy: The Case of New York University's College of Nanoscale Science and Engineering. In C. Van Horn, T. Edwards, and T. Green (Eds.), *Transforming US Workforce Development Policies for the 21st Century.* Kalamazoo: W.E. Upjohn Institute for Employment Research.

Semiconductor Industry Association. (2009, March). *Maintaining America's Competitive Edge: Government Policies Affecting Semiconductor Industry R&D and Manufacturing Activity.*

Skidmore College. MB107. http://www.skidmore.edu/management_business/mb107/index.php.

State University of New York. SUNY Works Campus Partnerships. https://www.suny.edu/suny-works/partnerships.

SUNY Adirondack. Academics. http://www.sunyacc.edu/academics.

SUNY Chancellor Promises More Internships in 2014. (2014, January 14). *Associated Press Newswire.*

SUNY Polytechnic Institute. (2016, June). *Strategic Plan.*

Sutton, R. I. (2013, October 3). Scaling: The Problem of More. *Harvard Business Review.*

United States Military Academy at West Point. A Brief History of West Point. http://www.usma.edu/wphistory/sitepages/home.aspx.

Wagner, A., Sun, R., Zuber, K., and Strach, P. (2015, May). *Applied Work-Based Learning at the State University of New York: Situating SUNY Works and Studying Effects.* Albany: The Nelson A. Rockefeller Institute of Government.

Zimpher, N. L. (2012). Foreword. In J. E. Lane and D. B. Johnstone (Eds.), *Universities and Colleges as Economic Drivers: Measuring Higher Education's Role in Economic Development* (p. xii). Albany: SUNY Press,

# Chapter 9
# The Changing Landscape of Tech Valley

**Abstract** The decade-long alignment of top state political leadership that supported the creation of Tech Valley has passed, and many of the original regional and local leaders that spearheaded the development effort are no longer engaged. The institutional disarray that followed one player's indictment and conviction, the shift in focus by state policymakers toward development of other upstate regions, and the continuing regulatory, financial, and operational travails of the Luther Forest Technology Campus underscore the fact that to be sustained, the successes achieved in the Capital Region will require a significant ongoing commitment by business, academic, and political leaders.

The creation of Tech Valley in New York's Capital Region is a success story, reflecting a broad regional effort involving hundreds of civic, business, and academic leaders and numerous institutions. The remarkable success achieved to date, however, does not guarantee that the region's gains during the past two decades will be sustained. Significantly, many if not most of the individuals involved in the creation of Tech Valley are no longer on the scene, and a number of key institutions have undergone restructuring, not always for the better. In a positive development for the state as a whole, other regions in New York are copying the Albany model, which means they are competing with the Capital Region for state funds, industrial investment, and talent. The Albany *Times Union* commented in 2013 in an article titled "Has 'Tech Valley' Peaked?" that—

> Now, with the NanoCollege breaking away from the University at Albany and expanding its outposts in every major city from here to Buffalo, some wonder if the [regional] momentum can be sustained.[1]

The challenge faced by the Capital Region is whether it can adapt to discontinuities and continue to build the institutions and infrastructure to sustain a high-tech manufacturing economy.

---

[1] "Has 'Tech Valley' Peaked?" Albany, *The Times Union* (November 11, 2013).

## 9.1 Changing Political Leadership

The troika of state political leaders who presided over the creation of Tech Valley in the decade after 1995 has passed from the scene. Governor George Pataki, Senate Majority Leader Joseph Bruno, and Speaker of the Assembly Sheldon Silver did not always see eye-to-eye, and the decision-making process that led to the success of Tech Valley was punctuated by bursts of acrimony and continual back-room horse trading by the three men.[2] But all of them came to share the vision that innovation-based economic development could reverse the erosion of New York's manufacturing base. They recognized what was needed to realize the vision and, at key points, put aside parochial concerns to achieve it. As one participant in the original outreach to Silicon Valley notes, the importance of the unusual "political confluence" between the three leaders which enabled the success of the effort: "three men in a room is the only way things got done."[3]

Governor Pataki did not seek re-election in 2006. One of his first acts as governor was his all-out, ultimately successful effort to reverse IBM's decision to move its headquarters out of New York. After that, by Joseph Bruno's account, Governor Pataki was skeptical about funding early initiatives at SUNY Albany but eventually changed his view and enthusiastically supported large state investments in research infrastructure in the Capital Region.[4] During Governor Pataki's tenure, SUNY Albany developed into the foremost center for applied nanotechnology in the world. Governor Pataki's commitment to the NanoCollege helped persuade IBM to build its first 300 mm wafer fab at East Fishkill. Governor Pataki was personally and deeply engaged in the successful effort to attract Sematech and later AMD to New York. It can be fairly said that during his administration, "Tech Valley" moved from an aspirational slogan to an established fact.

In July 2008 Senate Majority Leader Joseph Bruno announced his retirement.[5] Bruno used his power to direct state funds toward capacity-building in the Capital

---

[2] *See* Joseph L. Bruno, *Keep Swinging: A Memoir of Politics and Justice* (Brentwood, TN: Post Hill Press, 2016), pp. 138–149.

[3] Interview, Saratoga Springs, New York (September 16, 2015).

[4] According to Bruno, who asked the governor for funds to support SUNY Albany's nanotechnology research, Governor Pataki at first responded that "my staff doesn't think it's real. Not one of them. And we're not going to waste the money." But, as Bruno recalls, "The truth is, once [Pataki] got behind high tech he was a remarkable leader." Joseph L. Bruno, *Keep Swinging: A Memoir of Politics and Justice* (Brentwood, TN: Post Hill Press, 2016), pp. 135–136.

[5] "Bruno will Retire, End 32-year Career—'Time for Me to Ride off into the Sunset,' Senator Says in a Statement," Albany, *The Times Union* (July 16, 2008). Bruno left office in the shadow of an investigation of his business dealings which culminated in his acquittal by a federal jury in 2014. Bruno was convicted of honest services mail fraud in 2009. On appeal the conviction was thrown out when the Supreme Court ruled in another case that honest services must include proof of a kickback or bribe. Bruno was retried on the same charges and acquitted in 2014. "Bruno Acquitted," Albany, *The Times Union* (May 17, 2014). Bruno commented afterward that "The federal government spent 3 years investigating me, and I fought with the government in court for another five, all of which gave me a close look at a legal system overwhelmed by uncontrolled prosecutorial discretion and a judge who many thought enabled the prosecutors." Joseph L. Bruno, *Keep Swinging: A Memoir of Politics and Justice* (Brentwood, TN: Post Hill Press, 2016), p. xi.

## 9.1 Changing Political Leadership

Region to support high-tech research and manufacturing.[6] By his own account, he grew up in "a rundown duplex" in Glens Falls with no heat or hot water and watched his father, a blue-collar worker, forced into an austere retirement when he could no longer stay abreast of technological change in his workplace.[7] Bruno focused his efforts on building institutions in his region to educate and train the high-tech "blue collar men and women of the future."[8] Schenectady's *Daily Gazette* observed in a 2012 retrospective that "Bruno sank $100 million of state money into developing the enormous industrial park in Malta and Stillwater, the kind of money it takes to turn the middle of nowhere into somewhere."[9] In 2006 Bruno singled out the SUNY Albany NanoCollege and the planned AMD chip fab investment in Malta/Stillwater and said candidly—

> *Think that's a coincidence? That happened because as Majority Leader I sit at a table and it, three men in a room, works pretty well.*[10]

Assembly Speaker Sheldon Silver resigned in January 2015.[11] When Governor Mario Cuomo was defeated for re-election in 1994, Silver emerged as the "chief patron" of the NanoCollege at Albany, "the original investor of seed money when many could not spell nanotech."[12] Silver's support for nanotechnology continued over the years. In 2010 he brokered a deal involving $6.5 million in state incentives that the CEO of M+W Americas Inc., a major builder of high-technology manufacturing facilities, credits with his company's decision to relocate to the Capital Region in 2011, bringing $228 million in new investment.[13]

The exit of these three leaders has had palpable consequences in Tech Valley, perhaps the most visible manifestation of which is the financial plight of the Luther

---

[6] He recalled that "gradually, I began directing funds to HVCC, sometimes in dribs and drabs, other times sizable appropriations." Not long after becoming Majority Leader, he "got a quarter of a million dollars of seed money for the Center for Economic Growth. Their representatives used the funding to travel around the country and to Germany to learn about chip fabs and to drum up interest in putting one of them here." Joseph L. Bruno, *Keep Swinging: A Memoir of Politics and Justice* (Brentwood, TN: Post Hill Press, 2016), p. 138.

[7] Joseph L. Bruno, *Keep Swinging: A Memoir of Politics and Justice* (Brentwood, TN: Post Hill Press, 2016), pp. 3 and 137.

[8] Joseph L. Bruno, *Keep Swinging: A Memoir of Politics and Justice* (Brentwood, TN: Post Hill Press, 2016), p. 137.

[9] "Bruno, Others Did Hard Work for Nanotech," Schenectady, *The Daily Gazette* (May 12, 2012).

[10] "Mr. Bruno's Civics Lesson—He Defends 'Three Men in a Room' Budget Talks as Providing Benefits for the Capital Region," Albany, *The Times Union* (October 27, 2006).

[11] Silver resigned in the face of an indictment on federal corruption charges. Silver, who vowed to fight the charges against him, was convicted in May 2016 and sentenced to 12 years in prison. He is appealing his conviction based on the US Supreme Court's decision in *McDonnell v. United States*, which reversed the corruption conviction of a former governor of Virginia. "Silver Like Skelos, Can Remain Free While Appealing Graft Conviction," *The New York Times* (August 26, 2016).

[12] "Nano's Seeds Planted Long Ago," Albany, *The Times Union* (October 3, 2011); "Seed Cash for Tech Valley," Albany, *The Times Union* (February 23, 1999).

[13] "Nice Move," Albany, *The Times Union* (February 10, 2010).

Forest Technology Campus, the site which was created to attract semiconductor manufacturing to the region. Commenting on the nearly bankrupt tech campus in 2014, the Schenectady *Daily Gazette* observed that—

> *If George Pataki were still in the Executive Mansion and Joe Bruno in the Senate majority leader's seat instead of legal jeopardy* [a reference to charges Bruno was facing at the time], *there wouldn't be a problem. The two would simply have funneled money to the park and technology companies that occupy it, as they did with $100 million to initially develop the park and $1.2 billion in grants and tax credits to lure GlobalFoundries. But under three Democratic governors since 2007, the park has been left to fend for itself.*[14]

## 9.2 Events Under Governor Andrew M. Cuomo

In economic development policy, the governorship of Andrew M. Cuomo has seen a shift in the balance of power between the legislature and the executive in the direction of the latter. While "three men in a room" still make the final decisions with respect to the state budget, Governor Cuomo has largely supplanted the system pursuant to which individual legislators could steer funds to pet projects via "member items," a regime widely criticized as one in which "funding [is] distributed with minimal oversight in a murky process influenced by party and clout of local legislators."[15]

The current process is driven by the governor, who is advised by the state's professional economic development bureaucracy, most notably Empire State Development, and by outside think tanks and consultancies. Governor Cuomo has delivered six consecutive on-time budgets since taking office, but legislators complain that this involves ramming through legislation which is "passed on March 31, sometimes spilling into April 1, with deals coming together in the final hours and almost no time for public review."[16] Governor Cuomo has moved to significantly

---

[14] "One Troubled Tech Park in Malta," Schenectady, *The Daily Gazette* (April 6, 2014).

[15] "Development Councils in Trouble," Albany, *The Times Union* (March 19, 2011). During the 2017 budget negotiations, the "three men in a room" were actually four—Cuomo, Assembly Speaker Carl Heastie, Senate Majority Leader John Flanagan; and Senator Jeff Klein, who leads an eight-person breakaway group, the Independent Democratic Conference. "Whether three or four, the depiction of an exclusive club of legislative leaders and the governor alone in 'the room' mapping out the state's interests behind closed doors, is not entirely accurate. Often in the room are senior staffers—including budget directors and chiefs of staff—who are intensely involved in negotiation and planning, while outside voices—interest groups, lobbyists, rank-and-file legislators, think tanks, and others—also wield some, usually small degree of influence over the participants and their decisions." See "The 'Three-Men-in-a-Room' and Millions Outside," *Gotham Gazette* (March 30, 2017).

[16] New York law requires a 3-day waiting period between the time a bill is printed and when it is voted upon, affording time for review by lawmakers and the public. This requirement can be overridden; however, if the governor issues a "message of necessity" waiving the 3-day waiting period, to "speed the process along." One citizens' group representative objected in 2017 that "The message of necessity in the passing of the state budget has become as predictable as snow in January." See "The Three Men in a Room and Millions Outside," *Gotham Gazette* (March 30, 2017).

reduce the role of the legislature in deciding how and where state economic development funds are spent, with the executive assuming that role, to be exercised, in part, through newly created Regional Economic Development Councils originally administered by then Lieutenant Governor Robert Duffy, himself an upstate native (Rochester).[17] The new regime is far more inclusive than the one it replaced, engaging a broad range of local and regional leaders across the state in economic development but still, according to some critics, with limited transparency.

## 9.2.1 The Establishment of Regional Economic Development Councils

The hallmark of the Cuomo administration has been an effort to apply the Albany nanotechnology model in a systematic fashion in economically distressed upstate areas outside of the Capital Region. To his credit, Governor Cuomo has built on the previous work and experiences of past administrations and applied them during a time when he and the state have seen a set of unique challenges that have required—and forced—a more coherent approach in applying state support for economic development projects. In 2011 Governor Cuomo established ten Regional Economic Development Councils (REDCs) which were to coordinate investments by state agencies, benchmark performance by local governments before providing funding, and compete with each other for a part of $200 million in additional state funding to be used for regional growth and job creation.[18] Each REDC developed its own regional strategic plan and annually submitted project bids to the state for funds to implement the plans. The governor's office selected council members, which included at least one representative of labor and, as co-chair, the leader of a major university or college. The REDCs also served as conduits for myriad types of applications for state funding, ranging from sewer repairs to theater grants.[19] The

---

[17] In the 14 years prior to Cuomo's administration, governors and legislative leaders had vested in themselves the authority to borrow up to $7 billion "mostly for projects to be named later, at their discretion." The Syracuse *Post-Standard*, which ran a series of investigative reports into this tradition, commented in 2011 that "over the past 14 years, the state has authorized a total of $7 billion of borrowing under more than a dozen different names and acronyms. The money does not go through the normal channel—the state comptroller's office. Instead the money is handled by two independent public authorities—Empire State Development and the Dormitory Authority." See "Lawmakers still have their Pet Projects—Senate and Assembly Balk at Giving Governor Control of Unspent Funds," Syracuse, *The Post-Standard* (March 20, 2011). Kathy Hochul, a Buffalo native, succeeded Duffy in 2015. "Her Inheritance: An Eagerness to Serve," *New York Times* (May 29, 2011).

[18] "The Recipe for NY's Success," Albany, *The Times Union* (January 13, 2011); "Cuomo Names Local Board Members and Officials to 10 New Regional Economic Councils," Saratoga Springs, *The Saratogian* (July 28, 2011).

[19] "Cuomo Fund-Raisers Preceded Development Funding Awards," *Politico* (December 9, 2014).

ultimate selection of projects was undertaken by the governor, Empire State Development, and the governor's budget division, working as a team.[20]

The REDC process incorporated several concepts that worked well in the development of Tech Valley. The process was structured to nudge various local governmental units into working together on a regional basis, subordinating traditional internecine rivalries, as CEG was able to do, at least much of the time, in Tech Valley. This enabled economic development measures to be implemented in the context of coherent strategic plans for each region rather than distributing funds in scattershot fashion according to the preferences and clout of individual legislators and local jurisdictions. Equally importantly, the REDCs generated "bottom up" proposals from within each region enabling good ideas developed by local leaders—perhaps obscure but closely familiar with local conditions and potential—to be brought to the attention of state-level policymakers. Regardless of the level of success in obtaining state funding, through his Regional Council process, the Governor has been able to create a vehicle that has led to each region's private sector developing integrated, strategic, economic development plans aligned with their strengths and individual economies. This education process and collaborative effort which is in and of itself a best practice, has long eluded most regions of the state and, in fact, the nation. As the Saratoga Economic Development Corporation demonstrated in its seemingly quixotic pursuit of a massive investment in a chip fab, bottom-up initiatives can generate spectacular success, particularly when backed by state-level support at key junctures.[21]

Once regions had developed their strategic plans, Governor Cuomo showed "little tolerance for deviating from the plan." The governor "used his office and his control of state funds as a kind of bully pulpit." In cases in which regions were divided over some aspects of a particular project to be funded by the state, Governor

---

[20] "Real Regionalism Needs to be Restored," *The Times Union* (January 10, 2012). Critics of this arrangement charged that legislators' "member items" were merely being replaced by "governors' items." In 2011 Susan Lerner, the executive director of Common Cause New York, expressed concerns over vesting the ESD with authority to make final award decisions: "One official shouldn't have the final word." See "Economic Councils Come Under Fire," Schenectady, *The Daily Gazette* (April 24, 2011). ESD is advised by a Strategic Implementation Assessment Team (SIAT) which makes annual assessments of each region's progress and makes recommendations with respect to state investments in individual projects. The Assessment Team is comprised of New York State commissioners as well as an international trade expert from the US Department of Commerce and an expert on veterans' affairs. "ESD Team Reviews EDC Funding Proposals," Massena, *Daily Courier-Observer* (September 13, 2014).

[21] In 2011 Julie Shimer, chairwoman of the ESD, said in an interview that the regional councils were, in effect, screening mechanisms for ESD. In terms of picking the best projects, she observed that "despite its best efforts, making those decisions from Albany [is] always difficult. It's hard to have all the information on the ground and hear from all the stakeholders. So the governor feels, and I certainly agree with him, that getting the regional input is very important. Certainly all the state agencies did the best job they can with the information they have, and I think what's going to happen now is, we'll have much higher quality information and we'll be able to compare these proposals." See "How to Fix NY's Business Climate—It Starts with Regional Community-Based Planning, Says Development Corp. Chairwoman," Syracuse, *The Post–Standard* (August 18, 2011).

Cuomo made it clear that the localities had to work their differences out by a specified date or risk the loss of state funding for that project. The Governor's willingness and ability to exert this level of (firm) leadership has been seen by some as necessary and effective at a critical time when there is otherwise a leadership vacuum on many levels around the state—yet it is seen by others as overextending his authority. "When you're dangling millions of dollars, it is easier to get people to stick to the plan if you're willing, as Cuomo has been, to use the funding as a hammer."[22] Building coherent plans, providing serious funding, and supporting them over time is of course one of the lessons of the Tech Valley success.

Competition between the regions in the REDC process has been ruefully compared to "the Hunger Games," and the REDC system has attracted critics. A 2016 Glens Falls editorial complained that "instead of funding the most-needed projects"… the REDC awards "fund the projects with the best applications" and that entities that had "the connections and the cash to put together a knockout presentation" were more likely to prevail.[23] However, as the experience of Tech Valley demonstrates, in economic development, putting together good proposals is instrumental to successful outreach to the private sector, and preparing proposals is a necessary skill that the REDC process fosters.[24] There is no doubt that the regional development process forces communities to work together with stakeholders to develop comprehensive economic develop proposals, which may be considered in stark contrast of the former "member item" system where funding was often seen as going to pet projects based on the political pull of individual legislators.

In addition to the REDC process, Governor Cuomo sought to break up silos within the state government so that expenditures on infrastructure, housing, education, and other thematic areas could be integrated with economic development measures in a manner which cuts across agency lines to bring larger benefits to a given region. One former official who served in the governor's budget office under both Governors Pataki and Cuomo acknowledged that under the new system, politics continues to play a role in determining the allocation of development funds, but the new regime creates a significantly greater likelihood that meritorious ideas will be recognized and supported by the state.

At least one major apparent success story has emerged from the REDC process in Buffalo and Western New York. The economic development strategic plan developed by the Western New York REDC in 2011 has "shaped almost every major economic decision in the Buffalo Niagara region" in the 6 years that followed, contrasting sharply with the region's prior approach to economic development,

---

[22] "Cuomo Wields a Hammer to Nail Down Development for an Unsteady Region," *The Buffalo News* (April 13, 2017).

[23] "State Development Strategy is Incoherent," Glens Falls, *The Post-Star* (December 15, 2016).

[24] While discussing the reasons his company chose New York over other locations around the world for its next chip fab, AMD CEO Hector Ruiz said that, among other things, state and local officials had put together "the most well-crafted economic development package he could recall seeing." See "Tech Valley Vision Pays Off Big—Chip Maker AMD Hopes Rivals will Also Build Plants in Region," Schenectady, *The Daily Gazette* (June 24, 2016).

characterized as "nothing more than a random walk, just a random toss at the dartboard. No strategy, really, no goals."[25] *The Buffalo News* observed in February 2017 that—

> Under Gov. Andrew M. Cuomo, economic development has been focused through the lenses of the Regional Economic Development Councils that he created. Under its influence in Western New York, the impact has already been transformative. In addition to SolarCity, the money has funded the IBM Buffalo Medical Innovation Center, the Buffalo Medical Innovation and Commercialization Hub, Buffalo Manufacturing Works and the 43 North business competition, among other efforts.[26]

In a 2017 opinion piece, three past and one present co-chair of the Western New York REDC said that "The rest of the country has noticed. Buffalo is cool again, and the REDCs have become a national model."[27]

Another REDC-driven initiative appears to be gaining momentum, the Western New York Science and Technology Manufacturing Park (STAMP), which is being jointly advocated by the Finger Lakes and Western New York Regional Economic Development Councils as a potentially transformational multi-regional project. STAMP is a 1200-acre greenfield site at Alabama in Genesee County, a largely rural area between Buffalo and Rochester. The objective of the STAMP initiative is to attract nanotechnology manufacturing to the site, based on advantages such as low-cost hydropower electricity, the educational infrastructure of Buffalo and Rochester, and planned water and wastewater facilities. In 2016 ESD approved $46 million in funding for infrastructure in the park.[28] In 2015, 1366 Technologies, a US maker of solar energy equipment with a potentially revolutionary new technology, committed to invest $700 million at the STAMP site to create a manufacturing facility for solar wafers, but as of December 2017 the project was still pending final management go-ahead.[29]

---

[25] Howard Zemsky, former Co-Chair of the Western New York REDC and subsequently President and CEO of ESD, in "Instead of Finger-Pointing, Region Finds Focus with Council Strategy," *The Buffalo News* (April 13, 2017).

[26] "No More Bad Old Days—Legislators Should Give Up Their Wish to Dole Out the State's Development Cash," *The Buffalo News* (February 4, 2017).

[27] "Another Voice: Regional Councils—Economic Progress Under Threat from State Senate," *The Buffalo News* (February 28, 2017).

[28] "Empire State Development Approves $46 million for GCEDC's STAMP plan," *The Daily News* (August 18, 2016).

[29] "'Super Region' Marketing Part of $75K Contract," Batavia, *The Daily News* (December 2017). "Trump Weighs Heavily on 1366 Technologies Project—STAMP—Federal Loan, Competition from China will Factor in Manufacturers' Plans," Batavia, *The Daily News* (January 17, 2017). 1366, based in Bedford, Massachusetts, has pioneered a technology for producing silicon wafers by casting them in their ultimate shape in a mold, rather than the conventional method of slicing them from an ingot. The company believes that its technology will cut the cost of solar wafers by 50%.

## 9.2.2 The Transparency Issue

In 2017, in the wake of a procurement scandal involving some of the state's development projects, the state legislature demanded "what lawmakers call new transparency measures" with respect to the REDCs. The Assembly asked that members of an REDC be required to disclose information annually about their personal finances to avoid conflicts of interest.[30] Members of the Assembly sought creation of a public, searchable database of all aspects of economic development spending and a rule requiring 30 days' notice by the governor to the legislature of any intent to distribute economic development funds. The Senate has requested more comprehensive public reporting detailing how economic development funds were being spent.[31] Governor Cuomo, pushing back, says that legislators' real objective is "a quest for more control over pork-barrel spending" and that "state economic projects were marked by long delays when legislators had more say."[32] (See Box 9.1 for a discussion of achieving an appropriate level of transparency.)

> **Box 9.1 How Much Transparency?**
> Transparency is defined by the Non-Governmental Organization Transparency International as the "characteristic of governments, companies, organizations and individuals of being open in the clear disclosure of information, rules, plans, processes and actions."[33] As a general proposition, transparency in government fosters accountability of public officials and informed public participation in democratic processes and deters official corruption and conflicts of interest. "How could anyone be against transparency?" rhetorically asks Harvard Law School Professor Lawrence Lessig. "Its virtues and utilities seem so crushingly obvious."[34] Yet too much transparency can undermine democratic processes. Elimination of the secret ballot would limit citizens'

---

[30] The Schenectady *Daily Gazette* said in a March 2017 opinion piece that "most people don't want to publicly disclose information about their finances, but REDC members are not most people. They are well-connected business and community leaders who decide what to do with billions of dollars of taxpayer money." See "Time for Transparency with Economic Councils," Schenectady, *The Daily Gazette* (March 15, 2017).

[31] "Lawmakers Want Changes in How Cuomo Spends," *The Buffalo News* (March 14, 2017).

[32] "Lawmakers Battle Cuomo on Oversight of Job Programs," *The Buffalo News* (May 14, 2017).

[33] Transparency international, *Anti-Corruption Glossary* <http://www.transparency.org/glossary/terms/transparency>.

[34] Lawrence Lessig, "Against Transparency," *New Republic* (October 9, 2009).

exercise of the right to vote. Full disclosure of the deliberations of government agencies and advisory bodies would inhibit participants' willingness to speak frankly in making policy. Personnel decisions relating to hiring, firing, and promotion must be protected by confidentiality to enable evaluators to discuss and offer candid assessments.

In an economic development context, full transparency could actually bring much activity to a halt. In the creation of Tech Valley, high-tech companies in Silicon Valley only entertained overtures from state leaders seeking to persuade them to locate in New York on the condition that such discussions would be kept strictly confidential. Absent such secrecy in the early 2000s, there would be no GlobalFoundries in Malta/Stillwater today. Similarly, preparation of New York's incentive packages for companies considering investing in the state is necessarily kept under wraps to avoid tipping off competing regions and giving them an easy roadmap for putting together superior bids. The Saratoga Economic Development Council's 2002 sponsorship of visits by local civic leaders to other states where chip fabs were operating was instrumental in fostering local support for a fab in Malta and Stillwater but could have been derailed had publicity about the trips erupted into a furor over "junkets."

The abiding challenge for policymakers is to achieve a balance between the transparency necessary to keep the public well informed about the functioning of their institutions and to forestall corruption without bringing the entire process of governance to a halt—in other words, according leaders a degree of leeway to "get things done" that the public wants and needs. The late New York Senator Daniel Patrick Moynihan had this tension in mind when he addressed the Association for a Better New York in 1993, observing sardonically that 50 years previously in New York—

> We could do things in no time at all! In the dear old days of [Mayor] Jimmy Walker, we could build the George Washington Bridge in four years and one month, and think far enough ahead to make it structurally capable of carrying a second deck when the traffic grew.... We can do such things again. But it seems to me that we dare not lose the memory of what we have lost. [35]

Moynihan did not need to remind his audience that Mayor Walker, during whose administration a significant part of the infrastructure of modern New York City came into being, was forced to resign in the wake of a corruption scandal in 1932.

---

[35] Daniel Patrick Moynihan, "No Surrender: Toward Intolerance of Crime," Address to the Association for a Better New York (April 15, 1993).

## 9.2.3 Curbing the Legislature's Role in Economic Development

Governor Cuomo's budget for 2013–2014 eliminated so-called member items, which legislators had traditionally used to steer pet projects to their districts. The budget called for over $3 billion in new programs and discretionary funding which would be controlled solely by the governor, including $1.2 billion for economic development. The new arrangement prompted grumbling in the legislature with one legislative official asking "if the governor is given this authority, why is there even a need for a legislature? It's an awful lot of money to be allocated at the sole discretion of one person."[36] In 2014, John DeFrancisco, a Republican Senator who "had no clue" that Governor Cuomo would announce a new nanotechnology hub in his district and that the state had already picked a site, a developer, and a tenant, said that—

> It's very frustrating. There ought to be some collaboration. I just think the pendulum has gone too far with gubernatorial control. Maybe it was too far the other way with the legislature having too much control. I don't know. But it just seems right now one person should not be able to make same-day announcements of projects that affect the constituents of the legislators in that area.[37]

Sources in the Cuomo administration responded that the funds he wanted to control would be used to pay for "transformational projects, not a new roof on a social club" (see Footnote 36).

## 9.2.4 Spreading the CNSE Model

In May 2013, Governor Cuomo made it a priority to set up business incubators around the state, which the governor called "innovation hot spots." The announcement came as Governor Cuomo launched a third round of activity by the REDCs. The Governor wanted to help the state replicate the success of Tech Valley in other economically distressed regions in the state, using the same or similar developmental methods that had led to the success in Albany, expanding research activity into manufacturing.

Cuomo's vision was to establish a string of nanotechnology-oriented innovation clusters along an east–west corridor running from Albany to Buffalo, with sites in Rochester, Canandaigua, Utica, and Syracuse. If the plan worked—

> it will make the old Erie Canal route—which established the state as a major center of manufacturing and trade—into a modern-day tech corridor, rivaling Silicon Valley.[38]

---

[36] "Control Freaking: Legislators Vent at Cuomo for Hogging $3 B in Budget Power," *New York Daily News* (February 4, 2013).

[37] "Frustrated Lawmaker: 'There Ought to be some collaboration on Nano's Syracuse,'" Syracuse, *The Post-Standard* (March 9, 2014).

[38] "Big Project, Big Promises," Utica, *Observer-Dispatch* (October 12, 2014); "New York is Placing Big Bets in Upstate Cities," *Associated Press State Wire: New York* (September 12, 2015).

The Governor's vision was that state funds would primarily be used to invest in high-cost equipment and structures that would be owned by the state and made available to industrial users, serving as a basis for cooperation and attraction for high-tech firms.

"START-UP NY," a new incentive program, was a key element in Governor Cuomo's effort to "export" the CNSE development model to other upstate communities. START-UP NY offered 10 years of tax relief to early-stage and out-of-state companies that located their businesses on a SUNY campus, collaborated with the faculty, and created jobs. START-UP NY incentivized companies to locate on SUNY campuses to access the tax breaks, which also ensured "ready access to a trained work force."[39] Dean Fuleihan, CNSE's Vice President for Strategic Partnerships, said that START-UP NY is "really expanding the CNSE model throughout the state."[40] Existing and new CNSE sites across New York were pitched to potential industrial tenants as "START-UP NY eligible."[41]

### 9.2.5 Unease in the Capital Region

Governor Cuomo's new system of economic development posed a challenge to the Capital Region. Since the 1990s the foresight of the region's business and government leaders had enabled them to forge a consensus supporting a regional approach to economic development. Now, with the advent of the REDCs, the rest of New York was adopting such an approach, and the Capital Region faced nine competing regions rapidly learning how to apply the Albany model. In an application of the Albany precedent, innovation clusters grouped around research universities were proliferating across the state—good news, as a general proposition, but not necessarily an unmixed blessing for the Capital Region, which had to compete with these new innovation centers for state funding, new private-sector investment and skilled workers. Some questioned whether it made more sense to continue to focus support on a region that was proving to provide greater-than-expected ROI for the state taxpayer, or spread investments around the state by replicating the approach, which in turn might threaten the means to support the future success of the initial investments. On the positive side, the new research centers being established in the cross-state east–west axis resulted in increased economic activity and jobs in the Capital Region as well.[42] One example of this phenomenon is the GE-SUNY Poly

---

[39] "Big Project, Big Promises," Utica, *Observer-Dispatch* (October 12, 2014).

[40] "'Export' Plan Part of Larger State Strategy," Albany, *The Times Union* (November 10, 2013).

[41] "'Export' Plan Part of Larger State Strategy," Albany, *The Times Union* (November 10, 2013). START-UP NY has been "widely criticized for meager job creation despite heavy state spending to advertise it." In 2017 the governor was reportedly rebranding the program and revising the eligibility criteria. State officials continued to predict that the program would create about 4000 jobs in the next several years. "A Low Return Investment," Albany, *The Times Union* (March 26, 2017).

[42] Michael Liehr, who was named CEO of a new integrated photonics institute being established in Rochester, said in 2015 that "there is certain to be job creation in the Capital Region that comes

partnership—developing silicon carbide-based power electronics devices, which integrates a packaging operation in Marcy (near Utica) with silicon carbide wafer production at SUNY Poly in Albany.[43]

In the first two rounds of competitive bidding by the ten REDCs, the Capital Region REDC's proposals did not fare particularly well. James Barba, co-chair of the region's REDC, attributed these outcomes, in part, to the region's very success; during the years state funds were steered to local projects by Majority Leader Bruno:

> For over 15 years when Senator Bruno was majority leader this region got hundreds of millions of dollars. It's hard for me as a co-chair ... to say, "You know what? Give us even more."[44]

Many of the individuals interviewed for this study expressed the view that the Cuomo administration is steering state money away from the Capital Region, conceding that in light of the largesse the area had enjoyed under prior administrations, such a policy was politically astute and perhaps even essential. A 2013 commentary in *Albany Business Review* expressed this perspective:

> Cuomo is actively enticing business anywhere but Albany, the beneficiary of a state record $2.4 billion incentive package that recruited GlobalFoundries. Cuomo believes the region has had its fill, compared to other places upstate much more devastated by the demise of manufacturing, the popularity of offshoring and the allure of Texas, Florida and North Carolina.[45]

In fact, while under Governor Cuomo significant development funding was allocated to other areas of Upstate New York, the Capital Region continued to benefit from substantial state support as well as occasional personal intervention by the governor. In 2012 Governor Cuomo allocated $250 million toward new initiatives at the Albany NanoCollege, and another $250 million was approved in the FY 2018 budget.[46] In the third round of competitive bidding by the REDCs, the Capital Region REDC received the second-largest share of state development funding, $82.8 million.[47] SUNY Albany is getting $184 million in state funding to build a new College of Emergency Preparedness, Homeland Security, and Cyber Security.[48] In 2017, the state budget included $550 million in capital investment for SUNY Albany's College of Engineering and Applied Sciences.[49] In 2011, when the Trudeau Institute, a biomedical research located in Saranac Lake, was planning to relocate to

---

from the research [in Rochester] .... There's something in it for Albany." See "Institute to Bring More Jobs to Light," Albany, *The Times Union* (August 2, 2015).

[43] "What Happens at Quad-C?" Utica, *Observer-Dispatch* (March 25, 2017).

[44] "$50M is Back of the Pack," Albany, *The Times Union* (December 20, 2012).

[45] "Billions at Stake in Tech Arms Race as Luther Forest Flounders," *Albany Business Review* (October 18, 2013).

[46] "Nano fab X a 'Small Miracle,' McCall Says," Albany, *The Times Union* (February 15, 2012).

[47] "82.8M Win for Region," Albany, *The Times Union* (December 12, 2013).

[48] "A Slice of Funding Pie for U Albany," Albany, *The Times Union* (December 9, 2011).

[49] University at Albany, "State Budget Keeps Campus Moving Forward," Press Release (April 19, 2017).

North Carolina, Governor Cuomo "personally got involved to prevent the move," ultimately succeeding (see Footnote 48).

The Capital Region's comparatively meager share of state funding in 2011 ($62.7 million) reflected, in substantial part, the state's rejection of a single REDC project request, $25 million for a next-generation supercomputer at RPI. Rather than bias against the region by the governor's office, this setback may have reflected intra-regional squabbling which hurt prospects for state approval. The regional REDC was co-chaired by RPI President Shirley Jackson, and as reported in the press—

> The supercomputer proposal exposed a not-so-secret non-love affair between Dr. Jackson and Alain Kaloyeros ... [Kaloyeros] privately had been complaining that the supercomputer bid was a conflict of interest for Jackson. And two weeks ago, not so privately, Kaloyeros repeatedly ridiculed the supercomputer [proposal] during the public announcement of a new tech firm coming to the Watervliet Arsenal.[50]

The new supercomputer proposed by RPI would have served the entire state, not just the Capital Region. With REDC funding based on a 4:1 private/public match ratio, the recipient of state funding would have been required to cover 80% of the cost, and the project would have represented a very substantial commitment of funds by RPI.

## 9.2.6 Buffalo's Economic Renaissance

The first, largest, and to date most successful initiative by Governor Cuomo centered on Buffalo, the epicenter of Upstate New York's long economic decline. In 2006 Buffalo was characterized by the *New York Times* as "a once-mighty city reduced to a shadow of its former self."[51] The Erie Canal, which had enabled the city to become an economic powerhouse in the nineteenth century, had long ago been bypassed by rail, road, and air transportation, as well as by the St. Lawrence Seaway. Heavy industry, an economic mainstay of the city, collapsed in the latter part of the twentieth century. The city's population—which made it the tenth largest in the United States in 1920—shrank from 580,000 in 1950 to 290,000 in 2006.[52] At various points during that period, over half of the residents of Buffalo's Erie County were on some form of public assistance.[53] A 1995 opinion piece in the *Buffalo News* characterized the city as "an economic basket case and one of the nation's poorest urban centers."[54]

---

[50] "Cuomo Show Reveals the Big Winner," Schenectady, *The Daily Gazette* (December 16, 2011).

[51] "After a Half-Century of Decline, Signs of Better Times for Buffalo," *The New York Times* (September 18, 2006).

[52] "After a Half-Century of Decline, Signs of Better Times for Buffalo," *The New York Times* (September 18, 2006); "Can Buffalo Ever Come Back?" New York, *The Sun* (October 19, 2007).

[53] "How to Survive the Recession: From a City That Knows Some Guidelines on Stemming the Deluge," *The Buffalo News* (February 24, 1991).

[54] "Buffalo Must Build Industrial Base to Halt Economic Freefall," *The Buffalo News* (January 24, 1995). *The Buffalo News* observed in April 2017 that "For the better part of four decades, we chased

For decades journalists, economists, and business leaders have noted Buffalo's potential for an economic revival, citing abiding strengths such as proximity to Canada, numerous brownfield industrial sites with embedded infrastructure, a source of virtually unlimited renewable energy (Niagara Falls), and, most importantly, a concentration of excellent medical research centers and institutions of higher learning, including the largest school in SUNY's system, the University at Buffalo.[55] As Governor Andrew Cuomo observed, Buffalo had strong prospects despite its longstanding economic malaise:

> If Buffalo just got a fair shake, any reasonable assistance in a situation they did not cause on their own—they got caught in an economic transformation—if anyone could put out a hand, Buffalo could pull itself up. [56]

Yet for decades the city's development aspirations remained ephemeral. In the 1980s and 1990s Buffalo opened new retail centers, a sports arena, and a casino, initiatives which critics said generated low-paying, frequently part-time jobs and diverted public resources away from the local educational system. A promising initiative to establish a bioinformatics center at the University at Buffalo, launched in 2002, featuring recruitment of a "rock star" researcher to head the center, fizzled out within 3 years as the center failed to attract sufficient levels of private investment, and the "rock star" reportedly demonstrated that "social interaction was a weak spot in his personality."[57] Richard Azzopardi, spokesman for Governor Cuomo, pointed out in 2017 that "the legislature managed economic development funds for Buffalo for 20 years, spent billions, and accomplished nothing while Buffalo continued to decline."[58]

---

silver bullet solutions, from the Ghermezion brothers' mega-mall in Niagara Falls to its similarly misguided cousin, Benderson Development's Niagara Falls Factory Outlet mega-mall, and the most silvery bullet of them all, a Bass Pro at what is now Canalside. It was a parochial approach, guided by individual interests first and foremost. It was a 'What's in it for me?' approach that put the community's interests second, at best. It also was an object failure." See "Cuomo Wields a Hammer to Nail Down Development in an Unsteady Region," *The Buffalo News* (April 13, 2017).

[55] Ray Rudolph, chairman of the Albany-based engineering firm CHA companies, said in 2015 that Buffalo has "a history of manufacturing. The heritage is important because of the public acceptance—there is no NIMBYism. They have good bones. Lots of water, lots of sewers, power, all the infrastructure needed to support manufacturing, it's already there." Interview, Albany, New York (September 16, 2015). "Miracle Predicted for Area Economy—Editor of Forbes Sees City's Rebirth," *The Buffalo News* (May 13, 1989); "Buffalo Must Build Industrial Base to Halt Economic Freefall," *The Buffalo News* (January 24, 1995); "Upstate's Famously Defeatist Attitude Just Another Obstacle in the Way of a Resurgence—Are You a Defensive Pessimist?" Syracuse, *The Post-Standard* (February 18, 2007); "Buffalo Among Strongest Metro Areas," *Buffalo News* (September 15, 2009); "Can the Bio-Economy Succeed in Buffalo?" *The Buffalo News* (April 28, 2002); "Resuscitating the Heart of the City," *The Buffalo News* (May 13, 2001).

[56] "Why Cuomo is Devoted to Buffalo," *The Buffalo News* (January 22, 2017).

[57] "UB Hires 'Rock Star' of Bioinformatics," *The Buffalo News* (April 27, 2002); "A Buffalo Match Gone Wrong?" *The Buffalo News* (April 15, 2004); "The Bioinformatics Dream Gets a Wake-Up Call—Center Has Yet to Meet Expectations as an Engine of Economic Development," *The Buffalo News* (June 19, 2005).

[58] "Lawmaker Battle Cuomo on Oversight of Job Programs," *The Buffalo News* (May 14, 2017).

### 9.2.6.1 Planning an Economic Resurgence

In January 2012, Governor Cuomo pledged $1 billion in state funding over a 5-year paid period to revive Buffalo's economy, an initiative famously dubbed "The Buffalo Billion."[59] Governor Cuomo said in retrospect that—

> The billion dollars was important for shock and awe value. They were so down, and they had heard everything for so long, and they were so distrustful that I needed to say something that would actually get their attention and allow them to think "maybe it's different this time." [60]

In an expression of the governor's "bottom up" approach to economic development, the Buffalo Billion plan was "kept intentionally vague" to "let companies tell Albany what they need ... be it direct cash or tax credits or low-cost energy or regulatory relief—to move to Buffalo."[61] The governor was advised by McKinsey & Company and Bruce Katz, a long-time acquaintance of Governor Cuomo who was co-director of the Brookings Institution's Metropolitan Policy Program. Katz and Irene Baker, director of the state REDCs, emphasized the need to invest the state money gradually and wisely.[62] The state strategy focused on the development of a skilled local workforce with abilities matched to the needs of employers and potential in the region, encouraging start-ups and revitalizing depressed communities. As *The Buffalo News* reported—

> Rather than focusing on individual projects, the plan mainly concentrates on building a base of facilities and programs that will help create an economic environment attractive to private investors in ... targeted industries.[63]

The Buffalo Billion appeared to get off to a slow start as time was allocated for extensive planning by the REDC, assisted by the University of Buffalo, McKinsey, and Brookings. However, when the strategic blueprint was complete, funds were distributed relatively quickly to a succession of major projects which rapidly gathered momentum. In addition to initiatives funded by the Buffalo Billion, parallel initiatives were undertaken to expand the Buffalo Niagara Medical Campus and to redevelop the historic Canalside district in Buffalo.[64]

---

[59] "A Billion for Buffalo, is the City Ready?" *The Buffalo News* (January 29. 2012).

[60] "The Wind and Sun Are Bringing the Shine Back to Buffalo," *The New York Times* (July 21, 2015).

[61] "Cuomo Goes Out on a Limb by Pushing 'Buffalo Billion,'" *The Buffalo News* (January 29, 2012).

[62] "Cuomo Aide Offers Update on State Plans for Buffalo," *The Buffalo News* (June 14, 2012). Howard Zemsky, co-chairman of the Western New York REDC, said that "if we can make strategic choices, fact-based decisions, we can make progress. But I caution people that we're not going to look like Portland, Oregon in 2013 or 2014 .... We have people who are looking for jobs and jobs that are looking for people, and the two don't always match up .... We've had many decades of population decline and job loss. Unless we are successful in implementing these strategies, we will have continued decline." See "Fixing Our Economy is Slow Work," *The Buffalo News* (February 10, 2013).

[63] "Patient Approach Deemed Vital," *The Buffalo News* (February 8, 2013).

[64] "Buffalo Billion Goes Right Here Thanks—Area Reaping Benefits in Attitude, Economy," *The Buffalo News* (January 29, 2015).

The Buffalo initiative looked to the precedent of CNSE as a model for revitalizing western New York. *The Buffalo News* commented in February 2013 that "to see what state officials want for Buffalo, head 280 miles east, to the sprawling Nano Tech Complex in the University of Albany campus." As was the case with Albany, the state sought to build on Buffalo's existing strengths, which included excellent medical and other life sciences research and education institutions, and, as was the case with Albany, the lion's share of state investment was allocated to public research infrastructure to be owned by the state. The Buffalo initiative departed from the Albany model in one respect—while the Albany effort focused entirely on one theme, nanotechnology, the Buffalo initiative invested in three: biotechnology, clean energy, and advanced manufacturing.[65]

### 9.2.6.2 Biotechnology Center

In late 2012 the state launched an effort to replicate Albany's nanotechnology achievements in Buffalo in the field of biotechnology, featuring creation of a life sciences innovation center at the Buffalo Niagara Medical Campus aimed at creating a biotech industry cluster "fueled by University of Buffalo innovation, local medical expertise and private investment."[66] The state's investments were intended to leverage a much larger volume of private investment.[67] CNSE officials and the governor's office approached Albany Molecular Research, Inc. (AMRI), a global life sciences research and manufacturing enterprise, to establish a 250-person laboratory at the new innovation center. The state provided $50 million for the project, which involved $200 million in private investment. The state money was spent on new equipment ($35 million) and improvement of lab space ($15 million), with the equipment and lab resources owned by the state.[68] The investments were intended to create a critical mass of resources enabling the region to dominate the field; as was the case in nanotechnology in Albany, the vision of the Governor was to make Buffalo the center for medical innovation.

---

[65] "State Investment Follows Public-Private Partnership Model," *The Buffalo News* (February 25, 2014).

[66] "State Officials Want Buffalo to Become a Hub for Life-Sciences Innovation, and they are Investing $50 Million Toward Making that Vision a Reality," *The Buffalo News* (February 17, 2013). The Buffalo Niagara Medical Campus, established in 2001, is comprised of seven institutions, including the Roswell Park Cancer Institute, Buffalo General Hospital, and the State University of New York at Buffalo. Approximately 17,000 people are employed in the campus.

[67] Howard Zemsky said that "Our whole strategy for the medical campus, out whole focus, was to populate it with private-sector jobs." See "Companies to Complete Move to Permanent Home in 2015," *The Buffalo News* (December 8, 2013).

[68] "Albany Molecular to Open Buffalo Center," Albany, *The Times Union* (December 5, 2012); "A Model Partnership," Albany, *The Times Union* (December 8, 2012). At the end of 2015 Perkin Elmer joined AMRI as a partner in the initiative, supplying AMRI with equipment used in its research. "Companies to Complete Move to Permanent Home in 2015," *The Buffalo News* (December 8, 2013).

### 9.2.6.3 Clean Technology Center

The state's investments in life sciences in Buffalo were paralleled by significant investments in green technologies. In 2013 the state launched the RiverBend Project, investing $225 million to build new structures at a brownfield site along the Buffalo River, creating a clean energy and high-tech manufacturing research campus. Two initial industrial tenants, Silevo and Soraa, both moving from California, were lined up, forecast to create an initial 850 jobs at the site, with each firm investing $750 million.[69] The project was to be run by the SUNY Research Foundation assisted by officials from CNSE.[70] In March 2014 Empire State Development approved a tranche of $118 million drawn from to state's commitment of $225 million to fund initial development costs and machinery purchases for the RiverBend clean energy and high-tech manufacturing hub.[71]

In 2014, the state approved a budget that allocated the final $680 billion needed to fund the Buffalo Billion completely, ending questions whether "Governor Andrew M. Cuomo and the State Legislature would actually come up with $1 billion, or if the initiative would gradually fizzle and turn into something smaller."[72] An editorial in *The Buffalo News* stated that "Cuomo has been on a mission. And he had something to prove to Western New Yorkers who had rejected him in the last election. And he's done it. Huge ... now everything seems to be coming together." [73]

### 9.2.6.4 Genomic Medicine Center

In January 2014, the governor announced in his State of the State address that the state would invest $50 million from the Buffalo Billion to create a new genomic research facility, the Genomic Medicine Center, at the University of Buffalo's Center for Computational Research, with the university participating with the New York Genome Center in Manhattan and the Roswell Park Cancer Institute in Buffalo. This "bottom up" initiative arose out of a proposal made by the Western New York REDC and built on the region's existing strength in genomics research.[74] Four

---

[69] "Project Will be Located at Former Coke and Steel Plants," *The Buffalo News* (November 21, 2013). Soraa subsequently altered its plans and decided to locate its plant in Syracuse.

[70] "This is Like the Bills Winning the Super Bowl," *The Buffalo News* (November 22, 2013).

[71] "Medical Campus Gets $50 Million," *The Buffalo News* (March 29, 2014).

[72] "WNY Prospects 'Haven't Been this Bright in a Very Long Time,'" *The Buffalo News* (April 1, 2014).

[73] "$1 Billion Delivered—With the New State Budget, Governor has Met His Ambitious Goal for Buffalo," *The Buffalo News* (April 2, 2014).

[74] Roswell Park and the Hauptman-Woodward Medical Research Institute in the Medical Campus were established leaders in genetics, structural biology, and cancer research. In 2013 Roswell Park launched its Center for Personalized Medicine, where researchers decipher and analyze the genes of individual patients in order to develop improved tests and treatments for genetic abnormalities. "Cutting-Edge Science—Genome Center Can be Another Key Part of a Growing Medical Powerhouse," *The Buffalo News* (January 19, 2014).

out-of-town companies joined this effort as partners and began moving employees to Buffalo.[75]

In February 2014 IBM disclosed that it would become the first corporate member of the Genomic Medicine Center, bringing 500 employees to Buffalo. The state assisted this move with a $25 million commitment to build a "high-end software development center" in which IBM would be the first tenant and with $30 million to be spent on software, computers, and servers. All of these resources were to be owned by the state but available for use by industrial tenants.[76] *The Buffalo News* commented in an editorial that the IBM initiative and the other projects launched in Buffalo by the Cuomo administration—

> *follow the successful partnership model used in Albany, in which the state pays for the buildings, the laboratory equipment and computer systems, and recruits companies to staff and operate the facilities. It packs a powerful punch—one, we are sure, that some economic purists would protest. But it works and, what is more, the traditional model of economic development centered mainly on the private sector had done little for western New York after decades of decline. Some force was needed to break the pattern, and only New York, in the hands of a governor on a mission, had the clout and the passion to pursue that task successfully.*[77]

### 9.2.6.5 Advanced Manufacturing Center

In 2014 the state announced plans to form the Buffalo Niagara Institute for Advanced Manufacturing Competitiveness, a $54 million center on Buffalo's Main Street dedicated to themes such as flexible automation and controls, additive processing, advanced fabrication, and advanced materials and testing. The institute was to be operated by EWI, an Ohio consultancy specializing in helping companies develop innovative products. The concept behind the institute was to "create a center where manufacturers can work with engineers or use sophisticated equipment to help them turn ideas for new products into reality."[78]

---

[75] The industry partners included AESKU Diagnostics, a developer of diagnostic tests and instruments, and Lincagen, which uses DNA to test children with symptoms of autism and other forms of developmental challenge. "Cutting-Edge Science—Genome Center Can be Another Key Part of a Growing Medical Powerhouse," *The Buffalo News* (January 19, 2014).

[76] The new center was named the Buffalo Information Technologies Innovation and Commercialization Hub. IBM also planned to use SUNY Buffalo's large computing capability to translate genome research under way at a Manhattan facility into practical healthcare applications. "IBM to Bring 500 Jobs to Buffalo—Cuomo to Announce Plan for IT Center Downtown," *The Buffalo News* (February 24, 2014).

[77] "The place to be—IBM Becomes the Latest Trophy in Cuomo's Effort to Revive Buffalo," *The Buffalo News* (February 26, 2014).

[78] "Center to Develop Innovative Products," *The Buffalo News* (January 17, 2014); "Officials are Expected to Decide a Plan for Rainbow Center," *The Buffalo News* (April 5, 2014).

### 9.2.6.6 SolarCity

In June 2014 SolarCity, the largest solar energy services provider in the United States, disclosed plans to acquire Silevo, one of the first companies to establish a presence at RiverBend for $200 million. SolarCity, owned by entrepreneur Elon Musk, planned to buy solar panels manufactured by Silevo for incorporation into systems assembled by SolarCity.[79] With the acquisition, Silevo's original plan for a factory at RiverBend, which would employ 475 people grew to a factory five times the size of that originally planned and employing 3000 people. SolarCity planned to invest $5 billion in the site over the next 10 years.[80] The state supported the proposed manufacturing facility with a pledge to build the factory at a cost of $350 million and to purchase the equipment for $400 million, both of which the state would continue to own (see Footnote 80). The new factory would be three times longer than any other solar panel factory in the United States and one of the largest in the world.[81]

SolarCity has been buffeted by a global slump in demand for solar systems, mounting debt levels and a high-visibility procurement scandal involving RiverBend, but Elon Musk has reinforced the company through a series of deals, and the company's plans for a major factory in Buffalo remain on track, albeit with modifications of the original plan.[82] SolarCity's quarterly solar system installations plummeted from over 260 MW in the fourth quarter of 2015 to under 160 MW in the first quarter of 2017, reflecting declining global demand for new solar installations.[83] In November 2016 another Musk company, Tesla Motors, spent $2.6 billion to acquire SolarCity, "which has struggled financially despite revenue growth," a transaction which critics charged amounted to a bailout of SolarCity.[84] In January 2017 Tesla brought in Panasonic Corp. as an investor in SolarCity.[85] Panasonic, which has extensive experience in manufacturing solar cells, will operate the factory.[86]

---

[79] "SolarCity Strategy Reinforces Buffalo Niagara," *The Buffalo News* (June 19, 2014).

[80] "SolarCity's Riverbend Project packs a Major Economic Wallop: Promises 3,000 Jobs, $5 Billion Investment," *The Buffalo News* (September 24, 2014).

[81] "Yes, It's Buffalo ... City's Transformation is Augmented with Explosive Growth of RiverBend," *The Buffalo News* (September 25, 2014).

[82] Between March 2016 and March 2017, SolarCity eliminated 3000 jobs, or 20% of its workforce. SolarCity lost $820 million in 2016. "SolarCity Cut its Workforce by 2009 in 2016 ... 3,000 Positions Gone in Cost-Saving, More as Buffalo Plant Gears Up," *The Buffalo News* (March 3, 2017).

[83] "Solar Industry Slowdown Catches Up with SolarCity," *Investopedia* (May 5, 2017).

[84] "Tesla and SolarCity Merger Gets Approval from Shareholder," *CNBC* (November 17, 2016).

[85] Under the 10-year arrangement, Panasonic will help pay for capital costs at the SolarCity plant in Buffalo, while Tesla committed to buy Panasonic solar cells and modules to be used in Tesla's glass tile roofs and other products. "Panasonic's investment is significant because the Buffalo plant is expected to require more capital through the end of 2018 as it begins to ramp up production." See "Tesla and Panasonic Finalize SolarCity Deal—Japanese Tech Firm to Invest $256 Million," *The Buffalo News* (January 18, 2017).

[86] "Factory Plans Very Different from Original," *The Buffalo News* (March 8, 2017).

9.2  Events Under Governor Andrew M. Cuomo    297

By early 2017 construction of the SolarCity factory was nearly complete and the company planned to start installing equipment in the facility by early 2018. The company had roughly 40 employees on-site and was beginning to hold information sessions for people interested in entry-level jobs in the new factory including engineers, managers, manufacturing specialists, shipping and receiving clerks, and material handlers. SolarCity pledged to hire 1460 workers at the factory and create another 1440 new jobs in the region via supplier and vendors providing services to the factory.[87] "Manufacturing specialists," which will be needed in significant numbers, can be high school graduates but will need strong basic math capability and specialized skills to perform tasks such as use of computers for the automated processing of products.[88] The company planned to start production in Buffalo in 2017 or 2018, depending on the timing of completion of pilot production tests at a Tesla facility in California to "work out the kinks" in the process for making cutting-edge solar tiles.[89] *The Buffalo News* noted the potential impact of SolarCity on the region in April 2017, writing—

> If the SolarCity factory succeeds in creating its promised 2900 jobs at both the rooftop solar installer and its suppliers, that project alone could add about a half a percentage point to the region's job growth rate. And that would move the pace of hiring tantalizingly close to the national average—a prospect that only the biggest Buffalo Niagara booster would have dared to consider a decade ago.[90]

In January 2018 Tesla confirmed that production of solar roofing tiles had begun and that the workforce at the factory comprised about 500 people. It indicated that it was continuing to install and configure equipment at the site and that it envisioned the eventual annual output to consist of enough solar panels and roofing tiles to generate over 1 GW of electricity. Howard Zemsky, CEO of ESD, observed that "the ramp-up for production is well under way. Solar module production began last year, and Tesla is now producing tiles for its solar roof, with hundreds of people already hired."[91]

---

[87] "SolarCity Has Promised to Hire 1,460 Workers," *The Buffalo News* (January 18, 2017).

[88] "Hiring Time Has Arrived at SolarCity—Here's What Job Seekers Should Expect as They Apply for Entry-Level Positions at the Plant," *The Buffalo News* (January 29, 2017).

[89] Solar tiles are a premium product that look like regular roofing shingles—an alternative to "clunky solar panels" and are paired with energy storage capability. "In a Shrinking Market, Tesla Tackles SolarCity Changes," *The Buffalo News* (May 5, 2017).

[90] "Buffalo Billion II Ensures Against One-and-Done," *The Buffalo News* (April 7, 2017).

[91] "Tesla Confirms Production Start for Solar Roofing…Says Manufacture of Tiles in Buffalo Last Month as Hiring Continues," *The Buffalo News* (January 10, 2018).

#### 9.2.6.7 Economic Resurgence

With numerous major construction projects under way, Buffalo, having long provided "the punch line in jokes about snowstorms, also-ran sports teams and urban decline... is suddenly experiencing something new: an economic turnaround.... Buffalo is going like gangbusters."[92] Initiatives in health and life sciences, renewable energy, and advanced manufacturing, benefitting from the state's $1 billion investment, and a "comprehensive planning process" have helped to "spur the renaissance and reduce the region's unemployment rate to 5.3%." One consequence is "newly vibrant neighborhoods, many with reused buildings hosting apartments, shops and restaurants that ... attracted young workers from other cities."[93] One of many Buffalo out-migrants who has returned said in 2015 that "every day there's some hipster bar opening. I don't want to be in Brooklyn anymore."[94] *The Buffalo News* commented in 2015 that—

> The development under way in Buffalo is stunning for a region used to not much of anything at all happening. And the projects are [in] areas that put Buffalo at the forefront of various aspects of the high-tech economy. What was once a booming 19th century city (Eric Canal), and a thriving 20th century city (steel), is fast becoming a city of the 21st century.[95]

### 9.2.7 Other Upstate Initiatives

Governor Cuomo's launch of the Buffalo Billion initiative in 2012 left other upstate regions to "ask a simple question: What about us?"[96] A February 2012 opinion column in the Utica *Observer Dispatch* posed the following question:

---

[92] Interview with Darren Suarez, director of government affairs, Business Council of New York, Albany, New York (October 28, 2015).

[93] "The Wind and Sun are Bringing the Shine Backs to Buffalo," *New York Times* (July 20, 2016).

[94] "The Wind and Sun are Bringing the Shine Backs to Buffalo," *New York Times* (July 20, 2016); "Buffalo In Removable Economics resurgence," *Free Enterprises* (July 15, 2015); "New Day in Buffalo ... City's Revival is Gaining Momentum as Many Key Components Take Shape," *The Buffalo News* (January 24, 2015); "Millennials Find Reasons to Live in Buffalo," *The Buffalo News* (February 2, 2015); "Investments Helping Change New York's Anti-Business Tag," *The Buffalo News* (October 18, 2015); "A Long Time Counting ... Region's Positive Employment Numbers are the Product of Much Dynamic Effort," *The Buffalo News* (November 28, 2015); "Watershed Moment ... Buffalo is About to Reap Major Benefits From Years Of Planning and Investment," *The Buffalo News* (January 29, 2017).

[95] "Change in Attitude ... A Billion Dollars Later, Buffalo's Shedding its Reputation for Botching Development," *The Buffalo News* (February 9, 2015). Buffalo scored some successes in traditional manufacturing sectors in 2016 as well. General Motors revealed plans to invest $328 million in two Buffalo Niagara factories, including a new engine line in its Town of Tonawanda plant. Sumitomo Rubber said it would invest $87 million in its 93-year-old tire factory in Tonawanda. General Mills indicated it would shift some of its cereals production from other locations to its Buffalo plant. "Corruption Probes Have Wide Impact," *The Buffalo News* (January 18, 2017).

[96] "Cuomo Goes Out as a Limb by Pushing 'Buffalo Billion,'" *The Buffalo News* (January 29, 2012).

*What about us Governor? Like Buffalo, we have had a crisis in Central New York that's gone on too long. Over the past 20 years we've lost major players like Lockheed Martin, Bendix, Chicago Pneumatic, Oneida Limited, Rome Cable, Ethan Allen, Rite Aid and more. On top of all that, we lost an Air Force base that took away 5,000 good jobs.... [N]obody's talking about forking over $1 billion to us.*[97]

In fact, soon after these words were written, the state began unveiling major innovation-based economic development projects along the Albany–Buffalo axis, involving Rochester, Utica, and Syracuse. While it is too soon to assess the long-run prospects for these initiatives, some look promising, such as a new photonics hub in Rochester, and others have apparently fizzled, most notably a nanotech-themed film hub near Syracuse.

### 9.2.7.1 Rochester

In April 2013 the Albany *Times Union* reported that "the next frontier for expansion of New York State's nanotechnology economy is in the Rochester area."[98] CNSE officials had reportedly been acquiring property in the Finger Lakes area and developing plans for expansion of a technology park in Canandaigua, a lakeside town outside of Rochester. CNSE was also reportedly working to expand its Smart Systems Technology and Commercialization Center in Canandaigua, which was the site of a major DoE project to create a pilot light-emitting diode manufacturing line.[99]

In July 2015, federal, state, and local officials, including Vice President Biden, designated the city of Rochester as the national headquarters of a $610 million research and manufacturing hub dedicated to the emerging field of integrated photonics.[100] $110 million was being committed to the project by the Department of Defense, with the remainder coming from a combination of state and private investment. This hub, the American Institute for Manufacturing Integrated Photonics

---

[97] "Pumping Up State Economy Must Include US ... Let's See Some Regional Balance Coming from Albany," Utica, *Observer-Dispatch* (February 2, 2012).

[98] "Nano Heads out of Town," Albany, *The Times Union* (April 12, 2013).

[99] "Nano Heads out of Town," Albany, *The Times Union* (April 12, 2013). The Canandaigua facility was originally a Kodak clean room for solar electronic technology manufacturing. It came on the market during Kodak's Chapter 11 proceedings, which began in 2012. The state persuaded DoE to underwrite most of the cost of equipping the facility, transferring $19 million worth of equipment from a DoE site in Silicon Valley. "'Export' Plan Part of a Larger State Strategy," Albany, *The Times Union* (November 10, 2013).

[100] Photonics devices use light instead of electricity to perform processes such as transmission of data and sensor functions. "Integrated" photonics incorporate numerous functions on a single chip, such as sensors, wave guides, multiplex . . ." all that stuff goes into the chip, not the box. That reduces power by many orders of magnitude." These devices are incorporated in larger systems, such as automobiles, airplanes, communications systems at home appliances. Photonics devices hold the promise of enabling transmission of more information using much less data than conventional devices. Interview with Mike Fancher, associate professor of nanoeconomics, CNSE, Albany, New York (January 26, 2017).

(AIM Photonics) is a consortium of 124 organizations, including the Rochester Institute of Technology, the University of Rochester, MIT, the University of Arizona, the University of California and industrial members including IBM, Boeing, Raytheon, Corning and Texas Instruments.[101] In March 2017 AIM Photonics released a 400-page *Integrated Photonics Systems Roadmap* (IPSR), developed with input from roughly 700 photonics industry professionals, which forecasts anticipated market developments and technology needs, including training, tools, and standardization.[102] Four community colleges were part of the Consortium—HVCC, Schenectady County Community College (SCCC), Columbia-Greene Community College and Adirondack Community College.[103]

In 2016 two photonics companies committed to establish a presence in Rochester, invest $1.4 billion, and create 800 research and manufacturing jobs:

- *Photonica*, a California-based maker of visual display technology used in ultra-high definition televisions and large video displays, pledged to create 400 jobs at the photonics hub and at nearby Eastman Business Park.
- *Avogy*, a Silicon Valley-based developer of gallium nitride-based power electronics technology, committed to move its headquarters to Rochester and to employ 400 people at an average salary of $80,000 within 5 years.[104]

AIM Photonics is building a Testing, Assembly, and Packaging (TAP) facility for photonics devices in Rochester, with construction and tool installation beginning in 2017 and the facility expected to be fully operation in mid-2018. The State of New York has committed $250 million to this effort, which will cover the cost of renovating a former Kodak building as well as machinery, tools, technology licenses, and operations. The TAP facility will house the photonics industry's only open access 300 mm test site; a test, assembly, and packaging Manufacturing Execution System; and Wafer Fab Multiple Project wafer assembly tools and equipment.[105]

### 9.2.7.2 Utica

In 2013 CNSE announced plans to establish three semiconductor manufacturing plants at a site in the town of Marcy, outside of Utica. The site, called the Marcy Nanocenter, would have 8 million square feet and employ 5000 people when built

---

[101] "NY Wins $600 Million Hub for Photonics Research and Development," *Associated Press State Wire: New York* (July 27, 2015); "Region Selected for $600M Research Hub," Canandaigua, *Daily Messenger* (July 28, 2015).

[102] "A Roadmap for US Integrated Photonics," *Optics & Photonics* (March 23, 2017).

[103] "Forging Photonics Alliance," Albany, *The Times Union* (July 28, 2015).

[104] "Hub Aims to Bridge Gap Between Research, Product Development," *The Buffalo News* (March 18, 2016); "Cuomo: 2 Firms to Bring 1,400 Jobs," Canandaigua, *Daily Messenger* (March 31, 2016).

[105] "Governor Cuomo Announces Milestone Reached at AIM Photonics in Rochester," *US Fed News* (May 28, 2017).

out. The state and private-sector partners had reportedly committed $55 million for infrastructure improvements, site studies, planning, and marketing. SUNY Poly planned to build a $125 million semiconductor manufacturing facility, the Computer Chip Commercialization Center, or Quad C, adjacent to the Marcy site.[106] Ground was broken on Quad C in the summer of 2013.[107] In October 2013 Governor Cuomo announced that the state would invest $200 million in the site, which would be the home of a new semiconductor R&D consortium comprised of Advanced Nanotechnology Solutions, Inc., a company headed by former AMD CEO Hector Ruiz, IBM, Sematech, Tokyo Electron Atotech, and Lam Research.[108] Although Quad C was to have been finished by the end of 2014, the project was delayed by "changes in tenant requirements" and revisions of the original plan to provide "expanded infrastructure, cleanroom space and capabilities."[109]

In July 2015 AMS AG, an Austrian maker of sensors, announced plans to establish a wafer fabrication facility employing 1000 people at the Marcy Nanocenter.[110] AMS planned to build a three-story, 360,000–450,000 square foot, 300 mm wafer fabrication plant, as well as a 100,000 square foot administration facility.[111] Ground was broken for the AMS fab in April 2016.[112] But in December 2016, AMS disclosed that it was backing out of the project in the wake of the state bid-rigging scandal.[113]

In August 2015 General Electric announced a consortium in partnership with SUNY Poly to develop a manufacturing plant for silicon carbide semiconductor wafers for applications in power electronics.[114] GE committed to be the anchor ten-

---

[106] "High Hopes Built on Tiny Chips," Syracuse, *The Post-Standard* (October 20, 2013). Twenty million dollars in state funding was allocated to the project via the REDC process in 2011. Mohawk EDGE, an economic development group based in Rome, New York, began pitching Marcy as a site for semiconductor manufacturing in 1997. The Computer Chip Commercialization Center was conceptualized under Governor Paterson in 2009, with the original cost forecast at $45 million. "SUNY IT Tech Partnership to Bring 475 Jobs," Utica, *Observer-Dispatch* (July 15, 2009).

[107] "Ground Being Broken at Marcy Nanotech Site," Utica, *Observer-Dispatch* (June 27, 2013).

[108] "SUNY IT Investment Upped to $1.5 Billion—Governor Unveils Plan for Companies at Quad C," Utica, *Observer-Dispatch* (October 11, 2013).

[109] "Major Quad C Expansion on Horizon—Investment Doubling; More Jobs Possible," Utica, *Observer-Dispatch* (January 18, 2015); "Heastie: 'Have Some Faith' in Quad C—Assembly Speaker Stops in Utica During Upstate Tour," Utica, *Observer-Dispatch* (July 24, 2015).

[110] "Chip Plant Headed for Marcy," Utica, *Observer-Dispatch* (July 28, 2015).

[111] "Marcy Nanocenter Moving Forward—AMS Plant Construction to Begin in Spring," Utica, *Observer-Dispatch* (December 7, 2015).

[112] "Breaking Ground on Nano Promise," Utica, *Observer-Dispatch* (April 21, 2016).

[113] "Nano Utica Back to the Drawing Board—'Offers on the Table' After AMS Backs Out, Dimeo Says," Utica, *Observer-Dispatch* (December 20, 2016).

[114] Silicon carbide is a new material used in "wide bandgap" semiconductors which are more efficient than the silicon-based chips which are currently used to power devices and systems. According to the Department of Energy, silicon carbide-based devices could enable the shrinkage of a laptop computer power adapters to about one quarter of the current size. The new devices have broad potential industrial application, including a substantial reduction in the size of power-generating substations. "Cuomo: Chip Project Makes NY Competitive," Troy, *The Record* (July 15, 2014).

ant of the Quad C complex, creating an initial 470 jobs with the potential to employ 820 within 10 years.[115] The workers would include employees from GE, SUNY Poly, and other partner companies in the "Power Electronics Manufacturing Consortium."[116] While no GE business unit had natural skills in silicon carbide manufacturing, CNSE's know-how in this area was very strong, so the consortium gave GE something it could not have done on its own.[117] The State of New York contributed $135 million to support formation of the consortium.[118] Pursuant to this arrangement the state committed to build, equip, and operate a fabrication plant for producing 6-in. silicon carbide wafers at SUNY Albany which would eventually function as a foundry providing silicon carbide devices for GE and other customers. GE contributed platform silicon carbide intellectual property (IP) which was to be shared with collaborating industrial partners developing specific products and processes. The availability of world-class equipment plus GE's IP "gives a big head start to the partners."[119]

Then in March 2017 it was disclosed that Denmark-based Danfoss Silicon Power would take over the entire Quad C site in Marcy to package silicon carbide wafers being manufactured at SUNY Poly in Albany in collaboration with the GE-led consortium. The silicon carbide arrangement will work as follows:

- Six-inch silicon carbide wafers will be produced at SUNY Poly in Albany.
- The wafers will be shipped to the Danfoss facility in Marcy when it becomes operational.
- GE, which holds the IP for silicon carbide technology, will set up a dicing operation within the Danfoss facility to remove the chips from the wafers.
- Danfoss will package the chips into modules and assemblies capable of powering electronic devices.
- The modules will be sold, including some to GE which will use them to power its systems, including wind turbines, hybrid cars data centers.[120]

---

[115] "Huge in Nano," Syracuse, *The Post-Standard* (August 21, 2015).

[116] "GE, AMS Buildup to be Phased in," Utica, *Observer-Dispatch* (August 22, 2015). GE has a long history in the Utica area. It built a facility to make radio tubes in Utica in 1944 and expanded its operations dramatically after World War II. At its height, GE employed 6000 people in Utica. The company encouraged its employees to advance themselves and sometimes paid their tuition. One resident commented in 2015 that "Just about everyone in this community has had some relationship with GE at one time. It was the big employer in the area outside Griffiss Air Force Base." See "Will GE Bring Good Things to Life Again?" Utica, *Observer-Dispatch* (August 30, 2015).

[117] Interview with Mark Little, former head of GE Research, Schenectady, New York (April 7, 2017).

[118] "Cuomo: Computer Chip Project Makes NY Competitive," Troy, *The Record* (July 15, 2015). The state will own the buildings and equipment. "New Tech Investment," Albany, *The Times Union* (July 16, 2014).

[119] Interview at GE Global Research, Niskayuna, New York (April 6, 2016).

[120] "What Happens at Quad C?" Utica, *Observer-Dispatch* (March 25, 2017). "The Real Deal: Danish Company to Move into Quad C Promises 300 Jobs in Coming Years, Production Set to Begin in 2018," Utica, *Observer-Dispatch* (March 25, 2017).

In December 2017, GE announced that the construction of the wafer fabrication plant at SUNY Poly had been successfully completed. GE, which never planned to own or operate the facility, will withdraw from the consortium and can become a customer of the completed foundry, which will serve other companies as well. GE spokesman Todd Alhart said that the consortium "is moving from the first phase, which included the successful installation of tools and their qualification—with output exceeding everyone's expectations—to the second phase, featuring the production of power electronics chips as SUNY Poly works with GE and other potential industry partners on utilizing the line."[121]

### 9.2.7.3 Syracuse

In 2015, it was disclosed that California-based Soraa, a maker of light-emitting diode (LED) lighting with high-quality color, brightness, and efficiency, would invest $1.3 billion over 10 years in establishing a manufacturing facility in DeWitt, a suburb of Syracuse, near the film hub facility. The state committed $90 million to build and equip the facility, which it would own through SUNY Poly. Soraa would not pay rent for the facility but would invest in maintaining it and keeping the equipment up-to-date. The initiative was forecast to create 420 direct jobs in Syracuse.[122] The Central New York Hub for Emerging Nano Industries, announced by Governor Cuomo in 2014, was established in DeWitt, with the involvement of CNSE. The hub was to be devoted to nanotechnology-related R&D relevant to the film industry, including new technologies for post-production tasks such as sound, editing, and special effects. The state committed $15 million to the project and Onondaga County committed $1.4 million, with another $150 million in private sector investment anticipated.[123] Construction of the hub was completed in 2015, but prospective tenants delayed plans to locate there or to begin making films.[124] The *New York Times* reported in August 2016 that—

> [T]he building sits essentially vacant, and the hub has exactly two employees who work full time, including a cinematographer, Huayu Xu, recently hired to manage and promote the project. No films made it to the multiplex: Nearly two and a half years after the governor's announcement, the hub's anchor tenant, FilmHouse, has yet to release a production, and its president and other executives have been dogged by lawsuits, tax liens and seven-figure legal judgements.[125]

---

[121] "GE Funding Power Chip Partnership with SUNY Poly," Albany, *The Times Union* (December 26, 2017).

[122] "Lighting Lured—NY spends $90 million on a Facility to Attract a California Firm's 400 Jobs to Dewitt," Syracuse, *The Post-Standard* (November 1, 2015).

[123] "Nano Hub Update," Syracuse, *The Post-Standard* (February 25, 2015).

[124] "Update: Film House Again Delays Production of First Local Movies," Syracuse, *The Post-Standard* (February 26, 2015).

[125] "Cuomo's $15 Million High-Tech Film Studio? It's a Flop," *The New York Times* (August 22, 2016).

As of April 2017, the film hub stood vacant as the state reportedly searched for a new operator for the site, which was being "derided as a waste of taxpayer money."[126] In June 2018 the film studio building was sold for $1 to a newly created entity, Greater Syracuse Soundstage.[127]

The fate of the film hub does not demonstrate that the Albany model does not work but that, at least in this case, it was not tried. A 2017 retrospective observed that although SUNY Poly pledged to support the hub with university-based research and education, "no specific programs were developed." In contrast to the lavishly equipped NanoCollege in Albany, the film hub "opened without basic equipment" needed by most filmmakers, including equipment for film editing, sound mixing and other post-production work, and no carpentry shop in which to build sets. The sound stage had no ceiling grid lighting system or green screen. There was no trained workforce in the Syracuse area to support a film industry, and "without a steady stream of films to provide work, it's difficult to establish a pool of labor; without labor available, it's hard to attract filmmakers." IBM, which, as the NanoCollege's "anchor tenant" contributed hugely to its success, had no counterpart at the film hub, where the erstwhile anchor tenant "was in town for about a month," and shot one low-budget film.[128]

## 9.3 The Scandal

The scandal which erupted in 2016 with federal indictments of eight individuals, including Alain Kaloyeros, on charges of alleged bid-rigging, substantially set back New York's effort to promote innovation-based economic growth, delaying some projects and very likely playing a role in the collapse of several consortia. Many observers concluded that oversight of SUNY Research Foundation, which had eroded in recent years, needed improvement and that other institutional reforms are in order. That said, there is widespread sentiment that the broad, innovation-based economic development effort itself is well-conceived, that it is demonstrating results, and that it should continue. The scandal provides lessons on the importance of oversight and accountability, yet it also highlights the continued debate to define the proper balance between needed oversight and the ability to make decisions in a timely, cost-efficient fashion, so the United States can compete with the rest of the world, a need that is particularly acute in fields that involve fast-moving technology development and advanced manufacturing.

---

[126] "SUNY Poly in Loop for Filmmaking," Albany, *The Times Union* (April 2, 2017).

[127] "State's $15 million Film Hub Sold for $1 After Flopping," *The Buffalo News* (June 3, 2018).

[128] "Film Hub in DeWitt—New Award Category: Best Revival of a Failing Project," Syracuse, *The Post-Standard* (January 28, 2017).

## 9.3.1 Loosening Oversight of SUNY Research Foundation

At the center of the bid-rigging charges are several projects that were being administered by the SUNY Research Foundation through two of its nonprofits, Fuller Road Corporation and Fort Schuyler Management Corporation (FSMC). Such entities were the corporate vehicles through which the NanoCollege was established and expanded, making possible the rise of Tech Valley. The fact that the nonprofits were not bound by university rules is generally recognized as the key to their success.[129] At the same time, oversight of the nonprofits has diminished in recent years, a fact which ongoing investigations may conclude contributed to the current scandal. The question is thus whether oversight and accountability of the nonprofits can be strengthened to prevent abusive practices without destroying their ability to engage the private sector effectively. (See Box 9.2.)

The institutional constraints on SUNY Research Foundation, which were limited during the decade when the NanoCollege was formed, were loosened further in 2011. In 2011 the state legislature eliminated a requirement that the Office of the State Comptroller (OSC) had to pre-review every SUNY contract exceeding $250,000, a rule that was seen to be the cause of delays in implementing SUNY economic development projects. An editorial in *The Buffalo News* commented that "The relaxed oversight that had been sought by Gov. Andrew M. Cuomo will apply to potentially billions of dollars' worth of state and local government contracts in New York … [but] we agree with State Comptroller Thomas P. DiNapoli, who stresses that pre-audits are as important means for protecting taxpayer interests."[130] In the wake of the legislature's action, the comptroller retained post-audit oversight of all SUNY transactions, but an OSC spokesman commented in 2016 that "the comptroller's authority to review contracts is an important deterrent to waste, fraud, or abuse. The OSC spokesman said that over the past few years that oversight has been eroded, "noting that in 2015 state agencies had awarded nearly $7 billion in contracts" for which "our oversight was removed."[131]

> Meanwhile the governor and the legislature have created entities such as Fort Schuyler and Fuller Road that effectively remove our oversight for projects such as the Buffalo Billion. Such shadow entities are bad business for the state and for taxpayers.[132]

---

[129] Pradeep Haldar, Vice President of CNSE's Entrepreneurship, Innovation and Clean Energy Programs, observes that it takes 3–4 years to build a building using university bidding rules. "Fuller Road Corporation enabled it much faster." Interview, Albany, New York (November 30, 2016). Skidmore Professor Cathy Hill, who served as counsel to the Albany-Colonie Chamber of Commerce during the outreach effort to attract a chip fab to the region, called the Fuller Road-type nonprofits "brilliant" because the SUNY Research Foundation and its corporates do not operate under university rules and are much less constrained in forging deals with companies. Interview, Saratoga Springs, New York (September 16, 2015).

[130] "Restore DiNapoli's Oversight … Legislature Shouldn't Have Limited Comptroller's Authority Over Contracts," *The Buffalo News* (April 12, 2012).

[131] "Reform Groups Seek Transparency," Albany, *The Times Union* (October 3, 2016).

[132] "More SUNY Review Urged," Albany, *The Times Union* (September 28, 2016).

> **Box 9.2 Checks, Balances, and Results**
>
> Alain Kaloyeros is one of a long line of New York leaders reputably good at "getting things done" who eventually ran into legal trouble. Kaloyeros was convicted of bid-rigging by a jury in 2018. But there is no question that his responsibilities required him to navigate a dense thicket of public procurement laws, regulations, and procedures.
>
> In their magisterial 2010 book, *New York Politics*, Edward V. Schneier, John Brian Murtaugh, and Antoinette Pole observe that "Americans are more obsessed with official corruption than are the citizens of most other societies" and that "the laws of both Albany and Washington are larded with checks on graft." Public procurement usually requires competitive bidding with all major purchases and service contracts to be advertised in advance and awarded to the lowest bidder. Contracting is also subject to audits, disclosures by government employees of outside sources of income, and other forms of exposure. "Special investigations of various agencies and individuals are not uncommon, and . . . even private firms, and individuals who work for the state or local agencies are fair game for secret investigations of their finances and aspects of their private lives."[133]
>
> Such measures may limit corruption but "come at a price" in terms of direct costs and the accretion of red tape that stifles initiative and gives rise to various operational problems, up to and including institutional paralysis. "The number of steps that need to be taken before a purchase order can be processed is extraordinary." In the colleges of the City University of New York, the authors reported that it can take months to replace the ink supply of computer printers and that it takes so many weeks for most suppliers to be paid that many businesses refuse to take orders from CUNY. If a copy machine breaks down in a government office, an administrator may be required to competitively bid a replacement (see Footnote 133).
>
> The authors conclude that the public contracting system in New York is "mired in red tape and multiple levels of oversight." Contractors, subject to multiple controls, feel they are being treated as "quasi-criminals," become cynical and look for ways to "get things done," even if this involves dubious, gray-area practices. Public employees, unable to find contractors willing to work under the conditions specified by law, rewrite RFPs. Such moves are "in turn . . . likely to spawn more safeguards and greater suspicion" (see Footnote 133). David Frum wrote in *The Atlantic* in 2014 that—
>
>> When government seems to fail, Americans habitually resort to the same solutions: more process, more transparency, more appeals to courts. Each dose of this medicine leaves government more sluggish. To counter the ensuing disappointment,

---

[133] Edward V. Schneier, John Brian Murtaugh, and Antoinette Pole, *New York Politics: A Tale of Two States* (Armonk and London: M.E. Sharpe, 2010), pp. 246–249.

## 9.3 The Scandal

> *reformers urge yet another dose.... Reformers keep trying to eliminate backroom wheeling and dealing from American governance. What they end up doing is eliminating governance itself.*[134]

In 2015, media reports indicated that the US Attorney's office in Manhattan had served subpoenas on SUNY Poly in connection with a grand jury investigation of the bidding process associated with construction projects for the Buffalo Billion.[135] Governor Cuomo hired a special investigator, Bart Schwartz, to review all of SUNY Poly's projects in light of the federal grand jury investigation. SUNY Poly issued a statement to the effect that SUNY Poly and its affiliated nonprofit, Fort Schuyler Management Corporation had "followed every rule, regulation and law at every stage of the process and have been completely open and transparent in navigating through New York State's process for approving economic development projects, project contracts, and disbursement of economic development funds."[136] SUNY Poly pointed out that the contracting process is subject to so many checks and balances that the system "simply does not allow" its own unilateral approval of a project contract.[137]

Reinvent Albany, a watchdog group advocating increased transparency in government, challenged the SUNY position that checks and balances on its nonprofits were adequate, stating that "by using a mixture of non-profit groups . . . and academic institutions like SUNY Polytechnic, the state is blurring responsibility and reducing the accountability for decisions worth hundreds of millions of dollars."[138] Other state entities also indicated that procedures followed by some SUNY-affiliated nonprofits were problematic. In August, the state Dormitory Authority, which was funding the construction of the silicon carbide chip manufacturing line at SUNY Poly, acknowledged that it had not paid contractors on the job since April. The Authority said that it was responsible for overseeing SUNY Poly and its associated nonprofits and that—

> *documentation from the contractor that was recently provided to us by the Fuller Road Management Corporation was found to be incomplete, and we are working with Fuller Road Management to obtain the proper documentation as required.*[139]

---

[134] "The Transparency Trap," *The Atlantic* (September 2014).

[135] "Delay Upsets SUNY Poly," Albany, *The Times Union* (September 26, 2015).

[136] "SUNY Poly Details Oversight and Transparency Process Regarding Buffalo Projects," Memorandum by Jerry Gretzinger, SUNY Poly Vice President of Strategic Communications and Public Relations (September 26, 2015).

[137] Oversight bodies included Empire State Development, which was the designated funding agency, with additional financial oversight by the Division of the Budget (DoB) and, "as appropriate, the Office of the State Comptroller (OSC). "SUNY Poly Details Oversight and Transparency Process Regarding Buffalo Projects," Memorandum by Jerry Gretzinger, SUNY Poly Vice President of Strategic Communications and Public Relations (September 26, 2015).

[138] "More SUNY Review Urged," Albany, *The Times Union* (September 28, 2016).

[139] "Pay Delays Tied to Filings," Albany, *The Times Union* (August 4, 2016).

John Bacheller, a former Senior Vice President at ESD, observed that the SUNY-affiliated nonprofits were created as "private" entities, and therefore, they—

> did not operate with the full transparency and accountability that state entities must provide. As a result, board decisions were made in secret, and decision processes and criteria were not subject to public review. And, critically, contracting processes were not subject to review by the state comptroller. The comptroller plays a critical role in ensuring contracts issued by state entities meet legal requirements for fairness.[140]

In the wake of the indictments, Governor Cuomo transferred the functions of the Fuller Road and Fort Schuyler nonprofits to ESD, a more traditional state agency subject to freedom of information laws and "more transparent than the quasi-governmental nonprofits." Comptroller Tom DiNapoli argued that OSC's oversight powers should be restored with respect to pre-approval review of contracts, saying—

> If we had authority to look at SUNY construction projects in Albany, we might have weighed in. It might have gotten them to think twice before they came up with these alleged schemes. They should consider our office, which is set up to provide review, to give us a more direct role.[141]

While DiNapoli may be right, his critics observe that in 2012–2013 the comptroller's office conducted an audit of one of the SUNY Foundation's nonprofits, Fuller Road Corporation, at a time when alleged wrongdoing was occurring and concluded that the organization operated in an "ethical business climate" characterized by internal controls which safeguarded against fraud. The comptroller's office responded by stating that "when those at the top deliberately cover up their actions or engage in criminal activity, internal controls may not always flag the wrongdoing or raise red flags."[142]

## 9.3.2 Initial Perspective

While the effort to scale the Albany model throughout the state has naturally met with challenges, the degree of success is not yet known. Nonetheless, it is possible to draw some preliminary conclusions about the effort to spread the "Albany model" across the rest of Upstate New York. It is difficult to fault the aspirational aspect of this effort, that is, to stimulate innovation-based economic growth broadly and to bring thousands of high-quality jobs to long-distressed regions. Execution of such an ambitious and far-flung project was bound to encounter setbacks and errors, and in some cases, there appears to have been, at the very least, bad judgment by some individuals involved. Still, while some projects have been impacted, with respect to

---

[140] "Require Full Transparency for SUNY's 'Private' Nonprofits,'" Albany, *The Times Union* (September 28, 2016).

[141] "Reform Push in Scandal," Albany, *The Times Union* (October 15, 2016).

[142] "SUNY Poly's Plaudits Criticized," Albany, *The Times Union* (December 26, 2016).

the larger projects, the Albany model appears to have been closely followed and was yielding some impressive successes when the scandal broke.

In this context, it is important to note that while it is unfair to give too much credit to one individual in the process, it is a mistake to fault the vision that has involved many mid-level contributors through the years with many playing individual but important roles in the success of Tech Valley. One lesson is the danger of giving too much unchecked decision-making authority to one person, a lesson which the Governor and state lawmakers seem to recognize. By the time the scandal broke, the institutional checks and balances on Kaloyeros had eroded to the point that he was operating as a "one-man band, a lone ranger," as his defense attorney acknowledged, and some of his colleagues had expressed alarm at the degree of power he wielded over the state's investments.[143] The question will be to what degree policymakers will ultimately adjust the process and whether they will be able to strike the right balance between government oversight and control and the needed flexibility to enable the private sector to succeed.

Many staunch supporters of the model observe that in attempting to expand SUNY Poly's reach beyond the Capital Region at the same time that CNSE was itself growing rapidly, and merging with SUNY IT, may have caused the system to be over-extended. In addition, referring to the downside of one person's growing control (Kaloyeros), "The scope of his responsibilities got out of control, expanded to all of upstate."[144] Whereas previously he had worked in concert with institutions such as EDC, the Division of the Budget, and SUNY, which provided a variety of checks and balances, after 2013 he operated more like "a lone ranger," as several former state officials put it.[145] A Kaloyeros supporter spoke in 2016 of the "diseconomies of scale" that arose as CNSE's strategic partnerships grew in size and complexity ... [noting that] "75 partners are very different than 10 partners, too many actors, too many layers."[146] According to ESD's Howard Zemsky, as an institution CNSE did not grow sufficiently to keep pace with ever-larger responsibilities—

> As time went on the organization was ... asked to do a lot, and so while their portfolio of projects increased a lot over time, staff didn't increase in a commensurate way. They spread their footprint, they grew their portfolio, and it was hard to manage.[147]

One manifestation of CNSE's overstretch was the languishing of some projects that did not receive enough resources or attention, such as the film hub near Syracuse.

Governor Cuomo's upstate development drive appears to be encountering the same sort of scaling-up challenge as confronts the New York educational system (see Chap. 8), that is, how to replicate successful model programs on a statewide scale

---

[143] "Bid-Rigging Trial Begins for Ex-State University of New York Official," *Reuters* (June 18, 2018); "Alain Kaloyeros, Powerful Centerpiece of Buffalo Billion, Could Become a Household Word," *Gotham Gazette* (September 29, 2015).

[144] Interview with a former New York State senior official (May 2017).

[145] Interviews in Albany (2015 and 2016).

[146] Interview at CNSE, Albany, New York (November 2016).

[147] "Blame Grows, Creation Slows," Albany, *The Times Union* (February 19, 2017).

without an erosion of quality standards. The extraordinarily talented leaders required to scale initiatives cannot simply be multiplied. The numerous model educational programs surveyed in this study are often driven by unusually gifted educators and school administrators, and the question is whether the models themselves, properly applied by professionals who are competent but perhaps not as uniquely talented, can be scaled up and achieve comparable results on a broader scale. That challenge is being taken up by the state's economic development arm, Empire State Development.

### 9.3.3 Rebound

Soon after the indictments, Governor Cuomo announced that the economic revitalization of Upstate New York being undertaken by SUNY Poly would be turned over to ESD. ESD, headed by CEO Howard Zemsky, a Buffalo-based former developer who as co-chair of the Western New York REDC had played a major role in the effort to revive that city's economy, has a staff of 450 people operating out of 12 offices across the state.[148] Zemsky was characterized by *The Buffalo News* in 2017 as "[not] flashy . . . thoughtful and balanced . . . not aiming for the quick hit."[149] Zemsky was upbeat as ESD took over management of various SUNY Poly projects. Of SUNY Poly itself he said that "It's really important not to create a stigma over an institution that has done some extraordinary things. What is or isn't proven, what did or didn't happen, shouldn't really cloud the extraordinary work that so many people have done and the tremendous accomplishments that were realized over the decades."[150]

Zemsky quickly announced sweeping changes to the two SUNY economic development nonprofits, Fort Schuyler Management Corporation and Fuller Road Management Corporation. Board seats were added which were controlled by ESD, including a special nonvoting seat held by Zemsky himself. The boards of the nonprofits, which had operated in a relatively non-transparent manner, were subjected to more strict governance rules including new requirements with respect to state open meeting, freedom of information, and conflicts of interest laws. The bylaws of the nonprofits were rewritten to reflect these changes. The boards were required to establish an audit committee, hire a compliance officer, and create a whistleblower policy.[151] Robert Megna, the former state budget director, was assigned to oversee the finances of SUNY Poly's economic development projects across Upstate New York.[152]

The Fuller Road and Fort Schuyler nonprofits, under new management, moved to shake off the "Kaloyeros hangover." Megna, who became the new president of both organizations, said that they were trying to get "projects that either were on the back

---

[148] "State Unit Gets Key Revitalization Job," Albany, *The Times Union* (September 24, 2016).
[149] "Zemsky in the Driver's Seat for Buffalo Billion II," *The Buffalo News* (January 11, 2017).
[150] "Blame Grows, Creation Slows," Albany, *The Times Union* (February 19, 2017).
[151] "SUNY Poly Units Review," Albany, *The Times Union* (November 18, 2016).
[152] "SUNY Poly Spokesman Leaves Job," Albany, *The Times Union* (November 30, 2016).

## 9.3 The Scandal

burner or that were really only in development, finished.... [T]he state's made investments that we have to make the best of."[153] This entailed absorbing some losses. The Quad-C building in Marcy was leased on apparently concessional terms to Danfoss Silicon Power GmbH, thus becoming part of the silicon carbide consortium led by GE. This deal was originally concluded under Kaloyeros' tenure, and execution had been delayed in the wake of the scandal.[154] The announcement that Danfoss would create "at least" 300 high-tech jobs was seen as "a step in the right direction" by Utica Mayor Robert Palmieri, whose constituents had experienced years of uncertainty and disappointment with respect to attracting high-tech manufacturing.[155] A SUNY Poly lab in Canandaigua was sold at a loss to Akoustis, a North Carolina-based maker of smartphones, which planned to invest $20 million to create the "Smart Systems Technology Center," buoyed by $8 million in state tax credits.[156]

The state has assisted restructuring of some projects, such as the Advanced Pattern and Productivity Center (APPC), an advanced semiconductor lithography research effort involving GlobalFoundries, IBM, and Tokyo Electron. When the scandal broke, GlobalFoundries, which had already placed an order for a $120 million EUV lithography tool from ASML of the Netherlands, decided that it did not want to install the tool at SUNY Poly, effectively bringing the APPC collaboration to an end. However, GlobalFoundries continued discussions with ESD over how to undertake alternative projects that would fulfill the original APPC mission. As a result of these discussions, ESD offered GlobalFoundries a $7.5 million grant which covers about 8% of the cost of upgrading the ASML tool to a next-generation 7 nm version which will be installed in GlobalFoundries' Fab 8 in 2018 and used in the commercial production of semiconductors.[157]

In one of the first projects initiated after the scandal broke, SUNY Poly announced in February 2017 that it had won a $1.25 million US Commerce Department grant to create the Advanced Manufacturing Performance (AMP) Center, a high-tech training and R&D initiative. The AMP Center will partner with Edwards Vacuum, a semiconductor manufacturing equipment company, and with Infrcon, which makes equipment used in gas analysis and control. The AMP Center will work with Edwards Vacuum at SUNY Poly's Utica and Albany campuses to refine the company's vacuum and abatement systems used in semiconductor manufacturing.[158]

---

[153] "SUNY Poly Settles Low, Tries to Clear 'Kaloyeros Hangover,'" *Politico* (March 27, 2017).

[154] "Nano Utica Announcement Now Set for Friday," Albany, *The Times Union* (March 22, 2017).

[155] "Local Officials Excited, Relieved About Danfoss Announcement," Utica, *Observer-Dispatch* (March 25, 2017).

[156] The state invested $39 million in the Canandaigua lab and sold it to Akoustis for $2.75 million. Megna commented that "You could talk about the $39 million that was there and no jobs. Now we have a recovery of some of the facility and the promise of 200 jobs and $20 million in investment. We can look at some of these things as negatives or, take the view that it wasn't performing the way we wanted to and that now we have a private sector company." "SUNY Poly Settles Low, Tries to Clear 'Kaloyeros Hangover,'" *Politico* (March 27, 2017).

[157] "Scandal Roiled Chip Sector," Albany, *The Times Union* (May 31, 2017).

[158] "SUNY Poly Lands $1.75 Million in Grants," Albany, *The Times Union* (February 16, 2017).

### 9.3.4 New Leadership

In October 2017 incoming SUNY Chancellor Kristina Johnson appointed Megna to the post of Chief Operating Officer of SUNY's central administration.[159] In that capacity he continued to oversee the restructuring of the SUNY Research Foundations' corporations, Fuller Road and Fort Schuyler, until May 2018, when former GlobalFoundries CEO Doug Grose was chosen to head NY CREATES, a new entity which would absorb the two SUNY Poly real estate nonprofits, Fuller Road and Fort Schuyler, and manage SUNY Poly's research projects and economic development efforts. Johnson and Zemsky made clear that despite political pressure on the state to curb SUNY Poly's economic development activities, those efforts would continue.

> Johnson and Zemsky say these are incredible assets that have been created, especially locally. Johnson says that the Capital Region's own nanotech sector that is anchored by SUNY Poly and GlobalFoundries' Fab 8 chip factory is still evolving—and she believes there is a lot more to come since tech clusters generally take about 30 years to mature.[160]

Grose is an RPI graduate who worked most of his career for IBM in the Hudson Valley. He is tasked only with managing SUNY Poly's research and economic development activities, with the academic side to be run by another yet-to-be-determined individual. Megna commented that—

> There is no one happier in the State of New York than me that Doug has agreed to join us. There is no one more qualified than Doug. . . . The academic piece is still under discussion. And I'm sure the Chancellor will have some announcements to make in the near future. We expect that Doug and whoever comes in as interim president to SUNY Poly will work as partners to keep this moving forward.[161]

NY CREATES will be supervised by the SUNY Chancellor and will administer SUNY Poly's technology, administrative, legal, and human resource functions. It will interface with ESD and with the SUNY Research Foundation. NY CREATES will be created as a legal entity during 2019, and the old corporate nonprofit entities operating under the SUNY Research Foundation, such as Fuller Road and Fort Schuyler, will be dissolved and merged into NY CREATES.

### 9.3.5 New Initiatives

In 2019 SUNY Poly announced two major new initiatives in partnership with key semiconductor industry players. In November 2018 it was disclosed that semiconductor equipment maker Applied Materials would collaborate with New York State to establish an $880 million semiconductor manufacturing research center at SUNY

---

[159] "Budget Hand Megna Coming to SUNY Central," Albany, *The Times Union* (October 19, 2017).
[160] "Return to Trust in Tech School," Albany, *The Times Union* (May 6, 2018).
[161] "New Leadership at SUNY Poly," Albany, *The Times Union* (May 15, 2018).

Poly's NanoFabX, which previously housed the Global 450 consortium. The new center, the Materials Engineering Technology Accelerator (META Center) will open in 2019, featuring 24,000 square feet of cleanroom space. New York State will contribute $250 million to the 7-year project, with Applied Materials supplying the remainder. The project will reportedly create 400 new jobs, including individuals working for SUNY Poly, Applied Materials, and associated suppliers.[162]

Applied Materials' stated goal in opening the new center is to "speed customer availability of new chipmaking materials and process technologies that enable breakthroughs in semiconductor performance, power and cost."[163] The META Center will develop technology for not only applications in computer chips but artificial intelligence hardware and other devices that require large leaps in engineering, such as quantum computing (see Footnote 162).

Formation of the new consortium was facilitated by the fact that the top scientist at Applied Materials, Omkaram Nalamatsu, was an established figure in the Capital Region. Nalamatsu was formerly vice president for research at RPI, where he also taught materials science, engineering, and chemistry. He is also the president of Applied Ventures, his company's venture capital arm, which is planning to invest $20 million in Upstate New York startups.[164]

In 2019 New York State and IBM disclosed plans for a new collaboration to develop components for artificial intelligence and cloud computing systems. A $300 million artificial intelligence center will be created at SUNY Poly involving 326 new jobs over a 5-year period.

The Applied Materials and IBM collaborations will place SUNY Poly's research operations on a more solid financial footing, enabling the utilization of its semiconductor fabrication facilities for commercial activity. Those facilities cannot be shut down without damaging or ruining them, so the costs associated with continuous running must be incurred regardless. The Applied Materials and IBM deals will help to cover these costs.

## 9.4 The Saratoga Schism

While the high drama of Kaloyeros' indictment and its implications for the future of Tech Valley and the rest of upstate have occupied the attention of the media and the public, more prosaic challenges have arisen at the local level. The collaborative effort that saw the Capital Region's various jurisdictions working together in pursuit of a common objective—attracting a chip fab—has given way to a period of internecine controversy. Protracted local disputes have left the future of the Luther Forest Technology Campus uncertain and the original vision for that site unrealized.

---

[162] "$880M SUNY Poly Lab in the Works," Albany, *The Times Union* (November 16, 2018).
[163] Applied Materials, "New Applied Materials R&D Center to Help Customers Overcome Moore's Law Challenges," Press Release, November 15, 2018.
[164] "The Buzz," Albany, *The Times Union* (November 18, 2018).

As one frustrated economic development professional put it with respect to local jurisdictions, change came because at the local governmental level, "people retire. Nutjobs come in, trying to control."[165] Linda Hill, a former executive at National Grid who participated in the effort to attract the AMD fab to the region, recalled in 2015 that back then—

> We were rowers in a boat. We rowed in unison in an agreed direction. We had the need, the desire, the target. . . . Now, we don't row in the same direction. Communities don't know what they want, where they want to go. . . . The rowers in the boat don't have a clear target in mind. The coxswain is yelling but there's no teamwork.[166]

The drive to attract a semiconductor fab to the Capital Region was spearheaded by the Saratoga Economic Development Corporation and, in particular, its president Ken Green and founding member Jack Kelley—

> Green brilliantly and tactfully coaxed the complex project through town zoning approvals in Malta and Stillwater until final approval in 2004, and he was at the edge of the stage when Governor George Pataki announced a tentative deal for a chip plant in 2006, the deal that coaxed what was then AMD with an unprecedented $1.4 billion in state cash and future tax breaks.[167]

But Green resigned from SEDC in 2007 and subsequently left the region. Kelley resigned in 2008, moving on to a position in a local real estate firm.[168] Dennis Brobston, who had previously worked at SEDC for 10 years, succeeded Green as president in 2008, facing new circumstances.[169]

Tensions developed between Saratoga County officials and the SEDC under Brobston. In 2013 the county sought appointment to SEDC's board of an elected member of the County Board of Supervisors (the unelected County Administrator already sat on SEDC's board). SEDC, which characterizes itself as a private-sector entity, rejected the county's request for a role in governance, whereupon the county announced that it would not renew its annual $200,000 annual contract with SEDC.[170] The county complained of a lack of transparency at SEDC and criticized the lack of job creation "besides GlobalFoundries."[171] In 2014 the county created a

---

[165] Interview (October 2015).

[166] Interview with Linda Hill, Albany, New York (October 28, 2015).

[167] "Bruno, Others Did Hard Work for Nanotech," Schenectady, *The Daily Gazette* (May 12, 2012).

[168] "No Rush to Fill Leader Posts at Saratoga Business Unit," Albany, *The Times Union* (January 16, 2008); "Luther Forest Pioneer Leaving—Kelley was Founding member of SEDC," Albany, *The Times Union* (January 15, 2008).

[169] "Familiar Face at SEDC's Helm—Brobston, Who Spent 10 Years with Agency, Is Its New President," Albany, *The Times Union* (February 6, 2008).

[170] According to Brobston, SEDC refused to include a county official on its board because it would open SEDC's negotiations with businesses to the public. Such negotiations typically do not become serious until nondisclosure agreements are signed. Brobston said in 2016 that "We are a private organization, so we have no issue with confidentiality. They [the Prosperity Partnership] are required to report their minutes and everything online. FOIL doesn't apply to us, but it does apply to them." "New Program to Help Local Businesses," Glens Falls, *The Post-Star* (March 31, 2016).

[171] "SEDC Responds to Saratoga County's Decision Not to Renew Contract," Saratoga Springs, *The Saratogian* (May 2, 2013).

new economic development organization, the Saratoga County Prosperity Partnership, to take over SEDC's role.[172] As a result of this schism, Saratoga County emerged with two rival economic development organizations—a weakened SEDC and the largely untested Prosperity Partnership. The Schenectady *Daily Gazette*, commenting during the early days of the split, warned that—

> If the split does happen, efforts will be duplicated, wasting time and money. The County's economic development program will be fragmented and confused, rather than focused and clear, sending the wrong message to businesses. And the County's chances of winning state grants through Governor Cuomo's Regional Economic Development Council process could be hurt as well.[173]

In late 2015, Marty Vanags was designated president of the Prosperity Partnership and released a development plan for the county, "The Saratoga Strategy." The principal element of the plan was to use GlobalFoundries' presence in the county to generate more economic activity. The Prosperity Partnership "would work closely with the Luther Forest Technology Campus Economic Development Corporation to find suppliers that are interested in moving into the campus."[174] Concurrently, the SEDC under Brobston's leadership was "not going quietly into the night," launching a $3.5 million fundraising campaign in 2016 "that appears to send the message that the county will have to compete with SEDC well into the future for attracting new companies." "Neither Brobston, nor Marty Vanags ... have anything bad to say about one another, despite the messy split between the county and SEDC. . . ."[175] Each organization could point to some substantial initiatives.[176] However, as Brobston observed with respect to the split, "make no mistake, to people on the outside, they see a broken system. I have no doubt about that."[177]

In fact, competition between the two development organizations diverted time and resources away from the basic task of economic development. In 2015 the Prosperity Partnership sought an exclusive deal with the Saratoga County Industrial

---

[172] "Partnership Has a Home—But No Staff Yet," Schenectady, *The Daily Gazette* (January 10, 2015).

[173] "Out of Sync in Saratoga," Schenectady, *The Daily Gazette* (June 11, 2013).

[174] "Planting Seeds in Luther Forest," Albany, *The Times Union* (December 14, 2015).

[175] "SEDC Still Dealmaker After Split," Albany, *The Times Union* (January 24, 2016).

[176] The Prosperity Partnership is promoting the creation of the Next Wave Center, a physical location that would house semiconductor, supply chain, clean energy, and advanced manufacturing companies, provide education and training, and offer business support services. An advisory council was established to oversee the project in 2017, chaired by Gary Patton, chief technology officer of GlobalFoundries. SEDC, supported by National Grid, is coordinating a strategic study of energy use and needs in the Capital Region and claims credit for attracting new companies and supporting growth by established local companies. "Advisory Council to Help Area's Tech Industry," Saratoga Springs, *The Saratogian* (June 11, 2017); "SEDC Reports on its Impact," Albany, *The Times Union* (July 22, 2016); "Tax Deals Eyed for Kitware HQ," Albany, *The Times Union* (May 3, 2017); "Economic Developers Push for Technology Work Space" Schenectady, *The Times Gazette* (June 8, 2017).

[177] "Billions at Stake in Tech Arms Race as Luther Forest Flounders," *Albany Business Review* (October 18, 2013).

Development Agency (IDA), which would compel the IDA to sever its 36-year relationship with SEDC. Organizations that successfully bring projects to the IDA are awarded as much as $50,000 for a major project, so the financial stakes were substantial for both development organizations. IDA board members initially opposed the Prosperity Partnership's proposal, but by the spring of 2015 two of the IDA's seven board members had been replaced by the County Board of Supervisors.[178] In March 2016 the IDA board met and again raised concerns about cutting ties with SEDC. IDA Chairman Rod Sutton complained that competition between the two development organizations was "creating uncertainty" and that "it is time to let economic development take its course. You have a great team, but we're stagnant right now. Nobody knows where we're going." A member of the County Board of Supervisors effectively concurred, saying—

> *My frustration when I got on the [county board's] Economic Development Committee was who does what, and who are there? What it boils down to is, who is responsible for economic development in the county?*[179]

In July 2016 the IDA concluded agreements with SEDC and the Prosperity Partnership that the IDA would work with "both the county's rival development organizations ... following months of negotiations." Brobston, speaking for SEDC, expressed satisfaction with the arrangements and commented that "we just need to get back to business. We wasted months on this discussion."[180]

## 9.5 Continuing Financial Problems at Luther Forest

One very tangible consequence of Joseph Bruno's retirement has been protracted financial travail at the Luther Forest Technology Campus Economic Development Corporation. During his tenure Bruno steered state funds to LFTCEDC to support infrastructure projects, regulatory applications, and consultants' studies and other investments. LFTCEDC also secured loans from the state on the basis of Bruno's verbal assurance that at some point the loans would be converted to grants, e.g., forgiven. (See section on "Context" in Chap. 5.)[181] Unfortunately this commitment

---

[178] "IDA Sets Meeting to Sort Out SEDC, Prosperity Ties," Schenectady, *The Daily Gazette* (March 15, 2016).

[179] "Economic Development Groups Make Case to IDA—One Organization Wants to be Exclusive Marketer," Schenectady, *The Daily Gazette* (March 24, 2016).

[180] "Deal Reached With Economic Development Groups—IDA Says It Will Work with Both Agencies," Schenectady, *The Daily Gazette* (July 20, 2016).

[181] The Schenectady *Daily Gazette* observed in 2014 that "if George Pataki were still in the Executive Mansion and Joe Bruno in the Senate majority leader's seat instead of legal jeopardy, there wouldn't be a problem [at Luther Forest]. The two would have simply funded money to the park and technology companies that would occupy it, as they did with $100 million to initially develop the park and $1.2 billion in grants and tax credits to GlobalFoundries. But under three Democratic governors since 2007, the park has been left to fend for itself. And it has not done well." See "One Troubled Tech Park in Malta," Schenectady, *The Daily Gazette* (April 6, 2014).

## 9.5 Continuing Financial Problems at Luther Forest

was not committed to paper.[182] When Bruno retired, the flow of state funds ended. Instead of forgiving loans extended by the state, in 2010 the ESD, which held the mortgage on the Luther Forest property, threatened to foreclose.[183] The Schenectady *Daily Gazette* observed in 2012 that—

> The state has sunk better than $100 million into the Luther Forest Technology Campus, and without that outlay, GlobalFoundries wouldn't have built its $7 billion chip plant. But all indications are that without Joe Bruno on the tractor, that's all the state fertilizer the tech campus will get. The campus still needs big bucks for water and sewer extensions, and doesn't even have entrance signs for what's supposed to be a world-class technology center, but state officials are putting their cash elsewhere these days.[184]

The problem confronting LFTCEDC was the fact that apart from GlobalFoundries, no other companies were located in the Luther Forest campus, so the revenue anticipated from the sale or lease of sites did not materialize. By early 2012 more than a dozen companies that provided support for GlobalFoundries had established a physical presence in Saratoga County, but all of them located outside the campus.

### 9.5.1 The Tax Incentive Conundrum

When AMD concluded its deal with the state in 2006, it qualified for Empire Zone incentives which provided a state income tax credit for the company's entire local property tax bill. The Empire Zone program was phased out in 2010, and although GlobalFoundries was grandfathered and continued to receive those benefits, new companies could not. Although newcomers could request property tax abatements pursuant to payment-in-lieu-of-taxes agreements (PILOT) via local industrial development agencies, LFTCEDC's town approvals for the Luther Forest site prohibit such agreements, reflecting an assumption at the time those agreements were concluded that the Empire Zone benefits would continue throughout the campus.[185] The

---

[182] Interview with Dennis Brobston, president, Saratoga Economic Development Corporation. Saratoga Springs, New York (October 28, 2015).

[183] "State Takeover of Luther Forest Technology Campus Raising Concern," Saratoga Springs, *The Saratogian* (October 27, 2010); "Saratoga County Eyeing Luther Forest Campus Takeover," Saratoga Springs, *The Saratogian* (December 15, 2010). The prospect of an immediate state takeover receded when ESD head Dennis Mullen, who made the threat to the Luther Forest management in October 2010, was replaced by Kenneth Adams in January 2011, following the election of Governor Andrew Cuomo. "Head of Business Council Named to Lead State Economic Development Group," Glens Falls, *The Post-Star* (January 27, 2011).

[184] "Utica Getting the High Tech Grants Now," Schenectady, *The Daily Gazette* (December 1, 2012).

[185] "[I]ncentives—generally offered through an industrial development agency—are common throughout New York state and in much of the rest of the country. The town wrote a prohibition on using them in the tech park in 2004. At the time, it was thought to mean little since the state's Empire Zone program was giving property tax credits instead. But the program that offered those incentives has since ended, which economic development officials say has again made local incentives important." See "Officials Propose Takeover of Luther Forest Tech Park," Schenectady, *The Daily Gazette* (July 18, 2013).

bizarre result was that the tech campus was the only place in the entire State of New York where local laws prevented PILOT agreements.[186] SEDC's Brobston observed in 2012 that "there have been projects for the LFTC that we have talked about that have been deterred by lack of a PILOT." Mike Relyea, former president of LFTCEDC, said that—

> Marketing-wise, it is very difficult to enter into a conversation. The first thing they ask us is what our incentives are. When we can't offer incentives that ends the conversation pretty fast.[187]

In addition to the tax issue, LFTCEDC faced the problem of the zoning of the campus, which when approved in 2004 allowed establishment of a nanotechnology park, limited to nanotechnology and semiconductor support businesses. Relyea said that rezoning the land to allow a broader range of businesses could enable the campus to bid for "other potentially lucrative high-tech industries like photovoltaic energy."[188]

In March 2012 Malta Town Councilman John Hartzell introduced a resolution to the effect that the town would consider offering tax breaks to companies that located in the Luther Forest Technology Campus. Town Supervisor Paul Sausville said that he was not opposed to tax incentives per se but that they should be provided by the state or county, not the town. Sausville also said that there was "no reason to dismantle the current zoning."[189] In 2013 Sausville said Malta could not afford to offer tax abatement incentives for the tech park—

> It would be hard for me to go to the mom and pop residents and say "you don't have any exemptions, but we're going to incentivize more growth in the campus by giving them property and school tax exemptions."

## 9.5.2 Deepening Financial Woes

Former LFTCEDC President Relyea said frankly in 2012 that "we don't have a viable business model." The corporation had borrowed $9 million from the state to finance the original land purchase but had sold only one parcel that being to GlobalFoundries for $7 million. All of the GlobalFoundries money was used to improve infrastructure within the campus, and almost none of it was used to pay down the $9 million mortgage. In the meantime the corporation continued to pay property taxes on the remaining parcels and to finance its own increasingly

---

[186] "Changes would Help Luther Forest Realize Its Potential," Schenectady, *The Daily Gazette* (September 18, 2015).

[187] "Property Tax Breaks Urged for Tech Park," Schenectady, *The Daily Gazette* (March 12, 2012).

[188] "Luther Forest Tech Campus Official Says Zoning Change Necessary in Order to Market Land to Tech Firms," Saratoga Springs, *The Saratogian* (August 7, 2012).

[189] "Malta Councilman to Town Board: Consider Tax Breaks for Companies Interested in Luther Forest Tech Park," Saratoga Springs, *The Saratogian* (March 26, 2012).

cost-strapped operations (see Footnote 188). In 2013 the *Albany Business Review* summarized the dilemma facing Luther Forest in the following way—

> [T]he plight of the campus signals that the region is unprepared to fend for itself in the global competition for projects, much less the in-state battle developing at the direction of Gov. Andrew Cuomo.... No one works at Luther Forest any more. The board of directors cut all staff this year to conserve cash. And the campus owes $800,000 in road fees to its host town, Malta ... [w]ith Luther Forest functionally bankrupt, GlobalFoundries is on the hook for traffic studies and extending gas lines. So in addition to making computer chips, GlobalFoundries must pave its own roads and build its own water tower.[190]

Town Supervisor Sausville observed in 2012 that the town had agreed to build 5.5 miles of roads within the tech campus with the understanding that LFTCEDC would help finance their maintenance acknowledging that the corporation had paid in full 4 years previously but commenting that it had "come up short in 2010" and "this year and last year they said they couldn't pay at all" (see Footnote 188). By the spring of 2013 LFTCEDC owed the Town of Malta nearly $800,000 in overdue payments for road maintenance within the campus. The roughly $400,000 per-year payments were to be used for maintenance and snowplowing, but roughly half was required to be paid into a trust for future repair or maintenance. Relyea argued that the payments were too high, with more than half going to future costs, and observed that the town maintained 79 miles of road outside the campus for about $28,000 a mile but wanted to be compensated at $73,000 a mile for the roads inside the campus declaring "$400,000 a year is not a workable number." Some town board members were reportedly considering legal action against the corporation over the late payments. Sausville complained that "GlobalFoundries is looking out for GlobalFoundries. The LFTC is looking out for itself. The county is looking out for itself. Nobody is looking out for the town of Malta and its taxpayers."[191]

The rift between Saratoga County and SEDC "complicated" the prospects for attracting additional tenants to Luther Forest, leaving no local authority to serve as clearly designated "point" organization for recruiting companies. *The Saratogian* observed in 2015 that—

> The Albany-based Center for Economic Growth and Empire State Development are involved in the technology campus. But no one is clearly in charge, out front selling Luther Forest to prospective clients who are also considering sites from Hong Kong to the Rhine Valley, the way former SEDC president and vice president Ken Green and Jack Kelley landed GlobalFoundries.[192]

Pressure grew on Malta to change its stance. In 2013 Saratoga County offered to take ownership of the road system inside the tech park, but in return it wanted Malta and Stillwater to offer potential investors in the park sales, school, and property tax exemptions. The county hired a consultant, TIP Strategies, who agreed that tax

---

[190] "Billions at Stake in Tech Arms Race as Luther Forest Flounders," *Albany Business Review* (October 18, 2013).
[191] "Town Looking for Luther Forest Fees," Schenectady, *The Daily Gazette* (May 16, 2013).
[192] "Luther Forest Seeks Tax Breaks," Saratoga Springs, *The Saratogian* (March 7, 2015).

incentives—offered nearly everywhere else—were needed to attract new companies to the tech park.[193] TIP also recommended establishment of a full-time property manager for the campus "to keep tenants happy."[194] County Board of Supervisors Chairman Alan Grattidge said of incentive packages "it's the reality of New York state."[195] In November 2013 Congressman Paul D. Tonko convened an ad hoc group of regional leaders to discuss the tech park's problems, noting that "the park's problems are being followed in Washington."[196] However, "2 years of inconclusive discussions" of the tech park's problems by the town board and stakeholders followed.[197]

### 9.5.3 Zoning Breakthrough

In September 2014 GlobalFoundries and LFTCEDC filed petitions with the Town of Malta seeking zoning changes to enable the provision of local tax incentives and to reduce zoning regulatory requirements.[198] In October 2015, after a year of deliberations and hearings, the Malta Town Board unanimously approved a series of changes in the zoning ordinances sought by LFTCEDC, to take effect January 1, 2016. The tech campus was cleared to offer PILOT agreements to new tenants, the ordinances were relaxed to allow a broader range of businesses in the park, and an agreement was reached providing for transfer of ownership of the roads in the tech park from the town to Saratoga County. The arrearages in road maintenance payments by LFTCEDC to the town, which had grown to $1.5 million, were to be settled by a payment of $362,000.[199]

Following the zoning revisions, "Luther Forest Tech Campus officials … said they're optimistic about attracting new business to the campus."[200] The tech park got an added boost in March 2016 when the Saratoga County IDA agreed to buy 19 acres in Luther Forest for $743,000, which would "pump new money into the

---

[193] TIP Strategies, *Economic Development Strategic Plan Prepared for Saratoga County, New York* (March 2014), <http://www.saratogacountyny.gov/wp/wp-content/uploads/2013/11/Saratoga-Plan-FINAL.pdf>.

[194] "Consultant: Changes needed at Tech Park," Schenectady, *The Daily Gazette* (March 22, 2014).

[195] "Saratoga County Offers to Maintain Luther Forest Technology Campus Roads in Return for Tax Breaks," Troy, *The Record* (July 20, 2013).

[196] "Talks Focus on Offering Incentives in Luther Forest," Schenectady, *The Daily Gazette* (November 23, 2013).

[197] "Luther Forest Tax Breaks Get Nod," Schenectady, *The Daily Gazette* (April 22, 2015).

[198] "New Zoning sought for Tech Park," Schenectady, *The Daily Gazette* (September 20, 2014).

[199] "Tech Park Zoning Changes Get OK," Schenectady, *The Daily Gazette* (October 7, 2015); "Luther Forest Agrees to Pay Town $362 K for Road Work," Schenectady, *The Daily Gazette* (November 4, 2015).

[200] "Luther Forest Agrees to Pay Town $362 K for Road Work," Schenectady, *The Daily Gazette* (November 4, 2015).

struggling LFTC Economic Development Corporation."[201] However, a period of 5 full years had elapsed since the termination of the Empire Zone program had thrown the tech campus into financial distress and the apparent resolution of the problem through an accord with the town. As a local economic development professional noted—

> *The Town of Malta had a "no incentives" policy inside LFTC. They just reversed it in the last five weeks. So supply chain companies went elsewhere. . . . They now realize nothing's going into the park. They lost 5-6 years.*[202]

## 9.6 Fostering Startups

The Capital Region has grown high-tech manufacturing by recruiting well-established technology-intensive enterprises from outside the region. But as the Albany *Times Union* observed in 2013, unlike startups, such incumbent firms "are not the hyper wealth-creation engines that drive places like Silicon Valley and create millionaires who turn around and sink their earnings back into the economy." It pointed out that from April to June 2013, over 300 early-stage companies in Silicon Valley had raised $2.7 billion in venture capital, while in the same 3-month period, in all of Upstate New York, startups had raised only $9.7 million in venture funding, and only one of these deals was in the Capital Region, an employment-recruitment firm valued at $0.5 million.[203]

But in the 5 years since *The Times Union* made these observations, startup activity in the Capital Region has begun to pick up, including the launch of some promising nanotechnology companies, and within the region, "investment capital is getting easier to attract."[204] A 2017 Brookings report found that the Capital Region is "among the best places in the country for clean energy companies," citing factors such as the presence of GE and the growth of SUNY Poly.[205] Other regions have leveraged the recruitment of incumbent tech firms to foster an environment where innovative startups flourish and the early stages of the same phenomenon are observable in Tech Valley.

---

[201] "Country IDA Plans to Buy 19 Acres in Tech park—Purchase could give Luther Forest Technology Campus a Boost," Schenectady, *The Daily Gazette* (March 15, 2016).

[202] Interview with Brian McMahon, Executive Director, New York State Economic Development Council, Albany, New York (October 28, 2015).

[203] "Finding Funding to Stay Hot," Albany, *The Times Union* (October 2013).

[204] "Area Advantage," Albany, *The Times Union* (March 27, 2016); "Upstate Sees 67 Deals in Quarter," Albany, *The Times Union* (September 9, 2016); "Business Unicorns Are Not So Unusual," Schenectady, *The Daily Gazette* (February 1, 2018); "Startups Show Evolution of Capital Region's Tech Economy," Albany, *The Times Union* (November 16, 2016).

[205] "High Hopes for High Tech," Albany, *The Times Union* (November 20, 2016); "Clean Tech Cluster to Get Boost," Albany, *The Times Union* (October 23, 2016); "Area Tops in Clean Energy," Albany, *The Times Union* (June 2, 2017).

### 9.6.1 Traditional Sources of Regional Disadvantage

Stuart W. Leslie's pessimistic 2001 assessment of the Capital Region's potential to become a new Silicon Valley concluded that RPI's George Low, "had the right plan in the wrong place at the wrong time" and that for all of his own and the state government's success at refashioning RPI, "they could not overcome the regional disadvantage that kept them from competing effectively with emerging high-technology centers in other parts of the country."[206] Leslie noted that Low had sought to foster "home grown indigenous companies" in Troy just as Stanford had done in Silicon Valley and that the RPI business incubator had in fact proven surprisingly successful at spawning dynamic startups launched by RPI graduates. But as Leslie pointed out, the best of these new companies eventually relocated outside of the region—a phenomenon which continued long after Low's death in 1984.[207] Pradeep Haldar, CNSE's former vice president of entrepreneurship innovation, observed in 2016 that the main challenge facing the RPI and other regional business incubators was that they could not retain their own startups in the region: "companies move to California, Boston, companies are bought up."[208]

There is no simple cause or remedy for the problem identified by Leslie. In a 2015 interview, Craig Skivington, a Saratoga Springs-based entrepreneur, argued that the Albany area's "conservative" mindset prevented it from developing into a "hotbed for startups" and that "[E]ntrepreneurship is not what this area is focused on." He contrasted what he saw as the risk-aversion which characterized the business culture in the Capital Region with that of Silicon Valley and Austin where "there's not a stigma if someone starts a company and it doesn't make it. Here, people tend not to take the big swings."[209] A perception has long existed that "there isn't enough venture capital activity in the region to sustain an innovation economy."[210] The Capital Region "may also suffer from its proximity to much larger, perhaps more exciting metropolitan areas-- New York City and Boston"—which are centers of venture capital investment.[211] In 2015 the CEO of a Rochester-based venture fund, Theresa Mazzullo of Excell Partners, summarized why "Upstate New York startups struggle with early investment:"

---

[206] Stuart W. Leslie, "Regional Disadvantage: Replicating Silicon Valley in New York's Capital Region," *Technology and Culture* (2001), p. 237.

[207] One of the first and most successful tenants in the RPI incubator was Raster Technologies, founded by two RPI graduates in 1981 to develop technology for color graphics—a firm Low not-so-secretly hoped would become "the Hewlett-Packard of RPI." Instead the company "outgrew the incubator so fast that RPI could not hold onto it." Raster Technologies moved to Boston's Route 128, which Low characterized as "a great disappointment to us." Stuart W. Leslie, "Regional Disadvantage: Replicating Silicon Valley in New York's Capital Region," *Technology and Culture* (2001), p. 257.

[208] Interview, Albany, New York (November 30, 2016).

[209] "Tech Executive Are Why the Albany Region Does Not Need to be an Innovation Hub," *Albany Business Review* (February 24, 2015).

[210] "Focus on the Region's Potential," Albany, *The Times Union* (December 2, 2014).

[211] "There, But Not Quite There," Albany, *The Times Union* (September 7, 2014).

> *The struggles are in two particular areas, finding management talent to wrap around the tech so you have a company that's just emerging and they need a very strong CEO or COO. Talent that can help a company come to the next valuation in increasing milestones. We struggle with that because upstate's ecosystem has not been robust enough with serial entrepreneurs. The second one is capital, to meet the demand and help these companies in the later stages.*[212]

But a principal source of the Capital Region's "regional disadvantage" identified by Leslie in 2001—arguably correctly—was the speed at which Upstate New York's technology-based manufacturing industrial base unraveled in the 1980s and 1990s as mainstays like GE, IBM, and Kodak disinvested and moved research and manufacturing functions elsewhere. "Without a strong regional industrial base to capture and hold the innovations being generated by [RPI's Center for Industrial Innovation] and its other steeples of excellence, RPI ended up exporting its best ideas and best graduates to other places, including Silicon Valley itself."[213]

## 9.6.2 Perspectives from Other Regions

The Capital Region's high-tech entrepreneurial potential suffers from comparisons with Silicon Valley and Boston, where great research universities have fostered innovative new companies for over a century.[214] More apt comparisons are with North Carolina's Research Triangle Park and Austin, Texas, which evolved as centers of tech entrepreneurship through a drawn out two-step process which began with recruitment of established technology firms and was followed by a proliferation of startup activity. A 2015 study funded by the Kauffman Foundation, which promotes entrepreneurship, contrasted Silicon Valley and Boston with the more

---

[212] "Why Upstate New York's Startups Struggle with Early Investment," *Albany Business Review* (March 16, 2015).

[213] Stuart W. Leslie, "Regional Disadvantage: Replicating Silicon Valley in New York's Capital Region," *Technology and Culture* (2001), pp. 236–238.

[214] Since MIT's founding in 1861 it has encouraged "even (rather uniquely) faculty entrepreneurship since before the beginning of the 20th Century." Edward B. Roberts and Charles E. Eesley, *Entrepreneurial Impact: The Role of MIT* (Hanover, MA: Now Publishers Inc., 2011), p. 6. MIT and Stanford "were both committed to an endogenous strategy of encouraging firm formation from academic knowledge." Stanford, founded in 1891, looked to MIT as its model as an incubator of new firms. Stanford's founders believed that it could not achieve greatness unless it was surrounded by technology-based industries, which, because they did not then exist in California, would need to be created. In 1900, California depended on the East Coast for electrical equipment, so its president and a number of faculty members invested in new electrical firms being launched by Stanford graduates, an early successful example of which was Federal Telegraph (1909) which made significant contributions to the early development of radio communications. Similar Stanford-based important startups included Litton Engineering (1932), Hewlett-Packard (1937), and Varian (late 1930s). Henry Etzkowitz, "Silicon Valley: The Sustainability of an Innovative Region," *Social Science Information Journal* 52(4), 515–538. (2013); Timothy Sturgeon, "How Silicon Valley Came to Be," in Martin Kenney (Ed.), *Understanding Silicon Valley: The Anatomy of an Entrepreneurial Region* (Stanford, CA: Stanford University Press, 2000).

recent examples of the Research Triangle and Austin, which began to recruit high-tech manufacturing in the 1950s and achieved national prominence in the 1970s:

> *These regional entrepreneurial ecosystems [Austin and Research Triangle] have mainly benefited from the spawning of startup founders in both regions as large corporations have relocated here. . . . Our interviews, site visits, and data support conclusions from the entrepreneurial literature . . . that incumbent firms are a crucial source of entrepreneurial founders in both regions, and in some technology sectors, more so than regional universities, government facilities, or other anchor organizations.*[215]

Austin and Research Triangle involved de novo creation of a tech manufacturing base, whereas such a base originally existed in Upstate New York with the presence of such firms as GE, Kodak, and IBM—but as Leslie noted, by the 1990s New York's tech manufacturing base was declining so dramatically that the challenges presented resembled those facing regions where such manufacturing was not present.

In the mid-1950s Austin had one large home-grown technology-intensive manufacturer, the defense electronics contractor Tracor. Industrial recruitment efforts eventually attracted research and manufacturing firms like IBM (1967), Texas Instruments (1969), Motorola (1974), AMD (1979), Tandem Computers (1980), and Data General Corporation (1980). Startups followed, but not instantaneously. The first Austin-based tech startups which would become major companies included CompuAdd (1982) and Dell Computers (1984)—companies which were launched over 15 years after the region's first major recruitment successes and which did not achieve large scale for a number of years after startup.[216] North Carolina's Research Triangle began to flourish in its first decade by attracting large technology companies from outside the region, but startups eventually followed and, over time, snowballed, ultimately bringing more jobs to the region than the big companies still based in the park.[217] The thriving life sciences industry in and around the Research Triangle . . .

---

[215] Elsie Echeverri-Carroll, Maryann Feldman, David Gibson, Nichola Lowe and Michael Oden, *A Tale of Two Innovative Entrepreneurial Regions: The Research Triangle and Austin* (University of Texas at Austin and University of North Carolina at Chapel Hills, March 15, 2015), pp. 7–8.

[216] Elsie Echeverri-Carroll, Maryann Feldman, David Gibson, Nichola Lowe and Michael Oden, *A Tale of Two Innovative Entrepreneurial Regions: The Research Triangle and Austin* (University of Texas at Austin and University of North Carolina at Chapel Hills, March 15, 2015), p. 56

[217] The Research Triangle was launched in 1959 with an emphasis on recruitment of existing companies. George Simpson, a faculty member of University of North Carolina at Chapel Hill and director of the Research Triangle Institute, visited almost 200 companies in 1958–1959, finding significant interest in pharmaceutical, electronics, and chemistry firms interested in a "supply of graduates to staff future research projects." The first big tech tenant in the park was IBM in 1965, which over time brought in about 40 IBM organizations and which, 40 years later, in 2005 was still the park's largest employer, with about 11,000 workers. Fred M. Park, "Turning Poor Dirt into Pay Dirt," *METRO Magazine* <http://www.metronc.com/article/?id+421>; National Research Council, Charles W. Wessner (ed.) *Best Practices in State and Regional Innovation Initiatives: Competing in the 21st Century,* (Washington DC: The National Academics Press, 2013), pp. 231–240.

*is a story of attracting large multinational firms to locate their R&D operations and then encouraging startup firm formation in the wake of large-scale corporate mergers and acquisitions, layoffs, and restructuring.*[218]

Viewed against the background of Research Triangle and Austin, Tech Valley appears to be in the early stages of a similar trajectory in the wake of a highly successful industrial recruitment effort, which has augmented existing tech firms like GE and established a foundation for high-tech startups. Austin and Research Triangle were able to build on their newly-grown high-tech manufacturing bases to leverage the emergence of startups through strong university research programs and a variety of intermediate support organizations—perhaps most notably the North Carolina Biotechnology Center.[219] Similarly, in the Capital Region educational institutions and intermediate organizations are beginning to foster innovative technology-oriented startups.

### 9.6.3 CNSE Entrepreneurial Initiatives in Nanotechnology

The Capital Region is experiencing an uptick in nanotech-based startup activity, albeit starting from a miniscule base.[220] (Table 9.1 lists some of the nanotech startups in the Capital Region.) CNSE is helping to drive this process through a number of alliances and affiliated organizations. The Albany *Times Union* observed in July 2017 that—

> *With little fanfare SUNY Poly slowly but surely... [is] becoming a hot spot for clean energy and biotech startups that could become the Teslas of the future. You may not have heard much about them since they aren't publicly traded and have just a handful of employees— companies such as BessTech and Glauconix that have received a fair amount of publicity. These companies are being established by students and former students of Pradeep Haldar, SUNY Poly's Vice President of Entrepreneurship Innovation and Clean Energy. Haldar supports these companies through the SUNY Poly Advanced Research and Commercialization Initiative (SPARC), which works in partnership with two other initiatives affiliated with SUNY Poly, iCLEAN and the Tech Valley Business Incubator.*[221]

---

[218] Elsie Echeverri-Carroll, Maryann Feldman, David Gibson, Nichola Lowe, and Michael Oden, *A Tale of Two Innovative Entrepreneurial Regions: The Research Triangle and Austin* (University of Texas at Austin and University of North Carolina at Chapel Hills, March 15, 2015), p. 8. See also Nichola Lowe, "Beyond the Deal: Using Industrial Recruitment as a Strategic Tool for Manufacturing Development," *Economic Development Quarterly* 28(4) (2014).

[219] Elsie Echeverri-Carroll, Maryann Feldman, David Gibson, Nichola Lowe, and Michael Oden, *A Tale of Two Innovative Entrepreneurial Regions: The Research Triangle and Austin* (University of Texas at Austin and University of North Carolina at Chapel Hills, March 15, 2015), pp. 12–13 and 16.

[220] "Tech Startups Seek the Good Life in Saratoga Springs," *Politico* (December 2, 2014).

[221] "SUNY Poly Startups Created by Students Are Set to Launch," Albany, *The Times Union* (July 1, 2017).

Table 9.1 Nanotechnology startups in the Capital Region

| Company | Year founded | Technology focus | Academic origin | Location |
|---|---|---|---|---|
| BessTech | 2010 | Silicon nanostructures for lithium-ion batteries | CNSE | Albany |
| ThermoAura | 2011 | Thermoelectric nanocrystals | RPI | Colonie |
| Glauconix | 2014 | Nanostructures duplicating human eye tissue | CNSE | Albany |
| PBC Tech | 2008 | Nanomaterials with applications as batteries, supercapacitors | RPI | Troy |
| HocusLocus | 2008 | Nanobiotechnology for controlled expression of proteins in specific cells | CNSE | Albany |
| Eonix | 2013 | Ionic liquid electrolytes for ultracapacitors | CNSE | Albany |
| Lux Semiconductors | | Lightweight flexible solar cells` | CNSE | Albany |

#### 9.6.3.1 Teaching Nanotech Entrepreneurship

SUNY Poly has built entrepreneurialism into its nanotechnology curriculum. Professor Laura Schultz, a teacher of "nano economics," is responsible for creating an entrepreneurial environment at the NanoCollege that will foster the transfer of technologies into the commercial realm. A 2013 profile in the Albany *Times Union* characterized her as "in essence, a midwife who helps students deliver their nano research from lab to market." She said that the school wanted students "always to be thinking about how to create viable businesses out of the new technologies we are researching."[222] To date she has mentored over 30 technical teams at the school in exploring the commercial potential of their new technologies.[223] In addition, CNSE has fostered the creation of institutions for commercializing nanotechnology:

- **SUNY Poly Advancing Research and Commercialization (SPARC)**. SPARC is SUNY Poly's umbrella organization promoting entrepreneurship and innovation. It operates through two SUNY Poly affiliated incubators, iCLEAN and the Tech Valley Business Incubator. SPARC provides incubation and commercialization assistance to entrepreneurs including access to world-class nanotechnology equipment, prototyping, mentoring, and training. SPARC assists startups in raising funding and the development of strategic partnerships.
- *iClean*. In 2010 the New York State Energy Research and Development Authority (NYSERDA) awarded $1.5 million to CNSE to establish an onsite business incubator for startups utilizing nanotechnology in the field of clean energy—Incubators Collaborating and Leveraging Energy and Nanotechnology

---

[222] "Moving Nano Research from Lab to Market," Albany, *The Times Union* (June 25, 2013).
[223] SUNY Polytechnic Institute, Faculty Profile: Laura Schultz, <https://sunypoly.edu/faculty-and-staff/laura-schultz.html>.

("iCLEAN")—run in collaboration with the Hudson Valley Center for Innovation in Kingston, New York.[224] CNSE also raised $1.5 million in private funding from its industrial tenants, to be made available in the form of use of clean room equipment and other services.[225] By the end of 2010 about a half dozen companies were participating in iClean, with 25 local business executives volunteering to serve as mentors.[226] In 2014 the state renewed its 2010 award to iClean with another $1.5 million in funding, noting that iClean had graduated four startup companies, had worked with over 27 companies, and supported the creation and retention of 120 jobs.[227]

### 9.6.3.2 BessTech

BessTech, formed in 2010 as BESS (Battery Energy Storage Systems) Technologies, was the first nanotech-based startup to emerge from the NanoCollege. It was launched in 2010 by a group of Haldar's graduate students following their winning of the annual Tech Valley Business Plan Competition. The new company's focus was on using silicon nanostructures to increase the storage power of lithium-ion batteries, enhancing the quality of portable power storage systems. In 2012 the company, headed by entrepreneur Fernando Gomez-Baquero, struck a licensing deal with CNSE to take technologies developed in CNSE labs to market, boosted by access to the NanoCollege's labs and manufacturing facilities. CNSE would receive a share of the new company's future revenues.[228] Gomez-Baquero and Chief Technology Officer Isaac Lund characterized CNSE as a "powerhouse" on their side, including "being able to use all the great equipment" at the NanoCollege.[229]

In 2015 an AMD veteran and the former CEO of GlobalFoundries, Doug Grose, joined BessTech as chief technology officer, charged with bringing the Company's technology to market.[230] The same year BessTech's battery technology was awarded

---

[224] The Hudson Valley Center for Innovation is an incubator established in 2005 promoting economic development in the Hudson Valley. "High tech on the Hudson: Digital Dynamos Have Turned Kingston into Brooklyn North," *New York Daily News* (December 17, 2010).
[225] "Cleaning Up at Albany," Albany. *The Times Union* (March 19, 2010).
[226] "Hatching Clean-Tech Plans," Albany. *The Times Union* (December 8, 2010).
[227] SUNY Polytechnic Institute, "Energy-Focused Incubator at SUNY Polytechnic Institute and the Hudson Valley Center for Innovation," Press Release (October 10, 2014).
[228] "Whatever Happened To?" Albany, *The Times Union* (September 30, 2012).
[229] Haldar, who also serves as vice president for CNSE's clean energy program, pointed out that his institution had helped BESS raise funds via grants from NSF and NYSERDA, and the physical resources at CNSE obviated the need to raise "millions and millions" of dollars to demonstrate proof of concept. CNSE assets available to BESS included furnaces, deposition equipment, and measurement tools, as well as mentoring and investment, legal, and insurance contacts. "Company Seeking to Improve Battery Technology," Schenectady, *The Daily Gazette* (February 17, 2013).
[230] "Ex-GloFo CEO Joins Battery Startup," Schenectady. *The Daily Gazette* (July 15, 2015).

a patent in Japan, which was seen as providing "BessTech with a potential advantage in the Japanese market."²³¹ In 2015 BessTech became one of the handful of local startups to receive an investment from the regional angel fund Eastern New York Angels (ENYA) a commitment of $250,000.²³²

#### 9.6.3.3 HocusLocus

HocusLocus is commercializing a trans-RNA switching mechanism (SxRNA) using microRNA expression profiles to target and control expression of selected proteins in cells of specific tissues, disease states, and developmental stages. Developed by CNSE Professor Scott Tenenbaum, the technology is expected to have applications with respect to vaccines, therapeutics, molecular tools, and medical imaging. The company was launched in a partnership between Tenenbaum and local entrepreneur Ted Eveleth, who became aware of the technology at a SUNY Albany pre-seed workshop.²³³

HocusLocus was founded in partnership with CNSE and benefited from access to the NanoCollege's "unmatched nanobioscience-focused laboratories." It received initial funding from the SUNY Technology Accelerator Fund, the University of Buffalo Center for Advanced Biomedical and Bioengineering Technology and the regional angel fund ENYA. In 2015 it received over $1 million in grants from the NSF and NIH.²³⁴

#### 9.6.3.4 Glauconix

Glauconix originated as a team of three students from CNSE who won the top $100,000 prize at the New York State Business Plan Competition in 2014, featuring a new technology to create realistic eye tissue for drug screening to prevent and treat glaucoma.²³⁵ Soon afterward, the students, led by Dr. Karen Torrejon, launched Glauconix Biosciences, based at SUNY Poly in Albany. In 2017, Glauconix was awarded $750,000 from the National Science Foundation, a strong validation of the potential of the new company's technology, which the NSF grant would help to

---

[231] "Japan Patent Electrifies Lithium Battery Startup," Albany, *The Times Union* (June 24, 2015).

[232] "BessTech Draws in Venture Capital," Albany, *The Times Union* (May 14, 2015).

[233] State University of New York, *HocusLocus, LLC., the University at Albany at the College of Nanoscale Science and Engineering* (CICEP 2013 Case Study).

[234] "Startup Develops Genetic Switch Technology," Troy, *The Record* (July 20, 2015).

[235] "Nanocollege Takes Top Prize," Albany, *The Times Union* (April 29, 2014). The original research on this technology was conducted in a lab run by Dr. Susan Scharfstein, SUNY Poly professor of nanoscience, assisted by a $50,000 investment by the SUNY Technology Accelerator Fund.

commercialize.[236] Glauconix has received $500,000 in seed investment from the Eastern New York Angels and $975,000 in Phase I and II SBIR grants.

Glauconix offers platform services to reduce the cost of ophthalmic drug development, a 3D tissue system that reduces the risk of clinical trial failure.[237] CNSE's Pradeep Haldar observes that "Glauconix is one of our major successes." Torrejon was inspired to start the company while working in the lab of Dr. Susan Scharfstein, an associate professor of nanosciences at SUNY Poly, where the use of nanoscale scaffold was explored as a way of duplicating filter-like tissue in the human eye. Scharfstein's work was supported by a $50,000 investment from SUNY's Technology Accelerator fund to develop an ultimately successful commercial prototype. Buoyed by this success, Torrejon determined to start a company to launch the technology and enrolled in Haldar's class on entrepreneurship, which taught researchers how to transform their ideas into successful businesses. She met three collaborators in the class, who went on to win the state business plan competition in 2014.[238]

#### 9.6.3.5 Eonix

In 2013 four CNSE graduate students co-founded Eonix to commercialize an ionic liquid electrolyte to increase energy storage in ultracapacitors.[239] To date Eonix has attracted over $3.5 million in grants and investments, including $250,000 from NYSERDA and $50,000 from the NSF.[240]

### 9.6.4 RPI Entrepreneurial Initiatives

A number of individuals interviewed for this study singled out RPI as the main source of tech-oriented startups in the Capital Region, and it provides an infrastructure to facilitate them. According to Laban Coblentz, RPI chief of staff and associated vice

---

[236] Glauconix uses a synthetic meshwork that functions like the trabecular meshwork in the human eye, which can sometimes slow or halt the flow of aqueous fluid in the human eye, leading to partial or complete blindness. Glauconix' meshwork has been shown to be more cost-efficient and effective than the traditional use of cadaver eye tissue. SUNY Polytechnic Institute, "SUNY Poly Alumna, Founder and CEO of Glauconix Biosciences Awarded $750,000 by National Science Foundation for Commercialization of Technology Developed at SUNY Poly to Fight Eye Diseases," Press Release (July 25, 2017).

[237] Glauconix Biosciences, "What We Do," <http://www.http://glauconix.biosciences.com/whatwedo/>.

[238] SUNY Polytechnic Institute, "Grants, Investments Boost SUNY Poly CNSE Startup," SUNY Research Foundation News (September 4, 2015), <https://sunypoly.edu/news/suny-research-foundation-news-grants-investments-boost-suny-poly-cnse-start-0.html>.

[239] "Planting the Seeds of a New Company," *Siena News* (Summer 2013).

[240] NYSERDA, 2016 Clean Air Interstate Rule Annual Report on the New York Battery and Energy-Storage Technology Consortium (June 2017), p. 12.

president for policy and planning, as of 2011 RPI was launching 35–40 startups per year, employing "hundreds of people locally." In 2011 RPI unveiled its Emerging Ventures Ecosystem (EVE), an incubation program headed by entrepreneur and RPI professor Dick Frederick.[241] RPI's Severeno Center for Technological Entrepreneurship provides a variety of support services for startups, including mentoring, pursuit of funding opportunities, and training programs. While most RPI tech startups have involved software, artificial intelligence, gaming, and environmental and energy technology, several promising nanotech startups have also been launched.

### 9.6.4.1 ThermoAura

ThermoAura was established by a number of RPI faculty members in Troy in 2011 to develop a solid-state thermoelectric nanocrystal material, a bismuth telluride alloy, that converts heat into electricity. Its president, Rutvik Mehta, envisioned the company evolving into a multibillion dollar business. ThermoAura got a boost from a $393,000 NYSERDA grant in 2013 as well as a $750,000 SBIR award from the National Science Foundation.[242] In 2014 ThermoAura received $250,000 in investment by ENYA which enabled the company to start its first production facility for the nanocrystals in Colonie, New York.[243] In 2016 ThermoAura received an investment of $1 million by the Upstate Venture Association of New York.[244]

### 9.6.4.2 PBC Tech

In 2008 a multidisciplinary team of scientists (biotech, nanomaterials, and electronics) and an entrepreneur based on RPI's Severino Center for Technological Entrepreneurship began pursuing the vision of creating scalable, flexible structural sheets of energy storage material that could function as batteries and supercapacitors.[245] Founded as Paper Battery Company, the firm now operates as PBC Tech, based in Troy. In 2013 the company won a TIE50 "Top Startup" award at TIEcon 2013, a conference in Santa Clara, California, 1 of 50 out of 1142 firms nationwide

---

[241] "Network of Business Networking," Albany, *The Times Union* (February 8, 2011).

[242] "Growing Area Tech Firms Get Aid," Albany, *The Times Union* (February 15, 2013). One of the co-founders of ThermoAura is Ganpati Ramanath, an RPI professor and leading expert in the science and engineering of nanomaterials. His work has benefited from the availability of "state-of-the-art research equipment in his laboratory." Ramanath has over 145 peer-reviewed articles as well as numerous patents. "Nanoglue Cooked up in a $40 Microwave," Albany, *The Times Union* (May 21, 2013); "ThermoAura Receives Innovation Award," Albany, *The Times Union* (November 1, 2013).

[243] "High-Tech Firm Opens Up New Facility in Colonie," Saratoga Springs, *The Saratogian* (December 9, 2017).

[244] "Upstate Sees 67 Deals in Quarter," Albany, *The Times Union* (September 9, 2016).

[245] The CEO of the company, Shreefal Mehta, also serves as an adjunct professor of biomedical engineering at RPI. "They AIM to Help Students," Albany, *The Times Union* (November 17, 2016).

winning top honors.[246] In 2014 Paper Battery received $3.4 million from an out-of-state venture fund, Caerus Ventures.[247] In early 2018 PBC Tech announced a manufacturing partnership with KLA-Tencor in preparation for the commercial launch of an ultrathin supercapacitor later in 2018.[248]

### 9.6.5 Public Support Programs

The federal government and New York State administer programs which have proven instrumental in the launch of some nanotech startups in the Capital Region.

#### 9.6.5.1 Small Business Innovation Research Program

The federal Small Business Innovation Research (SBIR) program was created by a 1982 act of Congress requiring federal agencies with large research budgets to utilize a percentage (3.2% as of FY2017[249]) of their extramural research spending for grants or research contracts with small businesses. SBIR Phase I awards of $150,000 can be followed, in appropriate cases, by Phase II awards as high as $1 million, with additional funding possible. Federal agencies taking part in SBIR periodically release solicitations for Phase I, outlining the research themes eligible for contracts and grants. The awards enable proof of concept and prototyping and do not require surrender of intellectual property by recipients.[250] SBIR awards are not only a potentially critical source of funding for early stage companies but a form of "technology validation" in the eyes of angel investors, venture funds, and other sources of capital.

Two of the Capital Region's recent nanotechnology startups have benefitted from the SBIR program, ThermoAura and Glauconix. However, New York trails California and Massachusetts both in numbers of SBIR awards and total funding, suggesting that the state is not taking full advantage of a potentially significant source of capital for innovative startups (see Table 9.2).

---

[246] "Paper Battery of Troy Recognized," Albany, *The Times Union* (May 24, 2013).
[247] "Venturing Record Gains," Albany, *The Times Union* (August 24, 2014).
[248] "PBC Tech Readies Battery-Boosting PowerWRAPPER for Commercial Launch with Manufacturing Agreement and New U.S. Patent," *Nasdaq* (January 9, 2018). KLA-Tencor, a developer of process control systems for semiconductor manufacturing, opened a facility in Malta, New York, in 2011 concurrently with the completion of the GlobalFoundries Fab. "San Francisco Semiconductor supplier KLA-Tencor to open office Near GlobalFoundries," *Albany Business Review* (January 19, 2011).
[249] U.S. Small Business Administration, "About SBIR," <https://www.sbir.gov/about/about-sbir>, accessed February 21, 2018.
[250] National Research Council, Charles W. Wessner (ed.) *An Assessment of the SBIR Program* (Washington, DC: The National Academies Press, 2008).

**Table 9.2** California, Massachusetts, and New York SBIR Awards in 2017

| State | Total number of awards | Total value of awards (millions of dollars) |
|---|---|---|
| California | 628 | 307 |
| Massachusetts | 269 | 150 |
| New York | 144 | 71 |

SOURCE: U.S. Small Business Administration, SBIR website <http://www.sbir.gov>

#### 9.6.5.2 New York State Energy Research and Development Authority

The New York State Energy Research and Development Authority (NYSERDA) is a New York State public benefit corporation which provides funding, technical support, information, and analysis to promote energy efficiency and renewable energy. Its primary funding source is electric and gas ratepayers who are charged a fee by utilities (Systems Benefit Charge). Among other things NYSERDA provides early-stage funding and technological support for startups seeking to commercialize clean technology innovations in the Capital Region. It also provides financial support for clean tech incubators in the state, including iCLEAN in Albany, from which CNSE's first nanotech startup, BessTech, was successfully launched. Other Capital Region nanotech startups receiving early-stage financial and technical support from NYSERDA include Eonix, ThermoAura, and the Paper Battery Company (now operating as PBC Tech).

### 9.6.6 *Private Funding for Startups*

The Capital Region's startup climate has frequently been criticized for its paucity of early-stage private funding sources as well as the mentoring functions that angel investors and venture capital firms exercise with respect to promising but inexperienced young entrepreneurs.[251] An Upstate angel investor, Dick Frederick, commented in 2017 that to date the venture capital wave has not really reached the Capital Region: "Upstate is probably 5–10 years behind the curve for the rest of the venture world. Right now we have more potential deals than we have dollars to fund them."[252] In the past decade, however, a support network for tech-oriented startups has begun to emerge to the Capital Region. This includes business incubators operated under the auspices of RPI and CNSE and in-state angel and venture funds targeting local tech startups.

---

[251] "Venture Capitalists Aim to Help Local Startups," Glens Falls, *The Post-Star* (October 27, 2010).

[252] "Venture Funds Favor Robotics," Albany, *The Times Union* (January 31, 2017).

## 9.6 Fostering Startups

**Table 9.3** ENYA investments in prominent New York startups

| Startup | Technology Focus |
| --- | --- |
| Vital Vio | Biomedical and lighting design |
| Paper Battery Company | Ultrathin supercapacitors |
| ThermoAura | Nanotech process technology |
| Free Form Fibers | High performance fibers for ceramic matrix composites |
| Hocus Locus | Posttranscriptional regulation of RNA |
| BessTech | Lithium-ion batteries |
| Dumbstruck | Advanced video testing and optimization |
| Ener-G-Rotors | Economic conversion of low temperature heat to carbon free electricity |
| Glauconix | Ex vivo dynamic 3D human tissue models |
| Create Prosthetics | 3D printing for orthotists and prosthetists |

SOURCE: Eastern New York Angels, <http://www.easternyangels.com>

### 9.6.6.1 Accelerate 518

In 2011 a number of local colleges and universities in the Capital Region formed Accelerate 518 to provide funding, business incubation, and location and business guidance to new companies in the 518 area code.[253] It convenes entrepreneurs with capital providers from the angel and early stages to middle-market private equity investors and lenders.

### 9.6.6.2 Eastern New York Angels

ENYA, an "angel" investment fund, had its genesis in 2001 when four local entrepreneurs formed the Tech Valley Angel Network (TVAN), a group of wealthy investors who met periodically to assess potential investments in early-stage companies in Tech Valley. In 2005 CEG took over administration of TVAN from the founding individuals. TVAN was disbanded in 2010 but several members worked with CEG to form a fully managed angel fund, the Eastern New York Angels. ENYA's members manage the fund, participate in investment decisions, and provide support for portfolio companies.[254] Investments range from $50,000 to $250,000. Its portfolio includes many of the most prominent tech startups in the region (see Table 9.3).

In its first 4 years of operation, ENYA received 750 business plans from local businesses but funded only six, or less than 1%. Only 2% of the companies submitting business plans opted to submit to the rigorous due-diligence process ENYA required when it believed a business plan was worthy of serious consideration.[255]

---

[253] "Center Offers Help to Tech Startups," Schenectady, *The Daily Gazette* (October 6, 2011).

[254] Eastern New York Angels, <*http://www.easternnyangels.com*>.

[255] "How an Investment Group Helps Startups Take Flight," Albany, *The Times Union* (March 20, 2014).

### 9.6.6.3 Upstate Capital Association of New York (Formerly Upstate Venture Association of New York)

Upstate Capital is a statewide organization promoting venture capital and private equity investments in Upstate New York. Its members are investors, including venture, angel, and private equity funds. It convenes entrepreneurs with capital providers from the angel and early stages to middle-market private equity investors and lenders.

### 9.6.6.4 Upstate Venture Connect

Upstate Venture Connect (UVC) is a nonprofit networking organization that connects entrepreneurs with resources they need to build their companies. It is financed through donations and sponsorships and also offers advisory services to communities seeking to promote local entrepreneurship. UVC has helped create six angel funds involving 250 investors that have provided support totaling about $20 million to 80 startups.

In sum, the startup culture in Tech Valley is progressing but at a relatively slow rate. Some of the regions' institutions such as RPI have put in mechanisms to support entrepreneurship, yet the availability of funding remains a challenge. Perhaps most importantly, the culture in the region's universities lacks the focus on entrepreneurship and commercialization that characterizes centers of innovation elsewhere in the United States. To see more focus on commercialization, faculty need more internal incentives, involving both university funding and career encouragement. More broadly, greater attention needs to be paid by the region's educational institutions and governments to the opportunities presented by the federal SBIR programs. This could include training and mentoring for applicants, and successful applicants could receive additional funds through state matching grants as some states already are doing. Greater progress in firm formation and growth is certainly possible especially if it is facilitated by government–university–industry partnerships designed to address gaps in the region's innovation ecosystem.

## Bibliography[256]

After a Half-Century of Decline, Signs of Better Times for Buffalo. (2006, September 18). *The New York Times*.

Alain Kaloyeros, Powerful Centerpiece of Buffalo Billion, Could Become a Household Word. (2015, September 29).*Gotham Gazette*.

---

[256] As noted in the front matter of this book, the study also drew on interviews carried out by the authors and numerous articles from *The Times Union* (Albany), *The Daily Gazette* (Schenectady), the *Albany Business Review* (Albany), *The Post-Star* (Glens Falls), *The Record* (Troy), *The Saratogian* (Saratoga Springs) , *The Buffalo News* (Buffalo), The *Observer-Dispatch* (Utica), The *Daily Messenger* (Canandaigua), and the *Post-Standard* (Syracuse). These are not individually included in the bibliography.

# Bibliography

Applied Materials. (2018, November 15). New Applied Materials R&D Center to Help Customers Overcome Moore's Law Challenges. Press Release.

Bid-Rigging Trial Begins for Ex- State University of New York Official. (2018, June 18). *Reuters*.

Bruno, J. L. (2016). Keep Swinging: A Memoir of Politics and Justice. Franklin, TN: Post Hill Press.

Can Buffalo Ever Come Back? (2007, October 19). New York, *The Sun*.

Control Freaking: Legislators Vent at Cuomo for Hogging $3 B in Budget Power. (2013, February 4). *New York Daily News*.

Cuomo Fund-Raisers Preceded Development Funding Awards. (2014, December 9). *Politico*.

Cuomo's $15 Million High-Tech Film Studio? It's a Flop. (2016, August 22). *The New York Times*.

Eastern New York Angels. http://www.easternnyangels.com.

Echeverri-Carroll, E., Feldman, M., Gibson, D., Lowe, N., and Oden, M. (2015, March 15), *A Tale of Two Innovative Entrepreneurial Regions: The Research Triangle and Austin*. University of Texas at Austin and University of North Carolina at Chapel Hills.

Empire State Development Approves $46 million for GCEDC's STAMP plan. (2016, August 18). *The Daily News*.

ESD Team Reviews EDC Funding Proposals. (2014, September 13). Massena, *Daily Courier-Observer*.

Etzkowitz, H. (2013). Silicon Valley: The Sustainability of an Innovative Region. Social Science Information Journal. 52(4), 515-538.

Glauconix Biosciences. What We Do. http://www.http://glauconix.biosciences.com/whatwedo/.

Governor Cuomo Announces Milestone Reached at AIM Photonics in Rochester. (2017, May 28). *US Fed News*.

Gretzinger, J. (2015, September 26). SUNY Poly Details Oversight and Transparency Process Regarding Buffalo Projects. Memorandum. SUNY Poly.

High tech on the Hudson: Digital Dynamos Have Turned Kingston Into Brooklyn North. (2010, December 17). *New York Daily News*.

Lessig, L. (2009, October 9). Against Transparency. *New Republic*.

Lowe, N. (2014). Beyond the Deal: Using Industrial Recruitment as a Strategic Tool for Manufacturing Development. *Economic Development Quarterly* 28(4).

Moynihan, D. P. (1993, April 15). No Surrender: Toward Intolerance of Crime. Address to the Association for a Better New York.

National Research Council. (2008). *An Assessment of the SBIR Program*. C. W. Wessner (Ed.) Washington, DC: The National Academies Press.

National Research Council. (2013). *Best Practices In State and Regional Innovation Initiatives: Competing in the 21st Century*. C. W. Wessner (Ed.) Washington, DC: The National Academies Press.

New York is Placing Big Bets in Upstate Cities. (2015, September 12). *Associated Press State Wire: New York*.

NY Wins $600 Million Hub for Photonics Research and Development. (2015, July 27). *Associated Press State Wire: New York*.

NYSERDA. (2017, June). 2016 Clean Air Interstate Rule Annual Report on the New York Battery and Energy-Storage Technology Consortium.

Park, F. M. "Turning Poor Dirt into Pay Dirt. *METRO Magazine*. http://www.metronc.com/article/?id+421.

PBC Tech Readies Battery-Boosting PowerWRAPPER for Commercial Launch with Manufacturing Agreement and New U.S. Patent. (2018, January 9). *Nasdaq*.

Planting the Seeds of a New Company. (2013, Summer). *Siena News*.

A Roadmap for US Integrated Photonics. (2017, March 23). *Optics & Photonics*.

Roberts, E. B., and Eesley, C. E. (2011). *Entrepreneurial Impact: The Role of MIT*. Hanover, MA: Now Publishers Inc.

Schneier, E. V., Murtaugh, J. B., and Pole, A. (2010). *New York Politics: A Tale of Two States*. Armonk and London: M.E. Sharpe.

Solar Industry Slowdown Catches Up with SolarCity. (2017, May 5). *Investopedia*.
State University of New York. (2013). *HocusLocus, LLC., the University at Albany at the College of Nanoscale Science and Engineering.* CICEP 2013 Case Study.
Sturgeon, T. (2000.) How Silicon Valley Came to Be. In M. Kenney (Ed.), *Understanding Silicon Valley: The Anatomy of an Entrepreneurial Region*. Stanford, CA: Stanford University Press.
SUNY Poly Settles Low, Tries to Clear 'Kaloyeros Hangover.' (2017, March 27). *Politico*.
SUNY Polytechnic Institute. Faculty Profile: Laura Schultz. https://sunypoly.edu/faculty-and-staff/laura-schultz.html.
SUNY Polytechnic Institute. (2014, October 10). Energy-Focused Incubator at SUNY Polytechnic Institute and the Hudson Valley Center for Innovation. Press Release.
SUNY Polytechnic Institute. (2015, September 4). Grants, Investments Boost SUNY Poly CNSE Startup. *SUNY Research Foundation News*. https://sunypoly.edu/news/suny-research-foundation-news-grants-investments-boost-suny-poly-cnse-start-0.html.
SUNY Polytechnic Institute. (2016, June). *Strategic Plan*.
SUNY Polytechnic Institute. (2017, July 25). SUNY Poly Alumna, Founder and CEO of Glauconix Biosciences Awarded $750,000 by National Science Foundation for Commercialization of Technology Developed at SUNY Poly to Fight Eye Diseases. Press Release.
'Super Region' Marketing Part of $75K Contract. (2017, December). Batavia, *The Daily News*.
Tech Startups Seek the Good Life in Saratoga Springs. (2014, December 2). *Politico*.
Tesla and SolarCity Merger Gets Approval from Shareholder. (2016, November 17). *CNBC*.
TIP Strategies. (2014, March). *Economic Development Strategic Plan Prepared for Saratoga County, New York*. http://www.saratogacountyny.gov/wp/wp-content/uploads/2013/11/Saratoga-Plan-FINAL.pdf.
Transparency International. *Anti-Corruption Glossary*. http://www.transparency.org/glossary/terms/transparency.
The Transparency Trap. (2014, September). *The Atlantic*.
Trump Weighs Heavily on 1366 Technologies Project -- STAMP -- Federal Loan, Competition from China will Factor in Manufacturers' Plans. (2017, January 17). Batavia, *The Daily News*.
University at Albany. (2017, April 19). State Budget Keeps Campus Moving Forward. Press Release.
U.S. Small Business Administration. (2018). About SBIR. https://www.sbir.gov/about/about-sbir. Accessed February 21, 2018.
The Wind and Sun Are Bringing the Shine Back to Buffalo. (2015, July 21). *The New York Times*.

# Chapter 10
# Conclusion

New York's Tech Valley demonstrates that it is possible to reverse long-term economic decline in an old industrial region through the right mix of public policies and private-sector engagement. Tech Valley's success should dispel the notion that regional efforts to foster high-tech industry will necessarily fail because of embedded regional disadvantage. The experience has shown that regional competitive advantage can actually be grown and that public investments in research infrastructure can lead to a resurgence in well-compensated manufacturing jobs.

## 10.1  Is Tech Valley a Model?

The success to date of the Tech Valley effort raises the important question whether it is a model that can be applied elsewhere with appropriate adaptations. Governor Andrew Cuomo is overtly seeking to replicate Tech Valley in other areas of Upstate New York. But although Governor Cuomo's effort has already achieved some apparent success in the Buffalo/Niagara area, skeptics can point to the fact that the same intrinsic local advantages that worked in favor of Tech Valley are present throughout New York State and what worked in New York won't necessarily work in other states:

- **Financial resources**. New York is one of the biggest and wealthiest states in the United States, and few other regions are in a position to make risky financial outlays on the scale of the Tech Valley investments.[1]

---

[1] The risks associated with large-scale investments in advanced tools have been underscored by the winding down of the Global 450 consortium based at CNSE. That effort created the world's first and only 450 mm wafer fabrication plant, but further work has been suspended because the industry participants do not plan to make the transition from the 300 to 450 mm wafer size in the foreseeable future, an outcome that was not foreseen when the project was launched. As a result, SUNY Poly is "trying to get rid of $115 million worth of one-of-a-kind pieces of manufacturing equipment for computer chip factories that don't even exist yet." See "SUNY Poly's Tools on Sale," Albany, *The Times Union* (September 30, 2017).

- **Industrial foundation**. At the inception of the Tech Valley effort, New York already enjoyed "a collection of now priceless economic resources" in the form of large, long-established, technology-based manufacturers, including GE, IBM, Corning and Eastman Kodak, which provided the foundation upon which Tech Valley has been built.[2] Few other states have a comparable array of legacy firms.
- **Transportation**. New York enjoys significant advantages as a transportation center positioned between large nearby metropolitan areas, and most other states or regions do not occupy comparable geographic positions.

But while New York's intrinsic advantages were formidable, it should not be forgotten that many observers—perhaps a majority of academic economists who have looked at the region—concluded that, at least in the case of Upstate, embedded disadvantages made long-run economic recovery unlikely if not impossible. Regional disadvantages included a dense but fragmented collection of local and regional governmental units, relatively high taxes, high labor costs, a culture of defeatism, severe winters, and the longstanding reality that very substantial numbers of young adults from the region migrate to New York City and other urban centers.

Arguably practices and techniques that were instrumental in the creation of Tech Valley can be successfully applied in regions which lack some or all of New York's advantages. Tech Valley itself was predicated on careful study of other dynamic regions in which innovation drove growth. Beginning with Nelson Rockefeller's examination of Silicon Valley and George Low's initiatives at the Rensselaer Polytechnic Institute (RPI) based on Stanford's example, New York policymakers have long applied best practices drawn from here is no reason why techniques which New York borrowed and adapted from other regions cannot once again be borrowed, adapted, and applied elsewhere, including, in particular, other parts of the so-called Rust Belt. The result could be a substantial acceleration of the revival of US manufacturing.

## 10.2 Best Practices

"Best practices" are techniques that are widely accepted as superior to alternative methods because they predictably deliver better outcomes. New York's technology-based economic development efforts have been based on close study and application of best practices of other US states and regions as well as economic models developed in academia.[3] New York has also demonstrated some singular practices

---

[2] Martin Schoolman, "Solving the Dilemma of Statesmanship: Reindustrialization Through an Evolving Democratic Plan," in Martin Schoolman and Alvin Magid (eds.) *Reindustrializing New York State: Strategies, Implications, Challenges* (Albany: SUNY Press, 1986).

[3] Stuart W. Leslie and Robert H. Karagon, "Selling Silicon Valley: Frederick Terman's Model for Regional Advantage," *Business History Review* (Winter 1996); Laura I. Schulz, "Nanotechnology's Triple Helix: A Case Study of the University of Albany's College of Nanoscale Science and Engineering," *Journal of Technology Transfer* (2011).

## 10.2.1 Exceptional Policy Continuity

Every governor of New York since Nelson Rockefeller, whether Republican or Democrat, has shared a commitment to university-based, innovation-driven economic development. This commitment also extended to key legislative leaders, including Senate Majority Leaders Warren Anderson and Joseph Bruno and Assembly Speakers Stanley Fink and Sheldon Silver. The result has been a remarkable continuity in state policies and large-scale investments in university-based, industry-relevant innovation over a very long time horizon. Although each governor has introduced new initiatives and adjusted the direction of policies and practices of predecessor administrations, no governor has fundamentally reversed course or sought to erase the achievements of their predecessors. This policy continuity, rare in a democracy, is an important source of the region's success.

The sustained character of the state's investments and policies with respect to innovation-based economic development, spanning nearly a half century, is widely credited as a factor underlying the manifest success of Tech Valley. A 2013 study of the Albany Nanotech complex by Georgia Tech observed that—

> Some keys to this success include a state government, and particularly a succession of governors, that saw value in making capital investments that would lure cutting-edge researchers and companies to the area. ... [F]rom 2000-01 to the 2008-2009 fiscal year, the state invested $876.1 million in funding for nanotechnology research at the SUNY Albany College of Nanoscale Science and Engineering across 13 different projects. This level of support would not have been possible without the ongoing support of the state's top leadership.[4]

## 10.2.2 Strong Leadership at All Levels

The role played by senior state political leadership in the creation of Tech Valley is widely recognized and acknowledged. But business, academic, and local government leaders have made individual contributions to the effort that are less well-known but have been no less important. Local leaders shared a vision, saw what was needed to realize it, and took the initiative to bring it about. This study includes many examples, such as the Albany-area business leaders who fought local parochialism and pushed for a regional approach to economic development; the university, community college, and K-12 educators who devised and constantly refined

---

[4] Jason Chernock and Jan Youtie, "State University of New York at Albany Nanotech Complex," in Georgia Tech Enterprise Innovation Institute, *Best Practices in Foreign Direct Investment and Exporting Based on Regional Industry Clusters* (Atlanta: Georgia Tech Research Corporation, February 2013, Prepared for the Economic Development Administration, U.S. Department of Commerce). p. 66.

curricula relevant to high-tech manufacturing; and the locally-based economic development professionals who spearheaded the effort to draw high-tech manufacturing to the region and to create the infrastructure to support it.

Every community and region in the United States has individuals with the initiative, ideas, and skills necessary to drive local economic development, but such people may find that in the particular local context in which they find themselves, they are "voices in the wilderness," able to accomplish little. In the case of New York's Tech Valley, local leaders saw opportunities and acted on their own initiative to attract world-class high-tech manufacturing to the region. Importantly, however, state authorities supported them at key junctures, providing funding, arranging operational support from institutions like ESD, and providing legal and public policy support. The creation of Regional Economic Development Councils (REDC) represents an attempt to institutionalize the best practice of fostering leadership initiative at the local level, with the state encouraging brainstorming and strategic planning at the town, county, and regional level and providing substantial support for the best ideas and project proposals.

### 10.2.3 Preserving Key Parts of the Industrial Legacy

Rust Belt states have experienced massive erosion of traditional manufacturing industries and the contraction or disappearance of industry clusters. In the United Kingdom, deindustrialization in old manufacturing centers has proceeded so completely that entire supply chains have disappeared along with large manufacturers and the associated research capability, skills, and know-how.[5] New York saw extensive disinvestment by mainstay companies like IBM, GE, and Kodak, but in the case of IBM, the state took drastic and ultimately successful steps to keep the company from migrating elsewhere. The creation of Tech Valley built upon IBM's foundation, in the form of the company's expertise and know-how, as well as its continued presence in the state. There are other significant examples of Rust Belt regions building new industries with relevant knowledge, skills, and technology drawn from the heritage of failed manufacturing companies.[6]

---

[5] Centre for Research on Socio-Cultural Change, *Rebalancing the Economy for Buyer's Remorse* (working Paper No. 87, 2011), pp. 29–30; "Why Doesn't Britain Make Things Any More?" *The Guardian* (November 16, 2011).

[6] Ohio, for example, was devastated by the contraction of traditional manufacturing sectors like glass, rubber, and steel. However, a large number of small manufacturers remain who have proven able to parley traditional skills into new market areas. Machine tool makers, for example, moved into production of medical instruments and equipment. A number of universities and hospitals collaborated to establish the BioInnovation Institute to build on northeast Ohio's leadership in polymers to establish a biomaterials industry, including replacement parts for aging bodies and treatments for skeletal and joint ailments. See generally National Research Council, Charles W. Wessner (ed.), *Best Practices In State and Regional Innovation Initiatives: Competing in the 21st Century* (Washington, DC: The National Academies Press, 2013), pp. 111–142.

**Table 10.1** New York state funding of NanoTech facilities and equipment at SUNY Albany, 2000–2009

| Project | State funding entity | Year(s) | Budgeted amounts (Millions of Dollars) |
|---|---|---|---|
| CESTM building | State | 2000–2001 | 10.0 |
| CATN2 | NYSTAR | 2001–2002 | 9.7 |
| State Ctr excellence | ESD | 2003–2004 | 50.00 |
| Sematech facilities | ESD | 2003–2004 | 160.00 |
| Tokyo electron | ESD | 2004–2005 | 100.0 |
| CNSE | State | 2005–2006 | 8.3 |
| IMPL SE | FSD/SUCF | 2005–2006 | 5.0 |
| INDEX—Capital | ESD | 2006–2007 | 75.00 |
| INVENT | SUCF | 2006–2007 | 75.00 |
| Sematech—Machinery and equipment | ESD | 2008–2009 | 300.00 |

SOURCE: Office of the State Comptroller, *Fuller Road Management Corporation & The Research Foundation of the State of New York* (2010-5-5), p. 7

## *10.2.4 Large Public Investments in Research Infrastructure*

One of the principal factors underlying the emergence of Tech Valley has been a series of decisions by state policymakers to invest on an unprecedented scale in research equipment at universities in the region. As long ago as the mid-1990s, the nanotechnology research facilities of the University at Albany (SUNY Albany) were already differentiated from those of other US universities because of their high quality and scale. In 2001, as part of a cooperative effort, the state committed $50 million toward the cost of a 300-mm research wafer fabrication line at SUNY Albany, which was more than matched by $100 million from IBM.[7] A similar forward looking investment was made in 2006, with the state contributing $44 million to the establishment of a research supercomputer system at RPI, which at the time was the most powerful system of its type at any university in the world.[8] The then

---

[7] "$100 Million Boosts Tech Center," Albany, *The Times Union* (April 24, 2001). The state had previously funded the establishment of the world's first 200 mm research line, NanoFab 200, in 1996.

[8] IBM contributed $44 million and RPI and partner firms $34 million to the project. "A Magical Moment for Tech Valley—Many Await Chance to Use Supercomputers Which Will Link Region to a Powerful Network," Albany, *The Times Union* (May 12, 2006).

President of the College of Nanoscale Science and Engineering (CNSE), Alain Kaloyeros, emphasized that the scale of research facilities is a significant advantage:

> CNSE is the most advanced academic research enterprise in the academic world. The presence of cutting-edge, one-of-a-kind equipment and state-of-the-art cleanrooms allows scientists to conduct advanced research that is simply impossible without those capabilities. The large scale attracts corporate partners and a highly skilled workforce.[9]

As shown in Table 10.1, numerous other similar investments were undertaken. Such equipment, coupled with the expertise available at SUNY Albany and RPI, proved a powerful draw for co-location of research activities by global semiconductor companies, who could not find comparable sites anywhere else in the world.[10] A 2010 study by Professor Laura Schultz, a member of the CNSE faculty, noted that at the time that—

> The research facilities at the CNSE are four times greater than those available at the next largest center in Austin and house twice as many researchers. ... CNSE's main cleanrooms are designed to accommodate 300mm silicon wafers, the industrial standard, instead of 200mm, the academic research standard. ... CNSE is home of one of the two EUV lithography tools in the world.[11]

Investments in research infrastructure serving manufacturers need not entail vast financial outlays although in the case of New York the investments have been very substantial. The key is providing companies with research resources they could not otherwise afford, enabling them to introduce innovations into their products and industrial processes. This can involve relatively modest expenditures on specialized equipment, tools, and trained personnel that make relevant innovation research avail-

---

[9] Quoted in Laura I. Schulz, "Nanotechnology's Triple Helix: A Case Study of the University of Albany's College of Nanoscale Science and Engineering," *Journal of Technology Transfer* (2011), p. 564.

[10] The 300-mm research line established in the early 2000s at SUNY Albany was the only 300-mm facility at any university site in the world, and featured the most advanced manufacturing equipment available. In 2004, Albany NanoTech began using the world's first 193-nm preproduction immersion lithography system for 300-mm wafers, an ASML machine valued at $26 million, which was characterized by one industry expert as "bleeding edge" technology not even available yet on the market. Shonna Keogan, a spokesperson for Albany NanoTech, commented in 2004 that "basically we have every single lithography tool being used for research into chip development for commercial markets," including an extreme ultraviolet tool using technology that she characterized as "way, way off in the distance." Michael Tittnich, et al., "A Year in the Life of an Immersion Lithography Alpha Tool at Albany Nano Tech," in Proceedings of SPIE, Vol. 6151, *Emerging Lithographic Technologies* (2006).; "Albany NanoTech Fills Toolbox," Albany, *The Times Union* (August 26, 2004). IBM's John Kelly commented on the acquisition of the 193-nanometer immersion system, "that's the first in the world and the most advanced lithographic tool and it's sitting in Albany NanoTech." See "Brain Power Will Win NanoTech Wars," *Albany Business Review* (September 16, 2004).

[11] Laura I. Schultz, "Nanotechnology's Triple Helix: A Case Study of the University of Albany's College of Nanoscale Science and Engineering," *Journal of Technology Transfer* (2011), p. 561. "New Chips Mark UAlbany Milestone," *Albany Business Review* (September 16, 2004). In December 2003 the Albany NanoTech 300 mm research manufacturing line fabricated its first 300-mm silicon wafer, representing the first-ever such achievement by a university. A university official commented that this achievement by a university/industry team was unprecedented: "other universities don't even dream of doing something like this, much less doing it."

able to small businesses operating in market niches.[12] CNSE acts as a large and uniquely American version of Germany's Fraunhofer-Gesellschaft, a publicly supported system of generously funded, well-staffed research institutes that provide research tools and expertise to industry, not only large manufacturers but many small business (See Appendix B). Such research support has enabled small- and medium-sized German companies to become global leaders in their niche product areas.[13] The Fraunhofer research equipment addressing such technologies in most cases is far less costly than CNSE's semiconductor resources but is often best-in-class and provides the opportunity for students to have "hands-on" training and experience.

## 10.2.5 Industry-University Consortia

Even the best research equipment in the world is of little value without the organizational structure necessary to maximize the value and minimize the costs associated with that equipment with respect to research collaborations. The United States has evolved a vast array of institutional arrangements pursuant to which universities collaborate with industry (frequently with public financial support) to conduct basic and applied research, known variously as centers of excellence, joint laboratories, cooperative research centers, and engineering research centers.[14] A rich academic literature has analyzed this phenomenon and its potential for stimulating regional

---

[12] Youngstown, Ohio, for example, launched the Youngstown Business Incubator (YBI) in 1995, narrowly focusing on the goal of building a local cluster of business-to-business software firms. YBI invested in high-speed fiber-optics connections and a software-testing lab which would not have been cost-effective for small firms to acquire. The cluster grew dramatically, attracting, among other tenants, Turning Technologies LLC, which in 2007 was named the fastest-growing software company in the United States. PolicyLink, *To Be Strong Again: Renewing the Promise in smaller Industrial Cities* (2008), https://www.policylink.org/sites/default/files/ToBeStrongAgain_final.pdf; "Turning Technologies Rated Fastest-Growing," Youngstown, *Vindicator* (August 24, 2007).

[13] "[T]he research facilities of Fraunhofer serve as external, very well-equipped research departments of the Mittelstand [medium-sized] firms." In 2009, for example, the Fraunhofer began working with Roth & Rau, a firm with about 1100 workers, to develop manufacturing technology for the film photovoltaic cells. The Fraunhofer worked with the company to build a complete pilot manufacturing line at the institute to produce thin film PV cells efficiently—a project that resembled SUNY Albany's nanotechnology research initiatives, albeit on a much smaller scale. "Especially for the Mittelstand [highly competitive, medium-sized German firms] the Fraunhofer are very attractive. They find new equipment and can figure out what they never could on their own." See "German Innovation, British Imitation," *New Scientist* (November 21, 1992); Christian Homburg (ed.) *Structure and Dynamics of the German Misselstand* (Heidelberg and New York: Physica-Verlag, 1999), pp. 58–59; For a thorough review of the Fraunhofer system and its advantages, *see* National Research Council, Charles W. Wessner (ed.) *21st Century Manufacturing: The Role of the Manufacturing Extension Partnership Program* (Washington, DC: The National Academies Press, 2013), p. 252.

[14] Craig Boardman and Denis Gray, "The New Science and Engineering Management: Cooperative Research Centers as Government Policies, Industry Strategies and Organizations," *Journal of Technology Transfer* (February 2010).

economic growth.[15] The Albany Nanocomplex stands out against this background not as unique but as an unusually successful manifestation of university/industry/government collaboration in innovation.

CNSE was organizationally structured to minimize the cost and risk to semiconductor firms of the adoption of innovations, in the form of new equipment, materials, and processes. Cutting-edge semiconductor manufacturing equipment is not only very expensive—sometimes more than $60 million for a single tool—but in its first (alpha) generation is characterized by defects that only become apparent when the tools are used in a working manufacturing environment. The equipment maker observes the defects in an operational context and modifies the tool to create a more reliable beta version of the same tool. But a "bleeding edge" device manufacturer who has invested in the alpha version of the tool is stuck with expensive capital equipment that is less productive than beta versions available to its competitors and which may have little or no resale market value. The university-industry pilot lines at SUNY Albany have enabled manufacturers collectively to sidestep this trap. In addition, by participating in device production in a manufacturing research environment, industrial partners can move down the learning curve, gaining know-how that can be translated into better yields and lower costs in their own manufacturing lines.

CNSE has been characterized as a neutral, "Switzerland-type innovation lab." Its research facilities enable university researchers and industry partners to pool their efforts in a common area—

> *Ensuring the pooling of intellectual assets and physical resources to guarantee timely technology delivery, while allowing New York to act as the "referee" by providing the leveled playing field for each consortium participant to leverage its investments and protect its competitiveness.*[16]

The Nanocomplex rents large square-footages of space to industrial tenants who are participants in CNSE-industry research collaborations. These companies can take technology and know-how developed in the joint projects back to their own facilities in rented space and refine the knowledge into what may become its own proprietary technologies. While CNSE owns intellectual property developed in cooperative research facilities, industrial tenants own the intellectual property (IP) developed in their rented space.

## *10.2.6 Intermediary Organizations*

SUNY Albany and later SUNY Poly smoothed the traditionally-difficult interface between university research and the private sector by establishing not-for-profit corporate intermediaries which were the vehicles through which the university worked

---

[15] Henry Etzkowitz, *The Triple Helix: University-Industry Government Innovation in Action.* (New York and London: Routledge, 2008); Bruce Katz and Mark Muro, "The New Cluster Moment: How Regional Innovation Clusters Can Foster the Next Economy," *Brookings* (February 27, 2013).

[16] Alain Kaloyeros, "State is Well Poised for Nanotech Challenges," in Albany, *The Times Union* (December 26, 2012).

with industry.[17] The corporate entities were reportedly created because IBM wanted to deal with 501(c)(3) entities to ensure it was not dealing directly with the rules and governance structure typical of an academic campus. The corporate intermediaries were founded in collaboration with the SUNY Research Foundation, which provides university administrators "funds for programs and supplies not within the bounds of their regular budget authority" and which is not bound by university rules with respect to issues such as human resources and tenure.[18] The Nanocomplex was built under the auspices of the SUNY Research Foundation, and the faculty serving the complex were employees of the Foundation's not-for-profit entities. Without this autonomy, some observers believe that the Nanocomplex "never could have happened." Laura Schultz, a faculty member at CNSE, observes that—

> *When the CNSE was established, a new college was formed from the base up. The positions of faculty and staff were not constrained by the traditional academic expectations, but have been redefined to maximize technology transfer and economic development. Faculty and staff have been hired to enable the development of ties between companies. ... Faculty members experienced in industrial research better understand the needs of corporate partners, are able to identify potential collaborators, and can expedite the development of university and industry alliances.*[19]

More recently, the SUNY Research Foundation and its corporate intermediaries have become embroiled in the fall out arising out of Kaloyeros' indictment, with the organization criticized for inadequate levels of transparency. Significantly, the ESD, which stepped in to manage the consortia operating under the auspices of CNSE, has opted to implement institutional reforms at the intermediary organizations and keep them operating rather than shutting them down, a tacit acknowledgement of their importance as interfaces between the university and private companies. The existing intermediary bodies will be absorbed by the newly-established NY CREATES, which will continue the practice of interfacing between academia and the private sector through a corporate structure.

## 10.2.7 Forging a Regional Approach to Economic Development

The creation of New York's Tech Valley was enabled in large part by the Center for Economic Growth (CEG), an umbrella group of business organizations and leaders who recognized the extent to which political fragmentation was acting as a drag on

---

[17] These entities are the Fuller Road Management Corporation, Albany Nanotech, Inc., and Nanotech Resources, Inc.
[18] Edward V. Schneier, John Brian Mortaugh, and Antionette Pole, *New York Politics: A Tale of Two States*. (Aramark and London M.E. Sharpe, 2010), p. 46.
[19] Laura I. Schultz, "Nanotechnology's Triple Helix: A Case Study of the University of Albany's College of Nanoscale Science and Engineering," *Journal of Technology Transfer* (2011), p. 553.

growth. The CEG played an active role in advocacy, information sharing, financing, and convening in order to help create and sustain the cooperative momentum necessary for the success of the bid to attract a major semiconductor fabrication facility to the region. Its ability to work both above and with the individual political units of the region represented a key contribution. Similar successful regional efforts outside of New York in areas characterized by multiple small, competing jurisdictions demonstrate that this feature of the New York model, with adaptations, can be implemented in other US regions.[20]

CEG rapidly emerged as an effective advocate for regional collaboration as a way to combat economic decline. It commissioned studies, convened workshops and conferences, and served as a think tank, providing state policymakers with the intellectual foundation for high-tech-based economic development. CEG played a major role in securing state funding for rebuilding Albany International Airport and the Albany-Rensselaer's train station. It led the campaign to rebrand the region as "Tech Valley," an idea that was widely derided when first broached but which is now generally acknowledged. CEG engaged in a sustained effort to market the Capital Region to the global semiconductor industry and draw attention to the unparalleled research assets being created by the state's investments in local universities. A CEG official observed in 2013 that CEG and SUNY Albany's NanoTech Complex have always enjoyed a symbiotic relationship in drawing businesses to the Capital Region:

> As the NanoTech Complex was being developed, CNSE would exhibit with CEG at conferences under the banner of "NY Loves Nano" [a CEG marketing innovation]. These included key microelectronic conferences such as Semicon West in San Francisco and Semicon Europa in Dresden, Germany. [Today] CNSE's NanoTech Complex is internationally regarded, and as CEG highlights the region's strengths, the NanoTech Complex is one of the biggest components of the showcase.[21]

Beginning in 1997, CEG led the protracted effort to attract a new semiconductor manufacturer to the Capital Region, culminating in AMD's 2006 decision to build its next 300-mm facility in Malta/Stillwater. CEG worked closely with the Saratoga Economic Development Corporation (SEDC), a countywide economic development organization, to promote the Luther Forest site, providing it with funding at crucial junctures while also promoting alternative sites elsewhere in the Capital Region.[22] Business leaders in the region cite CEG as instrumental in the successful marketing of the Luther Forest site. The CEG model has been adopted by several other regions in New York and has been carried forward in Governor Cuomo's formation of Regional Economic Development Councils.

---

[20] See case study of Northeast Ohio in National Research Council, Charles W. Wessner (ed.) *Best Practices In State and Regional Innovation Initiatives: Competing in the 21st Century* (Washington, DC: The National Academies Press, 2013), pp. 111–141.

[21] Jason Chernock and Jan Youtie, "State University of New York at Albany Nanotech Complex," in Georgia Tech Enterprise Innovation Institute, *Best Practices in Foreign Direct Investment and Exporting Based on Regional Industry Clusters* (Atlanta: Georgia Tech Research Corporation, February 2013, Prepared for the Economic Development Administration, U.S. Department of Commerce).

10.2 Best Practices

**Table 10.2** SEDC engineering project team, 2006

| Company | Specialization | Tasks effort |
|---|---|---|
| M & W Zander | Builder of semiconductor wafer fabrication plants | Site planning, architecture, engineering, and construction |
| Abbie Gregg | Clean rooms | Vibration, electromagnetic, and radiofrequency measurements |
| LA group | Land planning | Initial site plans |
| C.T. Male | Engineering | Environmental, geographic analysis, and infrastructure planning |
| Creighton Manning Engineering | Engineering | Transportation studies and engineering |
| E/Pro | | Electrical transmission |
| National Grid | Power transmission | Electricity and gas |

SOURCE: "Project Case Study: High Tech Land Development," E News.com (May 2006)

### 10.2.8 *"Pre-Permitting" Potential Industrial Sites*

When Advanced Micro Devices decided in 2006 to build its next wafer fabrication facility at a site in Luther Forest, most of the substantive regulatory permitting had already been completed for a 300-mm fab of the type that AMD planned to build. The SEDC had secured the necessary approvals in advance through a "pre-permitting" exercise pursuant to which SEDC submitted an application for approval of a Planned Development District (PDD) zoned to accommodate 300-mm semiconductor manufacturing. The approval of the PDD by the Malta and Stillwater town boards in 2004 opened the way for the marketing of a "shovel-ready" site to semiconductor manufacturers. CEG President Kelly Lovell explained to town officials that creation of the PDD would—

> *give SEDC a shovel-ready site to market to a chip maker, who wouldn't have to jump through all the hoops the agency has already navigated—shaving two years off the time it usually takes to build a fab. The firm would only have to receive town approval for specific building site plans, a process that takes months rather than years.*[23,24]

The failure of an early pre-permitting exercise in North Greenbush provided lessons to regional leaders that enabled subsequent success in Malta and Stillwater. The rejection of a chip fab by the North Greenbush Town Board in the pre-permitting process enabled the weeding-out of this potential site before any commitment of resources, time and prestige had been invested by an out-of-state manufacturer, also avoiding the kind of reputational damage to the state that would have occurred with a high-visibility public rejection of an actual company. Pre-permitting enabled the

---

[22] "Task Force to Report on Efforts to Lure Chip Fab," Albany, *The Times Union* (September 13, 2000).
[23] "In a Forest, Two Roads Diverge," Albany, *The Times Union* (May 9, 2004).
[24] Martin Schoolman, "Solving the Dilemma of Statesmanship: Reindustrialization Through an Evolving Democratic Plan," in Martin Schoolman and Alvin Magid (eds.) *Reindustrializing New York State: Strategies, Implications, Challenges* (Albany: SUNY Press, 1986).

state to eliminate other seemingly promising sites in Upstate New York by identifying in advance potentially fatal flaws such as local land disputes, competing industrial projects, and hopelessly fragmented patterns of land ownership and control of access roads.[25] Most importantly, as state economic development professionals regrouped after the North Greenbush setback, lessons became apparent from the experience that helped guide the subsequent successful pre-permitting effort involving Luther Forest, including—

- Local worries over pollution, noise, and traffic needed to be addressed earlier in the process;
- Abundant "neutral information sources" needed to be made available to the public with respect to semiconductor manufacturing;
- Nomination of prospective sites was to be left to interested communities themselves rather than top-down site selection by senior state officials; and
- Before a site would be presented to industrial users, "all of the local officials have to be on board."[26]

## 10.2.9 Partnering with Industry Research Consortia

Although New York was passed over in favor of Texas as the location for Sematech's original research facility in 1988, New York continued its outreach to the semiconductor industry, reemphasizing state investments in local research universities and building relationships with a key semiconductor industry consortium, the Semiconductor Research Corporation (SRC), and engaging with Sematech.[27] As SUNY Albany built a critical mass of microelectronics research expertise and infrastructure and established corporate structures through which to interface with semiconductor industry research partners, the region's attraction for Sematech grew.[28] The facilities at SUNY Albany were featured in the 2001 state bid to become the site of Sematech's next research center. This was augmented by personal outreach by Governor George Pataki and other high-level state officials and a state incentive package valued at $210 million.[29] In July 2002 Sematech announced it would establish its next research center International Sematech North, at SUNY Albany. The most important decisional

---

[25] "North Greenbush Isn't the Only Site Having Trouble," Schenectady, *The Daily Gazette* (October 29, 1999).

[26] Kelly Lovell, president, Center for Economic Growth, "Task Force Heads West in Order to Lure Chip Fab Plant," Schenectady, *The Daily Gazette* (October 29, 2000).

[27] "Scientists Explore the Future of Computer Technology," Albany, *The Times Union* (December 11, 1990); Michael Fury and Alain E. Kaloyeros, "Metallization for Microelectronics Program at the University of Albany: Leveraging a Long-Term Mentor Relationship," *IEEE Explore* (1990).

[28] Robert W. Wagner, *Academic Entrepreneurialism and New York State's Centers of Excellence Policy*, Ph.D. dissertation, (SUNY Albany, 2007).

[29] "Sematech, SUNY Albany Seal EUV Lithography Program," *Solid State Technology* (January 29, 2003).

**Table 10.3** New York's successful incentives package for advanced micro devices

| Item | Amount (Millions of Dollars) |
|---|---|
| State grant for buildings and equipment | 500 |
| State grant for R&D | 150 |
| Empire zone tax credits/incentives | 250 (est.) |
| Infrastructure (includes some federal funds) | 300 (est.) |
| Total | 1200 |

NOTE: Commitment by Advanced Micro Devices was to: Create 1205 jobs by 2014; and Maintain 1205 jobs for 7 years (which was surpassed by GlobalFoundries' 3000 jobs)
SOURCE: "New York's Big Subsidies Bolster Upstate's Winning Bid for AMD's $3.2-Billion 300-mm Fab," *Site Selection* (July 10, 2006)

factors were reportedly "the caliber of work already being done at SUNY Albany and the enthusiastic backing it had from the governor and the legislature."[30]

## 10.2.10 A Professional Development Team

The Sematech success was followed by other successful outreach efforts by the state to semiconductor makers to establish a research presence in the state. However, the ultimate goal of New York policymakers was to attract inward investment by semiconductor producers in local manufacturing operations and the creation of thousands of new high-wage, high-skill manufacturing jobs. It was the large-scale commercialization of the R&D being conducted in the region that would provide the return on investment for taxpayers. To achieve this objective, New York undertook an effort that surpassed all previous bids, both successful and unsuccessful.

SEDC's choice of the Luther Forest site, a process which began in the mid-1980s, proved to be one of the most important aspects of the bid to attract a semiconductor manufacturer. Engineering studies revealed that the site was very "quiet," meaning that its geology sharply limits the transmission of vibrations which could interfere with the semiconductor manufacturing process. The presence of 60–200 ft of glacial sand deposits beneath the site protected it from vibrations. RF and electromagnetic field levels in Luther Forest are among the lowest in North America.[31] Luther Forest also held no federal or state wetlands, obviating the need for regulatory reviews that could delay construction.

SEDC's long and successful pre-permitting effort with respect to the potential semiconductor manufacturing site in Luther Forest has been described in the preceding section. This effort entailed the preparation of studies, most notably the Generic Environmental Impact Statement, which contained practical information of interest to site selection executives at semiconductor firms.

---

[30] "Albany No Longer a Secret in High Tech Chip World," *New York Times* (July 19, 2002).
[31] John S. Munsey, "Project Case Study: High Tech Land Development," *Civil and Structural Engineering* (May 2006).

Following pre-permitting and establishment of the PDD, SEDC retained a deep team of professional experts to develop proposals for semiconductor fabrication plants in Luther Forest, including M&W Zander, an engineering firm with extensive experience in building semiconductor manufacturing facilities (see Table 10.2). In addition to the bid itself, SEDC and other state economic development officials reached out to semiconductor manufacturers through numerous channels, most of them extremely discreet. Among other things, SEDC interviewed consultants who served large semiconductor makers and "word got back to their clients. Before we ever had any direct meetings with AMD, they had heard of us." Abbie Gregg, a semiconductor plant design expert who was part of SEDC's team of engineers, was a former colleague of AMD CEO Hector Ruiz and was able to communicate with him via email. "That got us the opportunity," commented SEDC President Ken Green in 2007.[32] Through this process, AMD emerged as a viable prospect interested in locating its next 300-mm wafer fabrication plant in Luther Forest.

### 10.2.11 Investing Big: An Internationally Competitive Incentives Package

Despite its many advantages, the main reason that New York won the competition for the new AMD fab is that it put forward an incentives package that was superior to that of other states and foreign countries. The state offered $1.2 billion worth of incentives, including a $500 million capital grant to AMD to pay for buildings and equipment and a $150 million grant for R&D, both with recapture clauses if the company didn't meet its commitments. The equipment grant was based on the proposals developed by SEDC's engineering team, which specified the types of equipment needed and calibrated the grant to cover their cost. AMD was eligible for tax credits and incentives worth as much as $650 million, which were formulaic based on the amount of capital investment and job creation by the company, via New York's Empire Zone program. Federal, state, and local funds estimated at $300 million would support infrastructural improvements. The approach took into consideration what is known in the economic development world: that investment in large, high-value anchor industries that require the support of regional research and development enterprises tend to stay around longer, create a higher-value supply chain of industries, and therefore stimulate the growth of an ecosystem of suppliers and small businesses underpinning the regional economy. The package was unique in that, by its terms, it built in time and flexibility to enable AMD to decide when the best moment would be to launch the actual construction effort.[33] (See Table 10.3.)

As a direct result of these combined efforts, in 2006, AMD, one of the world's leading semiconductor manufacturers, announced that it would build a chip fab at the Luther Forest Technology Campus in the towns of Malta and Stillwater, New York, to be operational by 2012. Key factors in its decision were the following:

- The site itself was one of the best in the world, including reliable power, and abundant water, and excellent underlying geology.

- A site in the United States was a hedge against global risks, e.g., SARS epidemics, earthquakes, and distance with respect to sites located outside the United States.
- The state's proposal was highly professional and accurately depicted the new fab's cost structure.
- A key consideration for AMD was its geographic proximity to the CNSE research infrastructure at SUNY Albany.[34]
- The state offered a substantial, internationally competitive incentives package. This was a critical factor in attracting the AMD/GlobalFoundries investment.
- State political leaders offered strong and sustained support, with both ad hoc funding for studies to further site development as well as the political will to offer a major incentives package.

In 2006, AMD's CEO told a New York audience that the site chosen was "superior to other locations the company considered in Germany and Singapore." He told the audience "you have collected tremendous possible sites for future selection" and complimented state and local officials for "the most well-crafted economic development package he could recall seeing."[35]

As previously mentioned, by all accounts the state's investment has paid off. The original Grant Disbursement Agreement between ESDC and AMD anticipated a 2.21-1 return on investment if AMD lived up to its commitment to create 1205 jobs and invest $3.2 billion to develop the project.

## 10.2.12 An Abiding Commitment to Relevant Educational Programs at All Levels

The success of the Tech Valley effort to date is, to a substantial degree, a testament to the state's commitment to relevant, high-quality educational institutional programs from K-12 through the post-graduate level. The large-scale state investments in the Albany NanoCollege have been paralleled by major public and private investments in universities, community colleges, and K-12 institutions within the region. A 2009 survey by the Semiconductor Industry Association of its members found that the most important factor considered by its member companies in choosing

---

[32] "Buzz Helped Attract AMD—Saratoga Economic Development Corp. Chief Says Targeting Behind-the-Scenes People Was Key," Albany, *The Times Union* (September 15, 2007).

[33] "New York's Big Subsidies Bolster Upstate's Winning Bid for AMD's $5.2 Billion 300-mm Fab," *Site Selection* (July 10, 2006); "GlobalFoundries—2010 Gold Shovel Project of the Year," *Area Development* (July 2010).

[34] John Frank, senior vice president of M&W Zander, the engineering company specializing in building semiconductor fabs, observed that CNSE was "a critical enabler in the eyes of a chip manufacturer. To be this close to a center of excellence in nanotech research, development, and manufacturing can be a major factor in the success of a new plant."

[35] "Tech Valley Vision Pays Off Big—Chip Maker AMD Hopes Rivals Will Also Build Plants in Region," Schenectady, *The Daily Gazette* (June 24, 2006).

sites for investment was the availability of a local workforce with the necessary education, training, and skills.[36] The Capital Region enjoyed strong K-12 schools and a number of excellent research universities engaged in nanotechnology programs and projects, such as RPI and SUNY Albany (now SUNY Poly). RPI and Union College also ran strong undergraduate engineering programs. Reflecting the caliber of RPI's undergraduate engineering programs, GlobalFoundries, the largest semiconductor manufacturer in the region, recruits more of its workforce from RPI than from any other institution.[37]

Semiconductor manufacturing requires larger numbers of technicians and operators with specialized skills than it does scientists and engineers.[38] Initiatives have been underway in the Capital Region since 1999, centered on the region's community colleges, to ensure that a pool of technician/operator workers exists with the requisite skills for semiconductor manufacturing. In 2003, Hudson Valley Community College (HVCC) in Troy established a 2-year degree program in semiconductor manufacturing technology "to train the technicians who work in the clean rooms," emphasizing "hands-on, applied nanotechnology and less-theoretical facets."[39] In 2006, HVCC concluded an agreement with CNSE enabling HVCC students to train on CNSE equipment, "the most advanced semiconductor equipment on the planet," to develop skills in mask preparation, clean room protocols, photoresist coating, etching, and other processes.[40] In 2007, HVCC established TEC-SMART, an extension located in the Luther Forest site adjacent to GlobalFoundries' manufacturing operations, featuring a real clean room and demonstration manufacturing line, to train workers with the skills needed to serve GlobalFoundries' operations.[41] Most of the courses required for HVCC's Semiconductor Technology Certificate are now taught at TEC-SMART, where students work in a manufacturing environment with workforce development partners.[42]

---

[36] Semiconductor Industry Association, *Maintaining America Competitive Edge: Government Policies Affecting Semiconductor Industry R&D and Manufacturing Activity* (March 2009).

[37] See summary of April 4, 2013 symposium keynote address by RPI President Shirley Ann Jackson in National Research Council, Charles W. Wessner (rapporteur), *New York's Nanotechnology Model: Building the Innovation Economy* (Washington, DC: The National Academies Press, 2013).

[38] In 2012 Pedro Gonzalez, GlobalFoundries' staffing manager for Fab 8, said in an interview that about 65% of the company's hires were technicians directly involved in the manufacturing process. "Talking About Fab 8's Work Force," Albany, *The Times Union* (September 9, 2012).

[39] "VCC Planning Nanotech Program," Schenectady, *The Daily Gazette* (January 18, 2003).

[40] "HVCC Training Goes Nanotech—Deal with Albany Gives Semiconductor Students Hands-On Experience on Most Advanced Equipment," Albany, *The Times Union* (December 15, 2006).

[41] "School to Boost Growing Sectors—HVCC to Build Center in Malta to Train Workers for Semiconductor, Alternative Energy Fields," Albany, *The Times Union* (December 14, 2007).

[42] "Tech Center Gets $2.15 Million Grant," Glens Falls, *The Post Star* (November 17, 2014). The Semiconductor Technology Certificate requires completion of 25 credits in courses which include Digital Electronics, Chemistry, Semiconductor and Nanotechnology, Semiconductor Methodology and Process Control, Semiconductor Manufacturing and Nanofabrication Processes, Vacuum and Power RF, and Electromechanical Devices and Systems. Hudson Valley Community College, *College Catalog* (2015–16), p. 122.

HVCC has shaped its curriculum to meet the actual needs of semiconductor manufacturers. HVCC's curricula are overseen by sectoral advisory committees of executives of local manufacturing firms. It constantly modifies and adapts its programs to meet the changing needs of local industry.[43] At the initiative of GlobalFoundries, HVCC also established training programs for GlobalFoundries employees and future employees based on the company's proprietary process technologies. This close and evolving interaction between a major high-tech manufacturer and the regional community colleges, combined with the training on cutting-edge equipment at SUNY Poly, represents an outstanding example of industry-university cooperation.

## 10.3 Challenges Facing the Region

During the interviews conducted for this study, regional leaders generally agreed that the effort to create Tech Valley has been a dramatic achievement but expressed concern as to whether it would remain sustainable over the long run. Worries included overdependence on a single industry known for its volatility and which is facing daunting global competitive challenges; uncertainty about the continuity of state support; organizational turmoil at CNSE/SUNY Poly; and problems associated with an abundance of local government authorities not always inclined to work in concert. Interviewees generally indicated that these problems can be surmounted but that such an outcome should not be taken for granted.

### 10.3.1 *Overdependence on One Industry*

Tech Valley's success is largely a function of the creation of research, manufacturing, and supply chain activities and jobs centered on a single industry, semiconductors. That industry has proven to be a driver of economic growth and prosperity in New York and other regions of the world. However, as noted above, the industry is prone to sharp cyclical swings and disruptive technological changes. Historical experience offers many examples of how regions heavily dependent upon one

---

[43] When AMD was considering establishing a wafer fabrication plant in Luther Forest, HVCC communicated with the company and sent two faculty members to AMD's Dresden fabs to map out the skill sets the company would require and develop an appropriate curriculum. After GlobalFoundries took over the Luther Forest project from AMD, it became apparent that AMD had a more highly developed skill set than GlobalFoundries sought. GlobalFoundries wanted technicians with "foundation skills" which the company would train itself. Accordingly, in 2015, HVCC created a new "Mechatronics" degree program, aligned with what GlobalFoundries wanted, combining mechanical and electrical engineering, motor control systems, and information technology to direct the systems that power semiconductor manufacturing equipment. "HVCC Offers New Engineering Degree in Mechatronics," Troy, The Record (March 8, 2015).

industry can be devastated by industrial contractions and restructuring.[44]

## 10.3.2 The Importance of Continuity in State Support

At present, semiconductor firms confront a number of daunting commercial risk factors, including the transition from 300- to 450-mm nodes; the looming end of Moore's Law, which has long guided industry R&D and investment patterns; global industry consolidation; and perennially cyclical and unpredictable patterns of market demand. At the same time, the united front of support for semiconductor manufacturing in the Capital Region on the part of New York's public authorities that existed in the mid-2000s is less evident today. Joseph Bruno retired in 2007 and the reliable flow of public funds to local projects has diminished. At the local level, the unity of purpose that characterized public authorities in the mid-2000s appears to have frayed. Saratoga County, for example, which led the extraordinary drive to attract a semiconductor manufacturer, now has two competing economic development authorities.

The state incentives package offered by New York State to AMD in 2006 is widely regarded in New York as a one-off investment needed to attract a manufacturer which will not need to be repeated to support further expansion of local semiconductor manufacturing. That perspective was a principal reason Austin lost Sematech to New York, which offered the consortium large incentives that Austin did not try to match. Daniel Armbrust, President and CEO of Sematech, commented in 2013 that—

> We came to Albany because of shared investments [a reference to state investments]. We share the infrastructure that's been put in here. In Texas we were on our own. R&D costs would have consumed all of our revenue. Most jurisdictions—except New York State and a little bit of federal—have concluded this industry is mature, just let it run.[45]

Today, in terms of state support, New York's established semiconductor manufacturers are increasingly on their own as Sematech found itself in Texas a decade ago. It is not at all clear what state support, if any, will be available with respect to GlobalFoundries' next round of investment for additional fabrication facilities (and the jobs and economic activity they engender), given the apparent view that the industry should now be able to stand on its own. To some extent, that assessment may have merit, but it does not reflect the realities of global competition or the positive effects of multiple rounds of investment on the region. The initial tax-related incentives provided to GlobalFoundries are due to expire beginning in 2019 and the associated

---

[44] Youngstown, Ohio, for example, was historically very heavily oriented toward steel manufacturing. At the end of the 1990s, the steel industry entered a deep recession and 31 plants in or near Youngstown shut down between 2001 and 2006. In 2007, the city's population of 82,000 was roughly half of what it was in 1967. In the 12 years between 2000 and 2012, Youngstown lost 40,500 jobs. A Cleveland-based economist estimated that it would take 52 years for the city to return to its employment levels of 2000. "Mahoning Valley Expert: Region's Recovery Will Take Decades" *Youngstown Vindicator* (September 23, 2012).

[45] Presentation by Daniel Armbrust, National Academies Symposium, *New York's Nanotechnology Model: Building the Innovation Economy*, Troy, New York, April 4, 2013.

reduced power rates shortly thereafter. Reduced taxes and competitive power rates are key in maintaining global competiveness. Will the economic development investment model shift to consider this reality and reflect an understanding that the presence of large, high-value manufacturers ultimately generate much more indirect tax revenue and economic impact and may warrant ongoing tax incentives? Of course, the key question is what will happen not if, but when, another jurisdiction offers a combination of major incentives, including a more stable and predictable public policy and infrastructural environment. In that case, the question arises as to whether GlobalFoundries would build its next fab in New York at all. Box 10.1 describes the relevance of political support to locational decisions in the semiconductor industry.

> **Box 10.1 The Relevance of Political Support to Locational Decisions in the Semiconductor Industry**
>
> Investment in semiconductor manufacturing entails commitment of billions of dollars in capital to build facilities over a span of years in the face of highly unstable market demand and constant, disruptive technological change. Building the massive infrastructure needed to support a wafer fabrication facility may also require years of planning, regulatory approvals, and construction. At the end of this process the manufacturer owns a facility which may or may not actually be utilized depending on unpredictable product decisions by semiconductor end users.[46] Given the normal market risks and uncertainties associated with the semiconductor business, it is not surprising that semiconductor manufacturers have tended to locate their fabrication plants in political jurisdictions where they can be assured of predictable, consistent, and comprehensive government support in terms of regulations, infrastructure, human resources, tax relief, and other financial support—locations such as Dresden, Hsinchu (Taiwan), Shanghai, Singapore, and Austin. A 2002 study of risk factors involved in building a semiconductor fab commented that "balancing enormous financial risk with cyclical market demands is like a no-limit poker game ... Delayed permits, incomplete tool hookups and similar problems can threaten the schedule and budget of the entire project."[47] Given these inherent risks, maintaining a supportive policy environment across the region may prove essential to sustain ongoing investment in one of the world's fastest evolving industries. For example, the current approach being employed by China to attract what they see as the world's most strategic industry incorporates all of the above: addressing regulatory approvals, the availability of infrastructure and human resources, supply chain support, creating a shared R&D ecosystem, reducing operational costs, and very substantial direct financial support.

---

[46] An AMD Board member once observed that building a wafer fab under such conditions is "like Russian roulette [but with a twist.] You pull the trigger and 4 years later you learn whether you blew your brains out or not." Hector Ruiz, *Slingshot: AMD's Fight to Free an Industry from the Ruthless Grip of Intel* (Austin, TX: Greenleaf Book Press, 2013), p. 8.

[47] Katherine Derbyshire, "Building a Fab—It's All About Tradeoffs," *Semiconductor Magazine* (June 2002).

## 10.3.3 A Federal Opportunity?

The fact that this study does not include a chapter on the federal role in the creation of Tech Valley reflects the reality that the effort was driven by state and local resources and leaders. On occasion, the federal government provided valuable and timely support, such as research grants and assistance in roadbuilding near the GlobalFoundries fab, but these contributions were modest relative to the effort as a whole. A question going forward is whether the state's go-it-alone approach will suffice to sustain Tech Valley's momentum, or whether a larger federal role should be encouraged.

Indeed, from the federal government's perspective, there seems to be a growing recognition of the importance of the national security advantages to be derived from these US-based, state of the art facilities. For example, during the development of GlobalFoundries, the offices of President, Vice President, and Commerce Secretary all reached out with an interest to visit the GlobalFoundries site, which led to several high-profile visits. Reflecting the innovative approach adopted by Global Foundries to address workforce needs (see above), GlobalFoundries was asked to give input into a 2010 report of the President's Council of Economic Advisors related to education and workforce development, which underscored the role of community colleges in demand-driven education.[48] This all points to the recognition early-on by the federal government of the importance of the ecosystem and GlobalFoundries, which is the only large, "pure-play" contract chip maker in the United States. Given that the regional development and advancements have occurred to a great degree without federal support, it would seem worthwhile to examine the possibility of greater federal investment as an active partner with the goal of securing and enhancing what is clearly a national asset. While the federal government has some mechanisms to invest in the private sector, establishing vehicles to invest to a much larger degree in commercial enterprises in this very strategic sector should be considered especially for agencies such as the Department of Defense and Department of Energy.

## 10.3.4 The Profusion of Local Government

New York "has one of the most complex networks of local governments of any state."[49] Largely a historical legacy, the entire state is a jigsaw of cities, counties, towns, villages, and a plethora of special-purpose jurisdictions (many overlapping) which include sewer districts, water authorities, highway and bridge authorities, school districts, fire districts, and so on. In 2007, the State Office of Comptroller put the total number of local governments at 53,177 plus 6658 special districts. The multiple approvals required by local governmental units pose a daunting challenge

---

[48] GlobalFoundries' Director of Government Affairs, Mike Russo, lead the national private-sector working group to provide guidance in the establishment of the National Network of Manufacturing Innovation (NNMI), now "Manufacturing USA".

[49] Edward V. Schneier, John Brian Murtaugh, and Antionette Pole, *New York Politics: A Tale of Two States* (Armonk and London: M.E. Sharpe, 2010), p. 24.

to the establishment and operation of large manufacturing plants with complex and extensive infrastructure needs.[50] This challenge was surmounted, in the case of GlobalFoundries, by a unified team of development officials and supportive political leaders, in the context of widespread public support for a semiconductor plant in Saratoga County. A similar broad based coalition is not assured to support future efforts, which potentially could be brought to a halt by one or more recalcitrant jurisdictions, as occurred in North Greenbush in 1999.

## 10.3.5 Other Challenges

In the local interviews conducted to develop this study, regional leaders identified a number of other issues of concern associated with Tech Valley, including distributional equity with respect to economic benefits both within the region and in the context of the state as a whole, the skills shortage (as described in Chap. 9) and the relative dearth of high-tech start-ups.

### 10.3.5.1 Addressing the Financial Difficulties of the Luther Forest Technology Campus

In other regions of the world, semiconductor manufacturing facilities are surrounded by supply chain firms, a phenomenon attributable in substantial part to the administration of the cluster by administering authorities with the resources and power to create the incentives to draw suppliers and service providers. The Luther Forest Technology Campus has neither. It has been struggling financially for years, with the result that efforts to attract supporting firms to the Luther Forest site have been impeded. Given the magnitude of the state's investments to date, administration of the site should be revamped. This could be achieved through state or federal takeover or by providing the existing administration with the additional authority and resources necessary to carry out its mission.

### 10.3.5.2 Ongoing Talent Loss

The recent influx of technology-intensive manufacturing and jobs, while welcome, and impressive, has not reversed broader longstanding trends of high unemployment and out-migration of talent even in the immediate Capital Region itself. The

---

[50] New York is not the only state in the United States characterized by balkanized local government, which is found in a number of older states. Northeastern Pennsylvania, for example, "is divided into hundreds of small political jurisdictions that often compete for residents, businesses, and property tax revenues." The average number of taxing districts in each of Ohio's 88 counties is 50, and as an Ohio foundation director observed in 2010, "the way we are fragmented governmentally impairs our ability to compete globally." <http://www.institutepa.org/PDF/indicators/jeedregionalization12.pdf>; "Philanthropy a Way of Life of Greater Clevelanders," *The Plain Dealer* (December 26, 2010). "Work Together, Or It Won't Work," Albany, *The Times Union* (February 28, 2003).

existence of a "skills gap" throughout the region is a major concern of local high-tech manufacturing firms. Local universities and community colleges confront financial pressures which threaten their ability to support necessary workforce development.

### 10.3.5.3 Distributional Equity

The geographic areas within the Capital Region experiencing the most dramatic economic and employment growth are relatively affluent. Economically distressed areas in Albany, Troy, and other communities in the Capital Region are largely isolated from the "Tech Valley" phenomenon, often reflecting prosaic problems such as the inadequacy or absence of transportation links. State leaders face understandable pressure for geographical diversification of economic development resources to other regions in New York, which may offer substantial returns in growth and employment. The question is not whether these investments are valuable but whether they will detract from efforts to sustain the continued development of the Tech Valley nanotech cluster.

### 10.3.5.4 A Slowly Emerging Start-Up Culture

Unlike Silicon Valley and Boston's Route 128, Tech Valley has not spawned start-ups which grow into major technology-intensive entities (e.g., Intel, Apple, Google). The region's growth trajectory resembles that of North Carolina's Research Triangle, which developed primarily through attracting established companies from outside of the state. The proximity of the Boston and New York metropolitan areas—in many ways an advantage for the Capital Region—undermines local efforts to foster start-ups through the gravitational pull they exert on young people, entrepreneurs, and venture-oriented investment. More focus and funding are needed to support and grow the nascent start-up culture, including early-stage capital through federal and state programs (e.g., SBIR) backed by best practice incubators, accelerators, and mentoring and networking programs.

### 10.3.5.5 The Challenge of Global Competition

In recent years a number of seemingly promising state and regional innovation-based economic development initiatives have foundered, reflecting the impact of global competition and technological change. As noted above, the semiconductor industry is seen as a strategic industry, and foreign governments are consequently prepared to make substantial investments in an attempt to develop this industry within their national economies. Such state-led efforts, backed by massive funding and not subject to market competition, can significantly distort both markets and investment decisions. It would be naive to think that Tech Valley is immune from similar challenges.[51] Global locational competition for semiconductor production facilities is fierce and will accelerate in the years ahead.

From a global perspective, a major source of this competition is the widely shared view that the semiconductor industry is a strategic industry, both as a key to regional economic development and also as a strategic national asset. Empirically, as this study has shown, the industry has a disproportionately positive impact on employment, wages, supply chain development, and national technological capabilities. Consequently, the location of new manufacturing facilities is actively pursued by governments around the world who are more than willing to provide major financial and other incentives.[52] China, for example, has launched a major effort with $150 billion war chest to develop a vertical supply chain for semiconductor production on its own soil.[53]

## 10.4 Looking Ahead

One of the great strengths of Tech Valley has been its continuity, leadership, and commitment, even as administrations have changed and the composition of the New York State Assembly has evolved. The change in leadership at SUNY Poly, the apparent absence of leaders in the Assembly such as Senator Bruno, and the disputes and duplication of economic development efforts in Saratoga have generated uncertainties about the region's commitment and ability to support this rapidly evolving high-tech industry. Fortunately, the new partnership initiative with Applied Materials, a leading semiconductor equipment supplier, and a $300 million partnership with IBM on artificial intelligence and cloud computing have reaffirmed the value of the SUNY Poly facilities and their attraction to leading high-tech companies. Similarly, the recent decision by the Cuomo Administration to support the partnership with Applied Materials in the creation of a $880 million center for manufacturing research has sent a clear signal of the state's ongoing commitment to this exceptional institution, which, along with GlobalFoundries' investment and the presence of GE and Regeneron, are the main pillars of the Capital region's ongoing growth. Nonetheless, the region is clearly facing a period of some turbulence. It will be important to send strong signals of continuity in both political commitment and investment in order to attract and reassure potential investors and maintain the region's reputational advantage.

---

[51] Toledo, Ohio, for example, led by University of Toledo and funded by the state, undertook an effort to transform itself from a city dependent on glass manufacturing into a center for photovoltaic research and manufacturing—but this effort was cut short in and after 2011 by massive Chinese dumping of PV modules which devastated the industry.

[52] State support for semiconductor investments is not a new phenomenon. *See,* for example, Laura Tyson's reference to the visible hand of the government in the development of the semiconductor industry, *Who's bashing whom: Trade conflict in high-technology industries,* (Institute for International Economics, 1993).

[53] President's Council of Advisors for Science and Technology (PCAST) "Report on Ensuring Long-Term U.S. Leadership in Semiconductors" (Washington, DC: Executive Office of the President, January 6, 2017).

## Bibliography[54]

Armbrust, D. (2013, April 4). Presentation at National Academies Symposium, *New York's Nanotechnology Model: Building the Innovation Economy*. Troy, New York.

Boardman, C., and Gray, D. (2010, February). The New Science and Engineering Management: Cooperative Research Centers as Government Policies, Industry Strategies and Organizations. *Journal of Technology Transfer*

Centre for Research on Socio-Cultural Change. (2011). Rebalancing the Economy for Buyer's Remorse. Working Paper No. 87.

Chernock, J., and Youtie, J. (2013, February). State University of New York at Albany Nanotech Complex. In Georgia Tech Enterprise Innovation Institute. *Best Practices in Foreign Direct Investment and Exporting Based on Regional Industry Clusters*. Atlanta: Georgia Tech Research Corporation. Prepared for the Economic Development Administration, U.S. Department of Commerce.

Fury, M. A., and Kaloyeros, A. E. (1993). Metallization for Microelectronics Program at the University of Albany: Leveraging a Long Term Mentor Relationship. *IEEE Xplore*. 59-60.

German Innovation, British Imitation. (1992, November 21). *New Scientist*.

Homburg, C. (Ed.) (1999). *Structure and Dynamics of the German Misselstand* (Heidelberg and New York: Physica-Verlag.

Hudson Valley Community College. *College Catalog* (2015-16).

Jackson, S. A. (2013, April 4). Keynote Address at National Research Council Symposium, "New York's Nanotechnology Model: Building the Innovation Economy." Albany, New York.

Leslie, S. W., and Karagon, R. H. (1996). Selling Silicon Valley: Fredrick Terman's Model for Regional Advantage. *Business History Review*.

Mahoning Valley Expert: Region's Recovery Will Take Decades. (2012, September 23). *Youngstown Vindicator*

National Research Council. (2013a). *21st Century Manufacturing: The Role of the Manufacturing Extension Partnership Program*. C. W. Wessner (Ed.) Washington, DC: The National Academies Press.

National Research Council. (2013b). *Best Practices In State and Regional Innovation Initiatives: Competing in the 21st Century*. C. W. Wessner (Ed.) Washington, DC: The National Academies Press.

National Research Council. (2013c). *New York's Nanotechnology Model: Building the Innovation Economy*. C. W. Wessner (rapporteur). Washington, DC: The National Academies Press.

New York's Big Subsidies Bolster Upstate's Winning Bid for AMD's $3.2 Billion 300-MM Fab. (2006, June 10). *Site Selection*.

Philanthropy a Way of Life of Greater Clevelanders. (2010, December 26). *The Plain Dealer*.

Office of the State Comptroller. (2010). *Fuller Road Management Corporation & The Research Foundation of the State of New York*. (2010-5-5).

PolicyLink. (2008). *To Be Strong Again: Renewing the Promise in Smaller Industrial Cities*. https://www.policylink.org/sites/default/files/ToBeStrongAgain_final.pdf.

President's Council of Advisors for Science and Technology. (2017, January 6). Report on Ensuring Long-Term U.S. Leadership in Semiconductors. Washington, DC: Executive Office of the President.

---

[54] As noted in the front matter of this book, the study also drew on interviews carried out by the authors and numerous articles from *The Times Union* (Albany), *The Daily Gazette* (Schenectady), the *Albany Business Review* (Albany), *The Post Star* (Glens Falls), *The Record* (Troy), *The Saratogian* (Saratoga Springs), *The Buffalo News* (Buffalo), The *Observer-Dispatch* (Utica), The *Daily Messenger* (Canandaigua), and the *Post-Standard* (Syracuse). These are not individually included in the bibliography.

Ruiz, H. (2013). *Slingshot: AMD's Fight to Free an Industry from the Ruthless Grip of Intel.* Austin, Texas: Greenleaf Book Group Press.

Schneier, E. V., Murtaugh, J. B., and Pole, A. (2010). *New York Politics: A Tale of Two States.* Armonk and London: M.E. Sharpe.

Schoolman, M. (1986). Solving the Dilemma of Statesmanship: Reindustrialization Through an Evolving Democratic Plan. In M. Schoolman and A. Magid (Eds.), *Reindustrializing New York State: Strategies, Implications, Challenges.* Albany: SUNY Press.

Schulz, L. I. (2011). Nanotechnology's Triple Helix: A Case Study of the University of Albany's College of Nanoscale Science and Engineering. *Journal of Technology Transfer.*

Tittnich, M., *et al.* (2006). A Year in the Life of an Immersion Lithography Alpha Tool at Albany Nano Tech. In Proceedings of SPIE, Vol. 6151, *Emerging Lithographic Technologies.*

Turning Technologies Rated Fastest-Growing. (2007, August 24). Youngstown, *Vindicator.*

Tyson, L. (1993). *Who's bashing whom: Trade conflict in high-technology industries.* Institute for International Economics.

Wagner, R. W. (2007). *Academic Entrepreneurialism and New York State's Centers of Excellence Policy*, Ph.D. dissertation. SUNY Albany.

Why Doesn't Britain Make Things Any More? (2011, November 16). *The Guardian.*

# Appendices

## Appendix A: Innovation-based Economic Development: Basic Concepts

The New York political, business, and academic leaders who have worked for a generation to revive the Upstate economy have been well aware of experiential reference points such as Silicon Valley and Boston's Route 128, where great research universities drove economic growth and regional prosperity. They also knew local examples of knowledge-based economic growth, such as the foundation of the great General Electric electrochemical research laboratories in Schenectady by an MIT professor, Willis Whitney, at the turn of the twentieth century.[1] In addition to such real-world models, in recent decades state leaders have been able to draw upon a growing body of learning developed in academia that systematically examines the dynamics and geography of knowledge-based regional economic development. These ideas have informed and guided the creation of what has come to be known as Tech Valley.

### Industry Clusters

The phenomenon of the industry cluster, in which enterprises in a given sector group themselves together in a particular location, thereby enhancing their competitiveness, is centuries old. The first systematic examination of clustering was undertaken in the late nineteenth century by British economist Alfred Marshall, who studied the specialty steel firms in Sheffield. He identified three elements that characterize a successful cluster, still referred to today as "Marshall's Trinity"[2]:

---

[1] George Wise, *Willis R. Whitney, General Electric, and the Origins of U.S. Industrial Research* (New York: Columbia University Press, 1985).

[2] National Research Council, Charles W. Wessner (ed.) *21st Century Manufacturing: The Role of the Manufacturing Extension Partnership Program* (Washington, DC: The National Academics Press, 2013), p. 303.

- **Supplier linkages**. *Companies supplying equipment, services, and materials to manufacturers are co-located in the cluster, reducing costs and fostering efficiency.*
- **Labor pool**. *The cluster is characterized by an abundant pool of laborers with the skills required by firms operating in the cluster.*
- **Knowledge spillovers**. *In Marshall's words, "secrets of trade are in the air" in a cluster, giving local firms access to market and technological intelligence, new designs, and new and better production processes.*

Marshall's ideas were carried forward, refined, and popularized in our own time by Michael Porter, who argued in his influential 1990 book, *The Competitive Advantage of Nations,* that in advanced economies, regional clusters of related firms and industries, rather than individual firms or sectors, were the principal source of economic competitiveness as well as rising regional employment and per capita income levels.[3] Porter's work was sufficiently convincing that it has come to dominate the approach to economic development in most advanced countries, including the United States, in which most state and regional development efforts associated with high technology are based on attempts at cluster formation and expansion.[4]

Marshall surmised that advances in transportation and communications would eventually render clusters obsolete, undermining the advantages associated with physical proximity. Instead, in today's high-technology industries, clustering has if anything become more pronounced. In substantial part, this reflects the fact that as manufacturing processes have become increasingly complex, major advantages are conferred by the local presence of key suppliers and service providers. While many manifestations of technology itself, such as designs, equipment, and parts, can be transmitted over long distances, so-called tacit knowledge, more colloquially known as "know-how," is best transferred on a face-to-face basis in a single location through demonstration and coaching, much as occurs in an industrial apprenticeship. In highly automated and complex manufacturing operations, involving the close interaction of many machines, "know-how" based on actual experience and observation is placed at a premium. For this reason, common instruments for development, refinement, and transmission of know-how are pilot manufacturing facilities in which new tools and processes can be tested and demonstrated and the know-how passed on to others in a face-to-face setting.[5]

---

[3] See also Michael Porter, "Clusters and the New Economics of Competition," *Harvard Business Review* (December, 1998).

[4] Mary J. Watts, "The Added Value of the Industry Cluster Approach to Economic Analysis, Strategy Development, and Service Delivery," *Economic Development Quarterly* 14(1) (February 2000). See also Stefano Breschi and Franco Malerba, *Clusters, Networks and Innovation* (Oxford: Oxford University Press, 2005).

[5] Michael Polanyi, a scientist who studied tacit knowledge, described it in 1958 as "an art which cannot be specified in detail, cannot be transmitted by prescription, since no prescription for it exists. It can be passed on only by example from master to apprentice. This restricts the range of diffusion to that of personal contacts. We find accordingly that craftsmanship tends to survive in closely circumscribed local traditions." Michael Polanyi, *Personal Knowledge: Toward a Post-Critical Philosophy* (Chicago: University of Chicago Press, 1958), p. 52. See also Ronald J. Gilson,

## The Key Role of Research Universities

In the high-technology industries, the most successful clusters are located near one or more high-quality research universities that educate and train students in disciplines relevant to local firms and conduct research closely linked to activities in the cluster. Graduates from these universities staff established companies and start up new ones. In Silicon Valley and Boston, Stanford and MIT, respectively, have for decades spun out streams of start-up companies that have evolved into extraordinarily successful clusters based on the information technologies and biotechnology, enhancing employment and raising living standards in both regions. North Carolina's Research Triangle Park, a centrally planned development based on three universities, succeeded in attracting numerous established high-tech companies to the region, ultimately transforming one of the poorest regions in the southeastern United States into one of the wealthiest.[6]

## Government and Innovation

In the United States the notion has long prevailed at the federal level that the government should not engage in industrial policy or pick "winners and losers" among industries and firms. However, exceptions have been made with respect to agriculture and industries associated with national defense, health, and energy security. The federal government has also supported basic research and innovation, particularly by small firms, through a broad array of decentralized programs. State governments have proven far less reticent in pursuing explicit industrial policies.[7] Although state economic development strategies once emphasized use of incentives to lure companies from other states and regions, these have been augmented by policies to encourage innovation and entrepreneurship, with state university systems functioning as economic development arms of state governments.[8]

---

"Legal Infrastructure of High Technology Industrial Districts: Silicon Valley, Route 128, and Covenants Not to Compete," *New York University Law Review* 74(3) (1999), p. 577; Eugene S. Ferguson, *Engineering and the Mind's Eye* (Cambridge, MA: The MIT Press, 1992), p. 56.

[6] See summary of remarks by Rick L. Weddle in National Research Council, Charles W. Wessner (ed.), *Understanding Research, Science and Technology Parks: Global Best Practices* (Washington, DC: The National Academies Press, 2009).

[7] Peter K. Eisinger, *The Rise of the Entrepreneurial State: State and Local Economic Development Policy in the United States* (Madison: University of Wisconsin Press, 1988).

[8] Walter H. Plosila, "State Science and Technology—Based Economic Development Policy: History, Trends and Developments and Future Directions," *Economic Development Quarterly*, 18(2) (2004); Irvin Feller, "Evaluating State Advanced Technology Programs," *Education Review* 12(3) (1988); Denis O. Gray, "Cross-Sector Research Collaboration in the USA: A National Innovation System Perspective," *Science and Public Policy* (March 2011); Jan Youtie and Philip Shapira, "Building on Innovation Hub: A Case Study of the Transformation of University Roles in Regional Technological and Economic Development," *Research Policy* 37(2008); Nathan Rosenberg and Richard R. Nelson, American Universities and Technical Advance in Industry," *Research Policy* 23 (1994).

## General Purpose Technologies

Nanotechnology is recognized as a general purpose technology (GPT)—that is, a technology that can dramatically affect an entire economy over time. Other examples include the internal combustion engine, electricity, computers, biotechnology, and the Internet.[9] GPTs affect many sectors, improve over time, and are available to users at progressively lower cost and enable invention of new products and processes. Marianna Mazzucato of the University of Sussex cites considerable academic analysis and empirical research in support of the proposition that large-scale, long-term government investment "has been the engine behind almost every GPT in the last century."[10] The private sector standing alone cannot be relied upon to develop GPTs because of the long developmental time frame involved, the level of risk, and the difficulty in finding sufficient volumes of "patient" investment capital.[11]

In 2000 President Clinton announced the National Nanotechnology Institute, signaling a major national commitment to a potentially revolutionary technology with implications across the economic spectrum. A number of state governments launched their own nanotechnology promotion efforts, but over time New York's has been the largest, most focused, and most successful.

## Triple Helix

In recent decades, the role played by universities in the development of regional high-technology industry clusters has been articulated in academia and refined into a formal model by Henry Etzkowitz, Loet Leydesdorff, and others—the so-called Triple Helix, which has been widely embraced by economic development professionals and policymakers.[12] A Triple Helix is a heavily networked combination of university, industry, and government organizations collaborating—usually through hybrid structures—in the promotion of innovation. In a Triple Helix each party takes on some of the roles of the other parties without abandoning their primary role. Thus, the university goes beyond its traditional duties of teaching and research and assumes an entrepreneurial function and the mission of regional economic development. Companies take on an educational role, training students, and work-

---

[9] The term "general purpose technology" was originated by Timothy F. Bresnahan and Manual Trajtenberg in 1995 and has found widespread applications in innovation studies. "General Purpose Technologies: Engines of Growth?" NBER Working Paper No. w4148. (Boston: National Bureau of Economic Research, 1995).

[10] Marianna Mazzucato, *The Entrepreneurial State: Debunking Public* vs. *Private Sector Myths* (London, New York and Delhi: Anthem Press, 2013), p. 62.

[11] Vernon W. Ruttan, "Is War Necessary for Economic Growth?" (Clemons Lecture, Saint Johns University Collegeville, MN, October 9, 2006), p. 30.

[12] Henry Etzkowitz, *The Triple Helix: University—Industry—Government Innovation in Action* (New York and London: Routledge, 2008); Henry Etzkowitz and Loet Leydesdorff, "The Dynamics of Innovation: From National Systems and 'Mode 2' to a Triple Helix of University—Industry—Government Relations," *Research Policy* (2000).

ers in the skillsets needed for advanced manufacturing. Regional and local governments may take partial responsibility for nontraditional functions such as ensuring that the necessary university research infrastructure exists, that promising startups have access to capital, and that manufacturers are supported with the specific, sometimes specialized infrastructure they need in order to operate.[13]

New York's development of a nanotechnology cluster centered on Albany's College of Nanoscale Science and Engineering (CNSE) was based on the Triple Helix model and is now cited as an example of successful application of Triple Helix principles. A professor at CNSE, Laura I. Shultz, observed in 2010 that—

> *The CNSE is an example of the implementation of the Triple Helix model. Since it was established in 2001, the CNSE has been successful in enabling the research of industrial partners such as IBM, Tokyo Electron, and SEMATECH as well as spurring economic development in NY's Capital Region.*[14]

University–industry–government collaborations to create regional economic growth have been studied extensively by leading policy-focused institutions. In 2007 the National Governors Association and the Council on Competitiveness released a report on cluster-based economic growth singling out successful Triple Helix-type collaborations.[15] The National Academies have convened a series of symposia and published reports on cluster-based economic development and competitive strategies.[16] In recognition that New York had re-emerged as a leading innovation region, these included a symposium and report on New York's nanotechnology-based development efforts in the Capital Region.[17] The Brookings Institution has published a number of studies of innovation clusters noting, among other things, the increasingly broad acceptance by policymakers of the validity of the approach:

> *Twenty years after Harvard Business School professor Michael Porter introduced the concept to the policy community and 10 years after its wide state adoption, clusters – geographic concentrations of interconnected firms and supporting or coordinating organizations – have reemerged as a key tool and rubric in Washington and in the nation's economic regions. After a decade of delay, the executive branch and Congress have joined state and local policymakers in embracing "regional innovation clusters" (RICs) as a framework for structuring the nation's economic development activities. At the state level,*

---

[13] Laura I. Schultz, "University Industry Government Collaboration for Economic Growth," in Jason E. Lane and D. Bruce Johnstone (Eds.), *Universities and Colleges as Economic Drivers: Measuring Higher Education's Role in Economic Development* (Albany: SUNY Press, 2012).

[14] Laura I. Schultz, "Nanotechnology's Triple Helix: A Case Study of the University of Albany's College of Nanoscale Science and Engineering," *Journal of Technology Transfer* (2011), p. 548.

[15] National Governors Association and Council on Competitiveness, *Cluster-Based Strategies for Growing State Economies* (Washington, DC: National Governors Association, 2007).

[16] National Research Council, Charles W. Wessner (ed.), *Understanding Research, Science and Technology Parks: Global Best Practices* (Washington, DC: The National Academies Press, 2009).

[17] National Research Council, Charles W. Wessner (rapporteur), *New York's Nanotechnology Model: Building the Innovation Economy* (Washington, DC: The National Academies Press, 2013).

*governors and gubernatorial candidates of both parties are maintaining or stepping up their longstanding interest.*[18]

The increasingly widespread effort to establish and grow innovation-based industrial clusters has demonstrated that there is no magic formula for doing so and that sustained, difficult, long-term efforts and investments are usually required. Successful clusters are rooted in the traditions, culture, and historical idiosyncrasies of a particular region, with the result that "replicating a successful cluster model elsewhere can be highly elusive."[19] In an influential 1994 study, buttressed by subsequent work, Annalee Saxenian contrasted the culture of clusters in Silicon Valley with that of Boston's Route 128 and concluded that Silicon Valley's traditions of risk-taking, informality, open exchange of information, and labor mobility enabled it to outpace Boston in the computer and information technologies, where these attributes were not present, at least to the same degree.[20]

## Bibliography[21]

Breschi, S., and Malerba, F. (2005). *Clusters, Networks and Innovation*. Oxford: Oxford University Press.

Bresnahan, T. F., and Trajtenberg, M. (1995). *General Purpose Technologies: Engines of Growth?* NBER Working Paper No. w4148. Boston: National Bureau of Economic Research.

Cortright, J. (2006, March). *Making Sense of Clusters: Regional Competitiveness and Economic Development*. Washington, DC: Brookings.

Eisinger, P. K. (1988). *The Rise of the Entrepreneurial State: State and Local Economic Development Policy in the United States*. Madison: University of Wisconsin Press.

Etzkowitz, H., and Leydesdorff, L. (2000). The Dynamics of Innovation: From National Systems and 'Mode 2' to a Triple Helix of University—Industry—Government Relations. *Research Policy*.

Feller, I. (1998). Evaluating State Advanced Technology Programs. *Education Review*. 12(3).

---

[18] Bruce Katz and Mark Muro, "The New 'Cluster Moment': How Regional Innovation Clusters Can Foster the Next Economy," (Washington, DC: Brookings, September 21, 2010); See also Mark Muro, "Regional Innovation Clusters Begin to Add Up," (Washington, DC: Brookings, February 27, 2013); Joseph Cortright, "Making Sense of Clusters: Regional Competitiveness and Economic Development," (Washington, DC: Brookings, March 2006).

[19] National Research Council, Charles W Wessner (rapporteur) *Growing Innovation Clusters for American Prosperity* (Washington DC: The National Academies Press, 2011), p. 8.

[20] AnnaLee Saxenian, *Regional Advantage: Culture and Competition in Silicon Valley and Route 128*. (Cambridge, MA: Harvard University Press, 1994).

[21] As noted in the front matter of this book, the study also drew on interviews carried out by the authors and numerous articles from *The Times Union* (Albany), *The Daily Gazette* (Schenectady), the *Albany Business Review* (Albany), *The Post-Star* (Glens Falls), *The Record* (Troy), *The Saratogian* (Saratoga Springs), *The Buffalo News* (Buffalo), The *Observer-Dispatch* (Utica), The *Daily Messenger* (Canandaigua), and the *Post-Standard* (Syracuse). These are not individually included in the bibliography.

Ferguson, E. S. (1992). *Engineering and the Mind's Eye*. Cambridge, MA: The MIT Press.

Gilson, R. J. (1999). Legal Infrastructure of High Technology Industrial Districts: Silicon Valley, Route 128, and Covenants Not to Compete. *New York University Law Review.* 74(3), 577.

Gray, D. O. (2011, March). Cross-Sector Research Collaboration in the USA: A National Innovation System Perspective. *Science and Public Policy.*

Katz, B., and Muro, M. (2010, September 21). The New 'Cluster Moment': How Regional Innovation Clusters Can Foster the Next Economy. Washington, DC: Brookings.

Mazzucato, M. (2013). *The Entrepreneurial State: Debunking Public* vs. *Private Sector Myths*. London, New York and Delhi: Anthem Press.

Muro, M. (2013, February 27). Regional Innovation Clusters Begin to Add Up. Washington, DC: Brookings.

National Governors Association and Council on Competitiveness. (2007). *Cluster-Based Strategies for Growing State Economies*. Washington, DC: National Governors Association.

National Research Council. (2009). *Understanding Research, Science and Technology Parks: Global Best Practices*. C. W. Wessner (Ed.) Washington, DC: The National Academies Press.

National Research Council. (2011). *Growing Innovation Clusters for American Prosperity*. C. W. Wessner (rapporteur). Washington DC: The National Academies Press.

National Research Council. (2013). *New York's Nanotechnology Model: Building the Innovation Economy*. C. W. Wessner (rapporteur). Washington, DC: The National Academies Press.

Plosila, W. H. (2004). State Science and Technology—Based Economic Development Policy: History, Trends and Developments and Future Directions. *Economic Development Quarterly.* 18(2).

Polanyi, M. (1958). *Personal Knowledge: Toward a Post-Critical Philosophy.* Chicago: University of Chicago Press.

Porter, M. (1998, December). Clusters and the New Economics of Competition. *Harvard Business Review.*

Rosenberg, N., and Nelson, R. R. (1994). American Universities and Technical Advance in Industry. *Research Policy* 23:326.

Ruttan, V. W. (2006, October 9). *Is War Necessary for Economic Growth?* Clemons Lecture, Saint Johns University, Collegeville, MN.

Schultz, L. I. (2012). University Industry Government Collaboration for Economic Growth. In J. E. Lane and D. B. Johnstone (Eds.), *Universities and Colleges as Economic Drivers: Measuring Higher Education's Role in Economic Development*. Albany: SUNY Press.

Watts, M. J. (2000, February). The Added Value of the Industry Cluster Approach to Economic Analysis, Strategy Development, and Service Delivery. *Economic Development Quarterly.* 14(1).

Wise, G. (1985). *Willis R. Whitney: General Electric and the Origins of US Industrial Research.* New York: Columbia University Press.

Youtie, J., and Shapira, P. (2008). Building on Innovation Hub: A Case Study of the Transformation of University Roles in Regional Technological and Economic Development. *Research Policy.* 37.

## Appendix B: The Foreign Innovation Challenge

Erosion of the US manufacturing base was one of the most important issues in the 2016 presidential election and is likely to remain at the center of the national debate for years to come. The Great Recession dealt a severe blow to the US manufacturing sector, which was already struggling for decades in the face of intensifying international competition.[22] While there have been recent indicators of recovery in the US manufacturing, reflecting, among other things, declines in natural gas prices and increasing factory automation, the challenges facing the US manufacturing remain stark. Virtually all analyses conclude that because the United States cannot and should not compete with foreign manufacturers on the basis of low wages, it must do so through innovation and the quantum jumps in labor productivity which innovation makes possible. However, despite an unparalleled research base and spectacular successes in some high-tech sectors, a recent study by the President's Council of Advisors in Science and Technology concluded that the United States was losing competitive leadership, not to China, but to other high-wage countries such as Germany and Japan in technology-intensive manufacturing industries employing high-skilled workers.[23] This reflects not only significant holes which have opened up in the US innovation infrastructure but also the impact of comprehensive, well-funded efforts by our leading competitors to institutionalize innovation throughout their advanced manufacturing industries.

---

[22] In mid-2013, roughly 12 million Americans were employed in manufacturing, down from about 20 million in the late 1970s. Martin Neil Baily and Barry P. Bosworth, "US Manufacturing; Understanding Its Past and Potential Future," *The Journal of Economic Perspectives* 28(1) (2014). Between December 2007 and December 2009 alone, the United States lost two million manufacturing jobs, or 17% of the total workforce, and employment fell to its lowest level since March 1941, when the United States was still mired in the Depression. Megan M. Barker, "Manufacturing Employment Hard Hit During the 2007–09 Recession," *Monthly Labor Review* (April 11, 2011).

[23] Presidents Council of Advisors on Science and Technology (PCAST), *Report to the President on Ensuring American Leadership in Advanced Manufacturing* (Washington, DC: Executive Office of the President, June 2011); Pamela M. Prah, "Has US Manufacturing's Comeback Stalled?" *USA Today* (July 30, 2013).

**US Vulnerability in Innovation**

The United States, with the world's foremost university system, simply has no peer anywhere in the world in the field of basic scientific research. However, as numerous studies have concluded with respect to the United States and other leaders in basic research such as the United Kingdom, France, and Canada, an abiding challenge has been the ability to translate scientific knowledge into commercially relevant products and processes which enhance the competitiveness of domestic industry and enable the retention and expansion of domestic employment.[24]

Historically, individual US companies have demonstrated great competence in converting scientific discoveries into commercial products and processes, but as a recent MIT study concluded, these firms are increasingly anomalies in the US industrial landscape.[25] The creation of the Albany Nano College represents one effort by a single state to address this challenge, but at the national level more is needed to enhance the US competitiveness. The recent initiative by the federal government to create a National Network for Manufacturing Innovation, now known as Manufacturing USA™, is an important step to address major gaps in the US innovation system. The promise of this network of public-private research institutes for applied research is reflected in the substantial private-sector participation and the industry contributions to the institutes.[26]

While this initiative shows great promise, other advanced countries already enjoy well-established institutional structures systematically promoting and facilitating innovation, which have been operating for decades. The best foreign public-private research organizations are not only well funded and well equipped but have also developed and refined the ability to transform scientific discoveries—whether domestic or foreign—into successful industrial products and processes. The existence, growth, and increasing sophistication of these foreign institutions is an important factor underlying the dynamic in which scientific discoveries are commercialized outside of the United States, resulting in little or no US employment while creating thousands of high-skilled jobs in other countries and leading to the

---

[24] Erich Bloch, "Seizing US Research Strength," *Issues in Science and Technology* (Summer 2003); Roli Varma, "Changing Research Cultures in US Industry," *Science, Technology and Human Values* (Autumn 2000); Tom Brzustowski, *Why We Need Move Innovation in Canada and What We Must Do to Get It* (Ottawa: Invenire Books, 2012), p. 247; Hermann Hauser, *The Current and Future Role of Technology Centres in the UK* (London: UK Department for Business, Innovation and Skills, 2010).

[25] Suzanne Berger, *Making in America: From Innovation to Market* (Cambridge, MA and London: MIT Press, 2013), p. 58.

[26] A federal FAQ with respect to NNMI characterized the issue as follows: "Q: Don't we already have many federal programs that achieve the goals of the NNMI? A: The simple answer is no. There are no current federal programs that have the required attributes to significantly influence the nation's competitiveness and successfully bridge 'the missing middle' in the manner and magnitude proposed for the NNMI." See "National Network for Manufacturing Innovation Frequently Asked Questions," *http://manufacturing.gov/docs/nnmi.faq.pdf* (accessed June 7, 2013).

tacit knowledge necessary for further advance.[27] The challenge thus posed by foreign public-private research organizations should be a stimulus as well as a source of best practice to be considered by US policymakers (both federal and state) in considering how to improve US manufacturing innovation.

## Connecting Research with the Market

All countries seeking to improve the relative competitiveness of their manufacturers must address the question of how to connect scientific discoveries taking place in their research base with the practical needs of industry. While cultural and regulatory factors always play a role in facilitating or impeding this process, successful innovating countries have developed a variety of institutional intermediaries which bridge the gap between the science base and the market. In the United States and the United Kingdom, this role was long performed by large corporate research laboratories such as Bell Labs (US) and AT&T Laboratories Cambridge (UK), but these have been radically downsized or eliminated over the past several decades. By contrast, some other advanced countries have developed large and highly successful public-private research intermediary organizations. Two of the most successful are located in Germany and Taiwan (see Table B.1):

- In **Germany** the Fraunhofer-Gesellschaft, a registered association under private law, is a network of over 60 institutes for applied research, with an average staff of 400 each, which performs research for and in collaboration with private industry, drawing on the knowledge base of the German universities and government basic research organizations.
- In **Taiwan** the Industrial Technology Research Institute (ITRI) is a not-for-profit research organization loosely supervised by the Ministry of Economic Affairs and which draws on the expertise of Taiwanese universities and foreign research partners to provide contract research for Taiwanese industries.[28]

Both institutions have had remarkable success. Indeed, it is difficult to overstate the impact and achievements of these two research organizations in the commercial realm:

---

[27] Such U.S.-invented "lost" technologies invented in the United States but produced mainly abroad include liquid-crystal displays, solar cells, wafer steppers, laptop computers, oxide ceramics, interactive video games, lithium-ion batteries, and various types of industrial robots. Presidents Council of Advisors on Science and Technology (PCAST), *Report to the President on Ensuring American Leadership in Advanced Manufacturing* (Washington, DC: Executive Office of the President, June 2011), pp. 4–5.

[28] Taiwan's Industrial Technology Research Institute (ITRI) is "arguably the most capable institution of its kind in the world in scanning the global technological horizon for developments of interest to Taiwanese Industry, and executing the steps required to import the technology—either under license or joint development—and then absorbing and adopting the technology for Taiwanese firms to use." John A. Mathews and Dong-Sung Cho, *Tiger Technology: The Creation of a Semiconductor Industry in East Asia* (Cambridge: Cambridge University Press, 2000).

Appendices

**Table B.1** Examples of successful public-private research intermediary organizations

|  | Fraunhofer | ITRI | IRAP | Catapult | Carnot |
|---|---|---|---|---|---|
| Country | Germany | Taiwan | Canada | United Kingdom | France |
| Direct supervisory authority | None | Ministry of Economic Affairs | National Research Council of Canada | Technology Strategy Board | National Agency Research |
| Form of entity | Private not-for-profit association | Government-owned research institute | Government program | Various private and public organizations | Public research institutions |
| Geographic footprint | Widely distributed across Germany | One main site in Hsinchu, one beta site in Tainan | Widely distributed but heavily concentrated in Quebec and Ontario | Plans for distribution across the United Kingdom | Distributed across France |
| Prototype development for companies | Yes | Yes | No | ? | Yes |
| Pilot lines/simulation platforms on premises | Yes | Yes | No | Yes | Yes |
| Company personnel can work on-site and use facilities | Yes | Yes | N/A | Yes | Yes |
| Spin-offs | Yes | Yes | No | ? | Yes |
| Number of institutes | 60 | 1 | 18 | 7 | 34 |
| Staff | 20,000 | 5728 | 4000 | Evolving | 19,000 |
| Patents | 6131 | 17,569 | N/A | N/A | 880/year |
| Annual "Core" Govt. Funding (millions of dollars) | 723 | 300 | 90 | 65 | 79 |

- The **Fraunhofer** system has been widely credited with Germany's reputation for high-quality engineering and manufactured products, for Germany's export performance, and for the country's singular ability to retain onshore manufacturing and manufacturing jobs in the face of intensifying competition from Asia.[29]

---

[29] *See generally* House of Commons, Committee on Science and Technology, "Second Report: Technology and Innovation Centres," February 9, 2011; Interview with Ekkehard Schulz, CEO of ThyssenKrupp, in "We are Not Driving the Price Hike," *Spiegel Online* (July 16, 2008); "Welcome to Berlin, Peter—Is it the Future?" *Sunday Telegraph* (February 7, 2010).

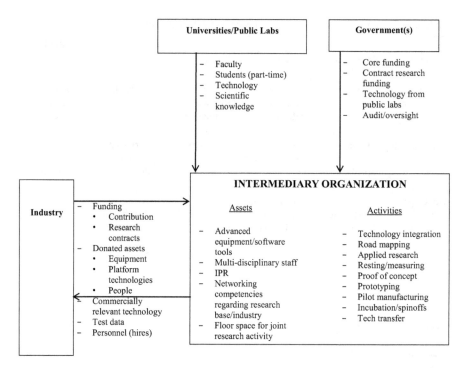

**Fig. B.1** Typical structure and functions of a public-private research intermediary organization. Source: National Research Council, *21st Century Manufacturing: The Role of the Manufacturing Extension Partnership Program* (Washington, DC: The National Academies Press, 2013), p. 144

- **ITRI** was arguably the principal factor in Taiwan's ability to transform itself from an agrarian, low-wage economy into a high-tech powerhouse within the course of a single lifetime.[30]

The Fraunhofer and ITRI differ in many respects, but they share the key role of integration—that is, bringing together both multiple fields of science and technology ("silo-breaking") and the different cultures of academia and industry to achieve common practical objectives in the form of commercially relevant products and industrial processes. In effect, they are successful exemplars of what has become known in academia as the "Triple Helix," a collaboration of universities, governments, and industry, usually through hybrid intermediary organizations like Fraunhofer and ITRI.[31] Figure B.1 illustrates the typical structure and functions of a public-private research intermediary organization.

---

[30] ITRI "has played an integral role in transforming Taiwan's economy from a low-tech, labor-intensive model to a high-tech, knowledge-based industrial core," Allen Hsu, "ITRI Pushes Technology Sector to New Frontiers of Innovation," *Taiwan Journal* (October 19, 2007).

[31] Henry Etzkowitz, *The Triple Helix: University-Industry-Government Innovation in Action* (New York and London: Routledge, 2008).

While it is generally acknowledged that successful innovation requires breaking down cultural and institutional barriers between scientific disciplines and academic research, on the one hand, and private industry, on the other hand, the task itself continues to vex policymakers in many advanced countries. The institutional structures and practices of Germany and Taiwan, which have achieved a degree of success in this area, are therefore relevant.

Technology Integration

ITRI has established internal technology integration centers which are tasked with integrating the research and technology generated by the institute's "core" laboratories, which specialize in discrete fields such as materials, measurement, and electronics. In emerging industries, these centers have overseen the development of capabilities with respect to materials, production equipment, process technology and integration, components, measurement, testing, and certification. These activities are conducted in collaboration with Taiwanese companies, many of them small- and medium-sized businesses, that are ultimately expected to specialize in one or more thematic areas as the emerging industry enters the commercial phase. In the words of one ITRI engineer, the centers' integrated research projects represent the early stages of what will become a complete industry chain. Taiwan is fortunate in that the ties between academia and industry are already very close and that industry can access the research base through multiple informal channels, such as alumni networks.

The situation is similar in Germany. Germany's FHG is internally structured to ensure that multiple technological disciplines are brought to bear on the problems of its industrial customers. Research contracts are served by pairs of "business units" and "research units," with each research unit specializing in a relevant technology area. The business unit ascertains the types of technology that are needed and which thematic Fraunhofer units should be called upon to jointly perform developmental research with respect to equipment, materials, processes, etc., while the technology unit reaches out to relevant expertise in the highly organized German research base. This system is reinforced by formidable internal communication structures that enable staff to ascertain which competencies exist throughout Germany and which relevant projects are under way throughout the Fraunhofer system.

University–Industry Integration

Fraunhofer and ITRI succeed as intermediaries in substantial part as a result of their deep ties with both the research community—primarily universities—and private companies. The extent of the interrelationships between the Fraunhofer and ITRI, on the one hand, and domestic universities, on the other hand, is such that the boundaries between institutions can become blurred. Each Fraunhofer institute is formally linked with one or more German universities with research competencies in the institute's thematic specialty. The director of every Fraunhofer is also a faculty member at the partner university. Such director/faculty members use their university roles to identify and steer the most promising students into working part-time at the Fraunhofer. Challenges arising out of the Fraunhofers' contract research for German industry become the basis for course offerings and dissertation topics at the university. Fraunhofer personnel teach university courses and may also enroll in them. Similarly, ITRI is located adjacent to two of Taiwan's best scientific research universities, Tsing Hua and Chiao Tung. The two universities supply many of ITRI's scientists and engineers, and researchers from the two universities frequently participate in ITRI's projects. Many leading ITRI managers serve as faculty at these institutions.

On the industry side, both ITRI and Fraunhofer develop commercial product prototypes for industrial clients. The Fraunhofer sponsors and supports offsite R&D centers located on company premises and may extend guest status to industrial clients who can operate research labs on company premises. Fraunhofer executives sit on company boards, and each Fraunhofer institute is advised by an advisory board that includes industry executives. ITRI's Technology Integration Centers work closely with local Taiwanese companies to establish industrial chains for emerging manufacturing industries, engaging in contract R&D services evaluation and verification of customers' materials, equipment, production processes and systems, and joint R&D. ITRI scientists and engineers commonly move on to Taiwanese companies, frequently rising to senior leadership positions.[32]

**Major Physical Sites for Shared Equipment and Cooperative Research**

German, Taiwanese, and British public-private applied research organizations operate at very substantial physical sites with relevant equipment, research infrastructure, and trained personnel, often operating research manufacturing lines as well as proof-of-concept and prototypic facilities. They are staffed ith individuals with superb engineering, process, and design competencies and possess deep intellectual property portfolios. In Germany's case, the Fraunhofer is a beneficiary of "the power and generosity ... of Germany's machine tool industry," which enables its laboratories to be equipped with state-of-the-art equipment made available on concessional terms.[33] For example, Germany's Fraunhofer Institut Fur Solare Energiesysteme (ISE) is Europe's largest solar energy research institute, with 1100

---

[32] The so-called "RCA 37" was a storied team of ITRI engineers that went to RCA's U.S. facilities in the 1970s to study and eventually transfer semiconductor technology to Taiwan, providing the genesis of Taiwan's semiconductor industry. Out of the RCA 37, "virtually the entire senior echelons of the subsequent semiconductor industry in Taiwan [were] formed." John A. Mathews, "The Hsinchu Model: Collective Efficiency, Increasing Returns and Higher Order Capabilities in the Hsinchu Science-Based Industry Park, Taiwan," Keynote Address, Chinese Society for Management of Technology 20th Anniversary Conference, Hsinchu, Taiwan (December 10, 2010).

[33] "Allied Research: The Fraunhofer Method," *Industries et Techniques* (October 20, 1987), JPRS-ELS-88-006.

employees and offices, test fields and laboratories occupying 27,000 square meters of floor space. It operates five accredited testing units for solar photovoltaic, solar thermal, and power electronics technologies. With regard to technological expertise, in 2015 this institute set a new world record for solar energy conversion and in 2014 set a world record for photovoltaic solar module efficiency.[34] There are over 60 similar Fraunhofers distributed across Germany and across the technology spectrum. In Taiwan, ITRI operates complete demonstration manufacturing lines and clean rooms on its premises; however, its efforts to expand to other sites apparently have not seen the same success.[35]

## Large and Sustained Public Investments

The most successful foreign manufacturing research organizations receive very substantial government financial assistance over a long time horizon. That fact shields them from the vagaries of the budgeting process and enables them to build up formidable cadres of professionals who do not need to devote a substantial proportion of their time searching for their next job in the shadow of prospective budget cuts. Such organizations receive between 10 and 70% of their funding in the form of public outlays that are not linked to the performance of research contracts.[36] In Germany a peculiar federal-state concordat reached in 1975 locks in federal and state core funding for the Fraunhofer on a 90:10 ratio every year, eliminating the existential uncertainties that characterize public funding of research in some advanced countries.[37]

This is important. As the United Kingdom's Hermann Hauser concluded in his landmark 2010 study, a need exists for "continued core funding for three functions: strategic high-risk research of medium- to long-term duration, competence development, and the acquisition and maintenance of large-scale facilities and specialist equipment." The lack of core funding was a principal factor underlying the failure of the United Kingdom's Faraday Centres, the country's first effort to emulate the Fraunhofer model, in the 1990s.[38]

---

[34] "Fraunhofer ISE Sets Another Solar Cell Efficiency Record," *Clean Technical* (September 25, 2015); "Fraunhofer ISE Sets PV Module Efficiency Record of 36.7 Percent," *Semiconductor Today* (July 14, 2014).

[35] ITRI's Display Technology Center, for example, operates a 3124 square meter second-generation laboratory pilot line (glass substrate size 370 × 470 mm) which in 2012 was producing flexible 20-in. thin film transistor liquid crystal displays (TFT-LCDs), a future-generation form of LCDs.

[36] The Fraunhofer received $723 million in "core" public funding in 2012, which was substantially augmented by publicly funded contract research and various types of funding from the European Union. Taiwan's ITRI received roughly $300 million in the same year.

[37] Markus Winnes and Uwe Schimack, *National Report*: "*Federal Republic of Germany*" (Institute for the Study of Societies, 1999).

[38] Andrew Webster "Bridging Institutions: The Role of Contract Research Organizations in Technology Transfer," *Science and Public Policy* (April 1994).

## Matching Industry Funding

Most successful foreign research intermediary organizations require funding or co-funding of research projects by participating companies. The Fraunhofer headquarters rewards individual Fraunhofer institutes according to a formula that assesses their relative success in obtaining private-sector funding: the more private money an institute is able to draw, the more funding the organization receives from headquarters. The effect of this practice is to direct government research funding toward thematic areas regarded as important by industry. On the other hand, some argue that this close relationship shifts Fraunhofer research toward research that, although valuable, is primarily incremental in nature, tending to exclude paradigm-shifting technologies.

## Multi-Generational Research Projects

Both ITRI and the Fraunhofer support research projects within certain thrust areas that provide the basis for follow-on projects over a span of many years. ITRI' s support for Taiwan's semiconductor industry, for example, began in the mid-1970s and has continued down to the present day, and it has branched off into related thematic areas like photovoltaics and liquid crystal displays. While most of the Fraunhofer's projects are 2 years or less in duration, many of them are related, mutually reinforcing, and conducted with the same core group of companies. For example, in the 1990s, ten Fraunhofer institutes with over 1000 employees were engaged in projects related to factory automation, addressing themes such as micromechanics, control systems, robotics, sensors, and CAO-CAM. This effort was a factor underlying German leadership in adopting factory automation.

## Support for Small Business

The Fraunhofer and ITRI regard support for small- and medium-sized enterprises (SMEs) as a key part of their mission. Most German SMEs cannot afford internal research departments and rely on external sources for technological improvements, with the Fraunhofer representing perhaps the single most important source of support. The resources of the Fraunhofer include not only the most advanced equipment, a large intellectual (IP) portfolio, and highly skilled and knowledgeable personnel but also include highly refined connections throughout the German, European, and global resource bases that can be brought to hear on behalf of small German companies.

The Fraunhofer has proven particularly important to the so-called Mittelstand, which are small- and medium-sized German firms which specialize in niche industrial sectors which they commonly dominate through superior quality and incremental innovation. The Mittelstand concentrate at the high end of comparatively obscure industrial products and make continuous incremental improvements to the products themselves and in the processes through which they are made, not enabling production of the cheapest products but rather those of the highest quality. A 2007 study found that roughly 1130 German SMEs held the number one or number two position in the world market for their products, or the number one position in the European market.[39] The Fraunhofer's research infrastructure underlies this leadership, with "the research facilities of the Fraunhofer [serving] as external, very well equipped research departments of the Mittelstand companies."[40]

In Taiwan, the economy is dominated by small businesses which cannot afford advanced R&D and which are vulnerable to competition from foreign multinationals. ITRI has addressed this challenge, in part, by transferring technology and skills to small firms which are sufficient to enable them to perform contract work within the supply chains of large multinationals and, in effect, to compete effectively with their foreign supply chain counterparts. ITRI also organizes the small firms to form "technology alliances" and complete supply chains for large Taiwanese enterprises, encouraging them to divide up and specialize in various research, design, and manufacturing tasks, so that they do not compete with each other. The end result has been the creation in Taiwan of unusually strong networks of domestic suppliers supporting large manufacturers.[41]

**Spinoffs**

Both ITRI and the Fraunhofer have deliberately spun off pieces of themselves to form new companies, not only a mechanism of technology transfer to industry but occasionally a way of creating entirely new industries. ITRI's venture capital subsidiary, the Industrial Technology Investment Corporation, screens proposed spinoffs from ITRI (which may be small or very large, involving personnel, equipment, technology, and intellectual property) and takes minority equity stakes in some of them. The Fraunhofer manages spinoffs through an internal division, Fraunhofer Ventures, which takes an equity stake of up to 25% in about half of the spinoffs.

ITRI can claim some of the most successful spinoffs from a public research entity in history. These include:

- **Taiwan Semiconductor Manufacturing Corporation (TSMC)**, the first and still the largest semiconductor foundry, a transfer involving 130 ITRI personnel

---

[39] Bernd Venohr and Klaus E. Meyer, "The German Miracle Keeps Running. How Germany's Hidden Champions Stay Ahead in the Global Economy," (Berlin School of Economics, May 2007).

[40] Christian Hamburg, "Structure and Dynamic of the German Mittelstand" (Heidelberg and New York: Physica-Verlag, 1999); "What's Behind the Success Story of German Manufacturing Industry?" *Xinhun* (February 23, 2012).

[41] See interview with Juney Shih, "Taiwanese IT Pioneers: Jouney (Chang-tang) Shih," recorded February 15, 2011 (Computer History Museum, 2011); "Wide Bandgap Alliance Formed in Taiwan," *Taiwan Economic News* (March 9, 2011); "ERSO to Set Up Flexible Electronics Alliance in July," *Taiwan Economic News* (May 19, 2005).

(including ITRI head Morris Chang) and a complete semiconductor manufacturing plant.
- **United Microelectronics Corporation**, a spinoff involving 31 ITRI personnel and a substantial quantity of semiconductor manufacturing equipment, creating a world-class semiconductor foundry.
- **Winbond Electronics**, the largest maker of branded semiconductors in Taiwan, created through the spinoff of about 200 personnel from ITRI.

The Fraunhofer's spinoffs have been more modest in scale, usually involving several of its staff, but they have historically enjoyed an extraordinary success rate, estimated at around 65–90% (in the United States, the failure rate for startups in the early 2000s was 90%).

## Branding

The Fraunhofer has an enviable reputation for excellence throughout the world, and its name is closely associated with the high quality of German products and engineering. A 2011 report by the UK House of Commons observed that "the name Fraunhofer resonates across the world and is widely associated with an impressive network of German technology and innovation centres."[42] ITRI is building a similar reputation, and some experts regard it as the best public institute for applied industrial research in the world. Both entities have devoted considerable effort to reputation-building, recognizing the importance of a strong brand in recruiting the best personnel, competing for industry research dollars, and pursuing international research collaborations. The UK's designation of its "Catapult Centres" is an aspirational attempt to build a similarly strong institutional brand.

## Maintaining Mission Focus

Successful government-supported innovation institutions like the Fraunhofer and ITRI have maintained a consistent focus on their core mission over time, which is the prosaic task of providing innovation support to industry. In part this reflects their comparative insulation from political pressure, which has averted wide swings in strategy with the change of government administrations. But these institutions have demonstrated their own abiding dedication to applied research and the ability to resist, for example, the gravitational pull exerted by basic research and the pursuit of scientific awards and academic distinction.

---

[42] "Second Report: Technology and Innovation Centres," Committee on Science and Technology, House of Commons, February 9, 2011.

## Autonomy

Although both Fraunhofer and ITRI are public organizations, they succeed in substantial part because of the relative autonomy they enjoy with respect to government authority. The Fraunhofer is a public association, and although on rare occasions it has been assigned special missions by the German government (such as upgrading the research infrastructure in the former East Germany), it is not normally subject to government direction. During the first decades of its existence, ITRI enjoyed considerable autonomy as a result of the ruling Kuomintang Party's longstanding tradition of according technocrats wide latitude and freedom from political interference.[43]

## Germany's Dual System

Germany is famous for its dual system of concurrent education and vocational training of students, to which is attributed the country's reputation for highly skilled industrial workers. This system is replicated within the Fraunhofer, which closely links university curricula with applied industrial research. University students and faculty comprise much of each institute's workforce, and the curricula and courses of study for graduate students at universities linked to the Fraunhofer are often closely correlated with technological themes emerging from the Fraunhofer's contract research for industry. Ultimately this system results in a flow of superbly qualified graduates into German industry.

## The Catapult Initiative in the United Kingdom

A number of countries seeking to improve their innovation performance have surveyed other advanced countries' research organizations to identify best practices that might be adapted and applied domestically. The Catapult initiative in the United Kingdom and the Carnot Institutes in France are recent examples of attempts to utilize the Fraunhofer model to improve the nexus between a strong research base and private industry.[44] The Catapult program is of particular interest to the United States because it is being implemented in a country with laissez-faire traditions similar to those of the United States, a comparable ambivalence among

---

[43] Robert Wade, *Governing the Market: Economic Theory and the Role of Government in Fast Asian Industrialization* (Princeton and Oxford: Princeton University Press, 1990): Markus Winnes and Uwe Schimack, *National Report*: "*Federal Republic of Germany*" (Institute for the Study of Societies, 1999).pp. 14 and 45.

[44] Jean-Michel Le Roux, *Carnot program* (2012), <*http://www.institute.carnot.edu/en/instituts-carnot*> (accessed in 2013).

policymakers at the national level toward "industrial policy," and the same willingness to accept major trade deficits in manufacturing.[45]

The Catapult program seeks to strengthen the United Kingdom's system of innovation intermediaries by taking some of the best existing applied research organizations, providing them with a base of government "core" funding over a 5-year timeframe, establishing a common system of governance and strategic direction and a building brand ("Catapult") intended to emulate the success of the Fraunhofer brand. Catapult Centres are being established in discrete technology areas in which the United Kingdom already enjoys a strong manufacturing base already. Overall direction and core funding is being provided by the Technology Strategy Board, a non-departmental public body largely staffed by individuals coming from careers in business rather than government. The ultimate objective is to create new manufacturing jobs in Britain.[46]

The oldest Catapult Centre, the High Value Manufacturing (HVM) Catapult, has only been operating since 2011, and other centers opened more recently. Allowing for the time necessary for initial stand-up of the centers and the development and launch of programs, not enough time has elapsed to judge whether their impact will be sufficient to characterize them as a success or to conclude that they rank with the Fraunhofer. Nevertheless, they already represent very substantial organizations closely linked with both academia and industry. The Offshore Renewable Energy Catapult, for example, based in Scotland, has 120 staff members and operates the largest concentration of multipurpose offshore renewable-energy technology test and demonstration facilities in the world. It is advised by one board comprised of 20 companies and another board made up of 10 research universities.[47]

The Catapults have already generated sufficient notable achievements that the Chief Technology Officer of the HVM Catapult, Dr. Phill Cartwright, could state in 2016 that "I think it's widely recognized that the Catapults have been successful." The HVM Catapult generated technologies for application in the construction sector which enabled early completion of projects such as the Manchester Metrolink and the City of London's Lendenhall Building. The technologies, which included off-site construction techniques and virtual design and production techniques, built on skills derived from other manufacturing sectors such as automobiles.[48] The HVM Catapult has achieved substantial advance in metal-cutting. A recent study concluded that the HMV Catapult has generated £15 for every £1 invested in it.[49]

---

[45] Trades Union Congress, *German Lessons: Developing Industrial Policy in the UK* (March 2011).

[46] "Fast Growth at Technology Centre Bodes Well for Manufacturing," *Professional Engineering* (June 14, 2012).

[47] "Economic Growth in Renewable Energy," *The Engineer* (January 7, 2014); "Offshore Renewable Energy Catapult is Ready for Action," *Herald Scotland* (January 3, 2014).

[48] "Interview: Dr. Phill Cartwright, Chief Technology Officer, High Value Manufacturing Catapult," *The Engineer* (March 7, 2016).

[49] "Interview: Dr. Phill Cartwright, Chief Technology Officer, High Value Manufacturing Catapult," *The Engineer* (March 7, 2016).

**Lessons for American Practice**

This review of some of the most successful foreign institutes to support manufacturing excellence highlights some of the key principles for a successful system to support manufacturing. These include:

- **Incentivizing cooperation** among universities, laboratories, and the private sector through federal funding with matching private-sector resources and the provision of common facilities to create an industrial commons for the cooperative development of new technologies.
- **Assuring a stable environment** with a continued commitment of both university and industry partners and the federal government. Manufacturing is a critical component of the American economy and is essential to the US defense base, and the resources and facilities should reflect that reality.
- **Providing substantial and sustained funding** is equally important for the effective operation of consortia designed to develop new materials and new processes and ultimately to support the emergence of new products.
- **Branding and outreach** are key elements in the success of a network of institutes designed to support researchers and manufacturers. The reputational benefits that accrue from previous success can act as a powerful incentive for continued collaboration.
- **Maintaining a focus** on applied research directly relevant to industry is a major source of success in foreign programs. The applied research should be complemented by an ability to address near-term manufacturing challenges where possible in cooperation with existing programs.
- **Focusing on the educational component** is essential. The centers need to contribute to workforce training—from the vocational to graduate to post doc levels. Providing students with hands-on experience of real-world problems and exposure to small companies, large corporations, and cooperative research focused on manufacturing serves to strengthen the talent pool.
- **Encouraging the creation of spinoffs** should be a priority, one which requires integration with existing funding sources at the state and federal levels (such as the $3 billion annual SBIR program) and in turn backed by policies to encourage entrepreneurial leave, seed funding, and, where appropriate, entrepreneurial training.
- **Incentivizing firms** to provide prototype equipment for testing and validation while also assuring effective IP protection can substantially enhance the value of common facilities and encourage the participation of a broad and diverse group of participants.

## Bibliography[50]

Allied Research: The Fraunhofer Method. (1987, October 20). *Industries et Techniques*. JPRS-ELS-88-006.

Barker, M. M. (2011, April 11). Manufacturing Employment Hard Hit During the 2007–09 Recession. *Monthly Labor Review*.

Berger, S. (2013). *Making in America: From Innovation to Market*. Cambridge, MA and London: MIT Press.

Bloch, E. (2003, Summer). Seizing US Research Strength. *Issues in Science and Technology*.

Brzustowski, T. (2012). *Why We Need Move Innovation in Canada and What We Must Do to Get It*. Ottawa: Invenire Books.

Economic Growth in Renewable Energy. (2014, January 7). *The Engineer*.

ERSO to Set Up Flexible Electronics Alliance in July. (2005, May 19). *Taiwan Economic News*.

Fast Growth at Technology Centre Bodes Well for Manufacturing. (2012, June 14). *Professional Engineering*.

Fraunhofer ISE Sets Another Solar Cell Efficiency Record. (2015, September 25). *Clean Technical*.

Fraunhofer ISE Sets PV Module Efficiency Record of 36.7 Percent. (2014, July 14). *Semiconductor Today*.

Hauser, H. (2010). *The Current and Future Role of Technology Centres in the UK*. London: UK Department for Business, Innovation and Skills.

House of Commons. Committee on Science and Technology. (2011, February 9). Second Report: Technology and Innovation Centres.

Hsu, A. (2007, October 19). ITRI Pushes Technology Sector to New Frontiers of Innovation. *Taiwan Journal*.

Interview: Dr. Phill Cartwright, Chief Technology Officer, High Value Manufacturing Catapult. (2016, March 7). *The Engineer*.

Le Roux, J. (2012). *Carnot program*. http://www.institute.carnot.edu/en/instituts-carnot. Accessed 2013.

Mathews, J. A. (2010, December 10). The Hsinchu Model: Collective Efficiency, Increasing Returns and Higher Order Capabilities in the Hsinchu Science-Based Industry Park, Taiwan. Keynote Address, Chinese Society for Management of Technology 20th Anniversary Conference. Hsinchu, Taiwan.

Mathews, J. A., and Cho, D. (2000). *Tiger Technology: The Creation of a Semiconductor Industry in East Asia* Cambridge: Cambridge University Press.

National Network for Manufacturing Innovation. (2013.) Frequently Asked Questions. http://manufacturing.gov/docs/nnmi.faq.pdf. Accessed June 7, 2013.

---

[50] As noted in the front matter of this book, the study also drew on interviews carried out by the authors and numerous articles from *The Times Union* (Albany), *The Daily Gazette* (Schenectady), the *Albany Business Review* (Albany), *The Post-Star* (Glens Falls), *The Record* (Troy), *The Saratogian* (Saratoga Springs), *The Buffalo News* (Buffalo), The *Observer-Dispatch* (Utica), The *Daily Messenger* (Canandaigua), and the *Post-Standard* (Syracuse). These are not individually included in the bibliography.

National Research Council. (2013). *21st Century Manufacturing: The Role of the Manufacturing Extension Partnership Program.* C. W. Wessner (Ed.) Washington, DC: The National Academies Press.
Offshore Renewable Energy Catapult is Ready for Action. (2014, January 3). *Herald Scotland.*
Prah, P. M. (2013, July 30). Has US Manufacturing's Comeback Stalled? *USA Today.*
President's Council of Advisors on Science and Technology (2011, June). *Report to the President on Ensuring American Leadership in Advanced Manufacturing.* Washington, DC: Executive Office of the President.
Taiwanese IT Pioneers: Jouney (Chang-tang) Shih. (2011, February 15). Recorded Interview. Computer History Museum.
Trades Union Congress. (2011, March). *German Lessons: Developing Industrial Policy in the UK.*
Varma, R. (2000, Autumn). Changing Research Cultures in US Industry. *Science, Technology and Human Values.*
Venohr, B., and Meyer, K. E. (2007, May). The German Miracle Keeps Running. How Germany's Hidden Champions Stay Ahead in the Global Economy. Berlin School of Economics.
Wade, R. (1990). *Governing the Market: Economic Theory and the Role of Government in Fast Asian Industrialization.* Princeton and Oxford; Princeton University Press.
Webster, A. (1994, April). Bridging Institutions: The Role of Contract Research Organizations in Technology Transfer. *Science and Public Policy.*
Welcome to Berlin, Peter—Is it the Future? (2010, February 7). *Sunday Telegraph.*
What's Behind the Success Story of German Manufacturing Industry? (2012, February 23). *Xinhun.*
Wide Bandgap Alliance Formed in Taiwan. (2011, March 9). *Taiwan Economic News.*
Winnes, M., and Schimack, U. (1999). *National Report: "Federal Republic of Germany."* Institute for the Study of Societies.

## Appendix C: Tech Valley Chronology

New York's research investments eventually led to manufacturing investments by industry but only over a long time horizon spanning the terms of six governors. From the launch of the Rensselaer Polytechnic Institute's (RPI's) Center for Integrated Electronics in 1981 to IBM's commitment to build a 300 mm fab at East Fishkill, a span of 19 years elapsed, and a total of 30 years elapsed before Global Foundries produced its first commercial wafers in 2012. During this period, if any single governor or the legislature reversed course and ended state support for the effort, it is unlikely that today's large and growing high-tech manufacturing base would exist. But that did not happen. The result is an extraordinary, bipartisan state achievement representing sustained commitments by multiple individuals and institutions over the course of a generation.

*1959–1973. Governor Nelson Rockefeller expands the State University of New York (SUNY) system and begins funding research in New York's private and public universities.* When Governor Rockefeller assumed the governorship in 1959, the SUNY system was hobbled by restrictions, including a prohibition on scientific research. Governor Rockefeller undertook a sustained effort to eliminate the restrictions, to vastly expand SUNY, and to employ the SUNY system as an instrument of economic development. In the 15-year period between the last year in office of Governor Rockefeller's predecessor and that of Governor Rockefeller himself, state outlays on higher education virtually exploded, from $43 million to $1052 million, and SUNY became the largest system of higher education in the United States.[51]

*1979–1981. Supported by Governor Hugh Carey, RPI President George Low attempts to replicate Silicon Valley in the Capital Region.* George Low modernized RPI's research infrastructure, created one of the first high-technology incubators, established an industrial park linked to RPI, and, with $30 million in state funding, created the Center for Integrated Electronics, a center for very large scale integration (VLSI) semiconductor research.

*1982–1983. Semiconductor Research Corporation (SRC) begins funding semiconductor R&D in New York universities.* The SRC is a consortium formed and funded by the US semiconductor companies (along with federal funding) which provides long-term support for R&D in industry-relevant basic microelectronics at the US universities. Initially, Cornell University and the University of Rochester received $50 million in research funding from SRC. SRC's research projects created an awareness in the semiconductor industry of New York's research resources and established a foundation for the state to bid for additional research projects and inward investment by the semiconductor industry.[52]

*1983. Governor Mario Cuomo begins establishing university-centered Centers for Advanced Technology (CATS) funded by the state.* State CAT funding enabled research universities to improve graduate programs, create new faculty positions, and modernize research equipment.[53]

*1987. Center for Economic Growth (CEG) formed.* CEG was formed by the Albany-Colonie Chamber of Commerce to promote a regional approach to economic development and to counteract the adverse effects associated with the fragmentation of governmental authority in the 16 counties of the Capital Region. CEG secured state funding to enable the rebuilding of Albany International Airport and sponsored initiatives to re-brand the region, including promotion of the concept of "Tech Valley" in 1998.

---

[51] Peter D. McClelland and Alan L. Magdovitz, *Crisis in the Making: The Political Economy of New York State Since 1945* (Cambridge: Cambridge University Press, 1981), p. 61.

[52] Robert M. Burger, *Cooperative Research: The New Paradigm* (Durham: Semiconductor Research Corporation, 2001).

[53] "Cuomo Tells TUE: High Tech Manufacturing Needed," Schenectady, *The Daily Gazette* (May 22, 1982).

*1988. Alain Kaloyeros appointed to faculty position at SUNY Albany.* Governor Mario Cuomo's Graduate Research Initiative, launched in 1987, allocated state funds to SUNY to enable it to offer higher salaries to faculty recruited outside the state and to pay for research equipment. Kaloyeros, who was interviewed for his position at SUNY Albany by Governor Cuomo himself, emerged as the leader of the College of Nanoscale Science and Engineering at SUNY Albany.

*1988. New York bids for Sematech.* In 1988, New York made a bid to the newly formed semiconductor manufacturing research consortium, Sematech, to be the site of the research facility. Although New York lost the bid to Texas, the bid was sufficiently strong—including an offer of $40 million in state incentives—that it underscored the fact that New York was emerging as a serious contender to attract high-tech manufacturing investment.

*1988. SRC and Sematech establish a research Center of Excellence in New York.* IBM—a major player in the Semiconductor Research Corporation, Sematech, and Semiconductor Industry Association—was particularly interested in research underway at SUNY Albany on chemical vapor deposition (CVD) technology for copper metallization. The creation of the Center for Excellence marked the beginning of a long and productive series of collaborations between SUNY Albany and IBM.

*1993. Governor Mario Cuomo establishes the CAT for Thin Films and Coatings at SUNY Albany under Kaloyeros.* In 1995, SUNY Albany disclosed that it would build a 75,000 square foot Center for Environmental Sciences and Technology (CESTM) which would house the CAT for Thin Films/Coatings. By the end of 1996, the CAT had worked with 50 companies, generated $32 million in industry funds and yielded over 20 new high-tech products.

*1995. Governor George Pataki forestalls move of IBM headquarters out of New York.* IBM, which has been a driving force behind the creation of Tech Valley, was on the verge of moving its headquarters out of New York in 1995. Governor Pataki made commitments to IBM which caused the company to shelve its plans to move.

*1997. CEG begins to concentrate on development of high-tech manufacturing in the Capital Region.* CEG launched a region-wide effort to attract a wafer fabrication facility to the region. CEG undertook marketing efforts directed at the semiconductor industry to highlight specific sites in the Capital Region.

*1998. National Grid begins funding to provide the informational and analytic basis for the development of a wafer fabrication site in Luther Forest.* National Grid provided funding for Saratoga County's development organization, Saratoga Economic Development Corporation (SEDC) to conduct studies of the site and develop strategic plans for seeking investment in the region by a global semiconductor manufacturer.

*1999. New York enacts the Jobs 2000 Act.* The legislation created the New York Office of Science, Technology and Academic Research (NYSTAR) with a first-year budget of $156 million, a major increase in state expenditures on research.

*1999. North Greenbush rejects semiconductor manufacturing.* "Chip Fab '98" was an initiative by New York State and regional economic development organizations to "pre-qualify" selected sites in the state by seeking regulatory clearance for generic semiconductor plants, which would enable the state to give assurances to potential investors that the regulatory hurdles had already been surmounted. This effort was set back when the Town Board of North Greenbush, where the front-running site was located, voted to block further review of the site as a potential chip fab location. The reverse provided useful lessons for the development officials involved, including the need to provide communities considering hosting a chip fab with abundant neutral information, and, as an alternative to site selection by state officials, to allow communities that had positive interest in attracting chip fab to self-select and pre-qualify their own proposed sites.

*2001. Governor Pataki begins establishing Centers of Excellence with a proposed investment of $283 million.* The new centers were intended to link university research with high-tech companies. One of the first three of these was established at SUNY Albany with a focus on nanotechnology. SUNY Albany was chosen over other institutions engaged in nanotechnology research because of its clear focus on industry relations.

*2001. New York and IBM commit funds for 300 mm research fab.* IBM and New York committed funds, $100 million and $50 million, respectively, for the establishment of a 300 mm wafer fabrication facility for research purposes at SUNY Albany. IBM, which was planning its own 300 mm manufacturing facility, would enjoy access to the research fab for its R&D projects.

*2001. The School of Nanosciences and Nanoengineering—the forerunner of the College of Nanoscale Science and Engineering—was established at SUNY Albany.*

*2002. SEDC presents plan for Luther Forest Technology Campus.* SEDC submitted plans to the Town Boards of Malta and Stillwater requesting rezoning of the Luther Forest site as a Planned Development District (PDD) which could accommodate up to four semiconductor wafer fabrication plants. The purpose of the initiative was to secure regulatory clearances in advance to enable SEDC to present semiconductor manufacturers with "shovel-ready" sites.

*2002. Sematech announces research center at SUNY Albany.* Ten months of discussions between senior New York leaders (including Governor Pataki) and Sematech culminated in Sematech's decision to base its next research center at SUNY Albany rather than Austin, Texas. New York pledged $160 million to the project and another $50 million for the construction work. Sematech reportedly chose Albany because of the quality of research underway at SUNY Albany, IBM's influence, and the enthusiastic support of the governor and legislature. In retrospect, Sematech's decision was recognized as part of a snowball effect in which semiconductor companies increasingly recognized the need to be present in the region themselves because of the presence of so many other leading industry players.

*2002. Tokyo Electron Ltd. (TEL) announces establishment of an R&D center at SUNY Albany.* TEL is one of the world's leading producers of semiconductor manufacturing equipment. It committed to invest $200 million in research at SUNY Albany over a 7-year period, with the state investing another $100 million. TEL made the investment because SUNY Albany had the only university-owned R&D

center focusing on 300 mm technology, and TEL would gain access to research equipment that would otherwise be prohibitively expensive.

*2004. The College of Nanoscale Science and Engineering (CNSE) is established at SUNY Albany.* CNSE had by far the world's largest and most advanced research facilities for nanotechnology, enabling research that could not possibly be conducted elsewhere. Over 90% of CNSE's PhD and MS graduates remain in New York State and are employed by leading nanotechnology companies. CNSE has arguably emerged as the foremost site for applied nanotechnology R&D in the world.

*2004. Malta and Stillwater approve creation of a PDD in Luther Forest.* The vote of the town boards of Malta and Stillwater reflected the major effort undertaken by development organizations to make information about semiconductor manufacturing available to the communities. The decision made it feasible for SEDC to purchase the land and begin efforts to market the "shovel-ready" site to the semiconductor industry.

*2005–2006. New York economic development professionals develop highly sophisticated site proposals for presentation to semiconductor manufacturers.* In a remarkable effort, SEDC assembled a first-class engineering project team of planners, engineers, and technical experts to create proposals that would resonate with semiconductor executives. State officials prepared a very substantial incentives package to make New York fully competitive with other regions for semiconductor manufacturing investment.

*2005–2007. New York invests $225 million in CNSE-industry research collaborations.* The state contributed to large-scale CNSE collaborations with industrial partners in research in immersion and extreme UV lithography (IMPULSE), nanolithography (INVENT), and nanoelectronics (INDEX). CNSE industry research partners included ASML, AMD, Micron Technologies, IBM, Texas Instruments, Intel, Applied Materials, Toshiba, and Sony. CNSE's expanding role in industry-relevant semiconductor R&D was an important factor underlying the 2006 decision by Advanced Micro Devices (AMD) to locate its next wafer fabrication plant in the Capital Region.[54]

*2005. Hudson Valley Community College (HVCC) begins offering 2-year degrees in semiconductor technology.* HVCC went on to expand its degree offerings in nanotechnology manufacturing, refining the curricula to meet the emerging needs of semiconductor producers.

*2006. RPI launches the Computational Center for Nanotechnology (CCNI).* CCNI, a partnership between IBM and RPI, houses the most powerful supercomputer system on any university campus in the world. The computational power of the system enables design and simulation of nanoscale electronic systems. The existence of the RPI center was credited as a factor behind AMD's decision to build its next wafer fabrication facility in the Capital Region.

---

[54] "RPI to Get Supercomputer—System Expected to Create 300–500 Jobs," Schenectady, *The Daily Gazette* (May 11, 2006).

*2006. Advanced Micro Devices (AMD) enters into agreement with New York to build a 300 mm semiconductor wafer fabrication plant at Luther Forest site.* The state pledged a grant of $500 million for buildings and equipment, a $150 million grant for R&D, infrastructure improvements valued at $300 million, and tax credits and other incentives worth an estimated $250 million. This effort was consistent with industry expectations and, critically, was superior to closely competing offers from other regions.

*2006. Governor George Pataki decides not to seek reelection.* Governor Pataki committed state resources on an unprecedented scale to developing the Albany area into the world's leading center of nanotechnology research. His efforts were instrumental in persuading IBM to retain its headquarters in New York and in securing commitments for investment in Tech Valley by Sematech and AMD.

*2007. Sematech announces it will move its headquarters from Austin to Albany.* Sematech committed to invest $150 million in cash and $150 million in cash equivalents over a 7-year period in research collaboration with CNSE. The state pledged $300 million. In 2010, Sematech announced it would move the remainder of its operations from Austin to Albany.

*2007. Tech Valley High School (TVHS) begins operations in temporary quarters in Troy.* TVHS offered area high school students intensive coursework in math, science, and technology and hands-on, project-based experience relevant to high-tech careers. TVHS eventually relocated to CNSE, where students are brought into close proximity to the research and technology development being conducted there.

*2008. Senate Majority Leader Joseph L. Bruno announces his retirement.* Leader Bruno was an enthusiastic and powerful advocate and patron for the development of institutions that made Tech Valley possible, including the NanoCollege, the Luther Forest Technology Campus, and training programs and infrastructure at the region's community colleges.

*2008–2009. Abu Dhabi's Mubadala Investment Company invests in AMD's semiconductor manufacturing operations.* The Mubadala investment, undertaken through its subsidiary, Advanced Technology Investment Company (ATIC), provided an owner with the financial resources necessary to undertake construction of a semiconductor wafer fabrication facility at the Luther Forest site. A new company, GlobalFoundries, was formed to build the facility and operate it as a foundry and to own and operate former AMD manufacturing facilities around the world. The initial ATIC commitment to the Luther Forest project of $4.2 billion would grow over time as new facilities were seen as necessary for the new company's competitive strategy. To assure this investment, New York approved transfer of the incentives package from AMD to GlobalFoundries.

*2009. GlobalFoundries reaches labor agreement with local construction unions.* Under the agreement union, wages would be paid by all firms working to build the GlobalFoundries fab, whether or not the employees were union members. The agreement ensured that the intricate, costly, and time-sensitive construction of the fab could proceed without labor disputes.

*2010. M + W Group announces it will move its US headquarters to Watervliet Arsenal near Albany.* M + W Group is one of the foremost constructors of semiconductor manufacturing facilities.

*2006–2011. All of the infrastructure projects necessary to enable GlobalFoundries to start operations on time are approved, planned, and completed.* Taken together, the infrastructure buildout itself, and the complex array of regulatory hurdles which were surmounted successfully to enable that buildout, are an extraordinary example of effective local and state government. New York State contributed about $80 million to infrastructure projects associated with the Luther Forest site.

*2009–2011. The GlobalFoundries wafer fabrication is constructed.* Construction proceeded smoothly, notwithstanding the unforeseen addition of new building projects to the original plan, including an expanded version of the original fab plan, a new R&D center, and an administration building. Disputes between GlobalFoundries and local governments were resolved to the apparent satisfaction of the parties concerned.[55]

*2010. HVCC starts operations of TEC-SMART, an extension campus in Luther Forest focusing on semiconductor manufacturing and alternative energy technology industries.* TEC-SMART features a simulated semiconductor manufacturing line and also trains students how to operate mechanical systems supporting semiconductor manufacturing, such as air and water handling.

*2011. Governor Andrew Cuomo discloses the formation of the Global 450 Consortium (G450C).* New York State entered into agreements with IBM, GlobalFoundries, Intel, Samsung, and TSMC to develop the technology necessary to enable manufacturing of semiconductors utilizing 450 mm wafers, a $4.8 billion project to be housed at CNSE. The state committed $400 million to the project.[56]

*2011. Governor Andrew Cuomo establishes ten Regional Economic Development Councils (REDCs) across New York State.* Each REDC drew up long-term economic development plans for its region, and the state government directed major funding to support the most promising proposals. Over time Governor Cuomo succeeded in gaining greater control over state economic development spending from the legislature.

*2012. Governor Andrew Cuomo launches the "Buffalo Billion" initiative to revitalize the economy of the Buffalo/Niagara region.* This effort, an attempt to replicate the "Albany model," saw the establishment of industry–academic collaborations in biotechnology, clean technology, genomic medicine, and advanced manufacturing. Buffalo has subsequently experienced a significant increase in economic activity and manufacturing-sector employment.

---

[55] "Fab 8 to Build Milestone New Chip This Year," Schenectady, *The Daily Gazette* (January 10, 2012).

[56] Office of the Governor, "Governor Cuomo Announces $4.4 Billion Investment by International Technology Group Led by Intel and IBM to Develop Next Generation Computer Chip Technology in New York" Press Release (September 27, 2011) <*http://www.governor.ny.gov/press/092711chiptechnologyinvestment*>

*2012. GlobalFoundries begins commercial production of semiconductors at the Malta/Stillwater site.* Initially fabricating chips using 32 nm design rules, it moved within the year to introduce 20 nm chips utilizing a process enabling 3D stacking of devices.

*2013. Governor Andrew Cuomo enlists Alain Kaloyeros to apply his model of innovation-driven economic development across Upstate New York.* Kaloyeros carried out this mandate and appeared to be achieving mixed results (apparent success in Buffalo with problems in other areas).

*2014. GlobalFoundries acquires IBM manufacturing facilities in East Fishkill, New York, and in Essex Junction, Vermont.* IBM and GlobalFoundries concluded a deal pursuant to which IBM transferred its semiconductor operations, $1.5 billion, and process technology and know-how to GlobalFoundries. GlobalFoundries would be IBM's exclusive supplier of 22-, 14-, and 10-nanometer semiconductors for 10 years. The IBM acquisition gave GlobalFoundries the world's leading market share in radio frequency (RF) devices, which are critical components in smartphones.

*2014. CNSE is transferred from SUNY Albany to SUNY at Utica/Rome to form SUNY Polytechnic (SUNY Poly).* In the wake of CNSE's departure, SUNY Albany launched a number of initiatives most notably the creation of a public college offering engineering training to undergraduates and a College of Emergency Preparedness, Homeland Security and Cybersecurity.,

*2014. Saratoga Country withdraws support from SEDC and establishes a competing development authority, the Saratoga County Prosperity Partnership.* Tensions between the two county economic development bodies impaired efforts to attract additional businesses to the area.

*2015. GE announced a consortium in partnership with SUNY Poly to develop a manufacturing plant for silicon carbide semiconductors for applications in power electronics.* A fabrication plant was established at SUNY Poly with state support, and a packaging facility for silicon carbide chips was established in Marcy, near Utica.

*2015. Assembly Speaker Sheldon Silver resigns.* Speaker Silver was one of the earliest New York political leaders to support the NanoCollege in Albany and was a consistent advocate for a strong state-level commitment to nanotechnology research.

*2016. Alain Kaloyeros is indicted for bid-rigging along with others.* Kaloyeros pleaded innocent to the charges, but he resigned from his academic and economic development posts and his talents were no longer available to support TechValley.

*2016. In the wake of Kaloyeros' indictment, several major industry collaborations in which he played a key role collapse, begin to wind down, or stall.* The Austrian semiconductor maker AMS backed out of a 2015 commitment to build a semiconductor fabrication plant in Marcy. The Global 450 Consortium ended, with none of the industrial partners moving to make the technological leap from 300 mm semiconductor wafers to 450 mm.

*2016. Under the leadership of Howard Zemsky, Empire State Development implements institutional reforms at SUNY Poly and moves to restart stalled projects.* Zemsky overhauled the management structure of SUNY economic development organizations and restructured stalled projects, absorbing losses where necessary to enable work to resume.

*2016. GlobalFoundries announces it will invest "billions of dollars" to develop technology for 7 nm chips at its facility in Malta/Stillwater.* Assigning a team of 700 employees to the task, the company in effect skipped the 10 nm node moving directly from 14 to 7 nm, which enhances performance by 30% and reduces production costs by 30%.

*2017. Denmark-based Danfoss Silicon Power announces it will take over the packaging operations for silicon carbide semiconductors at the Marcy site in collaboration with GE.* Silicon carbide wafers are being fabricated at SUNY Poly in Albany and sent to the Marcy site for packaging and incorporated into modules and assemblies capable of powering electronic devices.

*2017. GlobalFoundries announces a 20% capacity increase at Malta/Stillwater site.* GlobalFoundries' Malta/Stillwater investments reflect strong demand for advanced 14 nm devices which it only makes at the Malta/Stillwater site. The company announced its expansion plans as part of a global investment strategy which also includes capacity expansion in Germany and creation of a new fabrication plant in Chengdu, China.

*2017. A consortium comprised of SUNY Poly, GlobalFoundries, Samsung, and IBM announced a "major breakthrough" in fabricating 5 nm transistors using EUV lithography.* The breakthrough appeared to overturn predictions that nothing under 7 nm would be achievable for the foreseeable future.

*2018. New York and Applied Materials announced plans for an $880 million semiconductor equipment and materials center at SUNY Poly's NanoFabX.*

*2019. New York disclosed a new collaboration with IBM to develop component technologies for artificial intelligence, a $300 million project that will be housed at SUNY Poly.*

# Bibliography[1]

30 Great Small College Business Degree Programs 2016. (2016, July). *Online Accounting Degree Programs*.
450mm and Other Emergency Measures. (2016, September 22). *Semiconductor Engineering*.
450Mm/Copper/Low-K Convergence Report 2017. (2017, May 10). *Business Wire*.
500 Housing Units in Malta are Planned by Developers from Florida, Western NY. (2015, May 7). *Saratoga Business Journal*.
After a Half-Century of Decline, Signs of Better Times for Buffalo. (2006, September 18). *The New York Times*.
Alain Kaloyeros, Powerful Centerpiece of Buffalo Billion, Could Become a Household Word. (2015, September 29). *Gotham Gazette*.
Albany Medical Center. (2010, September 15). Albany Med and Saratoga Hospital Announce Joint Venture. Press Release.
Albany No Longer A Secret in High-Tech Chip World. (2002, July 19). *The New York Times*.
Allied Research: The Fraunhofer Method. (1987, October 20). *Industries et Techniques*. JPRS-ELS-88-006.
AMD, Intel Race to the Bottom. (2006, June 28). *Forbes*.
AMD's New Investors Make Big Bet on Chip Plants. (2008, October 8). Ocala, *Star Banner*.
AMD's Q4 Retreats Under Competition—ATI By Adds Debt, Intel's New Chip 'Inflicting Pain'. (2007, January 4). *San Jose Mercury News*.
Andy Plans Class Clone. (2013, February 27). *New York Daily News*.
Applied Materials. (2018, November 15). New Applied Materials R&D Center to Help Customers Overcome Moore's Law Challenges. Press Release.
Armbrust, D. (2013, April 4). Presentation at National Academies Symposium, *New York's Nanotechnology Model: Building the Innovation Economy*. Troy, New York.
Armour-Garb, A. (2017, January). *Bridging the STEM skills Gap: Employer/Educator Collaboration in New York*. Public Policy Initiative of New York State, Inc.
Arnoff Moving & Storage. (2017, June 17). Arnoff Company Opens to $11.6 Million Logistics Hub in Saratoga County. http://www.arnoff.com/blog.

---

[1] As noted in the front matter of this book, the study also drew on interviews carried out by the authors and numerous articles from *The Times Union* (Albany), *The Daily Gazette* (Schenectady), the *Albany Business Review* (Albany), *The Post-Star* (Glens Falls), *The Record* (Troy), *The Saratogian* (Saratoga Springs), *The Buffalo News* (Buffalo), *The Observer-Dispatch* (Utica), *The Daily Messenger* (Canandaigua), and the *Post-Standard* (Syracuse). These are not individually included in the bibliography.

As US Economy Races Along, Upstate New York is Sputtering. (1997, May 11). *The New York Times*.
Axcelis Technologies Lands in Clifton Park Aided by National Grid Infrastructure Grant. (2013, June 5). *Saratoga Business Journal*.
Babu Leads Clarkson's CAMP to Prominence. (2015, June 23). Massena, *Daily Courier-Observer*.
Bacheller, J. (2016, October 4). The Decline of Manufacturing in New York and the Rust Belt. *Policy by the Numbers*.
Bacon, D. *Organizing Silicon Valley's High Tech Workers*. http://dbacon.igc.org/Unions/04hitec2.htm.
Baily, M. N., and Bosworth, B. P. (2014, Winter). US Manufacturing: Understanding its Past and Potential Future. *Journal of Economic Perspectives*. 28(1).
Balanced College Concept. (2003). In W. Somers (Ed.), *Encyclopedia of Union College History* (pp. 83–88). Schenectady: Union College Press.
Ballston Spa Makes 'Top High Schools.' (2016, August 22). *The Ballston Journal*.
Ballston Spa Central School District. http://www.bscsd.org.
Barker, M. M. (2011, April 11). Manufacturing Employment Hard Hit During the 2007-09 Recession. *Monthly Labor Review*.
Beard Integrated Systems Moves Into Malta Offices Working With GlobalFoundries Plant. (2015, September 9). *Saratoga Business Journal*.
Belussi, F., and Caldon, K. (2009). At the Origin of the Industrial District: Alfred Marshall and the Cambridge School. *Cambridge Journal of Economics*.
Berger, S. (2013). *Making in America: From Innovation to Market*. Cambridge, MA and London: MIT Press.
Berman, E. P. (2012). *Creating the Market University: How Academic Science Became an Economic Engine* Princeton and Oxford: Princeton University Press.
Better Microchips Sought by Alliance. (2005, July 19). *Kansas City Star*.
Bid-Rigging Trial Begins for Ex-State University of New York Official. (2018, June 18). *Reuters*.
Big Blue to Stay in State, Build New Headquarters. (1995, February 17). *Watertown Daily Times*.
Black, M., and Worthington, R. (1986). The Center for Industrial Innovation at RPI. In M. Schoolman and A. Magid (Eds.) *Reindustrializing New York State: Strategies, Implications and Challenges* (pp. 261–265). Albany: SUNY Press.
Bloch, E. (2003, Summer). Seizing US Research Strength. *Issues in Science and Technology*.
Boardman, C., and Gray, D. (2010, February). The New Science and Engineering Management: Cooperative Research Centers as Government Policies, Industry Strategies and Organizations. *Journal of Technology Transfer*.
Breschi, S., and Malerba, F. (2005). *Clusters, Networks and Innovation*. Oxford: Oxford University Press.
Bresnahan, T. F., and Trajtenberg, M. (1995). *General Purpose Technologies: Engines of Growth?* NBER Working Paper No. w4148. Boston: National Bureau of Economic Research.
Bruno, J. L. (2016). *Keep Swinging: A Memoir of Politics and Justice*. Franklin, TN: Post Hill Press.
Brzustowski, T. (2012). *Why We Need Move Innovation in Canada and What We Must Do to Get It*. Ottawa: Invenire Books.
Buffalo Boondoggle. (2016, March 6). *New York Post*.
Building Permits in Serious Decline--Construction Firms Must Adapt or Go Out of Business. (2008 January 14). *Fort Wayne News Sentinel*.
The Bumpy Road to 450 mm. (2013, May 16). *Semiconductor Engineering*.
Burger, R. M. (2001). *Cooperative Research: The New Paradigm*. Durham: Semiconductor Research Corporation.
Buser, M. (2012, March 21). The Production of Space in Metropolitan Regions: A Lefebvrian Analysis of Governance and Spatial Change. *Planning Theory*. 288–289.
Business: Shirt Tale. (1938, February 21). *Time*.
Bypass Project Fast Tracked. (2010, January). *Professional Surveyor Magazine*.

Calhoun, D. H. (1960). *The American Civil Engineer: Origins and Conflict*. Cambridge, MA: Technology Press.
Camoin Associates. (2013, September 5). The Curious Case of GlobalFoundries and its Workforce: Setting the Stage. https://www.camoinassociates.com/curious-case-globalfoundries-and-its-workforce-setting-stage.
Can American Manufacturing Really Be Cornerstone of Economic Revival? (2012, February 8). *The Christian Science Monitor*.
Can Buffalo Ever Come Back? (2007, October 19). New York, *The Sun*.
Carayannis, E. C., and Gover, J. (2002). The Sematech-Sandia National Laboratories Partnership: A Case Study. *Technovation*.
Carroll, P. T. (1999, Spring). Designing Modern America in the Silicon Valley of the Nineteenth Century. *RPI Magazine*.
Centre for Research on Socio-Cultural Change. (2011). Rebalancing the Economy for Buyer's Remorse. Working Paper No. 87.
Chernock, J., and Youtie, J. (2013, February). State University of New York at Albany Nanotech Complex. In Georgia Tech Enterprise Innovation Institute. *Best Practices in Foreign Direct Investment and Exporting Based on Regional Industry Clusters*. Atlanta: Georgia Tech Research Corporation. Prepared for the Economic Development Administration, U.S. Department of Commerce.
Clark, J. B., Leslie, W. B., and O'Brien, K. P. (Eds.) (2010). *SUNY at Sixty: The Promise of the State University of New York* Albany: SUNY Press.
Clark, T. B. (1976, October-November). The Frostbelt Fights for a New Future. *Empire State Report II*. p. 332.
CNSE Working Group. (2013, June 13). *The SUNY College of Nanoscale Science and Engineering: A Vibrant Engine for Innovation, Education, Entrepreneurship and Economic Vitality for the State of New York*.
College of Nanoscale Science and Engineering. Landing Edge Research and Development Research Centers. http://www.sunycnse.com/LandingEdgeResearchandDevelopmentResearchCenters.
College of Nanoscale Science and Engineering. Quick Facts. http://www.sunycnse.com/AboutUs/QuickFacts.aspx.
College of Nanoscale Science and Engineering. (2008, May 5). *A Proposal for Undergraduate Academic Programs Leading to the B.S. in Nanoscale Science and B.S. in Nanoscale Engineering*. Submitted to SUNY Albany Senate.
Community Colleges Grow With Distinction. (1987, May 11). Albany, *Knickerbocker News*.
Construction Recovery will be Slow. (2009, December 9). *The Grand Rapids Press*.
Control Freaking: Legieslators Vent at Cuomo for Hogging $3 B in Budget Power. (2013, February 4). *New York Daily News*.
Cornog, E. (1998). *The Birth of Empire: Dewitt Clinton and the American Experience* Oxford: Oxford University Press.
Corona, L., Doutriaux, J., and Mian, S. A. (2006). *Building Knowledge Regions in North America: Emerging Technology Innovation Poles*. Cheltenham and Northampton, MA: Edward Elgar.
Cortright, J. (2006, March). *Making Sense of Clusters: Regional Competitiveness and Economic Development*. Washington, DC: Brookings.
Crane Operators Threaten Strike at Many Construction Sites. (2006, June 30). *The New York Times*.
Crazy Diamonds. (2013, July 20). *The Economist*.
Cross Subsidy in NY. (2007, January/February). *IEEE Power and Energy Magazine*.
Cryogenics Group Opens Facility in Malta, is Air Pump Supplier for GlobalFoundries. (2014, January 8). *Saratoga Business Journal*.
Cuomo, A. M. (2016, September 1). Rebuilding the Upstate Economy. *Huffington Post*.
Cuomo Fund-Raisers Preceded Development Funding Awards. (2014, December 9). *Politico*.
Cuomo Manages the Fallout from Corruption Scandal. (2016, September 29). WBFO.
Cuomo Pushes Technology to Kickstart State's Engine. (1994, January 27). *Watertown Daily Times*.

Cuomo's $15 Million High-Tech Film Studio? It's a Flop. (2016, August 22). *The New York Times*.
Defense Science Board. (2005, February). *Task Force on High Performance Microchip Supply*. Washington, DC: Department of Defense.
Derbyshire, K. (2002, June). Building a Fab—It's All About Tradeoffs. *Semiconductor Magazine*.
Development in Malta Prompts Delmar Optician to Open Saratoga County Office. (2013, August 6). *Saratoga Business Journal*.
Dhillon, H., Qazi, S., and Anwar, S. (2008). Mitigation of Barriers to Commercialization of Nanotechnology: An Overview of Two Successful University-Based Initiatives. *Proceedings of the ASEE 2008 Annual Conference*. Pittsburgh, Pennsylvania.
Dudley, W. C. (2014, October 7). The National and Regional Economy. Remarks at RPI, Troy, New York.
Eastern New York Angels. http://www.easternnyangels.com.
Echeverri-Carroll, E., Feldman, M., Gibson, D., Lowe, N., and Oden, M. (2015, March 15). *A Tale of Two Innovative Entrepreneurial Regions: The Research Triangle and Austin*. University of Texas at Austin and University of North Carolina at Chapel Hills.
Economic Growth in Renewable Energy. (2014, January 7). *The Engineer*.
Ehrlich, E. M. (2008). *Manufacturing, Competitiveness, and Technological Leadership in the Semiconductor Industry*.
Eisinger, P. K. (1988). *The Rise of the Entrepreneurial State: State and Local Economic Development Policy in the United States*. Madison: University of Wisconsin Press.
Empire State Development. (2001, December 14). *National Grid—Luther Forest Infrastructure Capital II—Upstate City-by-City (x044)*.
Empire State Department Corporation. (2010, March 25). *LFTCEDC—Luther Forest Infrastructure Capital II—NYSEDP and Update City-by-City (x043, x044)*.
Empire State Development Approves $46 million for GCEDC's STAMP plan. (2016, August 18). *The Daily News*.
ERSO to Set Up Flexible Electronics Alliance in July. (2005, May 19). *Taiwan Economic News*.
ESD Team Reviews EDC Funding Proposals. (2014, September 13). Massena, *Daily Courier-Observer*.
Etzkowitz, H. (2008). *The Triple Helix: University—Industry—Government Innovation in Action* (New York and London: Routledge).
Etzkowitz, H. (2012). *Silicon Valley: The Sustainability of an Innovation Region*. Triple Helix Research Group.
Etzkowitz, H. (2013). Silicon Valley: The Sustainability of an Innovative Region. *Social Science Information Journal*. 52(4), 515–538.
Etzkowitz, H., and Leydesdorff, L. (2000). The Dynamics of Innovation: From National Systems and 'Mode 2' to a Triple Helix of University—Industry—Government Relations. *Research Policy*.
EUV is key to 450mm Wafers. (2014, July 31). *Semiconductor Engineering*.
Fabless Future: Struggling AMD Spin-off Factories. (2008, October 7). *Associated Press*.
Fairweather, P., and Schnell, G. A. (1995). Reinventing a Regional Economy: The Mid-Hudson Valley and the Downsizing of IBM. *Middle States Geographer*. 28.
Fast Growth at Technology Centre Bodes Well for Manufacturing. (2012, June 14). *Professional Engineering*.
The Father of Silicon Valley. (2016, September 21). *TechHistoryWorks*.
Feds: Upstate New York Job Growth 'Flat'. (2016, August 18). *Syracuse.com*.
Fein, M. R. (2008). *Paving the Way: New York Roadbuilding and the American State, 1880–1956*. Lawrence, KA: University of Kansas Press.
Feller, I. (1998). Evaluating State Advanced Technology Programs. *Education Review*. 12(3).
Ferguson, E. S. (1992). *Engineering and the Mind's Eye*. Cambridge, MA: The MIT Press.
FMCC-HFM BOCES Collaboration Creates Career Pathway. (2011, December 24). *Targeted News Service*.
Foundry Sales Defy IC Decline. (2015, October 22). *EETimes*.

Fraunhofer ISE Sets Another Solar Cell Efficiency Record. (2015, September 25). *Clean Technical.*
Fraunhofer ISE Sets PV Module Efficiency Record of 36.7 Percent. (2014, July 14). *Semiconductor Today.*
Fred Terman, the Father of Silicon Valley. (2010, October 21). *Net Valley.*
From Pigs to Nanochips. (2013, August 19). *The Journal of Commerce.* http://www.techhistoryworks.com/silicon-valley-history/2016/9/21/the-father-of-silicon-valley.
Fury, M. A., and Kaloyeros, A. E. (1993). Metallization for Microelectronics Program at the University of Albany: Leveraging a Long Term Mentor Relationship. *IEEE Xplore.* 59–60.
Future Horizons. (2012, February 16). *Smart 2010/062: Benefits and Measures to Setup 450mm Semiconductor Prototyping and Keep Semiconductor Manufacturing in Europe.* Luxembourg: Office of Official Publications.
Gais, T., and Wright, D. (2012). The Diversity of University Economic Development Activities and Issues of Impact Measurement. In J. E. Lane and D. B. Johnstone (Eds.), *Universities and Colleges as Economic Drivers: Measuring Higher Education's Role in Economic Development* (p. 36). Albany: SUNY Press.
Geiger, R. L. (2010). Better Late Than Never: Intentions, Timing and Results in Creating SUNY Research Universities. In J. B. Clark, W. B. Leslie, and K. P. O'Brien, (Eds.), *SUNY at Sixty: The Promise of the State University of New York* (p. 172). Albany: SUNY Press.
German Innovation, British Imitation. (1992, November 21). *New Scientist.*
GF Closes on IBM Chip Business Purchase. (2015, July 1). *Semiconductor Engineering.*
Gillmore, C. S. (2004). *Fred Terman at Stanford: Building a Discipline, a University, and Silicon Valley.* Stanford CA: Stanford University Press.
Gilson, R. J. (1999). Legal Infrastructure of High Technology Industrial Districts: Silicon Valley, Route 128, and Covenants Not to Compete. *New York University Law Review.* 74(3), 577.
Glauconix Biosciences. What We Do. http://www.http://glauconix.biosciences.com/whatwedo/.
GlobalFoundries. (2011). *Reaping the Benefit of the 450mm Transition.* Semicon West.
GlobalFoundries. (2016, February 9). SUNY Poly and GlobalFoundries Announce New $500 M R&D Program in Albany to Accelerate Next Generation Chip Technology. Press Release.
GlobalFoundries—2010 Gold Shovel Project of the Year. (2010, July). *Area Development.*
GlobalFoundries on a Roll With 7nm and 5nm Announcements. (2017, June 15). *Forbes.*
GlobalFoundries: E. Fishkill Site Key to Overall Strategy. (2015, November 12). *Poughkeepsie Journal.*
GlobalFoundries Fires up its 7-nm Leading Performance Forges. (2017, June 16). *The Tech Report.*
GlobalFoundries: No Layoffs in IBM Chip Units. (2014, October 20). *Associated Press.*
GlobalFoundries Plans New York R&D Center. (2013, January 9). *EETimes.*
GlobalFoundries Plant Generates Demand for Site Inspectors. (2010, March 18). Albany, *Business Review.*
GloFo Shows Progress in 3D Stacks. (2014, March 19). *EE Times.*
Governor Cuomo Announces Milestone Reached at AIM Photonics in Rochester. (2017, May 28). *US Fed News.*
Governor Cuomo and Vice President Biden Announce New York State to Lead Prestigious National Integrated Photonics Manufacturing Institute. (2015, July 27). New York State Press Release.
Grads Exceed Average Earnings. (1987, March 12). Albany, *Knickerbocker News.*
Gray, D, Sundstrom, E., Tomasky, L. G., and McGowen, L. (2011, October 1). When Triple Helix Unravels: A Multi-Case Analysis of Failures in Industry-University Cooperative Research Centers. *SAGE Journals.*
Gray, D. O. (2011, March). Cross-Sector Research Collaboration in the USA: A National Innovation System Perspective. *Science and Public Policy.*
Gretzinger, J. (2015, September 26). SUNY Poly Details Oversight and Transparency Process Regarding Buffalo Projects. Memorandum. SUNY Poly.
Growth of Silicon Empire: Bay Area's Intellectual Ground Helped Sprout High Technology Industry. (1999, December 27). *The San Francisco Chronicle.*

Haldar, P. (2013). Pioneering Innovation to Drive an Educational and Economic Renaissance in New York State. In National Research Council, *New York's Nanotechnology Model: Building the Innovation Economy*. C. W. Wessner (rapporteur), p. 81. Washington, DC: The National Academies Press.

Hamdami, K., Deitz, R., Garcia, R., and Cowell, M. (2005, Winter). Population Out-Migration from Upstate New York. In Federal Reserve Bank of New York (Buffalo Branch), *The Regional Economy of Upstate New York*.

Hartley, D. (2013, May 20). Economic Decline in Rust Belt Cities. *Economic Commentary*.

Hastfed, D. M. (2003). *The Emergence of Entrepreneurship Policy: Governance, Start-ups and Growth in the US Knowledge Economy*. Cambridge: Cambridge University Press.

Hauser, H. (2010). *The Current and Future Role of Technology Centres in the UK*. London: UK Department for Business, Innovation and Skills.

Hi-Tech Funding Extended—Su Cater, Six Others to Benefit. (1987, March 16). *Syracuse Herald-Journal*.

High Tech Companies Team Up on Chip Research. (2012, August 27). *Wall Street Journal*.

High tech on the Hudson: Digital Dynamos Have Turned Kingston Into Brooklyn North. (2010, December 17). *New York Daily News*.

Hill, P. (2014, April 25). Preparing Middle-Class Workers for Middle-Skill Jobs. *Marketplace*.

Homburg, C. (Ed.) (1999). *Structure and Dynamics of the German Misselstand*. Heidelberg and New York: Physica-Verlag.

House of Commons. Committee on Science and Technology. (2011, February 9). Second Report: Technology and Innovation Centres.

How Did New York Become the Most Unionized State in the Country? (2014, September 3). *The Nation*.

How Semiconductors Are Made. *Intersil*. http://rel.intersil.com/docs/lexicon/manufacture.html.

Hsu, A. (2007, October 19). ITRI Pushes Technology Sector to New Frontiers of Innovation. *Taiwan Journal*.

Hudson Valley Community College. *College Catalog* (2015-16).

Hutcheson, G. H. (2017). Economics of Semiconductor Manufacturing. In Y. Nishi and R. Doering (Eds.), *Handbook of Semiconductor Manufacturing Technology* (2nd Ed.), p. 1137. Boca Raton, London, and New York: CRC Press.

IBM Braces for Additional Cuts, Plans to Implement First Layoffs. (1993, February 25). *Watertown Daily Times*.

IBM's Big New York Compute: $1.5 B Investment, 1,000 Jobs. (2008, August). *Site Selection*.

IBM, Partners Creating 1,000+ Jobs With $2.7 Billion in New York Projects. (2005, January). *Site Selection*.

IBM Pushes for Chip Consortium. (1987, January 7). Ft. Lauderdale, *Sun Sentinel*.

IBM Team Makes Atomic-Scale Circuitry Breakthrough. (2000, February 3). *Watertown Daily Times*.

Ideas in Action. (2006, Fall). *RPI Alumni Magazine*.

If You Build It, They Will Come. (2003, February 7). *The Chronicle of Higher Education*.

In IBM Country, A Kick in the Teeth. (1993, April 4). *Watertown Daily Times*.

In the Space of Five Years, It Looks Like 450mm Manufacturing Has Become Surplus to Current Requirements. (2016, June 28). *New Electronics*.

Income Inequality in the US by State, Metropolitan Area, and Country. (2016, June 16). *Economic Policy Institute*.

Income Surges for AMD—But Threat of Price War With Intel Raises Concern. (2006, July 21). *San Jose Mercury News*.

Inside GlobalFoundries' Fab 8. (2015, August 18). *EETimes*.

Interview: Dr. Phill Cartwright, Chief Technology Officer, High Value Manufacturing Catapult. (2016, March 7). *The Engineer*.

Irwin, D. A. (2004, September). The Aftermath of Hamilton's Report on Manufactures. *The Journal of Economic History*.

Is 450mm Dead in the Water? (2014, May 15). *Semiconductor Engineering*.

Is There a US Engineering Shortage? It Depends on Who You Ask. (2015, August 19). *Power Electronics*.

Jackson, S. A. (2013, April 4). Keynote Address at National Research Council Symposium, "New York's Nanotechnology Model: Building the Innovation Economy." Albany, New York.

Joint School of Nanoscience and Nanoengineering. http://jsnn.ncat.uncg.edu/.

Jindal, V. (2013, January). Getting up to Speed with Roadmap Requirements for Extreme UV Lithography. *SPIENewsroom*.

A Jobs Crisis? No It's a skills Crisis. (2013, January 16). *New York Daily News*.

Kaiser, D. (Ed.) (2010). *Becoming MIT: Moments of Decision*. Cambridge, MA, and London: The MIT Press.

Kalas, J. W. (1986). Reindustrialization in New York: The Role of the State University. In M. Schoolman and A. Magid (Eds.), *Reindustrializing New York State: Strategies, Implications, Challenges*. Albany: SUNY Press.

Kalas, J. W. (2010). SUNY Strides into the National Research Stage, In J. B. Clark, W. B. Leslie, and K. P. O'Brien, (Eds.), *SUNY at Sixty: The Promise of the State University of New York* (p. 161). Albany: SUNY Press.

Katz, B., and Muro, M. (2010, September 21).The New 'Cluster Moment': How Regional Innovation Clusters Can Foster the Next Economy. Washington, DC: Brookings.

Kich, L. B. (2002, December). End of Moore's Law: Thermal (Noise) Death of Integration in Micro and Nano Electronics. *Physics Letters*.

Kim, L. (1997, Spring). The Dynamics of Samsung's Technological Learning in Semiconductors. *California Management Review*.

Kingston: The IBM Years Gives a Peek into Tech Giant's History. (2014, June 4). *Hudson Valley Magazine*.

Korvink, J. G., and Greiner, A. (2002). *Semiconductor for Micro- and Nanotechnology: An Introduction for Engineers*. Wileys.

Krugman, P. (1991). *Geography and Trade*. Cambridge, MA: The MIT Press.

Lane, N., and Kalil, T. (2005, Summer). The National Nanotechnology Initiative: Present at the Creation. *Issues in Science and Technology*.

Le Maistre, C. W. (1989, October). Academia Linking with Industry—The RPI Model. *IEEE Xplore*.

Le Roux, J. (2012). *Carnot program*. http://www.institute.carnot.edu/en/instituts-carnot. Accessed 2013.

Leslie, S. W. (2000). The Biggest Angel of All: The Military and the Making of Silicon Valley. In M. Kenney (Ed.), *Understanding Silicon Valley: The Anatomy of an Entrepreneurial Region*. Stanford: Stanford University Press.

Leslie, S. W. (2001). Regional Disadvantage: Replicating Silicon Valley in the Capital Region. *Technology and Culture*.

Leslie, S. W., and Karagon, R. H. (1996). Selling Silicon Valley: Fredrick Terman's Model for Regional Advantage. *Business History Review*.

Lessig, L. (2009, October 9). Against Transparency. *New Republic*.

Lowe, N. (2011). Southern Industrialization Revisited: Industrial Recruitment as a Strategic Tool for Local Economic Development. In Daniel P. Gitterman (Ed.), *The Way Forward: Building a Globally Competitive South*. Chapel Hill: Global Research Institute.

Lowe, N. (2014). Beyond the Deal: Using Industrial Recruitment as a Strategic Tool for Manufacturing Development. *Economic Development Quarterly* 28(4).

*Luther Forest Technology Campus GEIS: Statement of Findings*. (2004, June 14). Draft adopted by Stillwater Town Board.

Made in America: Global Companies Expand in US Towns. (2012, April 30). *ABC News*.

Made in NY? Forget It, as State Loses to Others. (2017, March 10). Rochester, *Democrat & Chronicle*.

Mahoning Valley Expert: Region's Recovery Will Take Decades. (2012, September 23). *Youngstown Vindicator*

Major Technology Partnership Announced. (2009, July 21). *US Fed News*.

Malta and Stillwater Town Boards Approve PDD for Luther Forest Technology Campus. (2008, June 3). *New York Real Estate Journal*.

Malta Zones Research Site Industrial. (1999, December 28).

Manufacturers Bring Back Jobs to Central Mass. (2017, September 4). *WBJournal*.

Manufacturing Bringing the Most Jobs Back to America. (2016, April 23). *USA Today*.

Manville, M., and Kuhlmann, D. (2016, November 11). The Social and Fiscal Consequences of Urban Decline: Evidence from Large American Cities, 1980-2010. *Urban Affairs Review*.

Mathews, J. A. (2010, December 10). The Hsinchu Model: Collective Efficiency, Increasing Returns and Higher Order Capabilities in the Hsinchu Science-Based Industry Park, Taiwan. Keynote Address, Chinese Society for Management of Technology 20th Anniversary Conference. Hsinchu, Taiwan.

Mathews, J. A., and Cho, D. (2000). *Tiger Technology: The Creation of a Semiconductor Industry in East Asia* Cambridge: Cambridge University Press.

Mazzucato, M. 2013). *The Entrepreneurial State: Debunking Public vs. Private Sector Myths*. London, New York and Delhi: Anthem Press.

McClelland, P. D., and Magdovitz, A. L. (1981). *Crisis in the Making: The Political Economy of New York State Since 1945*. Cambridge: Cambridge University Press.

McKinsey & Company. (2015). *Capital 20.20: Advancing the Region through Focused Investment*. McKinsey & Company.

McNeil, R. D. et. al. (2007, September). Barriers to Nanotechnology Commercialization. Report prepared for U.S. Department of Commerce Technology Administration. Springfield: The University of Illinois.

Midnight in the Rust Belt. (2013, September 21). *Beltmag.com*.

Montgomery, D. (1989). *The Fall of the House of Labor: The Workplace, the State, and American Labor Activism 1865–1925*. Cambridge: Cambridge University Press.

Moore's Law: Past, Present and Future. (1997, June). *IEEE Spectrum*.

Moretti, E. (2013). *The New Geography of Jobs*. Boston and New York: Mariner Books.

Moynihan, D. P. (1993, April 15). No Surrender: Toward Intolerance of Crime. Address to the Association for a Better New York.

Munsey, J. S. (2006, May). Project Case Study: High Tech Land Development. *Civil & Structural Engineer*.

Muro, M. (2013, February 27). Regional Innovation Clusters Begin to Add Up. Washington, DC: Brookings.

National Governors Association and Council on Competitiveness. (2007). *Cluster-Based Strategies for Growing State Economies*. Washington, DC: National Governors Association.

National Grid. Spier Falls to Rotterdam 115kV Transmission Line Project. http://www9.nationalgridus.com/transmission/spier_rotterdam.asp.

National Grid. (2011, August). GlobalFoundries: New Gas Transmission Line. Article VII Application for a Certificate of Environmental Compatibility and Public Need.

National Grid. (2015, June 4). New Eastover Sub Station Provider Base for Continued Growth in New York Capital Region. Press Release.

National Grid Begins Construction on 115-kV Line in New York. (2011, November 2). *Transmission Hub*.

National Network for Manufacturing Innovation. (2013). Frequently Asked Questions. http://manufacturing.gov/docs/nnmi.fag.pdf. Accessed June 7, 2013.

National Research Council. (1984). *Competitive Status of the U.S. Electronics Industry: A Study of the Influences of Technology in Determining International Competitive Advantage*. Washington, DC: National Academy Press.

National Research Council. (1986). *New Alliances and Partnerships in American Science and Engineering*. Washington, DC: National Academy Press.

National Research Council. (2003). *Securing the Future: Regional and National Programs to Support the Semiconductor Industry.* C. W. Wessner (Washington, DC: The National Academies Press.

National Research Council. (2006). *A Matter of Size: Triennial Review of the National Nanotechnology Initiative* Washington, DC: The National Academies Press.

National Research Council. (2008). *An Assessment of the SBIR Program.* C. W. Wessner (Ed.) Washington, DC: The National Academies Press.

National Research Council. (2009). *Understanding Research, Science and Technology Parks: Global Best Practices.* C. W. Wessner (Washington, DC: The National Academies Press.

National Research Council. (2011a). *The Future of Photovoltaic Manufacturing in the United States.* C. W. Wessner (rapporteur). Washington, DC: The National Academies Press.

National Research Council. (2011b). *Growing Innovation Clusters for American Prosperity.* C. W. Wessner (rapporteur). Washington, DC: The National Academies Press.

National Research Council. (2013a). *21st Century Manufacturing: The Role of the Manufacturing Extension Partnership Program.* C. W. Wessner (Ed.) (Washington, DC: The National Academies Press.

National Research Council. (2013b). *Best Practices In State and Regional Innovation Initiatives: Competing in the 21st Century.* C. W. Wessner (Ed.) (Washington, DC: The National Academies Press.

National Research Council. (2013c). *New York's Nanotechnology Model: Building the Innovation Economy.* C. W. Wessner (rapporteur). Washington, DC: The National Academies Press.

Nation's First Security College Creates Student Opportunities. (2015, January 27). *Long Island Examiner.*

NEA Research. (2016, May). *Rankings & Estimates: Rankings of the States 2015 and Estimates of School Statistics 2016.* National Education Association. Table H-11.

New York Department of Transportation. *Round Lake Bypass Project.* Final Environmental Impact Statement.

New York is Placing Big Bets in Upstate Cities. (2015, September 12). *Associated Press State Wire: New York.*

New York State Office of the Comptroller General. (2013, January). *Fuller Road Management Corporation.* Report 2012-S-26.

New York's Big Subsidies Bolster Upstate's Winning Bid for AMD's $3.2 Billion 300-MM Fab. (2006, June 10). *Site Selection.*

NIMO, National Grid to Merge at End of Month. (2002, January 17). Syracuse, *The Post Standard.*

NY Wins $600 Million Hub for Photonics Research and Development. (2015, July 27). *Associated Press State Wire: New York.*

NY's High-Tech Hope. (2002, September 18). *New York Post.*

NYSERDA. (2017, June). 2016 Clean Air Interstate Rule Annual Report on the New York Battery and Energy-Storage Technology Consortium.

Office of the Governor. (2011, September 27). Governor Cuomo Announces $4.4 Billion Investment by International Technology Group Led by Intel and IBM to Develop Next Generation Computer Chip Technology in New York. Press Release. http://www.governor.ny.gov/press/092711chiptechnologyinvestment.

Office of the Governor. (2016, February 5). Governor Cuomo Announces New SUNY Emerging Technology and Entrepreneurship Complex at Harriman Campus. Press Release.

Office of the New York State Comptroller. (2010). *Fuller Road Management Corporation & The Research Foundation of the State of New York: Use of State Funding for Research into Emerging Technologies at the State University of New York at Albany: Nanotechnology.* (2010-S-4) http://www.osc.state.ny.us/audits/allaudits/093010/10s4.pdf.

Office of the State Comptroller. (2010). *Fuller Road Management Corporation & The Research Foundation of the State of New York.* (2010-5-5).

Offshore Renewable Energy Catapult is Ready for Action. (2014, January 3). *Herald Scotland.*

Oldenski, L. (2015, September). *Reshoring By US Firms: What Do the Data Say?* Peterson Institute for International Economics.

Ottman, T. (2010). Forging SUNY in New York's Political Cauldron. In J. B. Clark, W. B. Leslie, and K. P. O'Brien, (Eds.), *SUNY at Sixty: The Promise of the State University of New York* (pp. 15–29). Albany: SUNY Press.

Overseas Jobs are Coming Home--S.C. Business. (2013, September 8). Columbia, SC, *The State*.

Park, F. M. Turning Poor Dirt into Pay Dirt. *METRO Magazine*. http://www.metronc.com/article/?id+421.

Pawning Tools of the Trade: Construction Workers in Hock, Others Sell Gold. (2008, March 9). Riverside, *The Press-Enterprise*.

PBC Tech Readies Battery-Boosting PowerWRAPPER for Commercial Launch with Manufacturing Agreement and New U.S. Patent. (2018, January 9). *Nasdaq*.

Pecorella, R. F. (2012). Regional Political Conflict in New York State. In R. F. Pecorella and J. M. Stonecash (Eds.) *Governing New York State* (p. 14). Albany: SUNY Press.

Pendall, R. (2003). *Upstate New York's Population Plateau: The Third-Slowest Growing State*. Washington, DC: The Brookings Institution.

The Peridot Capitalist. (2007, January 27). AMD, Intel Price War Revisited. http://www.Peridotcapital.com/2007/01/amd-intel-price-war-revisited.html.

Phelan, T., Ross, D. M., and Westerdahl, C. A. (1995). *Rensselaer: Where Imagination Achieves the Impossible*. Albany: Mount Ida Press.

Philanthropy a Way of Life of Greater Clevelanders. (2010, December 26). *The Plain Dealer*.

Pillai, U. (2015, March). SUNY College of Nanoscale Science and Engineering. Monograph.

Planting the Seeds of a New Company. (2013, Summer). *Siena News*.

Platzer, M. D., and Sargent, J. F. (2016, June 27). *US Semiconductor Manufacturing Trends, Global Competition, Federal Policy*. Washington, DC: Congressional Research Service.

Plosila, W. H. (2004). State Science and Technology—Based Economic Development Policy: History, Trends and Developments and Future Directions. *Economic Development Quarterly*. 18(2).

Polanyi, M. (1958). *Personal Knowledge: Toward a Post-Critical Philosophy*. Chicago: University of Chicago Press.

PolicyLink. (2008). *To Be Strong Again: Renewing the Promise in Smaller Industrial Cities*. https://www.policylink.org/sites/default/files/ToBeStrongAgain_final.pdf.

Porter, M. (1998, December). Clusters and the New Economics of Competition. *Harvard Business Review*.

Prah, P. M. (2013, July 30). Has US Manufacturing's Comeback Stalled? *USA Today*.

Preparing Middle-Class Workers for Middle-Skill Jobs. (2014, April 25). *Market place*.

President's Council of Advisors on Science and Technology. (2003). *The National Nanotechnology Initiative at Five Years: Assessment and Recommendations of the National Nanotechnology Advisory Panel*. Washington, DC: Executive Office of the President.

President's Council of Advisors on Science and Technology (2011, June). *Report to the President on Ensuring American Leadership in Advanced Manufacturing*. Washington, DC: Executive Office of the President.

President's Council of Advisors for Science and Technology. (2017, January 6). Report on Ensuring Long-Term U.S. Leadership in Semiconductors. Washington, DC: Executive Office of the President.

Project Case Study: High Tech Land Development. (2006, May). *Civil and Structural Engineering*.

Quality Counts 2017: State Report Cards Map. (2017). *Education Week*. http://www.edweek.org.

Radosevic, S., et al. (Eds.) (2017). *Advances in the Theory and Practice of Smart Specialization*. London, San Diego, Cambridge, MA, and Oxford: Elsevier.

A Raise for the Record Books. (2007, June 5). *Inside Higher Ed*.

Reiert, E. (2007). *How Rich Countries Got Rich and Why Poor Countries Stay Poor*. London: Constable.

Rensselaer Polytechnic Institute. Center for Biotechnology & Interdisciplinary Studies. http://biotech.rpi.edu.

Rensselaer Polytechnic Institute Faculty Lauded. (2016, December 7). *RPI News*.
Replanting the STEM Common Core Should Help Students Master Necessary Skills. (2013, October 30). *Watertown Daily Times*.
Research Foundation of SUNY. (2006). *SUNY's Impact on New York's Congressional District 21*.
Reshoring: A Boost in American Manufacturing. (2017, August 30). *Machine Design*.
Reshoring Is an Issue for Europe Too. (2013, October 23). *Finanz & Wirtschaft*.
A Roadmap for US Integrated Photonics. (2017, March 23). *Optics & Photonics*.
Roberts, E. B., and Eesley, C. E. (2011). *Entrepreneurial Impact: The Role of MIT*. Hanover, MA: Now Publishers Inc.
Rockefeller Institute of Government. (2007, July). *The Supervisory District of Albany, Schoharie, Schenectady and Saratoga Counties: A Study of Potential Educational Reorganization in the Capital Region*. Prepared for New York State Department of Education.
Rockefeller Institute of Government of the University at Albany, and the University at Buffalo Regional Institute. (2011, June). *How SUNY Matters: Economic Impacts of the State University of New York*. State University of New York.
Rosenberg, N. (2003). America's Entrepreneurial Universities. In D. M. Hurt (Ed.), *The Emergence of Entrepreneurship Policy; Governance and Growth in the US Knowledge Economy*. Cambridge: Cambridge University Press.
Rosenberg, N., and Nelson, R. R. (1994). American Universities and Technical Advance in Industry. *Research Policy* 23:326.
Rosenberg, N., and Steinmueller, E. (2013, October). Engineering Knowledge. *Industrial and Corporate Change*.
Rudolph, F. (1977). *Curriculum: A History of the American Undergraduate Course of Study Since 1636*. San Francisco: Josey Bass.
Ruiz, H. (2013). *Slingshot: AMD's Fight to Free an Industry from the Ruthless Grip of Intel*. Austin, Texas: Greenleaf Book Group Press.
Ruttan, V. W. (2006, October 9). *Is War Necessary for Economic Growth?* Clemons Lecture, Saint Johns University Collegeville, MN.
Sá, C. M. (2011). Redefining University Roles in Regional Economies: A Case Study of University-Industry Relations and Academic Organization in Nanotechnology. *Higher Education* 61:193–208.
Samsung Licenses 3D Chip Manufacturing Tech to GlobalFoundries to Win More Orders. (2014, April 17). *Reuters*.
Samsung to Invest Additional $9.2 billion in its $14.44 billion fab. (2015, April 15). *KitGuru*.
Saxenian, A. (1994). *Regional Advantage: Culture and Competition in Silicon Valley and Route 128*. Cambridge, MA: Harvard University Press.
Schact, W. H. (2013, December 3). *Industrial Competitiveness and Technological Advancement: Debate Over Government Policy*. Washington, DC: Congressional Research Service.
Schaller, R. R. (2004). *Technological Innovation in the Semiconductor Industry: A Case Study of the International Technology Roadmap for Semiconductors*, Ph.D. Dissertation. Arlington, VA: George Mason University.
Schneier, E. V., Murtaugh, J. B., and Pole, A. (2010). *New York Politics: A Tale of Two States*. Armonk and London: M.E. Sharpe.
Schoolman, M. (1986). Solving the Dilemma of Statesmanship: Reindustrialization Through an Evolving Democratic Plan. In M. Schoolman and A. Magid (Eds.), *Reindustrializing New York State: Strategies, Implications, Challenges*. Albany: SUNY Press.
Schoolman, M., and Magid, A. (Eds.) (1986). *Reindustrializing New York State: Strategies, Implications, Challenges* Albany: SUNY Press.
Schultz, L., Wagner, A., Gerace, A., Gais, T., Lane, J. and Monteil, L. (2015). Workforce Development in a Targeted, Multisector Economic Strategy: The Case of New York University's College of Nanoscale Science and Engineering. In C. Van Horn, T. Edwards, and T. Green (Eds.), *Transforming US Workforce Development Policies for the 21st Century*. Kalamazoo: W.E. Upjohn Institute for Employment Research.

Schulz, L. I. (2011). Nanotechnology's Triple Helix: A Case Study of the University of Albany's College of Nanoscale Science and Engineering. *Journal of Technology Transfer.*

Schultz, L. I. (2012). University Industry Government Collaboration for Economic Growth. In J. E. Lane and D. B. Johnstone (Eds.), *Universities and Colleges as Economic Drivers: Measuring Higher Education's Role in Economic Development.* Albany: SUNY Press.

Scott, E., and Wial, H. (2013, May). *Multiplying Jobs: How Manufacturing Contributes to Employment Growth in Chicago and the Nation.* Chicago: University of Illinois at Chicago.

Sematech. (1991). *Annual Report.* Austin: Sematech.

Sematech, SUNY Seal EUV Lithography Program. (2003, January 29). *Solid State Technology.*

Sematech Touts the Benefits of its New York Alliance. (2002, July 19). *Austin American-Statesman.*

Semico Research Corporation. (2008a, February). *Economic Impact of the Semiconductor Industry on Upstate New York.*

Semico Research Corporation. (2008b, March). *Upstate New York: Assessing the Economic Impact of Attracting Semiconductor Industry.*

Semiconductor Industry Association. Policy Priorities: Tax. http://www.semiconductors.org/issues/tax/tax.

Semiconductor Industry Association. (2003, October). *China's Emerging Semiconductor Industry: The Impact of China's Preferential Valve Added Tax as Current Investment Trends.*

Semiconductor Industry Association. (2009, March). *Maintaining America's Competitive Edge: Government Policies Affecting Semiconductor Industry R&D and Manufacturing Activity.*

Semiconductor Industry Association. (2015, January). *US Semiconductor Industry Employment.*

Shapira, P., and Wang, J. (2007, November). Case Study: R&D Policy in the United States: The Promotion of Nanotechnology R&D. Atlanta: Georgia Institute of Technology.

Shermer, E. T. (2015). Nelson Rockefeller and the State University of New York's Rapid Rise and Decline. Rockefeller Archive Center Research Reports Online.

Silicon Valley, Meet Organized Labor. (2014, October 7). *New York.*

Simchi-Levi, D. (2012). US Re-Shoring: A Turning Point. *MIT Forum for Supply Chain Innovation 2012 Annual Re-Shoring Report.* Cambridge, MA: The MIT Press.

Skidmore College. MB107. http://www.skidmore.edu/management_business/mb107/index.php.

Small Manufacturers Feeling Squeeze—Shrinking of Industrial Giants Like IBM, Kodak Hurting State's Smaller Firms. (1993, May 17). *Watertown Daily Times.*

Solar Industry Slowdown Catches Up with SolarCity. (2017, May 5). *Investopedia.*

Spending Billions While Killing Jobs. (2017, March 28). *New York Post.*

Sperling, E. (2014, November 20). An Inside Look at the GlobalFoundries—IBM Deal. *Semiconductor Engineering.*

State Fights Silicon Valley for Computer Chip Site. (1987, July 29). Albany, *Knickerbocker News.*

State Investing $522 Million to Create High-Tech Jobs. (1999, November 11). *Watertown Daily Times.*

State to Up Ante in Sematech Pot to $80M. (1987, December 3). Albany, *Knickerbocker News.*

State University of New York. SUNY Works Campus Partnerships. https://www.suny.edu/suny-works/partnerships.

State University of New York. (2013). *HocusLocus, LLC., the University at Albany at the College of Nanoscale Science and Engineering.* CICEP 2013 Case Study.

Sturgeon, T. (2000.) How Silicon Valley Came to Be. In M. Kenney (Ed.), *Understanding Silicon Valley: The Anatomy of an Entrepreneurial Region.* Stanford, CA: Stanford University Press.

SUNY Adirondack. Academics. http://www.sunyacc.edu/academics.

SUNY Board of Trustees Resolution No. 2004-41. (2004, April 20).

SUNY Board of Trustees Resolution No. 2008-165. (2008, November 18).

SUNY Buffalo State. (2011, February 15). Research Foundation of SUNY Celebrates 60$^{th}$ Anniversary. Press Release.

SUNY Chancellor Promises More Internships in 2014. (2014, January 14). *Associated Press Newswire.*

SUNY POLY. (2015, October 21). SUNY Poly CNSE Announces Milestone as M+W Group Opens US Headquarters at Albany Nanotech Complex and Research Alliance Begins $105M Solar Power Initiative. Press Release.
SUNY Poly Settles Low, Tries to Clear 'Kaloyeros Hangover'. (2017, March 27). *Politico*.
SUNY Polytechnic Institute. Faculty Profile: Laura Schultz. https://sunypoly.edu/faculty-and-staff/laura-schultz.html.
SUNY Polytechnic Institute. (2014, October 10). Energy-Focused Incubator at SUNY Polytechnic Institute and the Hudson Valley Center for Innovation. Press Release.
SUNY Polytechnic Institute. (2015, September 4). Grants, Investments Boost SUNY Poly CNSE Startup. *SUNY Research Foundation News*. https://sunypoly.edu/news/suny-research-foundation-news-grants-investments-boost-suny-poly-cnse-start-0.html.
SUNY Polytechnic Institute. (2016, June). *Strategic Plan*.
SUNY Polytechnic Institute. (2017, July 25). SUNY Poly Alumna, Founder and CEO of Glauconix Biosciences Awarded $750,000 by National Science Foundation for Commercialization of Technology Developed at SUNY Poly to Fight Eye Diseases. Press Release.
SUNY Working Group Report. (2013, June 13). *The SUNY College of Nanoscale Science and Engineering*.
'Super Region' Marketing Part of $75K Contract. (2017, December). Batavia, *The Daily News*.
Sutton, R. I. (2013, October 3). Scaling: The Problem of More. *Harvard Business Review*.
Taiwanese IT Pioneers: Ding-Hua Hu. (2011, February 10). Recorded interview. Computer History Museum.
Taiwanese IT Pioneers: Jouney (Chang-tang) Shih. (2011, February 15). Recorded Interview. Computer History Museum.
Taiwanese IT Pioneers: Robert H.C. Tsao. (2011, February 17). Recorded Interview. Computer History Museum.
Tales of Silicon Valley Past: Legendary Founders Talk About Early Days at Fairchild. (1995, May 13). *San Jose Mercury News*.
Tajnai, C. (1996). From the Valley of Heart's Delight to Silicon Valley: A Study of Stanford University's Role in the Transformation. Stanford: Stanford University Department of Computer Science.
TD Economics. (2012, October 15). Onshoring, and the Rebirth of American Manufacturing.
Teaming Up to Get Workers Ready for the Tech of the Future. (2015, September 12). Columbia, South Carolina *The State*.
Tech Startups Seek the Good Life in Saratoga Springs. (2014, December 2). *Politico*.
Tesla and SolarCity Merger Gets Approval from Shareholder. (2016, November 17). *CNBC*.
TIP Strategies. (2014, March). *Economic Development Strategic Plan Prepared for Saratoga County, New York*. http://www.saratogacountyny.gov/wp/wp-content/uploads/2013/11/Saratoga-Plan-FINAL.pdf.
Tittnich, M., *et al.* (2006). A Year in the Life of an Immersion Lithography Alpha Tool at Albany Nano Tech. In Proceedings of SPIE, Vol. 6151, *Emerging Lithographic Technologies*.
Todtling, F., and Tripple, M. (2005). One Size Fits All? Towards a Differentiated Regional Innovation Policy Approach. *Research Policy*.
Tokyo Electron Plugging $300M R&D Center Into Albany, NY. (2002, December). *Site Selection*.
Top Ten Foundries 2017. (2017, December 1). *Electronics Weekly*.
The Town IBM Left Behind. (1995, September 10). *Business Week*.
Town of Malta Regulations. Chapter 167, Article VII, § 167-26.
Town of Stillwater Code. Art. XI § 211–162A(i)(a).
Town of Stillwater Code. Art. XI § 211–184.
Trades Union Congress. (2011, March). *German Lessons: Developing Industrial Policy in the UK*.
Transparency International. *Anti-Corruption Glossary*. http://www.transparency.org/glossary/terms/transparency.
The Transparency Trap. (2014, September). *The Atlantic*.

Trump Weighs Heavily on 1366 Technologies Project—STAMP—Federal Loan, Competition from China will Factor in Manufacturers' Plans. (2017, January 17). Batavia, *The Daily News.*
TSMC to Spend $10 Billion Building 450mm Wafer Factory. (2012, June 12). *Reuters.*
Tucker, F. M. (2008, Fall). The Rise of Tech Valley. *Economic Development Journal.*
Turning Technologies Rated Fastest-Growing. (2007, August 24).Youngstown, *Vindicator.*
Tyson, L. (1993). *Who's bashing whom: Trade conflict in high-technology industries.* Institute for International Economics.
UAlbany/CNSE/IT Implementation Teams. (2013). *Report to the SUNY Board of Trustees on the Potential Merger of the SUNY Institute of Technology and the SUNY College of Nanoscale Science and Engineering.*
United States Military Academy at West Point. A Brief History of West Point. http://www.usma.edu/wphistory/sitepages/home.aspx.
University at Albany. (2017, April 19). State Budget Keeps Campus Moving Forward. Press Release.
Upstarts and Rabble Rousers ... Stanford Fetes 4 Decades of Computer Science. (2006, March 20). *San Francisco Chronicle.*
The Upstate Economy is One of the Worst in the Country. (2016, September 16). *Politifact.com.*
Upstate New York Gets Nod For Sematech's $403M R&D Center. (2002, July). *Site Selection.*
The U.S. $4B Project that Got Away from Singapore. (2009, March 27). *The Business Times.*
U.S. Army and UAlbany NanoCollege Sign Agreement to Establish Unique Research Partnership. (2008, May 20). *Nanowerk.*
US Manufacturers 'Relocating' from China. (2013, September 23). *Financial Times.*
U.S. Paves Roads to Trusted Fabs. (2017, July 11). *EE Times.*
U.S. Small Business Administration. (2018). About SBIR. https://www.sbir.gov/about/about-sbir. Accessed February 21, 2018.
Utilities Can Play a Role in Attracting High-Tech Industry. (2006, November 1). *Electric Light & Power.*
van Agtmael, A., and Bakker, F. (2016). *The Smartest Places on Earth: Why Rustbelts are Emerging as Hotspots of Global Innovation.* New York: Public Affairs.
Varma, R. (2000, Autumn). Changing Research Cultures in US Industry. *Science, Technology and Human Values.*
Venohr, B., and Meyer, K. E. (2007, May). The German Miracle Keeps Running. How Germany's Hidden Champions Stay Ahead in the Global Economy. Berlin School of Economics.
Wade, R. (1990). *Governing the Market: Economic Theory and the Role of Government in Fast Asian Industrialization.* Princeton and Oxford; Princeton University Press.
Wadwha, V. (2013, July 3). Silicon Valley Can't Be Copied. *MIT Technology Review.*
Wagner, A., Sun, R., Zuber, K., and Strach, P. (2015, May). *Applied Work-Based Learning at the State University of New York: Situating SUNY Works and Studying Effects.* Albany: The Nelson A. Rockefeller Institute of Government.
Wagner, R. W. (2007). *Academic Entrepreneurialism and New York State's Centers of Excellence Policy,* Ph.D. dissertation. Albany: SUNY.
A Walk into 19th Century Troy, NY. (2012, June 27). *Upstate Earth.*
Watchdog Report: Upstate Sinks in a Sea of Legal Opioids. (2016, December 16). *Pressconnects.*
Watts, M. J. (2000, February). The Added Value of the Industry Cluster Approach to Economic Analysis, Strategy Development, and Service Delivery. *Economic Development Quarterly.* 14(1).
Webster, A. (1994, April). Bridging Institutions: The Role of Contract Research Organizations in Technology Transfer. *Science and Public Policy.*
Welcome to Berlin, Peter—Is it the Future? (2010, February 7). *Sunday Telegraph.*
What Happened to 450 mm? (2014, July 17). *Semiconductor Engineering.*
What We Can Learn from £100m and 10 Years Wasted on the Technique Programme. (2013, June 1). *WalesOnline.*

What's Behind the Success Story of German Manufacturing Industry? (2012, February 23). *Xinhun.*
Why 450 mm Wafer? (2012, August 8). *Semiconductor Engineering.*
Why Doesn't Britain Make Things Any More? (2011, November 16). *The Guardian.*
Wide Bandgap Alliance Formed in Taiwan. (2011, March 9). *Taiwan Economic News.*
The Wind and Sun Are Bringing the Shine Back to Buffalo. (2015, July 21). *The New York Times.*
Winnes, M., and Schimack, U. (1999). *National Report*: "*Federal Republic of Germany.*" Institute for the Study of Societies.
Wise, G. (1985). *Willis R. Whitney: General Electric and the Origins of US Industrial Research.* New York: Columbia University Press.
Youtie, J., and Shapira, P. (2008). Building on Innovation Hub: A Case Study of the Transformation of University Roles in Regional Technological and Economic Development. *Research Policy.* 37.
Zimpher, N. L. (2012). Foreword. In J. E. Lane and D. B. Johnstone (Eds.), *Universities and Colleges as Economic Drivers: Measuring Higher Education's Role in Economic Development.* (p. xii) Albany: SUNY Press.

# Index

**A**
Abu Dhabi, United Arab Emirates, 126, 127, 164, 390
Accelerate 518, 333
Adams, Ken, President and CEO, Empire State Development, 69
Advanced Manufacturing Performance (AMP), 223, 311
Advanced Micro Devices (AMD)
　agreement with New York, 77, 116, 127, 390
　asset-light strategy, 124–126
　financial difficulties, 121–127
　IBM collaboration, 119
　manufacturing expertise, 96, 119
　Materials Analysis Laboratory, Dresden, 119
　origins, 80, 127, 177, 351
　research presence in New York, 119
　search for site for 300mm water fabrication facility, 347
　spinoff of manufacturing operations, 126, 127, 187
Advanced Pattern and Productivity Center (APPC), 311
Advanced Technology Investment Company (ATIC), 126–128, 390
AIM Photonics, 82
Ainlay, Stephen, President, Union College, 250
Albany Business Review, xxvii, 32, 34, 35, 44, 54, 57, 62, 64, 66, 74, 75, 77, 80, 119, 174, 175, 199, 207, 224, 241, 244, 253, 254, 258, 259, 264, 289, 315, 319, 322, 331, 342, 368, 384, 395
Albany Business Review Schools Report, 264
Albany-Colonie Regional Chamber of Commerce, 42, 43, 97
Albany County, New York, 205
Albany Medical Center, 208, 209
Albany Molecular Research, Inc. (AMRI), 33, 293
Albany Nanotech Inc., 55, 63–64, 345
Albany Business Review, 100, 168
Albany, New York, 32, 165, 200, 217, 224, 235, 247, 253, 291, 298, 299, 305, 309, 314, 321, 322
Altes, Wallace, President, Albany-Colonie Regional Chamber of Commerce, 42
American Institute for Manufacturing Integrated Photonics (AIM Photonics), 82, 299–300
AMS AG, 301
Anderson, Warren, Majority Leader, New York Senate, 27, 339
Angelou, Angelos, economist, Austin, Texas, 66
Apollo programs, 12, 32
Applied Materials, xi, xx, 80, 81, 83, 202, 312, 313, 359, 389, 393
Apprenticeships, 220, 224, 229, 261, 268, 269, 273, 274, 364
Armbrust, Dan, CEO of Sematech, 67
Army Corps of Engineers, 140, 141, 148, 152
Arnoff Moving and Storage, 202
Arsenal Business & Technology Partnership, 72, 273
Arteris, 149, 173
Arthur, Chester, President of the United States, 248
Artificial intelligence, collaboration with SUNY Poly
　collaboration with Advanced Micro Devices, 80

Artificial intelligence, collaboration with SUNY Poly (*cont.*)
  collaboration with Professor Alain Kaloyeros, State University of New York at Albany, 85, 86, 88
  demonstration of molecular-level electronic circuits, 51
  employment impact, 188, 190
  financial contributions to semiconductor R&D in New York, 81
  first 64-bit microprocessor, 64
  immersion lithography process, 64, 83
  industrial legacy in New York, 6, 24, 340, 341
  Kelly, John E., Senior Vice President, 73, 249
  negotiations with Governor George Pataki, 44–46
  negotiations with Governor Mario Cuomo, 36–46
  Opel, John, CEO, 33
  participation in research consortia, ix
  plan to move headquarters out of New York, 44
  Semiconductor Industry Association activities, 37, 38
  Support for K-12 education, 264–266
  trusted foundry, 14
  wafer fabrication facility, East Fishkill, New York, 61
  wafer fabrication facility, Burlington, Vermont, 52, 191
ASML Holding NV, 64
Asodoorian, Alan, 164
Atmospheric Sciences Research Center, National Weather Service, 53
Austin, Texas, xi, xxi, 40, 58, 59, 95, 98, 100, 116, 232, 323, 388
Austria, 164
*Avogy*, 300
Axcelis Technologies, 201

**B**

Babu, S.V., Director, Center for Advanced Material Processing, Clarkson University, 235
Bacheller, John, Senior Vice President, Empire State Development, 308
Baldrey, Douglas, HVCC Assistant Dean of School of Engineering and Industrial Technology, 253
Ballston, New York, 161, 266, 270
Ballston Spa Central School District, xviii, 175, 177, 211, 265, 270
Ballston Spa, New York, 266, 270
BBL Hospitality, 207
Beard Integrated Systems, 200, 201
Berger, Ted, developer of the bionic brain, 248
BessTech, xxi, 325–328, 332, 333
Biden, Joe, U.S. Vice President, 82, 299
Bioengineering, 239, 249, 328
Blumberg, Samuel, Nobel Laureate of Medicine, 248
Board of Regents, New York, 29, 262
Boards of Cooperative Educational Services (BOCES), 261, 266–269
Boston, Massachusetts, 4, 7, 11, 25, 33, 55, 106, 195, 224, 239, 322, 323, 358, 363, 365, 366, 368
Boyer, Kim, Dean, College of Engineering and Applied Sciences, State University of New York at Albany, 245
Brobston, Dennis, President, Saratoga Economic Development Corporation, 208, 314–318
Brockway, Matt, Panalpina strategic accounts manager, 202
Brookhaven National Laboratory, 251
Brookings Institution, 23, 292, 367
Brooklyn Bridge, 31, 235
Brown's Beach, Stillwater, New York, 213
Bruggeman, Warren H., member of Board of Trustees, Rensselaer Polytechnic Institute, 32
Bruno, New York Senate Majority Leader Joseph
  "blue collar workers of the future", xvii
  Directs state funds to expansion of water infrastructure, 111
  Hudson Valley Community College, xviii, 218
  Luther Forest Technology Campus, 111
  retirement, 136, 278, 279, 316, 390
  support for nanotechnology, 279
  Tech Valley High School, 267
  "Three men in a room", 27, 279
Buffalo Billion initiative, 298, 391
Buffalo News, xxvii, 40, 44, 45, 88, 192, 283–285, 290–298, 300, 304, 305, 310, 368, 384, 395
Buffalo, New York, 2, 15, 106, 235, 283, 290, 298, 299, 392
Buffalo Niagara Institute for Advanced Manufacturing Competitiveness, 295
Buffalo Niagara Medical Campus, 292, 293
Build Program, RPI, 32

Index                                                                413

Bullard, Travis, Spokesman, GlobalFoundries, 118, 136, 177, 194, 225
Bullock, Quintin, President, Schenectady County Community College, 259
Burlington, Vermont, 14, 51, 191
Business Alliance for Tech Valley High School, 267
Business Council of New York, 220, 262, 298

C

California Institute of Technology, 82
Callanan, Ray, Chairman, Saratoga County Water Committee, 139
Camoin Associates, 264, 265
Canada, 291, 371, 373
Canandaigua, New York, 299, 300
Capital Region, vii, 4, 23, 57, 95, 137, 165, 182, 218, 277
Capital Region Building and Construction Trade Council, 272
Capital Region, New York, 3, 23, 25, 29, 32–35, 42, 235, 277, 322, 323
Capital Region Regional Economic Development Council (REDC), 27, 281–285, 287–290, 292, 294, 310, 315, 340, 346, 391
Capitol South Campus Center (CSCC), 263
Career and Technical Education (CTE), 266
Carey, Governor Hugh, viii, 22, 26, 27, 33, 36, 386
Carey, James, Clancy Moving Systems, 202
Castaldo, Joseph, General Manager, Crossgates Mall, 207
Caulfield, Tom, CEO of GlobalFoundries, 174
Center for Advanced Interconnect Science and Technology (CAIST), SUNY Albany, 58, 253
Center for Advanced Thin Film Technology, SUNY Albany, 84, 253
Center for Construction Trades Training (C2T), 273
Center for Economic Growth
  Albany airport modernization project, 97
  Albany-Rensselaer train stations modernization project, 42, 346
  Capital Region Semiconductor Task Force, 98
  creation (1987), 97
  high tech manufacturing emphasis, 97
  NY Loves Nanotech global outreach program, 97
  rebranding of Capital Region, ix
  regional approach to economic development, 345, 346, 386
  Support for Saratoga Economic Development Corporation, 96, 97, 160, 346
Center for Environmental Sciences and Technology Management (CESTM/aka/Nano Fab 200), SUNY Albany, 53
Center for Integrated Electronics, RPI, 33, 385, 386
Center for Manufacturing Productivity and Technology Transfer, RPI, 32
Center of Excellence in Nanoelectronics and Nanotechnology (CENN), 62, 70
Centers of Advanced Technology (CATs), viii, 26, 27, 36, 37, 45, 46, 52–55, 57, 58, 68, 239, 386, 387
Centers of Excellence, New York, 26, 46, 55, 60, 62, 348
Central New York Hub for Emerging Nano Industries, 303
Chandler, Arizona, 105
Chang, Morris, CEO of Taiwan Semiconductor Manufacturing Corporation, 125, 380
Charbonneau, Deborah, General Manager, Homewood Suites, 206
Chemical vapor deposition (CVD), 50–52, 54, 187, 226, 387
China, xx, 7, 13, 95, 123, 174, 284, 355, 359, 370, 393
Chip Fab, 98, 98, 101, 102, 113, 254, 387
Clancy Moving Systems, 202
Clarkson University, 235, 251, 265
Clifton Park, New York, 201
Clinton, Governor DeWitt, 16, 26
Clinton, President Bill, 9, 50
Clough Harbour & Associates, 138, 147
Coalition for Responsible Growth, 107
Cold Springs Road, 105, 112, 153, 196
College of Nanoscale Science and Engineering (CNSE)
  Collaborations with Community Colleges and Secondary Schools
  Collaborations with industry, 78
  creation, x, 78, 83, 85, 392
  curriculum, 75, 246
  employment effects in Capital Region, xi
  equipment, 352
  faculty, 75, 248, 342
  graduates, 329
  impact of Kaloyeros indictment, 345
  nanobioscience, 76, 328

College of Nanoscale Science and Engineering
    (CNSE) (*cont.*)
  nanoeconomics, 74, 76, 299
  nanoengineering, 68, 74, 75
  pilot manufacturing facilities, xi
  reputation, 234
  startups, 325, 326, 329, 332
College of Saint Rose, 218, 234, 251
Colonie, New York, 330
Committee on Foreign Investment in the
    United States (CFIUS), 127
Computer Chip Commercialization Center
    (Quad C), 301
Consortium, xxii, 12, 13, 15, 35, 37, 39, 41,
    44, 58, 60, 64–66, 78, 80, 82,
    87–89, 174, 207, 239, 251,
    300–303, 311, 313, 329, 337, 344,
    348, 354, 386, 387, 391–393
Construction industry, xv, 197
Construction jobs, xv, 22, 78, 186–190, 196,
    197, 200, 272
Continuous power supply (CPS), 178, 179
Copper metallization, 51, 387
Corbett, James W., co-founder of Joint
    Laboratories for Advanced
    Materials, 50
Cornell University, 37, 38, 40, 223, 386
Corning, 24, 82, 97, 300, 338
Cuomo, Governor Andrew
  budgets "on time", 280
  budget struggle with legislature, xix, 291, 391
  Buffalo Billion, 292, 294, 298, 307, 391
  Capital Region concerns over neglect, xix,
    262, 289
  Central New York Hub for Emerging Nano
    Industries, 303
  Kaloyeros, Professor Alain, 15, 41, 86
  "member items", xix, 280, 283, 287
  Regional Economic Development
    Councils, xix, 26, 281–284, 391
  silo breaking, 283, 374
  Trudeau Institute retention, 289
  upstate development projects, 309
Cuomo, Governor Mario
  bid for Sematech, 26, 39–41
  Centers of Advanced Technology
    (CAT), 52, 386
  continuity of economic development
    policies with those of Governor
    Hugh Carey, 41–43
  convenes working groups to combat
    recession, 44
  IBM, efforts to retain, 43, 44
  increased support for University-industry
    research collaboration, 36, 37
  interview with Alain Kaloyeros, 41, 387
Curriculum, 12, 15, 75, 221, 222, 233, 246,
    249, 253–255, 258, 260–262, 265,
    267–270, 273, 326, 353
Curry Road, 151, 153
Curtis, Carolyn, Academic Vice President,
    Hudson Valley Community College,
    219, 222, 230
CVD Equipment Corporation, 187

**D**

Daily Gazette, Schenectady, xxvii, 2, 32, 54,
    93, 133, 163, 182, 218, 279
Danfoss Silicon Power, 302, 311, 393
Dell Computers, 324
Delmar Opticians, 208
Denbeaux, Gregory, faculty member, College
    of Nanoscale Science and
    Engineering, 75
Department of Environmental Conservation
    (DEC), 140–143
Department of Housing and Urban
    Development, 263
DiNapoli, Thomas P., Comptroller, New York
    State, 305
Direct employment, xv, 188–193, 200, 208,
    209, 213
Division of the Budget (DoB), 307, 309
Dragone, Joseph, Superintendent, Ballston,
    Spa Central School District, 252,
    265, 266, 270
Dresden, Germany, xiii, 115, 117, 255, 346
Dudley, William C., President and CEO
    of Federal Reserve Bank of
    New York, 184
Duffy, Kristin, President SUNY Adirondack, 260
Duffy, New York, Lieutenant Governor
    Robert, 2, 281
Dunn, Kathleen, faculty member, College
    of Nanoscale Science and
    Engineering, 75
Dutchess County, New York, 61, 191, 193
Dyal, Gary, Executive Director, CVD
    Equipment Corporation, 187

**E**

Eastern New York Angels
    (ENYA), 328–330, 333
East Fishkill, New York, viii, ix, xvi, 14, 43,
    51, 172, 191, 219, 392
Eastman Business Park, 82, 300
Eastman Kodak, 23, 24, 338
Edison, Thomas, inventor, 223

Index 415

Edwards Vacuum, 201, 311
Ehrlich employment study, 188, 189, 191, 194
Ehrlich, Everett M., economist
    and statistician, 187
Ellsworth Commons, 204, 207, 208
Empire State Development (ESD), xix, 26, 45, 69, 70, 81, 96, 97, 99, 101, 114, 127, 134–136, 147, 159–161, 182, 188, 190, 194, 208, 209, 214, 241, 280–282, 284, 294, 297, 307–312, 317, 319, 340, 341, 345, 393
Empire Zone incentives, 317
Employment multipliers, 10, 185, 194
Engineering
    Albany Times Union profile, 325, 326
    description of IBM "[anchor] tenant" role at CNSE, 71
    mentoring role with technical teams at CNSE, 326
    nanoeconomics, 74, 76
    promotion of entrepreneurship of CNSE, 326
Environmental Facilities Corporation (EFC), 142
Environmental impact statement (EIS), xii, 107–111, 113, 140, 151, 152, 158
Eonix, 326, 329, 332
Erie Canal, 16, 21, 25, 26, 86, 287, 290
Essex Junction, Vermont, 172, 392
Etzkowitz, Henry, Professor, Stanford University, 5, 30, 323, 344, 366, 374
Excell Partners, 322
Extreme Ultra-Violet Light Lithography (EUVL), 65


**F**

Fab 8, Global Foundries, 98, 163, 169–171, 174, 177–179, 197, 199, 201, 205, 206, 208, 227, 228, 231, 311, 312
Fairchild Semiconductor, 31, 33, 116
Federal Reserve Bank of New York, 2, 184
Feldhan, Jim, Semico President, 186
FilmHouse, 303
FinFET, 172, 174
Finger Lakes, New York, 284, 299
Finishing Trades Institute, 274
Fink, Stanley, Speaker of New York Assembly, 27, 339
Focus centers, 55, 57, 58
Fort Schuyler Management Corporation (FSMC), 305, 307, 310
Foundry, 13, 14, 126–128, 164, 166, 173–175, 217, 302, 303, 379, 390
Foundry business model, 13, 125

Foundry Company, 126–128
Frederick, Dick, angel investor, 330, 332
Fuller Road Management Corporation (FRMC), 55, 56, 66, 68–70, 307, 310, 345
Fulton-Montgomery Community College (FMCC), xviii, 242, 246, 261, 268, 269
Fury, Michael, senior engineer, IBM, 50–52, 348


**G**

Gene Haas Foundation, 257
General Electric (GE), xxi, xxii, 4, 5, 24, 32, 33, 50, 68, 75, 82, 93, 184, 232, 235, 238, 241, 243, 255, 257, 259, 263, 267, 269, 301–303, 311, 321, 323–325, 338, 340, 359, 363, 392, 393
Generic Environmental Impact Statement (GEIS), 100, 104, 106, 108, 110, 111, 349
Georgia Institute of Technology (Georgia Tech), 50, 58, 78, 80, 339, 346
Georgia Tech Enterprise Innovation Institute, 78, 339, 346
Gerrard, Michael, Chairman, New York Environmental Law Section, 102
Gillibrand, Congresswoman Kirsten, xiv, 135, 167
Glauconix, xxi, 325, 326, 328, 329, 331, 333
Glens Falls, New York, xxvii, 138
GlobalFoundries
    Admin 2 building, 166
    agreement with construction unions, 167, 390
    construction of wafer fabrication plant, xiv, 137, 163, 273
    creation, 171
    employment effects in Capital Region, 183–189, 194
    Fab 8, 98, 144, 148, 163, 169–171, 174, 177–179, 197, 199, 201, 205, 206, 208, 227, 228, 231, 311, 312, 352
    IBM, acquisition of manufacturing operations, 14
    noise abatement efforts, 178–179
    Project Labor Agreement, 129, 167
    relations with Malta/Stillwater, xv, xviii, 9, 84, 94, 96, 98, 100, 102, 104, 105, 109, 184, 188–191, 200, 202, 205, 213, 225, 231, 269, 272, 273, 286, 392, 393
    sales tax exemption, 177, 178

GlobalFoundries (*cont.*)
  startup of operations, 205
  supply chain companies, 225
  Technology Development Center, 170–172, 178, 182, 196, 197
  tool installation, 169, 273
  use of local construction firms, 197
Gomez-Baquero, Fernando, CEO, Besstech, 327
Gonzalez, Pedro, 227, 352
Google, 224, 358
Gould, Gordon, inventor of the laser, 248
Grattidge, Alan, Chairman., Saratoga County Law and Finance Committee, 176, 320
Greater Austin Chamber of Commerce, 66
Greater Capital Association of Realtors, 204
Great Recession (2008), 22, 370
Green, Ken, President, Saratoga Economic Development Corporation, xii, 105, 106, 111, 115, 314, 319, 350
Gregg, Abbie, Consultant to Saratoga County Economic Development Corporation, xiii, 114, 115, 347, 350
Greyling, Werner, Project Manager, M+W Zander, 168
Grose, Doug, CEO, NY CREATES, Former CEO, Global Foundries, xix, 117, 129, 312, 327
Guilderland, New York, 165, 243

**H**

Haldar, Pradeep, Vice President of Entrepreneurship, Innovation and Clean Energy Programs, SUNY Polytechnic, 76, 305, 322, 325, 329
Halfmoon, New York, 205
Hamidi, Ken, engineer, Intel Corporation, 107
Hamilton, Alexander, U.S. Secretary of the Treasury, 21, 22
Harris, Richard, Halfmoon Planning Director, 205
Hartzell, John, Malta Town Councilman, 318
Helms, Robert, President, Sematech, 64
Hewlett-Packard, 31, 33, 35, 322, 323
Hewlett, William, co-founder, Hewlett Packard, 31
Higashi, Tetsuro, President and CEO, Tokyo Electron Ltd. (TEL), 67
Hill, Catherine, Clean Energy Leadership Institute, Skidmore College, 56, 61, 79, 101, 251
Hill, Linda, Lead Economic Developer, National Grid, 314
Hill, Penny, Associate Dean, TEC-SMART, Hudson Valley Community College, 228, 256, 257
Hilton Gardens, 207
Hilton House 2 Suites, 206
Hitchcock, Karen, President, University of Albany, State University of New York, 54, 74, 244
HMF BOCES-FMCC Apprenticeships, 261, 268, 269
HocusLocus, 326, 328
Homewood Suites, 206
Hotels, xv, 11, 199, 203, 205, 206
Hsinchu Science-Based Industrial Park, Taiwan, 6, 376
Hudson Falls, New York, 262
Hudson Valley, vii, xviii, 9, 15, 16, 42, 43, 97, 191, 218, 219, 221, 222, 228, 230, 233, 246, 252–254, 256, 257, 269, 312, 327, 352, 389
Hudson Valley Community College (HVCC)
  Advanced Manufacturing Program, 223, 257, 258
  Collaboration with College of Nanoscale Science and Engineering, 246–248
  employment of graduates, 257
  Gene Haas Foundation gift, 257
  history, 253
  introduction of new curriculum, 221
  Semiconductor degree program, 254–256, 264
  TEC-SMART, 255–257, 391
  training programs for semiconductor manufacturing, 218

**I**

Incentives
  competition between countries and regions, xxiii
  Empire Zone credits, 109, 117
  grants, 117, 127, 350, 390
  infrastructure support, 117, 129
  tax credits, 117, 349, 350, 390
  tax exemptions, 177, 178
Incubators, 33, 34, 44, 53, 61, 62, 238, 240–241, 287, 322, 323, 325–327, 332, 343, 358, 386
Indirect employment, 194, 199, 200, 208, 209
Induced employment, 203–210, 213
Industrial Design Corporation (IDC), 101–103, 106
Industrial Development Agency (IDA), xxvii, 110, 177, 178, 316, 317, 320, 321

Index                                                                                                                                              417

Industrial Technology Research Institute
    (ITRI), Taiwan, 125, 372–381
Infineon, xi, 60, 68, 80
Infrastructure
    controversy, 21
    Empire Zone credits, 129, 136, 349
    grants, 201
    infrastructure support, 62, 129
    local incentives, 109, 317
    tax benefits, 170
Infrastructure for semiconductor
        manufacturing
    electric power, xiv, 157–160
    regulatory approval
        requirements, xiv, 134, 355
    sewers, 133
    transportation, xiv, 149, 157
    water supply, xiv, 137, 145, 146
Innovation, vii, 4, 23, 49, 195, 222, 278
Innovation clusters, 4–6, 287, 288, 344, 367, 368
Integrated circuits (ICs), 10, 11, 37, 88, 120,
        126, 167, 173, 223
Intel Corporation, 39, 116, 120
International Business Machines (IBM), 9, 23,
        51, 93, 164, 188, 221, 278
    300 mm semiconductor processing node,
        x, 64, 188, 192
    anchor tenant, College of Nanoscale
        Science and Engineering, 69, 71, 304
International Sematech North, 65
International Union of Painters w/ Allied
        Trades, 274
Interstate 87, 112, 149


**J**

Jackson, Shirley Ann, President, Rensselaer
        Polytechnic Institute
    biotechnology, 72, 237, 239
    goals for RPI, 237
    information technology, 237
    support by Board of Trustees, 237
    tension with Alain Kaloyeros, 290
Janac, Charles, CEO of Arteris IP, 173
Japan, 13, 36, 37, 39, 67, 169, 328, 370
Jelinek, Lea, semiconductor industry
        analyst, 119
Jobs 2000 Act (J2K), 45, 387
John F. Kennedy International Airport,
        New York, 202
Johnson, Amy, President, Capstone, Inc., 267
Johnson, Kristina, Chancellor, State University
        of New York, 243, 312
Johnstown, New York, 261

Joint Laboratories for Advanced Materials
    (JLAM), 49
Jones, Robert, President, State University of
    New York at Albany, 85, 244, 245


**K**

Kaloyeros, Professor Alain, SUNY
        Polytechnic University
    attracts industry funds for semiconductor
        R&D, x, 51
    Center for Advanced Thin Film
        Technology, 52
    Center for Environmental Science and
        Technology Management, 53
    Chemical vapor deposition for copper
        metallization research, 51
    collaboration with IBM, 15
    College of Nanoscale Science and
        Engineering, 244
    conviction, 306
    formation of SUNY polytechnic, 15
    friction with Shirley Jackson, President,
        Rensselaer Polytechnic Institute,
        290
    indictment on bid-rigging
        charges, 15, 88, 305
    interview with Governor Mario
        Cuomo, 15
    investigation by federal authorities, 15
    Joint Laboratories for Advanced
        Materials, 49
    Karen Hitchcock, President, SUNY
        Albany, 54, 244
    metallization research, 51, 52, 387
    problems in extending Albany model
        across Upstate New York, 308
    resignation from SUNY Polytechnic, 392
Katz, Bruce, Co-Director of Brookings
        Institution Metropolitan Policy
        Program, 292, 344, 368
Kauffman Foundation, 323
K-12 education, xviii, 239, 262, 264–266
Kelley, Jack, Senior Vice President, Saratoga
        Economic Development
        Corporation, 100
Kelley, Jack, Vice President, Saratoga
        Economic Development
        Corporation, 319
Kelly, John E., Senior Vice President and
        Group Executive, IBM Technology
        Group, 73, 249, 342
Kennedy, Andrew, President, Center for
        Economic Growth, 32

Kilby, Jack, Texas Instruments, creator of first integrated circuit, 10
Killeen, Timothy, President, SUNY Research Foundation, 56
King, Michael, executive, National Grid, 161
King, Robert, Director, New York Governor's Office of Regulatory Reform, 101, 102
Kinowski, Edward, Stillwater Town Supervisor, 176
Korea, 13
Kors, Michael, Lord and Taylor designer, 207
Krishnamoorthy, Mukkai, Director, RPI Center for Open Source Software, 224

L

Land grant colleges, 4, 242
Leslie, Stuart W., Professor, Johns Hopkins University, 3, 5, 6, 12, 23, 29, 30, 32–35, 221, 235, 322, 324, 338
Leyden, Kevin, IBM, Co-founder of Business Alliance for Tech Valley High School, 267
Leydesdorff, Loet, Professor, University of Amsterdam, 5, 366
Lifshin, Eric, faculty member, College of Nanoscale Science and Engineering, 75
Linium Recruiting, 219
Lithography, 38, 40, 59, 60, 64, 65, 67, 76, 81, 83, 171, 174, 175, 255, 311, 342, 348, 389, 393
Litow, Stanley, IBM Vice President of Corporate Citizenship and Corporate Affairs, 221
Local governments, xxi, 41, 42, 44, 94, 95, 99, 113, 163, 281, 305, 314, 339, 353, 356, 367, 391
Lord & Taylor, 207, 208
Lovell, Kelly, President, Center for Economic Growth, 104, 106, 347, 348
Low, George
    Apollo program leadership, 32
    Center for Industrial Innovation (CII), 33, 40, 238
    fundraising efforts, 33
    incubator created at RPI, 240–241
    Rensselaer Polytechnic Institute (RPI) leadership, 33
    Silicon Valley, attempts to replicate, 32, 33, 386
    "three steeples of excellence", 33
Luther Forest, New York, 100, 102, 127, 135, 169

Luther Forest Technology Campus (LFTC), xx, 96, 100, 101, 104–113, 134, 135, 141, 144, 147, 148, 150, 152, 157, 158, 161, 162, 201, 202, 209, 280, 285, 313, 317–321, 350, 357, 388, 390
Luther Forest Technology Campus Economic Development Corp (LFTCEDC), xx, 96, 99, 109, 136, 144, 146, 149, 155, 156, 212, 315–320

M

M+W Group/aka/M+W Zander
    contract with Saratoga Hospital, 208
    oversight of construction of GlobalFoundries water fabrication plant, xiii, 350
    participation in Saratoga Economic Development Corporation engineering project team, 114, 347
    relocation of headquarters to Albany, 72, 391
    tool installation at GlobalFoundries, 168, 169
    Watervliet Arsenal training center, 72, 164, 273, 391
Mackay, Alexander, Owner, Saratoga Water Services, Inc., 138, 140, 142, 143, 147, 148
Malta Med Emergent Care, 209
Malta, New York, 165, 202, 223, 225, 228, 229, 235, 257, 331
Malta Planning Board, 166, 170, 197
Malta Town Board, 104–109, 130, 320
Marcy Nanocenter, 300, 301
Marcy, New York, 15
Marshall, Alfred, economist, 5, 363, 364
Massachusetts Institute of Technology (MIT) Stanford University, 4, 5, 7, 8, 30, 32, 58, 60, 77, 80, 235, 236, 300, 323, 363, 365, 371
Matonak, Andrew, President, Hudson Valley Community College, 221, 223, 255, 256
Mazzone, Angelo, owner and founder of Mazzone Hospitality, 165, 168, 199, 201, 207
Mazzone Hospitality, 201, 207
Mazzucato, Marianna, Professor, University of Sussex, 366
McKinsey & Company, 219, 292
Mead, Carver, Professor of Engineering and Applied Science, California Institute of Technology, 125
Meager, David R., Malta Town Supervisor, 109

Index 419

Megna, Robert, Chief Operating Officer, State University of New York Central Administration, 312
Megna, Robert, Senior Vice Chancellor and Chief Operating Officer, State University of New York, 310
Mehta, Rutvik, President, Thermo Aura, 330
Metallization, 38, 50–53, 348, 387
Microelectronics Advanced Research Corporation (MARCO), 57, 58
Microelectronics and Computer Technology Corporation (MCC), 35, 39, 94
Microfluidics, 173
Micron Technologies, 60, 389
Minuteman program, 12
Mohawk Valley Edge, 186
Monroe Community College, 235
Moodys, 149, 211
Moore's Law, 11, 13, 88, 123, 313, 354
Moreau Industrial Park, 100, 102
Moreau, New York, xiv, 100
Moretti, Enrico, Professor of Economics, University of California at Berkeley, 10, 195, 208, 209
Morrill Act of 1862, established Land Grant Colleges, 242
Motorola, 53, 54, 263, 324
Moynihan, Senator Daniel Patrick, 286
Mubadala Investment Company, 126, 164, 390
Mullen, Dennis, Empire State Development Chairman, 147, 317
Musk, Elon
  Factory in Buffalo, 296
  SolarCity, 296–297
  Tesla Motors, 296

N
Namometer, viii, 8, 9, 50, 64, 81, 171, 172, 175, 392
Nanobioscience, 76, 328
Nanoeconomics, 74, 76, 270, 299
Nanoengineering, 68, 74, 75, 388
Nanoscience, 50, 68, 75, 164, 246, 272, 328, 329, 388
Nano Tech Complex, CNSE, 293
Nanotechnology, vii, 3, 23, 49, 109, 136, 164, 181, 224, 278
Nanotech Resources Inc., 55, 345
National Academy of Engineering, 235
National Business Incubation Association, 240
National Grid
  Connect 21 initiative, 160
  economic development efforts, 97
  expertise, 96
  natural gas pipeline to support Global Foundries, 161
  new power lines to support Global Foundries, xiv
  new substations, 158, 160
  participation in outreach to Silicon Valley, 96
  sponsorship of technical studies National Nanotechnology Initiative (NNI), 96
National Nanotechnology Initiative (NNI), 9, 50, 51
National Science Foundation, xxvii, 50, 221, 239, 268, 328–330
Netherlands, 169, 311
New York Battery and Energy-Storage, 251, 329
New York City, New York, 16, 21, 25, 26, 42, 106, 163, 202, 223, 224, 286, 322, 338
New York Department of Labor, 191
New Yorkers for Fiscal Fairness, 187
New York Exec we Union Energy Leadership (NY EXCEL), 251
New York Genome Center, 294
New York Office of Science, Technology and Academic Research (NYSTAR), viii, 26, 42, 45, 70, 341, 387
New York Route 9, 149, 150
New York State Department of Transportation (DOT), 150, 151, 156
New York State Electric & Gas Corp, 158
New York State Energy Research and Development Authority (NYSERDA), 100, 102, 104, 251, 256, 270, 326, 327, 329, 330, 332
New York State Environmental Quality Review Act (SEQRA), 102, 107
New York State Health Department, 137
New York State Science and Technology Foundation (NYSSTF), 29, 36, 45
New York State Supreme Court, 143
New York State Thruway, 16, 25
North Greenbush Town Board, 104, 347
North Greenbush, New York, 15
Northeast Technology Corridor, 191
Noyce, Robert, Co-Founder Intel Corporation, 10, 11, 167
Nuclear Regulatory Commission, 237

O
Obama, President Barak, 42, 243, 262
O' Connor, Kevin, President, Center for Economic Growth, 42, 45, 97
Office of Regulatory Reform, 101, 102

Office of the State Comptroller
    (OSC), 69, 305, 307
O'Leary, Vincent, President, State University
    of New York at Albany, 49
Oneida County, New York, 186, 299
Onondaga County, New York, 303
Optics, 46, 269, 300, 343
Orange County, New York, 183, 188, 254
Oregon, 93, 292

**P**
Pace Energy and Climate Center, 251
Packard, David, Co-Founder,
    Hewlett - Packard, 31, 33
Panalpina, 202
Pataki, Governor George
    Centers of Excellence, viii, 26, 45, 62, 388
    collaboration with Senate Majority Leader
        Joseph L. Bruno and Speaker of the
        Assembly Sheldon Silver, viii, 27,
        99, 278
    designation of SUNY Albany as Center of
        Excellence, 46, 62
    increase in R&D spending, 283, 388
    Jobs 2000 Act, 45
    New York Office of Science, Technology
        and Academic Research
        (NYSTAR), 45
    outreach to Sematech, 66, 348, 388
    retirement, 279
    retention of IBM, 390
Paterson, Governor David, 26, 71, 87, 118,
    128, 129, 301
Pathways in Technology Early College High
    School (P-TECH), 220, 261, 262,
    269, 271
PBC Tech, 326, 330–332
Pearl Street Station, New York, 223
PeroxyChem, 201
Peterson, Thomas, attorney for The Town of
    Malta, 108, 109
*Photonica*, 300
Pilot manufacturing, xi, 59–62, 79–80, 343, 364
Porter, Michael, Professor, Harvard Business
    School, 5, 364, 367
Power Electronics Manufacturing
    Consortium, 302
Pre-permitting, xi, xxii, 94, 98, 101–103, 116,
    134, 150, 347
Project India, 104
Project Labor Agreement (PLA), xv, 129, 167,
    171, 189

Public Policy Institute of New York, 221, 223
Putnam County, New York, 202
Pyeongtaek, 123

**Q**
Questar III BOCES, 267, 268

**R**
Range, Michael, 244
Recovery Sports Grill, 207
Regional Economic Development Councils
    (REDC), xix, 26, 281–285,
    287–290, 292, 294, 301, 310, 315,
    340, 346, 391
Reilly, Emily, Global Foundries, 223, 225,
    229, 235
Relyea, Michael, Director, Luther Forest
    Technology Campus Economic
    Development Corporation
    (LFTCEDC), 42, 136, 318
Rensselaer County, New York, xxvii, 102, 103
Rensselaer Polytechnic Institute (RPI)
    Blue Gene Supercomputer System, 73
    Build Program, 32
    Center for Biotechnology and
        Interdisciplinary Studies
        (CBIS), 239
    Center for Computational Innovations
        (CCI), 239
    Center for Directed Assembly of
        Nanostructures, 239
    Center for Integrated Electronics, 33, 385
    Center for Lighting Enabled Systems and
        Applications (LESA), 239
    Center for Manufacturing Productivity and
        Technology Transfer, 32
    Computational Center for Nanotechnology,
        xiii, 71, 73, 120, 239, 241, 389
    Emerging Ventures Ecosystem
        (EVE), 240, 330
    George. M Low Center for Industrial
        Innovation (CII) graduates,
        employment by local companies,
        33, 40, 238
    incubator, 33, 34, 238, 240–241, 322, 386
    Jackson, Shirley Ann, President, 32, 72,
        237, 290, 352
    origins, 35
    Rensselaer Technology Park, 39, 240
    Schmitt, Roland W., President, 238
    undergraduate education strengths, 237

Index 421

Rensselaer, New York, viii, 9, 30, 31, 51, 183, 184, 198, 243, 252, 267, 338, 385
Research Triangle, North Carolina, xxi, 4, 55, 185, 323, 324, 358, 365
Restaurants
   employment effects of GlobalFoundries, CNSE, 207
   Global Cafe, 201
   Mazzone Hospitality activities, 201
   Recovery Sports Grill, 207
   Wheatfields Restaurants, 207
Rice, Steven, 249
Rifenburg Construction, 153
RiverBend, 294, 296
Rochester Institute of Technology, 82, 235, 300
Rochester, New York, 15, 82, 284, 299, 322
Rockefeller, Governor Nelson
   Advisory Council for the Advancement of Industrial Research and Development, 29
   concern over stagnant economy in Upstate New York, 23
   expansion of State University of New York, 26
   interest in "Silicon Valley" model, 5, 29, 30
   meetings with Fredrick Terman, "father of Silicon Valley", 5, 29, 30
   New York State Science and Technology Foundation (NYSSTF), 29
   support for research-based economic development, 5, 25, 28–35
Rockefeller Institute
   of Government, 242, 243, 266
Rome, New York, 3, 82, 246, 299, 301
Ross, James, New York State Department of Labor analyst, 184
Roswell Park Cancer Institute, 293, 294
Round Lake bypass
   construction, 133, 152–154, 156, 196
   delay in wetland disturbance permit, 152
   New York Department of Transportation study, 150
   opening for vehicular traffic, 153
   regulatory issues, 150, 151
   Saratoga County sponsorship of engineering design, 151
   traffic impact, 150
   Zoning legislation, 150
Round Lake, New York, xiv, 150
Route 128, Massachusetts, 4, 31, 33, 55, 358, 368
RPI Incubator
   closure, 240
   creation by George Low, 240
   criticism of closure, 240
   National Academies study, 34
Ruiz, Hector, CEO, Advanced Micro Devices, xiii, 96, 115–118, 122–124, 126, 128, 283, 301, 350, 355
Russo, Michael, Global Foundries
   Electric power requirements for Global Foundries, 157–160
   Project Labor Agreement (PLA), 129, 167
   spokesman for Global Foundries, 129
Rust Belt, 1–3, 7, 23, 28, 224, 338, 340
Ryan, James, faculty member, College of Nanoscale Science and Engineering, 75

S
Saburro, Rich, executive, Center for Advanced Technology, SUNY Albany, 54
Sacks, Dixie Lee, Mayor, Round Lake, New York, 112, 150
Sales tax, 104, 177, 178, 200
Samsung, xi, 13, 80, 81, 87, 89, 94, 98, 100, 115, 123, 172, 174, 391, 393
Sanders, Jerry, founder, Advanced Micro Devices, 116
S&P, 149
Santa Cruz, California, 178
Saratoga County Board of Supervisors, 136–138, 140, 141, 151
Saratoga County Chamber of Commerce, 188, 203, 205
Saratoga County, New York, 100, 137, 144, 184, 320
Saratoga County Prosperity Partnership, xxvii, 315, 392
Saratoga County Sewer Commission, 113, 148
Saratoga County Water Authority (SCWA), 137–139, 141–144, 147, 148
Saratoga Economic Development Corporation (SEDC)
   attracts companies to Saratoga, 96, 160, 315, 319
   Brobston, Dennis, 136, 208, 314–316, 318
   engineering project team, 114, 347, 389
   environmental approvals, xii, xiii
   foundation (1978), 99
   goals, 349
   Green, Ken, xii, 100, 105, 111, 115, 150, 314, 350
   infrastructure initiatives, 111, 158
   job creation, 314
   Kelly, Jack, xii, 101, 182
   leadership, xii, 315

Saratoga Economic Development Corporation
(SEDC) (*cont.*)
    pre-permitting effort, 150, 158, 347, 349
    proposal to Advanced Micro Devices, 99,
        148, 347
    purchase of land in Luther Forest, xii, 124
    selection of Luther Forest site, 349
    support from National Grid, Center for
        Economic Growth, 96, 99, 160
Saratoga Hospital, 208
Saratoga Lake, 138, 141, 213
Saratoga PLAN, 122, 151, 152, 155, 157
Saratoga Springs, New York, vii, xxvii, 56, 61,
    72, 101, 104, 130, 135, 136, 138,
    139, 142, 143, 145–147, 153, 154,
    157, 159, 162–165, 170, 176–179,
    184, 196, 201, 203, 204, 206, 209,
    211, 212, 219, 232, 248, 251, 260,
    265, 267, 269–271, 278, 281, 305,
    314, 315, 317–319, 322, 325, 330,
    368, 384, 395
Saugerties New York, 187
Sausville, Paul, Malta Tow Supervisor, 133,
    170, 318, 319
Sawyer, Diane, ABC journalist, 182
Scharfstein, Professor Susan, SUNY Poly,
    328, 329
Schenectady County Community College
    (SCCC)
    grant from Bill and Melinda Gates
        Foundation, 260
    grant from U.S. Department of Labor, 184,
        191, 220
    Nanoscale Materials Technology Program,
        258–259
    Smart Scholars Early College Program, 260
    training for Global Foundries, 259
Schenectady, New York, 2, 32, 54, 93, 122,
    133, 163, 182, 217, 279
Schmitt, Roland W., President RPI, 238
Schubert, David Captain, United States
    Navy, 85
Schultz, Laura, Assistant Professor of
    Nanoeconomics, College of
    Nanoscale Science and, 69, 71,
    74, 326
Schumer, Senator Charles, xiv, 135, 256, 259
Schwartz, Bart, investigator appointed by
    Governor Andrew Cuomo, 307
Sematech, ix, xi, xxiii, 12, 14, 39–41, 51,
    64–67, 74, 232, 254, 348, 354, 367,
    387, 388, 390
Semico employment study, 186–189, 195,
    197, 200

Semiconductor industry
    associations, xvi, 11, 12, 37, 57, 59, 80,
        123, 161, 191, 194, 209, 217, 231,
        351, 352, 387
    complexity, 12
    disaggregation, 13, 14
    employment, xvi
    equipment, xx, 52, 67, 71, 81, 83, 165,
        202, 253, 311, 344, 352, 353, 359,
        380, 388
    fabless firms, 13
    foundries, xi, 14, 125, 126
    government policies, 12, 352
    history, 12, 66
    instability, xx, 11–14
    manpower needs, 220
    manufacturing process, 67, 83, 119–121,
        125, 148, 349
    materials, 191
    Moore's law, 13, 88, 123
    rising costs, 123
    Sematech consortium, 64
Semiconductor Industry Association (SIA),
    xvi, 11, 12, 37, 57, 59, 80, 123, 161,
    191, 194, 209, 217, 231, 351,
    352, 387
Semiconductor manufacturing equipment, 52,
    59, 63, 64, 67, 81, 83, 165, 178,
    202, 311, 344, 353, 380, 388
Semiconductor manufacturing process
    assembly, 121
    deposition, 120
    doping, 121
    etching, 121
    packaging, 121
    photolithography, 121
    testing, 121
Semiconductor materials, 191
Semiconductor Research Corporation (SRC),
    ix, xxiii, 37, 38, 40, 57, 80, 348,
    386, 387
Semiconductors, xvi, 9, 10, 13, 37, 43, 59, 68,
    84, 100–102, 105, 106, 108, 110,
    111, 113–116, 119–126, 130, 161,
    172, 184, 191, 222, 225, 280, 300,
    301, 311–314, 318, 353, 359, 380,
    391–393
Semiconductor wafers
    200 mm, 89
    300 mm, 89
    450 mm, 89
    fabrication process, x, xiii, 116, 219, 347,
        388, 390
    Global 450 Consortium, 87, 89, 391, 392

Index 423

Increasing cost, 123
Research and development, 52
  risks, 52
  transition in size, 88
SEMICON Europe, 161
SEMICON West, 124, 161, 217
Semico Research Corporation, 186, 194, 197
Sensor-on-a-chip, 85
Seward, William, US Secretary of State, 248
Shimkus, Todd, President, Saratoga County Chamber of Commerce, 203, 205, 206
Siena College, 234, 250
Silicon carbide, 289, 301, 302, 307, 311, 392, 393
Silicon germanium (SiGe), 173
Silicon on Insulator (SOI), 173
Silicon photonics, 173
Silicon Valley, California, 3–6, 11, 15, 23, 29, 30, 32–35, 39, 43, 55, 58, 72, 80, 116, 122, 161, 167, 185, 191, 195, 223, 224, 235, 236, 278, 286, 287, 300, 321–323, 338, 358, 363, 365, 368, 386
Silver, Sheldon, Speaker of New York Assembly, viii, xiv, xix, 27, 53, 59, 62, 87, 99, 278, 279, 339, 392
Simcoe, William, Executive Director, Saratoga County Water Authority, 140
Singapore, 94, 117, 125, 166, 172–174, 351, 355
Skidmore College, 218, 234, 251
Skills gap, xviii, 218–221, 223, 232, 358
Small Business Innovation Research Program (SBIR), 329–331, 358, 383
Smith, David, spokesman, Sienna College, 250
Smith, Zimri Lice ("Zim"), member, Saratoga Design Review Commission, 110, 151, 154–157
SolarCity, 284, 296–297
Solar wafers, 284
Soraa, 294, 303
Spencer, James, Director of New Venture Development, Rensselaer Polytechnic Institute, 241
Spitzer, New York Governor Elliot, 26
Standard Oil, 235
Stanford University
  Department of Electrical Engineering, 12, 223
  graduate programs, 32
  Honors Cooperative Program, 31
  role in the creation of Silicon Valley, 5, 29, 30, 33

Stanford Industrial Park, 31, 33, 240
Stanford Research Institute, 31
Terman, Fred, Provost, Support for University-based economic development, 5, 29, 30, 32, 240
Use of MIT as a model for university's role in local economy, 4, 30, 32, 60
Stark, Jeff, President, Capital Region Building and Trades Council, 272, 273
START-UP NY, 288
Startups, xxi, 33, 45, 50, 81, 83, 313, 321–326, 328, 330–334, 357, 358, 380
State Education Department, 221, 260, 270
State Environmental Quality Review Act (SEQR), 102, 107, 108, 158
State University of New York (SUNY), vii–viii, 26, 29, 68, 85, 242–243, 252, 309, 386
State University of New York at Albany
  Alain Kaloyeros initiatives, x, 50, 387
  College of Computing and Information, 245
  College of Emergency Preparedness, Homeland Security and Cybersecurity, 245, 246
  College of Engineering and Applied Sciences, 245, 246, 289
  curriculum, x
  graduates, 235, 244, 246
  investment shortfall, xx, 300–304, 316–321
  Karen Hitchcock leadership, 54, 244
  Kim Boyer recruited as dean of engineering college, 245
  Robert Jones leadership, 85, 244
  Spinoff of College of Nanoscale Science and Engineering, x, xvii, 68, 119, 244, 339, 367, 387–389
State University of New York at Stony Brook, 37, 54
State University, of New York at Utica-Rome, 82, 86, 87, 246
Stillwater, New York, xii, xiv, xviii, 9, 84, 94, 96, 98, 100, 102, 104, 105, 108–111, 130, 133, 139, 148, 150, 155, 156, 158, 174–176, 178, 184, 188–191, 196, 200, 202, 203, 205, 208, 210–213, 225, 231, 269, 273, 279, 286, 319, 346, 350, 388, 389, 392, 393
Stillwater School District, 175
Stillwater Town Boards, 100, 104–106, 108, 109, 111, 148, 158, 347
Stokes, Julia, Chairman, Saratoga PLAN, 152, 155

Suarez, Darren, Director of Government
    Affairs, Business Council of
    New York, 220, 262, 298
Sumitomo Cryogenics of America (SCAI), 200
SUNY Adirondack, xviii, 242, 243, 246, 260
SUNY Empire State College, 234, 248
SUNY Polytechnic Institute (SUNY POLY)
    collaborations with Community Colleges
        and Secondary Schools, 264–266
    College of Nanoscale Science and
        Engineering, xix, 82, 86, 221, 247,
        268, 301, 303
    creation, 83, 85, 175
    declining enrollment, 244
    employment impact in Capital Region,
        xvii, 15, 235
    iClean, 325, 326
    internships, 63, 243, 248, 250, 263, 272
    Kaloyeros departure, xix, 309
    startups, 325
    SUNY Poly Advancing Research and
        Commercialization (SPARC), 326
SUNY Research Foundation, 56, 85, 294,
    304–308, 312, 329, 345
Supercomputers, xiii, 73, 120, 237, 239, 241,
    249, 250, 290, 341, 389
Superfund, 100
Supply chain, xi, xv, xvi, xx, xxiii, 5, 7, 10,
    22, 24, 78, 81, 83, 86, 185, 189,
    193–195, 200–203, 210, 225, 231,
    315, 321, 340, 350, 353, 355, 357,
    359, 379
Supply chain jobs, xv, 193, 194
Switzerland, x, 79, 202, 344


T
Taiwan, 6, 13, 95, 123, 125, 174, 355, 372,
    373, 375–377, 379
Taiwan Semiconductor Manufacturing
    Corporation (TSMC), 87, 89, 124,
    125, 174, 379, 391
Tappan Zee Bridge, 16
Technology and Education Center for
    Semiconductor Manufacturing and
    Alternative Renewable
    Technologies (TEC-SMART),
    255–257, 352, 391
Technology Development Center (TDC),
    GlobalFoundries, 170, 171, 178
Tech Valley, vii, xxi, xxii, 3, 7, 14, 22, 26–28,
    43, 183, 190, 208, 266–268,
    277–334, 337, 338

Tech Valley High School (TVHS), xviii, 267,
    268, 390
Tech Valley STEMsmart Alliance, 269
Templeton, Richardson, CEO, Texas
    Instruments, 248
Tenenbaum, Professor Scott, College of
    Nanoscale Science and Engineering,
    328
Terman, Frederick, Provost, Stanford
    University, 5, 6, 29, 30, 32, 33, 240,
    338
Tesla Motors, 296
Texas Instruments, 10, 80, 115, 125, 248, 300,
    324, 389
ThermoAura, xxi, 326, 330–333
Thiel, Brad, faculty member, College of
    Nanoscale Science and
    Engineering, 75
1366 Technologies, 284
3D chip stacks, 171
Through Silicon Vias (TSVs), 171
Times Union, Albany, xxvii, 368, 384, 395
Tokyo Electron Ltd. (TEL), 64, 67, 74, 81,
    83, 388
Tonko, Congressman Paul D., 320
Tool installation, 169, 273, 300
Torrejon, Dr. Karen, founder, Glauconix
    Biosciences, 328
Tozzi, Tony, Town of Malta Planning and
    Development Director, 128, 166, 168
Tracor, 324
Training, xvii, xviii, xxiii, 8, 21, 32, 36, 44,
    61, 63, 72, 192, 205, 211, 217–274,
    300, 311, 315, 326, 330, 334, 352,
    353, 366, 381, 383, 390, 392
Trinity Alliance, 263
Triple Helix, 5, 6, 30, 70, 74, 76, 78, 338, 342,
    344, 345, 366–370, 374
Troy, New York, vii, xviii, xxvii, 3, 5, 28, 31,
    39, 41, 74, 81, 100, 102, 104, 114,
    117, 126–128, 135, 142, 143, 156,
    164, 168, 184, 192, 218, 219,
    221–224, 230, 233, 235–238, 240,
    241, 250, 251, 253, 256, 257, 262,
    263, 265, 267, 269, 271, 272, 301,
    320, 322, 326, 328, 330, 331,
    352–354, 358, 368, 384, 390, 395
Troy Riverfront P-TECH, 269
Trudeau Institute, 289
Trump, President Donald, 2, 284
Tucker, Michael, President, Center for Economic
    Growth, xxviii, 25, 97, 186, 252
Turner Construction, 169, 199

## U

Unemployment, 183, 184, 219, 298, 357
Union College, 15, 218, 233, 234, 248–250, 258, 352
Unions, 26, 94, 128, 129, 167, 189, 219, 272–274, 390
United Association of Plumbers and Steam Fitters, 274
United States Army, 72, 85, 100, 104, 273
United States Military Academy at West Point, 15, 233
University of Minnesota, 238
Upstate Capital Association of New York, 334
Upstate New York, vii, xix, xxi, xxvii, 2, 3, 6, 15, 21–46, 93, 100, 159, 160, 186, 194, 196, 197, 289, 290, 310, 313, 321–324, 334, 337, 348, 392
Upstate Venture Connect (UVC), 334
U.S. Attorney's Office, Manhattan, 307
Utica, New York, xii, xxvii, 3, 15, 82, 86, 227, 261, 287, 289, 298–303, 311, 317, 368, 384, 392, 395

## V

Vanags, Marty, President Saratoga County Prosperity Partnership, 315
Van Patten, Ronald Wayne, resident of Malta, opposed easement for Zim Smith Trail, 157
Venture capital, 313, 321, 322, 328, 332, 334, 379
Vibration Engineering Consultants, 178

## W

Walker, New York Mayor Jimmy, 286
Wallingford, David, 138
Warren County, New York, 183, 266
Watervliet Arsenal, 71, 164, 273, 391
Watervliet, New York, 72, 83, 164, 165, 198, 238
Welch, Jack, CEO, General Electric, 33
Wencis, Ed, project manager, National Grid, 162
Westchester County, New York, 2, 183
Western New York Regional Economic Development Council, 284
Western New York Science and Technology Park (STAMP), 284
Westinghouse, 235
Wheatfields Restaurants, 207
Whitney, Rick, President of U.S. Operations, M+W Group, 5, 363
Whitney, Willis, founder of General Electric research laboratory, 5, 363
Williams, C. Wayne, President, The College of Saint Rose, 251
Wood, Thomas, Chairman, Saratoga County Board of Supervisors, 136
World War II, 22, 28, 29, 32, 33, 104, 236, 253, 302

## Z

Zemsky, Howard, CEO of Empire State Development, xix, 284, 292, 293, 297, 309, 310, 312, 393
Zimpher, Nancy, Chancellor, State University of New York, 85, 223, 240, 242, 243, 263
Zim Smith Trail, 110, 151, 154–157

CPSIA information can be obtained
at www.ICGtesting.com
Printed in the USA
BVHW010746201219
567308BV00010B/2/P